VOLUME THREE HUNDRED AND FORTY ONE

INTERNATIONAL REVIEW OF CELL AND MOLECULAR BIOLOGY

Biology of T Cells - Part A

INTERNATIONAL REVIEW OF CELL AND MOLECULAR BIOLOGY

VOLUME THREE HUNDRED AND FORTY ONE

International Review of CELL AND MOLECULAR BIOLOGY

Biology of T Cells - Part A

Edited by

LORENZO GALLUZZI
Weill Cornell Medical College, New York, NY, United States

NILS-PETTER RUDQVIST
Weill Cornell Medical College, New York, NY, United States

ELSEVIER

ACADEMIC PRESS
An imprint of Elsevier

Academic Press is an imprint of Elsevier
125 London Wall, London EC2Y 5AS, United Kingdom
525 B Street, Suite 1650, San Diego, CA 92101, United States
50 Hampshire Street, 5th Floor, Cambridge, MA 02139, United States
The Boulevard, Langford Lane, Kidlington, Oxford OX5 1GB, United Kingdom

First edition 2018

ISBN: 978-0-12-815385-7
ISSN: 1937-6448

For information on all Academic Press publications visit our website at https://www.elsevier.com/books-and-journals

Publisher: Zoe Kruze
Acquisition Editor: Ashlie M. Jackman
Editorial Project Manager: Teresa Pons-Ferrer
Production Project Manager: James Selvam
Cover Designer: Alan Studholme

Typeset by TNQ Technologies

CONTENTS

CONTRIBUTORS

Adrienne E. Borrie
Department of Physiology and Pharmacology, Western University, London, ON, Canada; Division of Clinical Pharmacology, Western University, London, ON, Canada

Jolanda Brummelman
Laboratory of Translational Immunology, Humanitas Clinical and Research Center, Milan, Italy

Fanny Chalmin
Department of Medical Oncology, Centre Georges-François Leclerc, Dijon, France; Centre de Recherche INSERM LNC-UMR1231, Dijon, France; Univ. Bourgogne Franche-Comté, Dijon, France

Ana A. Fernández-Ramos
Institut National de la Santé et de la recherche Médeicale (INSERM) U1147, Paris, France; Paris Descartes University, Paris, France

Hareem Ghazanfar
St. John Hospital and Medical Centre, Detroit, MI, United States

François Ghiringhelli
Department of Medical Oncology, Centre Georges-François Leclerc, Dijon, France; Centre de Recherche INSERM LNC-UMR1231, Dijon, France; Univ. Bourgogne Franche-Comté, Dijon, France; Platform of Transfer in Cancer Biology, Centre Georges-François Leclerc, Dijon, France

John W. Hickey
Department of Biomedical Engineering, Johns Hopkins University School of Medicine, Baltimore, MD, United States; Institute for NanoBiotechnology, Johns Hopkins University School of Medicine, Baltimore, MD, United States; Cell Engineering, Johns Hopkins University School of Medicine, Baltimore, MD, United States

Etienne Humblin
Department of Medical Oncology, Centre Georges-François Leclerc, Dijon, France; Centre de Recherche INSERM LNC-UMR1231, Dijon, France; Univ. Bourgogne Franche-Comté, Dijon, France

Uqba Khan
St. John Hospital and Medical Centre, Detroit, MI, United States

Alyssa K. Kosmides
Department of Biomedical Engineering, Johns Hopkins University School of Medicine, Baltimore, MD, United States; Institute for NanoBiotechnology, Johns Hopkins University School of Medicine, Baltimore, MD, United States; Cell Engineering, Johns Hopkins University School of Medicine, Baltimore, MD, United States

Marie-Anne Loriot
Institut National de la Santé et de la recherche Médeicale (INSERM) U1147, Paris, France; Paris Descartes University, Paris, France; Clinical Chemistry Department, Georges Pompidou European Hospital, APHP, Paris, France

Enrico Lugli
Laboratory of Translational Immunology, Humanitas Clinical and Research Center, Milan, Italy; Humanitas Flow Cytometry Core, Humanitas Clinical and Research Center, Milan, Italy

Saman Maleki Vareki
Department of Oncology, Western University, London, ON, Canada; Cancer Research Laboratory Program, Lawson Health Research Institute, London, ON, Canada

Nicolas Pallet
Institut National de la Santé et de la recherche Médeicale (INSERM) U1147, Paris, France; Paris Descartes University, Paris, France; Clinical Chemistry Department, Georges Pompidou European Hospital, APHP, Paris, France

Karolina Pilipow
Laboratory of Translational Immunology, Humanitas Clinical and Research Center, Milan, Italy

Jonathan P. Schneck
Institute for NanoBiotechnology, Johns Hopkins University School of Medicine, Baltimore, MD, United States; Cell Engineering, Johns Hopkins University School of Medicine, Baltimore, MD, United States; Department of Pathology, Johns Hopkins University School of Medicine, Baltimore, MD, United States

Frédérique Végran
Centre de Recherche INSERM LNC-UMR1231, Dijon, France; Univ. Bourgogne Franche-Comté, Dijon, France; Platform of Transfer in Cancer Biology, Centre Georges-François Leclerc, Dijon, France

T CELLS: FRIENDS AND FOES

Nils-Petter Rudqvist*[,1] and Lorenzo Galluzzi*[,§,¶,1]

*Department of Radiation Oncology, Weill Cornell Medical College, New York, NY, United States
[§]Sandra and Edward Meyer Cancer Center, New York, NY, United States
[¶]Université Paris Descartes/Paris V, Paris, France
[1]Corresponding author: E-mail: deadoc@vodafone.it

Keywords: Antigen presentation; Autoimmunity; Cancer; Dendritic cells; Immune checkpoints; Immunosurveillance; Macrophages; TCR; Viral infections

Although the late 19th century witnessed several groundbreaking discoveries that laid the foundation of modern immunology, including the identification of "germs" as the cause of infectious disease (by Robert Koch and Louis Pasteur) and the first descriptions of phagocytosis (by Élie Metchnikoff) (Smith, 2015; Cohn, 2017), it was not until the mid-1960s that the essential role of lymphocytes in adaptive immunity was established, owing to the pioneering work of Jim Gowans (Gowans and et al., 1962). Since then, an extensive experimental effort unveiled not only the remarkable phenotypic and functional diversity of lymphocytes but also their implication in a vast array of physiological and pathological processes (Smith and et al., 2017; Ebbo and et al., 2017; Rawlings and et al., 2017; Qi, 2016; Bjorkstrom and et al., 2016). $CD19^+$ lymphocytes expressing the so-called B-cell receptor (BCR, a monomeric transmembrane variant of a type M immunoglobulin) are commonly referred to as "B cells," reflecting their main anatomical site of maturation (the ***b***one marrow, in humans, and the ***b***ursa of Fabricius in birds). Similarly, $CD3^+$ lymphocytes expressing the so-called T-cell receptor (TCR), in any of its variants, are generally dubbed "T cells" or "T lymphocytes," with direct reference to the ***t***hymus (Murphy and Weaver, 2016). Both B and T cells employ their exclusive receptors for reacting to antigenic entities as they proliferate to establish an adaptive immune response. However, while activated B cells differentiate into plasma cells that secrete mature immunoglobulins (de facto, soluble variants of the BCR), the expansion of T cells is not accompanied by TCR secretion (Murphy and Weaver, 2016). Moreover, while the BCR can recognize both particulate and soluble antigens, most TCRs recognize only small peptides presented in the context of MHC molecules. That said, a small percentage (approx. 2%) of human T cells expresses a poorly diverse set of TCRs composed of γ and δ chains that recognize specific antigens irrespective of MHC presentation, de facto operating as an innate immune effector (Nielsen and et al., 2017; Mondragon and et al., 2016). Of note, the entire combinatorial diversity of the BCR and TCR repertoires in a single individual is (at least theoretically) capable to provide protection against virtually all possible antigenic challenges (Murphy and Weaver, 2016; Rudqvist and et al., 2018).

The cellular and molecular biology of T cells has major pathophysiological relevance. On the one hand, mature T cells are quintessential for the control of infections, not only as they can directly eliminate cells bearing intracellular pathogens—function performed by $CD8^+$ cytotoxic T lymphocytes (CTLs) (Tscharke and et al., 2015; Voskoboinik and et al., 2015)—but also as they secrete a large panel of cytokines that support innate and humoral immunity against extracellular pathogens—a function performed by both $CD4^+$ helper T (T_H) cells and $CD8^+$ CTLs (Bevan, 2004; Crotty, 2015; Gasteiger and Rudensky, 2014). Moreover, T cells have a major role in both natural and therapy-driven anticancer immunosurveillance (Kroemer and et al., 2015; Shankaran and et al., 2001). Thus, T cells are potent weapons that control organismal homeostasis and can (at least potentially) be harnessed for therapeutic purposes. Accordingly, when T cells are the target of viral infections, such as in the case of HIV-1, the profound immunosuppressive state that ensues T-cell depletion may have catastrophic consequences (in the absence of adequate clinical management) (Barouch and Deeks, 2014; Walker and Yu, 2013). On the other hand, T cells underlie the pathogenesis of various conditions including most (if not all) autoimmune disorders, graft rejection, and some hematological tumors (e.g., cutaneous T-cell lymphoma) (Espinosa and et al., 2016; Chong and Perkins, 2014; Suarez-Fueyo and et al., 2017; Petrelli and van Wijk, 2016; Tripodo and et al., 2009). Thus, in some clinical settings, the deregulated activity or proliferation of T cells constitutes a prominent therapeutic target.

In *The biology of T cells* (a thematic issue of the successful *International Review of Cell and Molecular Biology* series), world-leading experts in the field offer an overview of the major molecular and cellular aspects of T-cell biology and their links to human pathophysiology. Each of the chapters of *The biology of T cells* focuses on one specific facet of the development, phenotype, function, study, or pathophysiological relevance of T lymphocytes. The book (which consists of two parts, part A and part B) is equally addressed to expert investigators, who may wish to expand their knowledge on the biology of T cells, and to newcomers to this exciting and rapidly expanding area of research.

ACKNOWLEDGMENTS

LG is supported by an intramural startup from the Department of Radiation Oncology of Weill Cornell Medical College (New York, US), and by Sotio a.s. (Prague, Czech Republic).

REFERENCES

Barouch DH, Deeks SG. Immunologic strategies for HIV-1 remission and eradication. Science 2014;345:169–74.

Bevan MJ. Helping the CD8(+) T-cell response. Nat. Rev. Immunol. 2004;4:595–602.

Bjorkstrom NK, et al. Emerging insights into natural killer cells in human peripheral tissues. Nat. Rev. Immunol. 2016;16:310–20.

Chong AS, Perkins DL. Transplantation: molecular phenotyping of T-cell-mediated rejection. Nat. Rev. Nephrol. 2014;10:678–80.

Cohn M. Learning from a contemporary history of immunology. Immunol. Res. 2017;65:573–91.

Crotty S. A brief history of T cell help to B cells. Nat. Rev. Immunol. 2015;15:185–9.

Ebbo M, et al. Innate lymphoid cells: major players in inflammatory diseases. Nat. Rev. Immunol. 2017;17:665–78.

Espinosa JR, et al. Memory T cells in organ transplantation: progress and challenges. Nat. Rev. Nephrol. 2016;12:339–47.

Gasteiger G, Rudensky AY. Interactions between innate and adaptive lymphocytes. Nat. Rev. Immunol. 2014;14:631–9.

Gowans JL, et al. Initiation of immune responses by small lymphocytes. Nature 1962;196:651–5

Kroemer G, et al. Natural and therapy-induced immunosurveillance in breast cancer. Nat. Med. 2015;21:1128–38.

Mondragon L, et al. Immunosuppressive gammadelta T cells foster pancreatic carcinogenesis. OncoImmunology 2016;5:e1237328.

Murphy K, Weaver C. Janeway's Immunobiology. ninth ed.). Garland Science; 2016.

Nielsen MM, et al. Gammadelta T cells in homeostasis and host defence of epithelial barrier tissues. Nat. Rev. Immunol. 2017;17:733–45.

Petrelli A, van Wijk F. CD8(+) T cells in human autoimmune arthritis: the unusual suspects. Nat. Rev. Rheumatol. 2016;12:421–8.

Qi H. T follicular helper cells in space-time. Nat. Rev. Immunol. 2016;16:612–25.

Rawlings DJ, et al. Altered B cell signalling in autoimmunity. Nat. Rev. Immunol. 2017;17:421–36.

Rudqvist NP, et al. Radiotherapy and CTLA-4 blockade shape the TCR repertoire of tumor-infiltrating T cells. Cancer Immunol. Res. 2018;6:139–50.

Shankaran V, et al. IFNgamma and lymphocytes prevent primary tumour development and shape tumour immunogenicity. Nature 2001;410:1107–11.

Smith MJ, et al. B cells in type 1 diabetes mellitus and diabetic kidney disease. Nat. Rev. Nephrol. 2017;13:712–20.

Smith KA. Editorial: a living history of immunology. Front. Immunol. 2015;6:502.
Suarez-Fueyo A, et al. T cells and autoimmune kidney disease. Nat. Rev. Nephrol. 2017;13:329–43.
Tripodo C, et al. Gamma-delta T-cell lymphomas. Nat. Rev. Clin. Oncol. 2009;6:707–17.
Tscharke DC, et al. Sizing up the key determinants of the CD8(+) T cell response. Nat. Rev. Immunol. 2015;15:705–16.
Voskoboinik I, et al. Perforin and granzymes: function, dysfunction and human pathology. Nat. Rev. Immunol. 2015;15:388–400.
Walker BD, Yu XG. Unravelling the mechanisms of durable control of HIV-1. Nat. Rev. Immunol. 2013;13:487–98.

CHAPTER ONE

Transcriptional Programs Underlying Cd4 T Cell Differentiation and Functions

Fanny Chalmin[*,§,¶], **Etienne Humblin**[*,§,¶], **François Ghiringhelli**[*,§,¶,||,1] **and Frédérique Végran**[§,¶,||]

[*]Department of Medical Oncology, Centre Georges-François Leclerc, Dijon, France
[§]Centre de Recherche INSERM LNC-UMR1231, Dijon, France
[¶]Univ. Bourgogne Franche-Comté, Dijon, France
[||]Platform of Transfer in Cancer Biology, Centre Georges-François Leclerc, Dijon, France
[1]Corresponding author: E-mail: fghiringhelli@cgfl.fr

Contents

International Review of Cell and Molecular Biology, Volume 341
ISSN 1937-6448
https://doi.org/10.1016/bs.ircmb.2018.07.002

Abstract

Understanding the basis of cellular differentiation is a fundamental issue in developmental biology but also for the comprehension of pathological processes. In fact, the palette of developmental decisions for naive CD4 T cells is a critical aspect of the development of appropriate immune responses which could control infectious processes or cancer growth. However, the current accumulation of data on CD4 T cells biology reveals a complex world with different helper populations. Naive CD4 T cells can differentiate into different subtypes in response to cytokine stimulation. This stimulation involves a complex transcriptional network implicating the activation of Signal Transducer and Activator of Transcription but also master regulator transcription factors allowing the functions of each helper T lymphocyte subtype. In this review, we will present an overview of the transcriptional regulation which controls process of helper T cells differentiation. We will focus on the role of initiator transcriptional factors and on master regulators but also on other nonspecific transcriptional factors which refine the T helper polarization to stabilize or modulate the differentiation program.

1. INTRODUCTION

CD4 T cells are initially produced in a "naive" form having not yet encountered any antigen. Activation and acquisition of their functions require several signals. The first signal consists in the recognition of the antigen-major histocompatibility complex (MHC II) complex by the T cell receptor (TCR) in the presence of the CD4 coreceptor and the CD3 complex allowing the transmission of the signal. The antigen-MHC II complex is found on cells called antigen-presenting cells (APC) which may be macrophages, dendritic cells, or even B cells. On APCs, the binding of costimulation molecules such as CD80 and CD86 with CD28 ligand on T lymphocytes, constitutes the second essential signal for activation of CD4 T cells. These two signals allow the survival and proliferation of T lymphocytes but are not sufficient to obtain a functional differentiation. A third signal is needed and is provided by cytokines secreted by the cells from the microenvironment. The diversity of this latter signal explains the multitude of functions played by CD4 T cells. CD4 T cells proliferate and polarize into different effector CD4 T cells depending on their interaction with the recognized antigen and cytokines in the environment. Cytokines participate in the differentiation of helper T cells by the activation of the *JAnus tyrosine Kinase* (JAK) and *Signal Transducer and Activator of Transcription* (STAT) signaling pathway.

Initially, immunologists described two types of immune responses that required the action of CD4 T cells. The first response is mediated by the humoral response and the second by cell-mediated immunity. Mosmann and Coffman first recognized that CD4 T cells could be subdivided into two distinct populations according to their secretions and functions. CD4 lymphocytes are thus separated according to their inflammatory properties or their ability to secrete interleukin (IL)-4 (Mosmann and Coffman, 1989). The first description of in vitro lymphocyte differentiation was reported simultaneously by the groups of Dr. Susan Swain and Dr. William Paul in 1990. They demonstrated that stimulation of naive CD4 T cells with TCR ligands and IL-4 allowed to cells, after a few days of culture, to acquire the ability to secrete IL-4 (Swain et al., 1990a, 1990b; Le Gros et al., 1990). These cells, called Th2, are characterized by a high production of IL-4, IL-5, and IL-13 and to a lesser extent IL-10 and IL-9. Th2 lymphocytes are necessary for protection against extracellular parasites. For in vitro differentiation of Th2 cells, IL-4 and IL-2 are required. In 1993 Anne O'Garra, Ken Murphy, and their colleagues demonstrated that naive CD4 T cells could acquire ability to secrete interferon γ (IFNγ) in vitro (Hsieh et al., 1992). IL-12 secreted by dendritic cells and macrophages plays a central role in the polarization of these CD4 T cells producing IFNγ, called Th1. These lymphocytes produce, in addition to IFNγ, IL-12, IL-2, and lymphotoxin α. Th1 cells coordinate the immune response against intracellular microorganisms.

Immunologists have attributed the development of autoimmune diseases to the action of Th1 cells. However, even if the neutralization of IL-12 reduces the induction of experimental autoimmune encephalomyelitis (EAE), a murine model for multiple sclerosis, mice deficient for IFNγ are paradoxically more sensitive to the development of EAE. In 2000 Kastelein's team discovered IL-23, which shares the same subunit (IL-12p40) as IL-12, which led to a reassessment of the relative contributions of IL-12p70 and IL-23 in the induction of EAE (Oppmann et al., 2000). In 2003 the demonstration is made that IL-23, and not IL-12p70, plays the leading role in induction of EAE (Becher et al., 2003). The link between IL-23 and IL-17 expression permitted to identify Th17 cells as different from conventional Th1 and Th2 cells. Indeed, Th17 cells express low levels of T-bet (T-box protein expressed in T cells) and GATA-binding protein 3 (GATA3) and produce neither IFNγ nor IL-4, cytokines which suppress the differentiation of Th17 cells. In 2006 Stockinger, Weaver, Kuchroo, and their colleagues each showed that Th17 cells could be induced in vitro from naive mouse CD4 T cells by TCR stimulation in the presence of IL-6 and transforming growth factor (TGF-β; Bettelli et al., 2006; Mangan et al., 2006; Veldhoen et al., 2006). Th17 cells play a role in immune responses against extracellular bacteria and fungi. They are involved in the induction of many organ-specific autoimmune diseases.

The immune system uses many mechanisms to maintain immunological self-tolerance and protects the host against exacerbated immune responses. The concept of "suppressive" T cells was issued since the early 1970s. The authors observed that thymectomy 2—3 days after birth induces the development of an autoimmune disease thus suggesting that some immunosuppressive cells are generated in the thymus after birth to prevent the development of an autoimmune disease (Nishizuka and Sakakura, 1969). It is only in 1995 that an important study conducted by Sakaguchi and colleagues described a unique $CD4^{+}CD25^{+}$ T cell population with potent regulatory activity (Sakaguchi et al., 1995). Regulatory T cells (Treg) that actively suppress the function of conventional T cells are a key regulator by which the immune system limits inappropriate or excessive responses. There are two types of $CD4^{+}$ Treg cells, the "natural" Treg (nTreg) and the induced Treg (iTreg). nTreg cells develop in the thymus during positive and negative selection, whereas iTreg cells are induced in the periphery, after antigenic stimulation. In 1997 a major breakthrough was achieved by Dr Roncarolo's team. Based on the study of patients with severe combined immunodeficiency, they were able to highlight the immunosuppressive role of IL-10, capable of inhibiting a graft versus host reaction following transplantation (Bacchetta et al., 1994). Subsequently, they were able to show that the exogenous addition of this cytokine to human or murine CD4 T cells leads to the development of highly immunosuppressive type 1 (Tr1) regulatory cells secreting large amounts of IL-10 and TGF-β (Groux et al., 1997). In 2001 the study of patients with immunodysregulation polyendocrinopathy enteropathy

X-linked 9 (IPEX) syndrome and the study of Scurfy mice, which develop an X-linked lymphoproliferative disease, allowed to link the development of severe autoimmunity to *Forkhead box 3* (Foxp3) protein mutations (Bennett et al., 2001; Brunkow et al., 2001). It was in 2003 that Foxp3 protein was identified as the transcription factor responsible for the Treg polarization (Walker et al., 2003). Foxp3 is important in the development of nTreg and some iTreg cells, but it is important to note that it is not required for Tr1 cell polarization. Both nTreg and iTreg contribute to a balance between peripheral tolerance and immune responses to infections.

In 1994 Schmitt et al. described for the first time the production of IL-9 by naive CD4 T lymphocytes in the presence of IL-2 under TGF-β stimulus (Schmitt et al., 1994). Similarly, the authors found that adding IL-4 improved the release of IL-9. Overall, this suggests that these cytokines act synergistically and have a dose-dependent effect on IL-9 production. This study also highlighted the inhibitory effect of IFNγ on IL-9 production by neutralizing the effects of IL-4. However, it is only 15 years later, in 2008 that Veldhoen et al. discovered a subpopulation of T-lymphocytes specialized in the secretion of IL-9 (Veldhoen et al., 2008b). This study demonstrated the emergence of IL-9 "preferentially" producing T cells in response to TGF-β in the presence of IL-4 and in the absence of IL-6. It appears that TGF-β is capable of profoundly altering the transcriptional and secretory profiles of Th2 cells, leading to the "reprogramming" of this subtype into secretory IL-9 cells. This confirmed the identification of a new CD4 T subpopulation, different from those already described, with high expressions of IL-9 and IL-10 and thus called Th9 lymphocytes. At the same time, the study by Dardalhon et al. confirmed that a combination of IL-4 and TGF-β allowed to polarize CD4 T cells into a subset of IL-9 secretory cells (Dardalhon et al., 2008).

The name « Tfh (*T follicular helper*) cells » was firstly used in 2000 to describe a subpopulation of CD4 T cells present in human lymphoid tissues and expressing the chemokine receptor C-X-C chemokine receptor (CXCR5). The role of this cell population is mainly to allow the humoral response involving B cells. The expression of CXCR5 together with a weak expression of the CCR7 (*C—C chemokine receptor*) allows the Tfh cells to move from the T cell zone to the B cell follicles where they will directly support B lymphocyte expansion and differentiation (Breitfeld et al., 2000; Schaerli et al., 2000).

Current knowledge reveals a differentiation pattern common to the different T helpers (Fig. 1). CD4 T cells differentiate depending on the

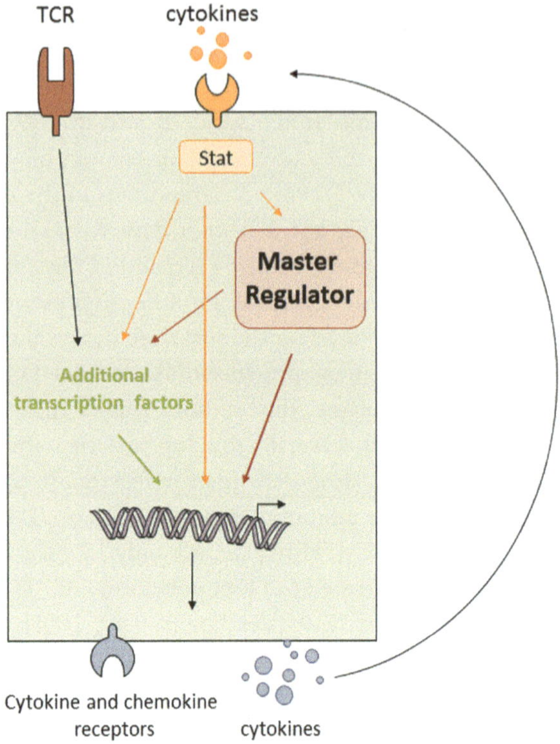

Figure 1 After antigenic stimulation, TCR engagement associated with the costimulation signals contributes to the activation and proliferation of a $CD4^{+}$ T lymphocyte. The orientation of the $CD4^{+}$ T cell differentiation is then a reflection of the cytokinic environment. STAT proteins (in orange) are critical mediator of extracellular stimuli, especially cytokines. These proteins are directly responsible for master regulator expression (in red). The functional specialization of $CD4^{+}$ T lymphocytes is finally coordinated by a specific transcriptional program built around this master regulator. In addition to Stat and master regulator, this transcriptional network required additional transcription factors (in green). Cooperation between all these factors leads to a subset-specific gene expression (cytokines, cytokines, and chemokines receptors; in blue). Cytokines produced by T cells contribute to the maintenance of polarization.

microenvironment and cytokines present. A true transcription factor that initiates CD4 T polarization, STAT proteins are activated following the recognition of a cytokine by its receptor (Fig. 1). Depending on the cytokine(s) encountered, the landscape of activated STAT differs and leads to the development of different T helper. These initiating factors are responsible, among other things, for the expression of a "master regulator" (Fig. 1). These factors support a central role in CD4 T cell polarization. Specific to each T subtype, these factors in collaboration with the STAT proteins are necessary and often sufficient for T cell polarization. These master regulators support the identity of the T helper subtypes and are important for stabilizing the cellular phenotype, in particular by limiting the polarization toward another T subset. Together the STAT proteins, the master regulators and the TCR commitment allow the expression of a multitude of transcriptional factors (Fig. 1) involved in the survival of cells, the maintenance of their phenotype, the establishment of their functions, or their migratory capacity. All these actors are organized in a complex transcriptional network specific to each subtype and resulting in the production of cytokines, chemokines, and receptor specific to T helper (Fig. 1) and supporting their biological functions. Some subset however makes exceptions to this scheme, in fact, although the development of Th9 cells requires the activation of STAT5 and STAT6, the expression of its PU.1 master regulator dependents on the signaling pathways induced by the TGF-β. Moreover, although many transcriptional factors are important to the development of Tr1 cells, none of them appears as the master regulator due to their lack of specificity for the subtype.

2. STAT FAMILY IN CD4 DIFFERENTIATION

STAT proteins are a common and essential signaling pathway for the T helper lymphocyte differentiation program (O'shea and Plenge, 2012). These proteins are latent cytoplasmic transcription factors until their activation by extracellular signaling proteins, mainly cytokines bind to specific receptors on the cell surface. These extracellular signaling proteins cause activation of various tyrosine kinases, including JAKs (Darnell et al., 1994) capable of activating STAT proteins by phosphorylation (Table 1). Activated STAT proteins thus form dimers that will accumulate in the nucleus of cells, bind to DNA, and induce the expression of genes important for the development and functions of differentiated T cells, thus establishing a transcription program unique to each T helper subset. To induce effective transcriptional activity, STAT proteins are able to recruit additional transcription factors (Levy and Darnell, 2002) and induce local chromatin remodeling. STATs modify the acetylation status of histones through the recruitment of histone acetyltransferases (HAT) such as CBP/p300 but also to limit the recruitment of histones methyltransferases in order to facilitate access to chromatin to transcription factors (O'shea et al., 2011).

Table 1 STAT biology in T cells: description of cytokines that induce STAT activation and consequences of STAT deficiency in mice

STAT	Cytokines	Phenotype of Stat Deficient Mice	References
STAT1	IFNγ, IFNα/β, IL-6, IL-21, IL-27, IL-5, IL-7, IL-9, IL-10, IL-35	• Susceptible to viral disease • Unable to respond to IFN	Durbin et al. (1999), Meraz et al. (1996)
STAT2	IFNα/β	• Susceptible to viral disease	Park et al. (2000)
STAT3	IL-6, IL-21, IL-23, IL-10, IL-5, IL-7, IL-9, IL-15, IL-27	• Early embryonic lethality Conditional deficiency: • Resistant to EAE, defect in IL-17 production • Enhance Treg development	Takeda et al. (1997), Yang et al. (2007), Harris et al. (2007)
STAT4	IL-12, IL-35	• Defect in Th1 responses	Harris et al. (2007), Kaplan et al. (1996b)
STAT5a/b	IL-2, IL-3, IL-7, IL-15, IL-5, IL-9, IL-21	• Perinatal lethality (99%) • Defect in B and T cell development and proliferation	Yao et al. (2006)
STAT6	IL-4, IL-13, IL-3	• Defect in Th2 response and B cell activation • Reduced airway response to allergen • Aberrant Th1 and Th17 response	Kaplan et al. (1996a), Summers et al. (2011)

2.1 Th1: STAT4/STAT1 and STAT5

The differentiation of Th1 cells involves the presence of IL-12p70 in the microenvironment. STAT4 is a critical mediator of IL-12–stimulated gene regulation in Th1 cells. IL-12 activates the Janus family tyrosine kinases JAK2 and Tyk2, which in turn phosphorylate STAT4 on tyrosine 693 (Y693) (Higashi et al., 2005). The recognition of this cytokine by its receptor, IL-12R, initiates the activation of the STAT4 signaling protein, essential for the polarization of Th1 cells. Indeed, mice deficient for *STAT4* present an impaired Th1 development in response to intracellular pathogens (Kaplan et al., 1996b; Thierfelder et al., 1996). STAT4 is directly responsible for the expression of many specific Th1 cell genes, including *Tbx21*, encoding the master regulator T-bet, but also the *Ifng* and *Il12rb2* genes (Lund et al., 2004; Hoey et al., 2003). The secreted IFNγ, a specific cytokine of Th1 cells, also participates in the maintenance of the phenotype by a positive feedback loop. IFNγ signaling results in STAT1 activation by phosphorylation of tyrosine 701 (Y701) by Janus kinases JAK1 and JAK2 (Darnell et al., 1994). The STAT1 activation also contributes to *Tbx21* and *Ifng* expression (Zhu et al., 2012; Lighvani et al., 2001). Together STAT1 and STAT4 are necessary for the development and maintenance of the Th1 phenotype.

While STAT5 is an important regulator of Th2 and Th9 cell development, it is also necessary for the polarization of Th1 cells. Indeed, the blockade of JAK3, the kinase upstream of the STAT5 activation inhibits the production of IFNγ in Th1 cells (Shi et al., 2008). Similarly, IL-2 inhibition suppresses polarization of Th1 cells (Liao et al., 2011). Activation of the IL-2/STAT5 pathway is necessary for the expression of *Il12rb2* and *Tbx21*, making it possible to sensitize cells to IL-12 and differentiate into Th1 (Liao et al., 2011).

2.2 Th2: STAT5/STAT6

IL-4, the cytokine characteristic of Th2 cells, is also responsible for their polarization. The IL-4 binding to its receptor results in the activation of Janus kinase JAK1 and JAK3 responsible for the STAT6 activation. STAT6 is phosphorylated on Y641 and translocated to the nucleus in response to IL-4 (Sherman et al., 2002). STAT6 is necessary and sufficient to induce the expression of the GATA3 key transcription factor (Zhu et al., 2001). STAT6 is also required for the expression of many additional transcription factors such as interferon regulatory factor (IRF4), c-Maf, or Gfi-1 (*growth factor independent 1 transcriptional repressor*) that are necessary for the development of Th2 cells. That is why STAT6-deficient mice present a defective Th2 response (Kaplan et al., 1996a; Shimoda et al., 1996). IL-2 is involved in the proliferation and survival of Th2 cells but also in maintaining their phenotype and functions. In the absence of IL-4, IL-2 is able to induce a modest polarization of naive T cells into Th2 (Shimoda et al., 1996). In response to IL-2, JAK3 induce the phosphorylation of tyrosine 694 of STAT5a and tyrosine 699 of STAT5b (Lin et al., 1996). STAT5 regulates IL-4 receptor (IL-4Rα) expression and improves cell sensitization to IL-4 (Liao et al., 2008). In addition, STAT5 participates in maintaining the expressions of GATA3 and IL-4. The activation of STAT5 also induces the expression of *NLR Family Pyrin Domain Containing 3* (NLRP3), which plays an important role as a transcription factor in Th2 cells and participates in the production of Th2-like cytokines (IL-4, IL-5, and IL-13; Bruchard et al., 2015).

2.3 Th17: STAT3

The Th17 cell polarization requires the presence of IL-6 but also TGF-β. IL-21 which is secreted by the Th17 cells ongoing their differentiation amplifies it, and IL-23 stabilizes the phenotype. STAT3 is the main signal transducer for IL-6, IL-21, and IL-23 (Zhou et al., 2007; Korn et al., 2007; Nurieva et al., 2007). STAT3 activation requires phosphorylation on tyrosine 705 (Y705) by Janus kinase JAK1/2 and serine 727 (S727) by *mitogen-activated protein kinase* (MAPK) (Wen et al., 1995). STAT3 is indispensable for the production of IL-17 and, its deletion leads to the loss of IL-17 production (Zhou et al., 2007). STAT3 is directly responsible for the expression of *Rorc* and *Rora*, encoding the RORγt and RORα (*Retinoic acid receptor—related Orphan Receptor* $\gamma t/\alpha$) master regulators of Th17 cells (Zhou et al., 2007; Yang et al., 2007). STAT3 is also involved in the expression of the c-Maf factor required for IL-21 production and the regulation of IL-23 receptor expression (IL-23R) by Th17 cells (Durant et al., 2010). In addition, IL-6 and the activation of STAT3 are also necessary to limit the action of TGF-β, in particular, by blocking the expression of Foxp3 factor and thus the development of Treg cells (Laurence et al., 2012). Finally, in the presence of TGF-β, STAT3 is also involved in the development of immunosuppressive functions of Th17 cells. First, TGF-β inhibits the expression of the Gfi-1 repressive factor, and then STAT3 induces the expressions of *Entpd1* and *Nt5e* genes encoding, respectively, for the CD39 and CD73 ectonucleotidases (Chalmin et al., 2012).

2.4 Treg: STAT5

Induced regulatory T cells (iTreg) that differentiate at the periphery are obtained in the presence of TGF-β and IL-2. While TGF-β is directly responsible for the expression of Foxp3, STAT5 is also necessary for the regulation of its expression, notably, by improving the accessibility of chromatin to other transcription factors by epigenetic modifications like demethylation (Burchill et al., 2007; Lal et al., 2009). Activation of STAT5 after binding of IL-2 to its receptor CD25 is essential for the development of Tregs. STAT5 enhances Treg cell functionality by the regulation of the *Foxp3* gene on which it binds and stabilizes the suppressor phenotype. In particular, the conserved noncoding sequence (CNS2) enhancer region, which binds several transcription factors, including STAT5, has been found to be required for the maintenance of *Foxp3* expression (Feng et al., 2014). One study has shown that STAT5 binding to CNS2 enhances the stability of Treg cells in inflammatory contexts (Liu et al., 2015). Like STAT3 that limits the polarization of Treg cells, STAT5 limits the development of Th17 cells and suppresses the IL-17 secretion (Laurence et al., 2007). In the same way, it prevents the Tfh polarization (Johnston et al., 2012; Nurieva et al., 2012).

2.5 Tr1: STAT3/STAT1

IL-27, a cytokine belonging to the IL-12 family, is a critical regulator of Tr1 cell development. Its receiver consists of the two subunits WSX1 and gp130 responsible for the respective activation of the channels JAK1/STAT1 and JAK2/STAT3, all necessary for the Tr1 polarization. The STAT3 activation plays a central role in the Tr1 cell development. As in the Th17 cells, STAT3 allows the expression of the c-Maf factor, itself responsible for regulating the cytokines characteristic of Tr1 cells, IL-10, and IL-21. STAT3 also induces IL-21 receptor expression improving Tr1 cell sensitivity to IL-21 (Pot et al., 2009). In response to this cytokine, STAT3 is also activated allowing the phenotype maintenance. STAT3 also participates in the expression of the *Icos* costimulation molecule necessary for the amplification and maintenance of the Tr1 phenotype. When Tr1 cells are polarized, STAT3 is also required to limit *Foxp3* expression and iTreg polarization. Following its activation, STAT3 is capable of binding to the *Foxp3* promoter in a negative regulatory region. This fixation reduces the acetylation of the promoter and the accessibility to the Smad3 transcription factor involved in Foxp3 expression (Huber et al., 2008; Xu et al., 2010). STAT1 activation by IL-27 is essential for Tr1 cell development. The STAT1 signaling pathway leads to the expression of the T-bet factor by Tr1 cells. T-bet leads to the production of IFNγ by Tr1 cells but also limits Th17 polarization. STAT1 via the expression of T-bet represses the expression of the RORγt factor (Lazarevic et al., 2011), thus the mice deficient for *Stat1* or *Tbx21* (encoding for T-bet) present an increased number of Th17 cells in particular during systemic inflammation (Villarino et al., 2010). Similarly, in the absence of STAT1, IL-27 induces the development of Th17 cells (Peters et al., 2015).

2.6 Tfh: STAT3/STAT1

IL-6 is the key cytokine for inducing polarization induction of Tfh in a STAT3-dependent manner. STAT3-deficient CD4 T cells have a profound defect in Tfh cell differentiation, accompanied by decreased *germinal center* (GC) B cells and antigen-specific antibody production (Ray et al., 2014). Activation of STAT3 by IL-6 is responsible for the expression of the Bcl6 (*B-Cell CLL/Lymphoma*) master gene and IL-21, a cytokine produced by Tfh cells. The activation of STAT3 is also important to limit the expression of the IL-2 receptor and thus the activation of STAT5, a Tfh polarization suppressor (Choi et al., 2013). IL-21 secreted by Tfh cells is also capable of activating the STAT3 signaling pathway and maintaining the Tfh phenotype (Liu et al., 2013). While STAT3 plays a major role in the Tfh cell differentiation, the absence of STAT1 has also been shown to limit the development of Tfh cells. More modestly than STAT3, STAT1 can be activated downstream of IL-6 and IL-21—induced signaling pathways. STAT1 is also necessary for early expression of Bcl6 and secretion of IL-21 by Tfh. Type I interferons, through the activation of STAT1, improve the expression of the *Bcl6* master gene, as well as the *Cxcr5* and *Pdcd1* genes characteristic of the Tfh signature (Ray et al., 2014; Choi et al., 2013).

2.7 Th9: STAT5/STAT6 and STAT1

Initially considered as a particular subtype of Th2 cells, Th9 cells also require the activation of STAT5 and STAT6 for their development. STAT6 is necessary for the expression of the transcription factors GATA3, IRF4, and basic leucine zipper ATF-like transcription factor (BATF), essential for the IL-9 production. However, in Th9 cells, STAT6 also limits the expression of key factors T-bet and Foxp3 that could be responsible for inhibiting the Th9 polarization. As in the Th2 cells, STAT5 participates in the expression of IL4Rα. STAT5 intervenes directly in the expression of IL-9 by Th9 cells and participates in the expression of the IRF4 factor, necessary for the development of this subtype. STAT5 also limits the action of Bcl6 factor, a repressive factor of polarization Th9 (Bassil et al., 2014; Liao et al., 2014).

STAT1 plays an ambivalent role in the development of Th9 cells. On the one hand, IL-1 is capable of improving IL-9 production by CD4 T cells (Schmitt et al., 1991), and activation of STAT1 plays an important role in this induction. Indeed, naïve human CD4 T cells differentiated in Th9 in the presence of STAT1 activating cytokines (IFNα, IFNβ, IL-6, or IL-21) develop a higher IL-9 production (Wong et al., 2010). Similarly, in murine Th9 cells, the activation of STAT1 by IL-1β is responsible for the induction of the IRF1 expression, which greatly improves the expression of *Il9* and *Il21* (Vegran et al., 2014). On the other hand, IFNγ and IL-27 are two cytokines capable of inducing the polarization of Th1 cells, thereby inhibiting Th9 polarization in a STAT1 and T-bet–dependent manner (Murugaiyan et al., 2012). Similarly, IFNγ induces STAT1 binding to *Il9* promoter during in vitro Th9 cell differentiation drastically suppresses IL-9 secretion, notably, by reducing the ability of cells to respond to IL-4 (Schmitt et al., 1994).

3. MASTER REGULATORS OF CD4 T CELLS SUBSETS

The so-called "master regulator" transcription factors are indispensable for the determination of the helper T cell phenotype and the cytokine production. These transcriptional factors are specifically expressed in a particular T cells population and play a major role in stabilizing the phenotype of the fine tune of polarization. The transcriptional activity of these factors is mainly determined by their high level of expression within the cell. STATs and activated Smads proteins are directly responsible for the expression of the master regulators during the CD4 T cell differentiation process. Thus, just like the activation of STAT proteins that reflect the microenvironment, the expression of these factors depends on the cytokines present (Fig. 2). These master regulator factors, together with STAT-activated proteins, are directly responsible for the production of key cytokines produced by the various T helper subtypes.

Many years of studies have allowed the identification of the critical transcription factors necessary for the development of different T helper subsets. Thus T-bet, GATA3, RORγt, Foxp3, Bcl6, and PU.1 are the master regulators of the subtypes Th1, Th2, Th17, Treg, Tfh, and Th9, respectively.

3.1 T-Bet: Th1 Master Regulator

Th1 differentiation is initiated by IL-12 and IFNγ (Trinchieri et al., 2003). IL-12 present in the cellular environment induces the production of IFNγ by natural killer (NK) cells. This extracellular IFNγ binds to its receptor (IFN-γR) leading to the activation of STAT1 in the lymphocyte cytoplasm. Activated STAT1 can then act as a transcription factor and induce T-bet

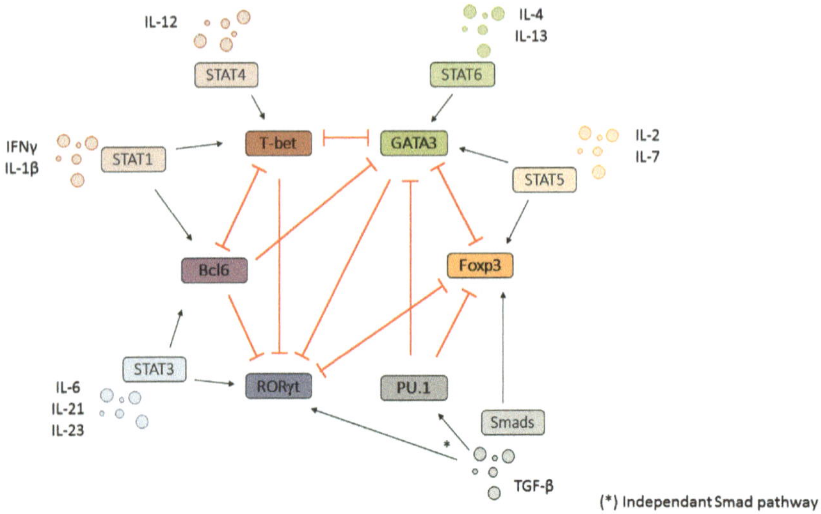

Figure 2 Through STAT proteins activation, cytokines present in the microenvironnement dictates master regulator expression in each T cell subset. T-bet, GATA3, RORγt, Foxp3, Bcl6, and PU.1 are the master regulators of the subtypes Th1, Th2, Th17, Treg, Tfh, and Th9, respectively. These factors are not only essential for the orientation of the transcriptional program toward a particular T-subtype helper but they also limit the polarization of cells toward another subset.

expression, directly regulating the transcription of *Ifng* and the *il12 receptor* (*Il12r*) in Th1 cells (Lighvani et al., 2001). In parallel, IL-12p70 binds to its receptor inducing STAT4 phosphorylation that will then translocate into the nucleus to transactivate *Ifng*, *Il12rb2,* and *Tbx21* genes (Lovett-Racke et al., 2004). Importantly, IFNγ produced by the Th1 cells acts in an autocrine way to support their differentiation through a positive control loop. T-bet also significantly suppresses expression of IL-2 (induced by TCR stimulation). IL-2 is an essential cytokine for activation, proliferation, and differentiation of T helper cells. This role of T-bet on IL-2 production is mediated by the inhibition of nuclear factor κB (NF-κB) p65 in Th1 and Th2 differentiation conditions (Hwang et al., 2005a).

The T-bet induction is also responsible for the repression of Th2 and Th17 differentiation (Lazarevic et al., 2011; Lugo-Villarino et al., 2003). Indeed, T-bet represses the transcription of the gene coding for IL-4 (Djuretic et al., 2007) and inhibits GATA3, the central transcription factor of Th2 lymphocytes (Hwang et al., 2005b). Consistently, T-bet suppression spontaneously leads to the induction of Th2 both in vitro and in vivo (Szabo et al., 2000, 2002). Moreover, T-bet blocks the transcription of *Rorc* that encodes the RORγt protein, the main transcription factor of Th17 differentiation (Lazarevic et al., 2011). Other proteins such as runt-related transcription factor 3 (RUNX3; Djuretic et al., 2007) or Eomes (*Eomesodermin*) (Fujio et al., 2010) interact with T-bet to promote or inhibit Th1 cell differentiation respectively. Finally, it has been described that T-bet plays a transcriptional repressing role by interacting directly on the promoter regions of genes encoding SOCS1 and SOCS3 (*suppressor of cytokine signaling*) and thus inhibiting their expression and increasing the IFNγ production by Th1 cells (Oestreich et al., 2011). During the advanced stage of the Th1 polarization, T-bet recruits and associates with the Bcl6 transcriptional repressor to negatively regulate the activity of the *Ifng* locus. This mechanism limits the overproduction of IFNγ and thus prevents autoimmune disorders. T-bet expression also blocks the Tfh differentiation by favoring Th1 differentiation (Nakayamada et al., 2011). These data suggest that T-bet is a transcription factor allowing a tight regulation of Th cell development.

Thus, in patients, the *Tbx21* mutations, gene coding for T-bet, are often associated with a Th1 response defect and a dysregulation of Th1/Th2 or Th1/Th17 balance. *Single-nucleotide polymorphisms* are described in diseases such as asthma (Chung et al., 2003), rheumatoid arthritis (Chae et al., 2009), or systemic sclerosis (Gourh et al., 2009), suggesting that T-bet is a transcription factor allowing a tight regulation of Th cell development.

3.2 Gata3: Th2 Master Regulator

The exposure of naive CD4 T cells to IL-4 and IL-2 allows them to acquire Th2-like functions. IL-4 initiates a signaling cascade leading to phosphorylation of STAT6 (Kaplan et al., 1996a). Phosphorylated STAT6 is responsible for the transcription of the gene encoding GATA3, the master regulator of Th2 differentiation (Zhu et al., 2006b). The GATA3 expression is required for Th2 differentiation since its absence leads to the abrogation of their differentiation (Zhang et al., 1997). The transcriptional activity of GATA3 is responsible for the production of IL-5 and IL-13 and reinforces the expression of IL-4 (Agarwal et al., 2000).

Once induced, GATA3 increases its own expression directly (Ouyang et al., 2000) and via Dec2 transcription factor (Yang et al., 2009). GATA3 is essential for the early steps of Th2 differentiation and for already differentiated Th2 activation (Pai et al., 2004; Zhu et al., 2004). Indeed, even after their differentiation, the loss of GATA3 expression decreases cytokine production as well as proliferation and increases the *Ifng* expression.

GATA3 cooperates with other transcription factors such as Runx, Ets, *Activator Protein 1* (AP-1), *nuclear factor erythroid 2–related factor 1* (Nrf1), and *friend leukemia integration 1* (FLI1). Using a ChIP-Seq analysis, this study showed the presence of GATA3 binding in other TCD4 subtypes. For example, the binding sites located within the loci of *cytotoxic T-lymphocyte–associated antigen 4* (*Ctla4*) and *Icos* (*inducible T-cell COStimulator*) are shared by different types of T lymphocytes including Treg, Th1, Th2, and Th17. These results indicate that even though GATA3 targets many genes within Th2 cells, it could also be involved in the regulation of other cell subtypes by acting within larger transcription factor complexes (Wei et al., 2011).

Although the expression of GATA3 is weaker in Th9 than in Th2 cells (Veldhoen et al., 2008b; Goswami et al., 2012), it is an important STAT6 target gene implicated in Th9 differentiation. Indeed, Dardalhon et al. showed that in vitro Th9 polarized cells deficient in GATA3 lose their ability to produce IL-9 (Dardalhon et al., 2008). However, the GATA3 functions in the Th9 polarization are still badly characterized with divergent data regarding its impact on IL-9 secretion (Goswami et al., 2012; Tamiya et al., 2013; Veldhoen et al., 2008b). Indeed, in the absence of GATA3, a high expression of FoxP3 is observed in the Th9 cells leading to the inhibition of the IL-9 secretion (Dardalhon et al., 2008). Thus a balanced expression of GATA3 within Th9 could be responsible for the FoxP3 inhibition (Mantel et al., 2007) without activating the transcriptional program of Th2 cells. However, too strong expression of GATA3 tends to repress the transcriptional activity of PU.1 that is essential for Th9 polarization, leading to the expression of Th2 genes.

Intrinsic expression of GATA3 in Treg is required for their accumulation in inflammatory sites and to maintain FoxP3 expression. However, GATA3 limits the Treg polarization to an effector T cell phenotype producing inflammatory cytokines (Rudra et al., 2012). Indeed, in vivo and in vitro, under inflammatory conditions, GATA3-deficient Treg cells produce a high level of IL-17 (Wang et al., 2011). Therefore GATA3 seems to be able to suppress a Th17 differentiation program within the Treg cells. The observation made by Wei et al. that GATA3 binds to the *Il17* and *Rorc* locus suggests that GATA3 could directly regulate the Th17 differentiation program independently of FoxP3 expression (Wei et al., 2011). GATA3 is also involved in Th1 differentiation repression through reduction of the STAT4 expression (Usui et al., 2003). Similarly, by interacting with RUNX3, it inhibits the Eomes expression and the IFNγ production

(Yagi et al., 2010). It has also been reported that GATA3 recruits Gfi-1. This transcription repressor is early and transiently induced after IL-4/STAT6 signaling and TCR stimulation. Gfi-1 promotes selective proliferation of overexpressing GATA3 cells (Zhu et al., 2002, 2006a).

3.3 RORγt: Th17 Master Regulator

RORγt (or RORC2 in humans) is a nuclear receptor that plays an important role in the early development of Th17 subset (Sun et al., 2000). In the small intestine lamina propria, T cells expressing RORγt are IL-17 producing unlike non-RORγt expressing cells. In vitro, Th17 lymphocytes polarized by IL-6 and TGF-β need the expression of RORγt to produce IL-17. Furthermore, it has been also shown that a forced expression of *Rorc* is sufficient for CD4 T lymphocytes to produce IL-17 even in the absence of polarizing cytokines (Ivanov et al., 2006). RORγt could directly bind to the *Il17* promoter and thus favoring its expression. The fundamental role of RORγt has been also highlighted in several experiments using RORγt-deficient mice. These mice develop less severe autoimmune diseases. RORα, another member of the retinoid nuclear receptor family, is also expressed similarly to RORγt in Th17 cells. A study by Yang et al. (2008b) has indicated that RORα could play a similar but not identical role to RORγt in Th17 differentiation, suggesting that Th17 polarization could be triggered by these two specific transcription factors. RORα binds to the CNS2 of *Il17*. Overexpression of both RORα and RORγt increases *Il17* expression in a synergistic way. On the contrary and consistent with this finding, their deletion blocks the Th17 differentiation. Although RORγt is clearly required for IL-17 production, it also coordinates the activity of many important transcription factors (Ciofani et al., 2012) including Batf, Irf4, and STAT3.

A previous study has linked an alteration of IL-17—mediated immunity to the development of chronic mucocutaneous candidiasis disease (CMCD; Puel et al., 2011) characterized by recurrent or persistent infections of skin, nails, and oral and genital mucosa caused by *Candida albicans*. A recent study revealed that a *Rorc* mutation leads to a weakened immunity against *Candida* and *Mycobacterium* responsible for the development of CMCD (Okada et al., 2015). The functional loss of RORγt results in an absence of IL-17 secretory T cells and a weakening of mucosal immunity due in part to the absence of development of *mucosal-associated invariant T cells*. This study highlights the importance of RORγt factor for IL-17 secretion and Th17-mediated immunity.

RORγt is also involved in the inhibition of Treg differentiation since the loss of RORγt expression leads to an increase in the FoxP3 expression, the main transcription factor of Treg cells (Burgler et al., 2010). Indeed, the development programs of Th17 and Treg are tightly interconnected. During TCR stimulation, a naïve T cell can express FoxP3 and become Treg cell under TGF-β stimulation. However, in the same condition but in the presence of IL-6 or IL-21, the Treg differentiation is abrogated in favor of the Th17 differentiation (Chen et al., 2003; Kretschmer et al., 2005). IL-6 plays a central role in the balance between Th17 and Treg differentiation by favoring Th17 (Bettelli et al., 2006). The mechanism of action through which IL-6 and IL-21 control this balance is by the *Foxp3* and *Rorc* expressions (Zhou et al., 2008). In line with this, conditional suppression of FoxP3 in adult mice allows an increase in *Rorc* and *Il17* expression (Li et al., 2007b). It is clear that TGF-β is required for both FoxP3 and RORγt expression although the downstream signaling cascade leading to one or the other has to be different (Yang et al., 2008a). RORγt, RORα, and FoxP3 can bind to each other and reciprocally inhibit their functions (Zhou et al., 2008; Du et al., 2008). This association is probably at the basis of the balance between the Treg and Th17 cells. This relationship is corroborated by the results obtained in *Il6*-deficient mice presenting a severe default in generating Th17 cells and simultaneously increasing the number of Treg in the peripheral repertoire (Bettelli et al., 2006; Korn et al., 2007).

3.4 Foxp3: Treg Master Regulator

With TGF-β and IL-2 in the environment, naive CD4 T cells differentiate into iTreg. In response to TGF-β, iTreg expresses the *Foxp3* transcription factor that favors both the development and immunosuppressive functions of these cells (Chen et al., 2003). These cells produce TGF-β and act in an autocrine manner to organize a feedback loop and support their expansion. The Smad2 and Smad3 proteins are activated downstream of TGF-β signaling to induce *Foxp3* expression. Smad3 also plays a role in the differentiation of iTreg by blocking *Rorc*'s expression (Martinez et al., 2009).

FoxP3 expression is crucial in the gene regulation involved in the suppressive functions of Treg. Scurfy mice, which have a truncated FoxP3 transcription factor, harbor abnormalities in suppressive functions (Brunkow et al., 2001; Ramsdell and Ziegler, 2014). Indeed, these mice quickly die because of a severe lymphoproliferative syndrome. The consequences of the FoxP3 mutation in mice clearly show this transcription factor importance in immune homeostasis. Similarly to scurfy mice, mutations in the Foxp3 gene have been identified in patients with IPEX syndrome who develop severe autoimmunity (Bennett et al., 2001). Depending on the Foxp3 mutation, this leads to a more or less severe alteration of the development, the phenotype, and the functions of Treg cells. In addition, $CD4^{+}CD25^{+}$ cells found in periphery are anergic and effector T cells present a defect in cytokine production (Bacchetta et al., 2006). Despite the existence of $Foxp3^{-}$ regulatory cells, this transcriptional factor appears as a central regulator, essential for homeostasis and immune tolerance.

Foxp3 is able to bind approximately 700 genes. It can act both as an activator of transcription and as a repressor. Separate transcriptional programs driven by Foxp3 organize the proliferation and functions of regulatory T cells in the periphery (Zheng et al., 2007). FoxP3 forms a large protein complex (about 400–800 kDa) with over 360 different factors that may be transcription factors or chromatin-modifying factors (Rudra et al., 2012; Hori, 2012). In Tregs, FoxP3 physically interacts with RUNX1, suppresses IL-2 and IFNγ production, and upregulates Treg-associated molecules, thereby promoting suppressive activity (Ono et al., 2007).

FoxP3 blocks Th2 and Th17 differentiation by inhibiting the *Gata3* and *Rorc* expressions and thus suppressing the IL-4 and IL-17 production (Zeng et al., 2009). By binding to the RORγt/RUNX1 complex, FoxP3 inhibits *Il17* expression (Zhang et al., 2008). Foxp3 also suppresses the IL-6 function by binding to the phosphorylated STAT3 transcription factor, thereby repressing Th17 differentiation (Chaudhry et al., 2009). FoxP3 can also directly bind to RORγt to inhibit *Il17* expression by decreasing RORγt binding to the *Il17* promoter (Zhou et al., 2008; Ma et al., 2008).

The transcriptional activity of FoxP3 also depends on its ability to facilitate the epigenetic remodeling of its targets. For example, the FoxP3 binding to the genes that encode IL-2 and IFNγ leads to deacetylation of H3 histone reducing gene expression. In contrast, FoxP3 promotes *Il2ra*, *Ctla4*, and *Tnfrsf18* expressions by inducing histone acetylation close to the promoters (Chen et al., 2006; Li et al., 2007a). The blocking of the FoxP3 interaction leads to an increased acetylation of the *Il2* promoter thus reducing Treg functions (Bettini et al., 2012).

3.5 Bcl6: Tfh Master Regulator

Bcl6 was initially described as an inhibitor of Th2 responses by blocking STAT6 binding to DNA (Dent et al., 1997). The transcription factor Bcl6 is the master regulator of Tfh lymphocytes (Chtanova et al., 2004). Mice deficient in Bcl6 develop multiorgan inflammatory diseases, an increase in immunoglobulin (IgE) production and a defect in the GC reaction (Ye et al., 1997). Bcl6 is preferentially expressed in the Tfh cells defined by the CXCR5 and programmed cell death 1 (PD-1) markers (Chtanova et al., 2004). In vitro, the Bcl6 expression can be upregulated by IL-6 and IL-21. In addition, Bcl6-deficient T cells lose their ability to differentiate into CXCR5+ PD-1+ Tfh, while CD4 T cells that overexpress Bcl6 have an increase in Tfh differentiation in the absence of exogenous cytokines (Yu et al., 2009; Nurieva et al., 2009).

In parallel, the Bcl6 expression inhibits the Th1, Th2, or Th17 cell development by decreasing the *Tbx21*, *Gata3*, and *Rorc* expressions, respectively (Nurieva et al., 2009). Bcl6 binds to the *Tbx21* and *Rorc* promoters which respectively determine the Th1 and Th17 cell development, resulting in IFNγ and IL-17 repression. In addition, Bcl6 suppresses the expression of many microRNAs involved in the Tfh regulation, including miR-17-92 which blocks the CXCR5 expression (Yu et al., 2009). Bcl6 is repressed by the B lymphocyte–induced maturation protein-1 (Blimp-1) transcription factor. Overexpression of Blimp-1 in CD4 T cells prevents Bcl6 expression and significantly reduces Tfh cell differentiation without affecting Th2, Th17, and Treg differentiation. In addition, Blimp-1 suppression in CD4 T cells shows an increase in Tfh cell differentiation (Johnston et al., 2009). Blimp-1 is expressed in non-Tfh T cell subsets in response to stimulation indicating a mechanism by which other subsets inhibit Tfh differentiation (Crotty et al., 2010). In Th9 cells, Bcl6 is weakly expressed. The Bcl6 ectopic expression within Th9 cells leads to an alteration of the differentiation process and a decrease in IL-9 production. Bcl6 is able to bind the *Il9* promoter to adjacent regions to STAT5, thus limiting the *Il9* expression. Reciprocally, STAT5 or IL-2–deficient T cells show an increase in Bcl6 expression. Thus there is a negative correlation between the Bcl6 expression and the *il2* and *il9* expressions (Liao et al., 2014; Bassil et al., 2014).

3.6 PU.1: Th9 Master Controller

PU.1, an Ets family member, was the first transcription factor described to play a role in the IL-9 production by Th9 cells. The PU.1 expression in

Th9 cells is induced by TGF-β independently from Smad signaling (Tamiya et al., 2013) and IL-4 pathways (Chang et al., 2010). In Th9 cells the mechanisms leading to the PU.1 expression are yet to be elucidated. T cells lacking PU.1 factor exhibit inhibition of IL-9 production, while ectopic expression of PU.1 induces IL-9 production. These results suggest that PU.1 may be a switching factor between CD4 T cell subsets (Chang et al., 2010). Similarly, the degradation of PU.1 by p62-dependent autophagy selectively represses Th9 cell differentiation (Rivera Vargas et al., 2017). PU.1 directly binds to the *Il9* promoter and is responsible for HAT recruitment (Chang et al., 2010) inducing chromatin remodeling at *Il9* locus. PU.1 interacts directly with the Gcn5 HAT (Goswami and Kaplan, 2012). In PU.1-deficient cells, Gcn5 is not associated with the *Il9* promoter, whereas ectopic expression of PU.1 allows the Gcn5 binding to the *Il9* promoter (Goswami and Kaplan, 2012). PU.1 function is not strictly limited to Th9 cells. In T cells, PU.1 was initially described for its ability to regulate the heterogeneity of cytokine expression in Th2 cells. The presence of PU.1 is thus associated with a Th2 subpopulation characterized by its weak expression in *Il4* (Chang et al., 2005). Ectopic PU.1 expression in Th2 cells results in a significant decrease in specific cytokines expression (Chang et al., 2010). Through its physical interaction with GATA3 or IRF4, PU.1 is able to block their binding to DNA and thus limits their transcriptional activity. Through the HDAC recruitment, PU.1 also acts as a transcriptional repressor of *Gata3* expression (Yashiro et al., 2015). PU.1 also represses some genes that are important for Tfh cell functions including *Il21, Pdcd1* (encoding PD-1), *Cd40lg* (encoding CD40L; Awe et al., 2015). Compared with wild-type mice, mice with PU.1-deficient T cells show an increase in Tfh cell number. Thus PU.1 could also indirectly favor the Th9 cell differentiation by limiting the differentiation programs of other T helper subsets.

4. ADDITIONAL TRANSCRIPTION FACTORS

While the activation of STAT proteins and the expression of master regulators are essential for the specialization of T cells, they are not sufficient for the development and establishment of the functions of these cells. Indeed, the differentiation of helper T cells involves many additional factors required for the development and maintenance of the cellular phenotype. These factors can be induced by various stimuli such as activation of TCR, present cytokines, or metabolic changes (Fig. 3).

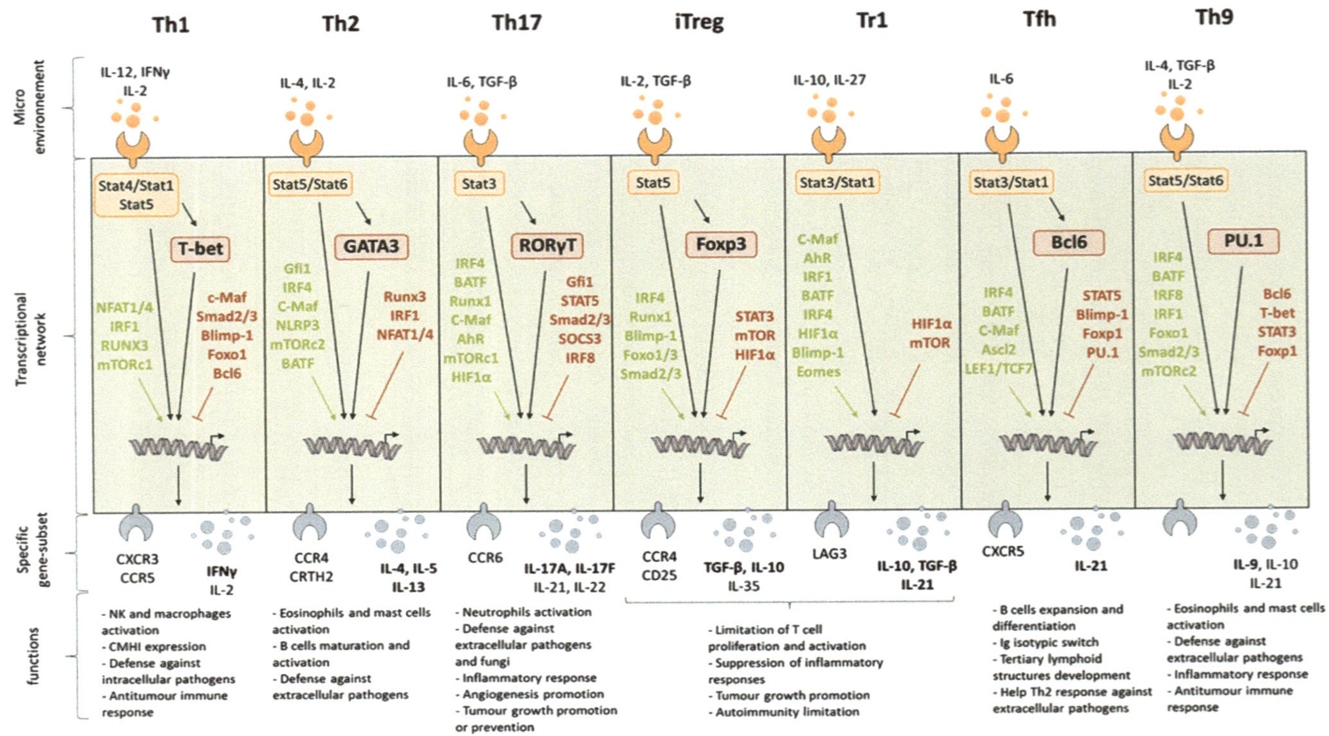

Figure 3 Distinct microenvironments initiate different transcriptional network. In addition to the central role of STAT proteins (orange box) and master regulator (red box), these networks require the intervention of various additional transcription factors (in green). The cooperation between these factors leads to a subset-specific gene expression (in blue) which support the functions specific to each T helper. These networks are also the target of many repressive factors restricting the development of some helper subtypes in favor of another (in red).

4.1 Th1: IRF1, RUNX3, NFATc1

In addition to the presence of IFNγ and IL-12, a strong involvement of the TCR is also a prerequisite for the development of Th1 cells. This strong activation of the TCR leads, among other things, to the expression of the nuclear factor of activated T-cells (NFAT1) factor, which is able to bind the *ifng* promoter to improve its activity (Sica et al., 1997). NFAT1 deficiency results not only in decreased expression of *ifng* but also in an aberrant increase in IL-4 production (Kiani et al., 2001; Kiani et al., 1997). Similarly, the deletion of NFAT1 and four factors results in an important expression of Th2-associated cytokines, IL-4, IL-5, and IL-10 at the expense of IFNγ expression (Rengarajan et al., 2002b; Ranger et al., 1998). Thus the NFAT family factors are necessary for the development of Th1 cells by their ability to modulate the IFNγ expression but also their ability to suppress Th2 polarization.

The activation of NF-κB by TCR engagement is also necessary to initiate polarization of Th1 cells. Indeed, deletion of the RelB factor results in a decrease in the IFNγ expression and an early blocking of Th1 differentiation with lack of T-bet expression and a strong decrease in STAT4 phosphorylation.

At the CD4 T cell level, the activation of STAT1 by IFNγ leads to a strong increase in the IRF1 expression. IRF1 appears to be indispensable for Th1 differentiation. Indeed, IRF1 directly modulates cell sensitivity to IL-12p70. IRF1 is essential by acting on *Il12rb1*, the gene encoding the beta-1 subunit of the IL-12 receptor (IL-12Rβ1). IRF1 directly interacted with and activated the *Il12rb1* promoter in CD4 T cells (Kano et al., 2008). IRF1 participates in the suppression of polarization of Th2 cells in favor of Th1 cells because of its ability to bind to the *Il4* promoter, thus repressing its expression (Maruyama et al., 2015).

The transcription factor RUNX3 is also induced during the differentiation of Th1 cell in a T-bet—dependent manner (Djuretic et al., 2007). RUNX3 in cooperation with the master regulator T-bet participates in the expression of IFNγ (Djuretic et al., 2007; Yagi et al., 2010). Its deficiency results in a decrease in the Th1 key cytokine expression but also in an increase in IL-4 production (Yagi et al., 2010). RUNX3 is also an important suppressor of Th2 polarization as evidenced by its ability to limit the expression of *Gata3* and *Il4* (Yagi et al., 2010; Kohu et al., 2009).

As a true metabolic sensor of the microenvironment, the mTOR signaling pathway also participates in the differentiation of Th1 cells. IL-2 and the CD28 costimulation molecule initiate the activation of phosphatidylinositide 3-kinase (PI3K) and Akt kinases, which participate in the activation of the mammalian target of rapamycin complex (mTORc1) (Ray et al., 2015). T cells deficient for mTOR are unable to maintain T-bet expression and fail to differentiate into Th1 (Delgoffe et al., 2009). The mTORc1 complex participates in the phosphorylation of the T-bet protein, which is necessary to induce the production of IFNγ by Th1 (Chornoguz et al., 2017).

4.2 Th2: Gfi-1, IRF4, c-Maf, NLRP3, Batf, Nfatc2

Polarization of Th2 cells requires a lower TCR signal compared with Th1. This difference in commitment leads to an activation of the NFAT2 factor, which improves IL-4 (Brogdon et al., 2002) production by being directly attached to its promoter. These results were confirmed by the deletion of NFAT2 which results in a decreased IL-4 secretion (Yoshida et al., 1998) and a decreased Th2 cell proliferation. As for Th1 cells, the blocking of NF-κB, in particular of the p50 subunit, blocks very early the polarization of Th2 cells by preventing the expression of *Gata3* and thus cytokines *Il4*, *Il5*, and *Il13* (Das et al., 2001).

The activation of STAT6 by IL-4 is responsible for the expression of the IRF4 factor, which plays a critical role in the differentiation of Th2 cells (Savitsky et al., 2010; Taniguchi et al., 2001). Mice deficient for IRF4 develop a defective Th2 response and reduced sensitivity to IL-4, the cytokine responsible for Th2 polarization. IRF4 is capable of modulating the polarization of Th2 cells at many levels. IRF4 improves activation of the Il4 promoter and IL-4 production (Rengarajan et al., 2002a). Its transcriptional activity depends on various factors that also participate in the polarization of Th2 cells such as NFATc2 or NLRP3. Indeed, the interaction between NFATc2 and IRF4 allows their fixation on the *Il4* promoter and the initiation of transcriptional activity (Rengarajan et al., 2002a). Far from its initially described function as an inflammasome actor, the NLRP3 receptor also exhibits a transcriptional activity in T cells. NLRP3 is indeed capable of binding to the *Il4* promoter and of inducing its expression in cooperation with IRF4 (Bruchard et al., 2015). In addition, NLRP3-deficient Th2 cells show severe alteration of their phenotype resulting in decreased IL-4, IL-5, and IL-13 secretion, decreased CCR4 membrane receptor expression and decreased GATA3, and c-Maf transcription factors (Bruchard et al., 2015).

The activation of STAT5 by IL-2 also contributes to the functional development of Th2 cells. The Gfi-1 transcription factor induced by IL4/STAT6/IRF4 pathway is also required to sensitize Th2 cells to IL-2 (Zhu et al., 2006a). The absence of Gfi-1 in Th2 cells leads to a partial failure of activation of the IL-2/STAT5 pathway leading to a decrease in their proliferation and an increase in their death by apoptosis (Zhu et al., 2002, 2006a). Gfi-1 can also be independently induced by IL-4 through the extracellular signal-regulated kinases (ERK)/MAPK pathway upon TCR activation. Gfi-1 is also involved in stabilizing the GATA3 protein and maintaining the Th2 phenotype (Shinnakasu et al., 2008).

C-Maf contributes to Th2 polarization in several ways. The c-Maf transcriptional factor is specifically involved in controlling IL-4 expression and not other Th2-related cytokines (Kim et al., 1999). On the one hand, its expression in Th1 cells allows a reduction of the IFNγ production even if it does not initiate the IL-4 secretion (Ho et al., 1998). Thus c-Maf also contributes to the polarization of Th2 cells by limiting the differentiation of Th1 cells.

mTOR deficiency prevents the development of Th2 cells by limiting the expression of *Gata3*. However, if mTORc1 complex is necessary for the differentiation of Th1 cells, its absence does not affect the development of Th2 cells, while the mTORc2 complex is essential for differentiation of Th2 cells (Delgoffe et al., 2011).

4.3 Th17: IRF4/Batf, Gfi-1, RUNX, mtor, AhR, c-Maf, SOCS3, HIF1α

IRF4 is essential for the differentiation of Th17 cells and its functions in EAE (Brustle et al., 2007). The IRF4 contribution to the Th17 program requires cooperation with the BATF and JUN factors (Li et al., 2012; Zheng et al., 2009) Thus BATF deficiency also causes a defect in the development of Th17 cells (Schraml et al., 2009). Together these factors (IRF4/BATF/JUN) are responsible for the expression of *Il17*, *Il21*, *Il23r*, and *Rorc* (Li et al., 2012). The IRF4/BATF transcriptional module contributes to the remodeling of chromatin during polarization of Th17 cells. IRF4 participates in the recruitment of HAT p300, which promotes access to chromatin and the recruitment of transcription factors such as RORγt (Ciofani et al., 2012).

Aryl hydrocarbon receptor (AhR) is a dependent ligand transcription factor, which can sometimes promote polarization of Treg (Quintana et al., 2008), sometimes that of Th17 (Quintana et al., 2008). Activation of AhR improves the development of Th17 cells and their production of IL-17 and IL-22 (Quintana et al., 2008; Veldhoen et al., 2008a). However, AhR is not essential for the implementation of Th17-like responses. In contrast, AhR deficiency results in the inability of Th17 cells to secrete IL-22 (Veldhoen et al., 2008a).

The commitment of the ICOS costimulation molecule leads to the expression of c-Maf in CD4 T cells. Blocking this pathway causes a failure of Th17 cells to produce IL-21 (Bauquet et al., 2009). The c-Maf deficiency in Th17 cells does not alter the early stages of polarization, but these cells lose their ability to respond to IL-23 (Sato et al., 2011; Bauquet et al., 2009). The c-Maf factor is necessary for the expression of IL-23R and therefore required for the development of memory Th17 cells (Sato et al., 2011). C-Maf also cooperates with the transcription factor Sox5 (*SRY-Box 5*) to participate in the expression of the master controller RORγt, independently of STAT3 (Tanaka et al., 2014).

The transcription factor RUNX1 controls the expression of the master regulator RORγt, then in cooperation with it, promotes the expression of *Il17* in Th17 cells. However, in cooperation with the Foxp3 factor, RUNX1 participates in the suppression of the Th17 polarization (Zhang et al., 2008).

As in the development of Th1 cells, the Th17 polarization requires the activation of the PI3K/Akt/mTOR pathway and the mTORc1 complex. The mTORc2 complex does not play a major role in the development of Th17 cells (Lee et al., 2010; Sauer et al., 2008) but participates indirectly by limiting the development of Treg cells and the expression of Foxp3 (Kerdiles et al., 2009; Ouyang et al., 2009). Activation of the mTORc1 complex is necessary for the expression of the hypoxia-inducible factor 1 alpha (HIF1α) factor. As a true metabolic sensor, HIF1α promotes the glycolytic activity of cells, favoring the expansion of Th17 cells at the expense of Treg cells (Shi et al., 2011; Dang et al., 2011). HIF1α actively participates in the expression of the RORγt factor, then in cooperation with it, initiates the recruitment of HAT p300 on the *Il17* promoter, inducing its expression. In addition, HIF1α is capable of interacting with Foxp3 leading to its degradation by the proteasome (Dang et al., 2011). HIF1α is thus a major player in the regulation of the balance between the development of Treg and Th17 cells (Shi et al., 2011; Dang et al., 2011). The TGF-β required for the differentiation of Th17 cells also allows the activation of the mTORc1 complex. This activation is responsible for inhibiting the SOCS3 factor, which is a negative regulator of STAT3 activation

(Qin et al., 2009). MTORc1 thus supports the phosphorylation of STAT3 and the development of Th17 cells (Delgoffe et al., 2011). mTORc1 also participates in suppressing the Th17 Gfi-1polarization repressive factor (Kurebayashi et al., 2012). The fixation of the Gfi-1 repressive factor on the *Il17*promoter prevents the binding of RORγt (Kurebayashi et al., 2012), and its presence on the promoters of the genes encoding the CD39 and CD73 ectonucleotidases limits the fixation of STAT3 and their expressions (Chalmin et al., 2012).

4.4 Treg: IRF4, Runx, Foxo, Blimp, Smad

IRF4 is also highly expressed in Treg cells. If its presence is not essential to the development of these cells, it is essential for the implementation of immunoregulatory properties. In contrast to the role of IRF4 in Th2 cell differentiation, IRF4 expression gives Treg cells the ability to suppress Th2 responses. Mice with a conditional deletion of IRF4 in Treg cells develop aberrant Th2 responses although the number of Foxp3 cells is not altered. In the absence of IRF4, Treg cells present a defect of ICOS activation marker expression and a strong decrease of their IL-10 secretion (Zheng et al., 2009).

IRF4 is also responsible for the expression of the Blimp-1 repressive factor (Cretney et al., 2011) essential for the establishment of Treg cell functions (Kallies et al., 2006; Martins et al., 2006). Blimp-1 in cooperation with IRF4 regulates IL-10 production in Treg cells, as well as their activation and migratory capacity. Blimp-1 is also involved in Treg cell survival by regulating Bcl2 expression (Cretney et al., 2011). The Blimp-1 factor also suppresses the expression of *Tbx21* and *Bcl6*, thus limiting the polarization of Th1 and Tfh cells (Cimmino et al., 2008; Johnston et al., 2009). IRF8 is also expressed in Treg cells where it is necessary for development and functions. Indeed, its deletion leads to a decrease in the expression of the CXCR3 receptor required for the Treg cell migration at inflammatory sites. In addition, IRF8-deficient Tregs have aberrant production of IL-4 and IL-17 cytokines (Lee et al., 2016). However, the molecular mechanisms underlying this observation are not explained.

The mTOR pathway plays an important role in the development of Treg cells. Activation of the PI3K/Akt/mTOR pathway is essential for the development of effector T cells, but inhibition of this channel leads to the default development of Treg cells (Delgoffe et al., 2009). Activation of the Akt/mTOR pathway promotes the nuclear exclusion of Foxo1 (*Forkhead Box O*) factor. Foxo1 is an important factor in the survival of naive T cells, which tends to limit their polarization in effector T cells, thus contributing to the control of homeostasis. Foxo1 is essential in the development of Treg cells and their functions (Kerdiles et al., 2010; Ouyang et al., 2012). Foxo1 deficiency causes a decrease in the CTLA4 receptor and also promotes the secretion of IFNγ. Foxo1 is required for Treg cells to respond to TGF-β. Foxo3 is also expressed in Treg cells and appears to play a similar role to Foxo1 (Kerdiles et al., 2010). Together, Foxo1 and 3 factors participate directly in the expression of the master gene of Treg cells, Foxp3 (Ouyang et al., 2010; Harada et al., 2010).

Proteins of the RUNX family also participate in the development of Treg cells. In their absence the regulatory functions of Treg cells are greatly diminished leading to the development of lymphoproliferative syndrome. RUNX deficiency results in lack expression of Foxp3 (Rudra et al., 2009) since RUNX1 and three proteins are able to initiate expression of the Foxp3 master regulator by binding to its promoter (Bruno et al., 2009; Kitoh et al., 2009; Klunker et al., 2009). RUNX1 also requires the presence of nuclear receptor subfamily 4 group A (Nr4a) factor to induce Foxp3 expression (Sekiya et al., 2011). In addition, RUNX1 is an interaction partner of Foxp3, their interaction tends to limit the transcriptional activity of RUNX1 thus limiting the production of IL-17 and IL-2 by Treg cells (Zhang et al., 2008; Ono et al., 2007).

4.5 Tr1: c-Maf, Ahr, Batf/IRF1, IRF4, HIF1a, Blimp-1, Eomes

STAT3 activation by IL-27 or IL-21 allows the expression of the c-Maf factor that plays a central role in Tr1 cell development. Indeed, mice deficient for c-Maf or the IL-21 receptor have a defect in IL-10 secretion by CD4 T cells and also a defect in Tr1 cell development in response to IL-27 (Pot et al., 2009). The ICOS costimulatory molecule is a key player in the secretion of IL-21 by CD4 T cells (Bauquet et al., 2009). In Tr1 cells the ICOS-induced signaling participates in the expression of c-Maf and the production of IL-21 thus maintaining the phenotype (Pot et al., 2009). The c-Maf transcription factor is a direct regulator of IL-10 and IL-21 production by Tr1 cells (Apetoh et al., 2010).

IL-27 induces a strong expression of the AhR transcription factor in Tr1 cells. Its absence results in a decrease in IL-10 secretion by Tr1 cells and conversely the presence of its 2,3,7,8-tetrachlorodibenzo-p-dioxin or 6-Formylindolo[3,2-b]carbazole ligands allows an increased secretion of IL-10. AhR interacts with the c-Maf factor, their cooperation is necessary to induce the expression of IL-10 and IL-21 by Tr1 cells (Apetoh et al., 2010).

In addition, AhR participates in the suppressive function of Tr1 cells independently of IL-10. Indeed, the activation of AhR by its ligands improves the expression of GzmB by Tr1 cells (Gandhi et al., 2010), leading to a specific killing of myeloid cells (Magnani et al., 2011).

Similarly to the development of Th17 cells, the BATF factor regulated by STAT3 activation is required to early initiate the transcriptional program of Tr1 cells (Karwacz et al., 2017). On the other hand, if BATF requires the presence of IRF4 to induce Th17 cell development, BATF preferentially cooperates with IRF1 to induce early chromatin remodeling in Tr1 cells (Karwacz et al., 2017). BATF-deficient Tr1 cells lose the ability to limit EAE development. The BATF factor regulates the development of Tr1 cells at various levels, in fact if it participates directly in the expression of *Il10*, *Icos*, and *Prf1*, it also regulates the expression of factors important for the differentiation of Tr1 cells such as c-Maf, Ets1, Eomes or Hif1α. IRF1 participates, on the one hand, in the development of Tr1 cells thanks to its cooperation with BATF but also directly regulates the expression of certain genes such as *Il10* or *Prf1*. However, if BATF-deficient Tr1 cells are unable to control EAE development, IRF1-deficient Tr1 cells accelerate the disease development. IRF1's interaction with BATF essential to Tr1 cells also appears to limit the cooperation between IRF4 and BATF, an important transcriptional module for the Th17 cell development. IRF1 therefore promotes the polarization of Tr1 cells at the expense of Th17 cells.

Just as IRF4 is an important factor in the development of the immunosuppressive functions of iTreg Foxp3^{+} cells, IRF4 is necessary for the functions of Tr1 cells. Activin-A participates in the development of Tr1 cells in humans through the expression of IRF4. IRF4 is a key partner of AhR, involved in the regulation of *Icos* and *Il10* (Tousa et al., 2017). The activation of the TCR and in particular of the Itk kinase is also necessary for the development of Tr1 cells, by activation of the Ras/IRF4 pathway. Indeed, Itk-deficient cells present a defect of IL-10 secretion and a loss of suppressive functions (Huang et al., 2017). Itk-deficient murine and human cells show a strong decrease in IRF4 and AhR expression. If AhR ligands are ineffective at restoring Tr1 phenotype, IRF4 overexpression can restore AhR expression, differentiation, and Tr1 cell functions (Huang et al., 2017). On the other hand, the role played by IRF4 on c-Maf expression remains ambiguous, notably due to the contradictory results obtained between the murine and human cells. Inhibition of Itk in human cells has no impact on c-Maf expression, underscoring the importance of AhR and IRF4 factors in the development of the Tr1 phenotype

(Huang et al., 2017). On the other hand, the Itk deficiency in murine cells profoundly alters the expression of c-Maf, making it difficult to distinguish effects of, on the one hand, the absence of IRF4 and, on the other hand, the absence of c-Maf.

HIF1α is induced early in Tr1 cells development and supports a bivalent role during their development. Indeed, HIF1α expressing CD4 T cells are unable to differentiate into Tr1, and the early blocking of this factor after 6 or 12 h of polarization limits the development of Tr1 cells (Mascanfroni et al., 2015). Similarly to Th17 cells, aerobic glycolysis is required to initiate Tr1 cell polarization. Inhibition of HIF1α expression after 24 h of differentiation significantly improves Tr1 cell polarization (Mascanfroni et al., 2015). However, in the longer term, HIF1α initiates the degradation of the AhR protein by the proteasome and limits the polarization of the Tr1 cells. Thus HIF1α is essential for the early control of glycolysis during the early stages of Tr1 cell differentiation, while at a later stage, AhR limits the expression of HIF1α and supports the regulation of Tr1 cell metabolism (Mascanfroni et al., 2015). It was showed that 25-hydroxycholesterol acts as a negative regulator of Tr1 cells in particular of IL-10 secretion via liver X receptor signaling (Vigne et al., 2017).

IL-10 secreted by Tr1 cells can be regulated in different ways depending on the context. Thus, in an inflammatory context, IL-12 allows the expression of the factor Blimp-1. Blimp-1 then directly regulates IL-10 production in Tr1 cells and is thus a major regulator of inflammation induced by TNF (Montes De Oca et al., 2016). Blimp-1 is also involved in the expression of the Eomes factor, as evidenced by the lack of expression of Eomes and the decrease in IL-10 and granzyme B production by CD4 T cells deficient for Blimp-1 (Zhang et al., 2017). Overexpression of Eomes in Blimp-1—deficient Tr1 cells partially restores their IL-10 and granzyme B production and limits the expression of *Il2*, *Il4*, *Il17*, and *Foxp3* (Zhang et al., 2017). The T-bet factor is also required for Eomes expression in Tr1 cells. Like T-bet, the Eomes factor helps maintain the Tr1 phenotype (IFNγ^+, IL-10$^+$, GZMB$^+$) by limiting the polarization of Th2 or Th17 cells (Zhang et al., 2017).

Inhibition of the Akt/mTOR signaling pathway leads to the development of regulatory T cells. Inhibition of this pathway by rapamycin in the presence of IL-10 leads to the generation of IL-10 secreting Tr1 cells (Battaglia et al., 2006a). If the mechanisms remain poorly described, the IL-10 is indeed able to maintain its own expression through the activation of the p38/MAPK channel (Brockmann et al., 2017).

4.6 TFh: Ascl2, TCF-1, LEF-1, Foxp1, BATF1, c-Maf, IRF4

Since the discovery of Bcl6 as a transcription factor required for Tfh cell differentiation, other studies have shown the role of transcriptional regulators and revealed different aspects of Tfh cell development, migration, and function. Achaete-scute family BHLH transcription factor 2 (Ascl2) is a transcription factor that is crucial for the development and functions of Tfh cells. Ascl2 acts as a specific transcriptional regulator of Tfh cells because it is highly expressed in these cells and its induction precedes Bcl6. Like Bcl6, Ascl2 inhibits the differentiation of Th1, Th2, and Th17 cells and directly controls the genes involved in Tfh development, such as induction of CXCR5 and CXCR4, inhibition of CCR7 and P-selectin glycoprotein ligand-1 expression, and signaling of IL-2. In addition, Ascl2 promotes T cell migration into follicles and stimulates Tfh cell development in vivo. Overexpression of Ascl2 increases the expression of *Cxcr5*, but not *Bcl6*, *Prdm1*, *Batf*, and *Il21* (Liu et al., 2014).

T cell—specific factor 1 (TCF-1) and lymphoid enhancer binding factor 1 (LEF-1) participate in the differentiation of Tfh cells. Two studies show that both transcription factors intervene early in development of Tfh cells (Choi et al., 2015; Xu et al., 2015). The selective loss of LEF-1 and TCF-1 in T cells affects Tfh differentiation and formation of GCs. On the contrary, when the expression of LEF-1 is forced, the differentiation of Tfh is increased. These two factors regulate the balance between Bcl6 and Blimp-1 in Tfh. TCF-1 acts as a positive and negative regulator of Bcl6 and Blimp-1, respectively. However, the activity of LEF-1 remains unclear. In addition, TCF-1 could also enhance IL-6 receptor expression promoting Bcl6 expression and suppress IL-2 receptor expression by decreasing Blimp-1 expression. Expression of TCF-1 is suppressed by IL-2 via Blimp-1 repression. TCF-1 can act as a regulator upstream of the Bcl-6-Blimp-1 axis since it can bind to the *Prdm1* regulatory site to inhibit Blimp but also to the *Bcl6* promoter to form a complex and improve its transcription (Xu et al., 2015).

Foxp1 is also involved in the differentiation of Tfh cells as a critical negative regulator. Foxp1-deficient naïve CD4 T cells can differentiate preferentially into Tfh cells allowing GC formation and antibody responses (Wang et al., 2014). Foxp1 directly and negatively regulates *Il21* expression in CD4 T cells (Linterman et al., 2010). Moreover, in the absence of Foxp1, ICOS expression is increased, and Tfh differentiation is promoted in an ICOS-dependent manner (Wang et al., 2014).

In Tfh cells, BATF appears to directly control the expression of *Bcl6* and *c-Maf* which are important genes for Tfh subset commitment. BATF-deficient T cells drastically express decreased Bcl-6 and c-Maf expressions. When the Bcl6 expression is restored in BATF-deficient T cells, the Tfh activity is partly restored in vivo. However, in the absence of Batf, Bcl6 is not sufficient for the development of Tfh cells. By regulating both *Bcl6* and *c-Maf* expression, BATF is at top of the transcriptional hierarchy that controls Tfh differentiation (Ise et al., 2011). This factor is also important in *Il4* expression by Tfh lymphocytes (Sahoo et al., 2015). Indeed, BATF forms a complex with IRF4, STAT3, and STAT6 and induces the IL-4 production by Tfh by directly binding and activating CNS2 region of *Il4* locus (Vijayanand et al., 2012).

C-Maf factor positively regulates the production of IL-4 by Tfh lymphocytes. In the Bauquet et al. study, c-Maf—deficient chimeric mice show defective Tfh cell differentiation in vivo. This alteration of the Tfh response is due to a defect in IL-4 and IL-21 induction and maintenance in an ICOS-dependent manner (Bauquet et al., 2009). However, in Tfh cells, c-Maf interacts with NFAT and JunB to regulate *Il4* expression (Zhu et al., 2010), which could explain why IL-4 induction is independent of Gata3 in Tfh cells (Yusuf et al., 2010).

The IRF4 transcription factor has been identified as a prerequisite for inducing Tfh differentiation and for GC formation. In IRF4-deficient CD4 T cells, the Bcl6 expression and other Tfh-related molecules are greatly reduced. Although $IRF4^{-/-}$ cells have a weak Bcl6 expression, they are completely unable to differentiate into Tfh cells due to the lack of interaction between IRF4 and Bcl6 (Bollig et al., 2012).

4.7 Th9: IRF4, Batf, Irf8, ETV5, mTORc2, Foxo1, IRF1, NFAT1

The IRF4 transcription factor, widely described for its importance in the polarization of Th2 and Th17 cells (Brustle et al., 2007; Lohoff et al., 2002), is also essential for the Th9 cell differentiation. Indeed, the Th9 lymphocytes differentiated from the naïve CD4 T cells deficient for IRF4 significantly decrease their IL-9 secretion (Staudt et al., 2010). Conversely, the ectopic expression of IRF4 in Th9 cells improves IL-9 production (Jabeen et al., 2013). IRF4 is present on the *Il9* promoter and its fixation improves the transcription of the *Il9* gene. IRF4 inhibit the expression of factors that limits the Th9 development, such as T-bet and IFNγ (Jabeen et al., 2013; Staudt et al., 2010).

BATF, transcription factor of the AP-1 family, forms with IRF4 a transcription module (Glasmacher et al., 2012; Li et al., 2012) necessary for the development of Th9 cells. Naïve CD4 T cells isolated from BATF-deficient mice show profound alteration of the polarization process of Th9 cells. In the same way as IRF4, BATF binds to the *Il9* promoter at the CNS1 regulatory region, and allows activation of transcription. However, unlike IRF4 which regulates only a portion of the Th9-associated genes, BATF deficiency results in a lack of expression of almost all genes associated with the Th9 phenotype (Jabeen et al., 2013). The difference in the impact of IRF4 and BATF transcription factors on Th9 development indicates that although these factors form a functional transcriptional module (Glasmacher et al., 2012; Li et al., 2012), they also constitute a unique and necessary contribution.

Structurally close to IRF4, its IRF8 counterpart actively participates in the polarization of Th9 cells. The transcriptional function of IRF8 is dependent on these interaction partners. On the one hand, IRF8 participates in a multiprotein complex involving the PU.1, IRF4, and BATF factors, necessary for the induction of cytokines characteristic of Th9 cells, notably *Il9* and *Il21*. On the other hand, IRF8 supports the polarization of Th9 cells by limiting their IL-4 production. Indeed, in the presence of the protein Ets transcription variant (ETV6), IRF8 is capable of forming a complex which initiates the suppression of the transcriptional activity of *Il4* (Humblin et al., 2017).

In addition to PU.1 and ETV6, another member of the Ets factor family is involved in the polarization of Th9 cells. Expression of ETV5 is dependent on IL-4—induced signaling and on expression of the IRF4 transcription factor. ETV5 is capable of binding to the *Il9* promoter to participate in its expression. In its absence the IL-9 production is diminished while its ectopic expression increases its production (Koh et al., 2016).

Another factor of the IRF family, IRF1 is expressed in Th9 cells following the activation of STAT1 in the presence of the proinflammatory cytokine IL-1β in the microenvironment. IRF1 participates in combination with IRF4 and PU.1 factors in the production of IL-9 by Th9 cells but also greatly improves their secretion of IL-21 (Vegran et al., 2014).

Smads proteins are also necessary for the development of Th9 cells, the Smad2/3/4 complex induces *il9* expression (Tamiya et al., 2013; Wang et al., 2013). In addition to the direct effects on the transcriptional regulation of *Il9*, the Smad2/Smad3 deficiency also causes increased expression of the

T-bet and Rorγt transcription factors, both of which can suppress IL-9 production (Goswami et al., 2012; Tamiya et al., 2013).

At the *Il9* promoter level, the NFAT1 factor is required to allow chromatin accessibility and binding of the NF-κB p65 factor (Jash et al., 2012). Indeed, a significant decrease in the IL-9 secretion is observed in CD4 T cells deficient in NFATc1 or NFAT2 (Koch et al., 2016). These transcription factors contribute to the rapid induction of IL-9 when polarizing naïve CD4 T cells into Th9. APCs expressing OX40L or OX40 target agonist antibodies are powerful inducers of IL-9 production in in vitro differentiated Th9 cell (Xiao et al., 2012). OX40, a member of the TNF family, activates the TNF Receptor Associated Factor 6/NF-κB RelB-p52 signaling pathway, leading to the fixation of the RelB-p52 heterodimer to the *Il9* promoter. While the induction of IL-9 by OX40 depends on STAT6, it does not require the presence of the PU.1 transcription factor (Xiao et al., 2012).

Close to Th2, the development of Th9 cells requires the presence of the mTORc2 complex. In fact, a deficiency of the Rictor factor, a member of the mTORc2 complex, leads to a decrease in the activation of STAT6 and a decrease in the expression of the IRF4 factor (Chen et al., 2017). In addition, activation of the PI3K/AKT/mTOR pathway by IL-7 also contributes to the development of Th9 cells. Indeed, the activation of this pathway improves the expression of p300 HAT leading to a higher acetylation of the *Il9* promoter and expression of Foxo1 factor (Bi et al., 2017). Foxo1 therefore competes with the Foxp1 inhibitor to bind to the *Il9* promoter (Bi et al., 2017; Buttrick et al., 2018).

5. FUNCTIONS

These complex transcriptional networks support the polarization of CD4 T cells and enable them to adapt to the microenvironment. These networks are directly responsible for the expression of a cytokine panel specific to each subtype and thus support the functions of the T helper cells (Fig. 3).

5.1 Th1: Cell-Mediated Immune Responses

Th1 cells orchestrate cell-mediated immune responses to intracellular pathogens such as viruses, bacteria, or tumor antigens (Del Prete, 1992; Mosmann and Coffman, 1989). They exert their effects indirectly, through the cytokines they secrete. IFNγ Th1 secretion activates NK cells, stimulates the maturation of dendritic cells (Pan et al., 2004), and supports the phagocytic and bactericidal functions of macrophages (Suzuki et al., 1988). IFNγ also causes an increase in MHCI molecules on the surface of APC, improving antigen presentation to CD8 T lymphocytes (Seliger et al., 2008). The IL-2 they secrete is involved in the proliferation of cytotoxic CD8 T lymphocytes and the acquisition of a memory phenotype (Bourgeois et al., 2002; Janssen et al., 2003). Th1 cells play a key role in helping CD8 T cells. Through their ability to recruit and activate CD8 and NK T cells, Th1 cells promote antitumor immune response (Shankaran et al., 2001).

However, Th1 responses may cause tissue immunopathological lesions, especially in the presence of chronic infection. This lymphocyte subtype is indeed associated with many chronic inflammatory diseases such as multiple sclerosis, insulin-dependent diabetes, and rheumatoid arthritis (Chen and Liu, 2009). Similarly, an unmanaged Th1 response from the recipient to the transplant is a major obstacle to successful transplantation.

5.2 Th2: Humoral Immune Response

Th2 cells direct a humoral response directed against extracellular pathogens (parasites, bacteria, toxins) by cooperation with B lymphocytes secreting antibodies or cellular effectors of innate immunity. IL-4 and IL-5 are particularly important for IgE production by B lymphocytes (Li et al., 2011). These secreted IgE will be recognized by the Fc receptors on the surface of mast cells and basophilic polynuclear cells leading to their activation (Steinke and Borish, 2001). Furthermore, IL-13 is the cytokine responsible for helminth expulsion and airway hypersensitivity related to asthmatic pathology (Wynn, 2003).

Th2 cells thus play a major role in the induction and persistence of asthma and other allergic pathologies (Del Prete, 1992; Sokol et al., 2009). IL-5 and IL-9 promote the recruitment of eosinophils and mast cells, respectively (Li et al., 2011). IL-4 and IL-13 may also have proatherogenic effects and support the development of atheromatous plaques. Th2 cells and IL-4 tend to promote tumor growth either through a direct effect of IL-4 on cancer cells or by inhibiting cellular immunity (Hallett et al., 2012; Kobayashi et al., 1998; Powrie et al., 1993).

5.3 Th17: Fungal Bacteria Defenses

The marked inflammatory properties of Th17 cells, formerly attributed to Th1 responses, stem from their ability to secrete IL-17A, a highly proinflammatory cytokine. Th17 cells are involved in anti-infectious defenses against extracellular pathogens and fungal bacteria (Puel et al., 2011; Weaver et al., 2007). They contribute in particular to the recruitment and the activation of the neutrophils on the infectious sites. The clearance function of certain pathogens by Th17 cells requires a massive inflammatory response. IL-22 participates in this response and thereby strengthens host defenses by promoting secretion of antibacterial peptides and promoting barrier self-renewal (Aujla et al., 2008). IL-21 has pleiotropic effects: it activates NK cells and T lymphocytes and supports the differentiation of B lymphocytes into effector and memory cells (Korn et al., 2007; Leonard and Spolski, 2005). IL-17 induces the production of inflammatory cytokines such as IL-6, IL-1, TNFα, and promotes angiogenesis via vascular endothelial growth factor.

Th17 is also the main effector of autoimmune pathologies such as psoriasis, rheumatoid arthritis, multiple sclerosis, and chronic inflammatory bowel disease. Th17 promotes local inflammation by recruiting myeloid cells and promote stroma reaction via proangiogenic secretion.

Th17 cells have a controversial role in anti-tumor immunity. On the one hand, Th17 cells promote tumor progression. First, IL-17 stimulates the production of angiogenic factors, favoring a vascular network feeding the tumor (Berger et al., 2013; Chung et al., 2013). Moreover, differentiated in the presence of IL-6 and TGF-β, Th17 cells secrete IL-10, an immunosuppressive cytokine (Mcgeachy et al., 2007). In addition, these Th17 cells express on their surface the CD39 and CD73 ectonucleotidases, responsible for the production of immunosuppressive adenosine (Chalmin et al., 2012). Conversely, Th17 cells may have a powerful antitumor activity (Muranski et al., 2008, 2011), particularly by promoting infiltration of Th1, T CD8, and NK cells (Kryczek et al., 2009) but also through their ability to differentiate into IFNγ-producing cells (Martin-Orozco et al., 2009).

5.4 Treg: Homeostasis and Immune Tolerance

Regulatory T cells are involved in maintaining cellular homeostasis and immune tolerance. There are many regulatory T cells subtypes, and while their biology, cytokines, and suppression mechanisms may be different, all these subtypes participate in a common function in regulating immune responses. Treg cells perform their functions through the secretion of immunosuppressive cytokines such as TGF-β (Powrie et al., 1996), IL-10 (Asseman et al., 1999), or IL-35 (Collison et al., 2007). These cells are also capable of initiating cytolysis of effector T cells or APCs following granzyme and perforin secretion (Cao et al., 2007; Magnani et al., 2011). Treg also modulates the microenvironment leading to metabolic disturbances. Indeed, the strong expression of the IL-2 receptor (CD25) participates in the sequestration of this cytokine, limiting the proliferation and activation of Nk, T CD4, and T CD8 cells and promoting their death by apoptosis (Pandiyan et al., 2007). In addition, Treg cells have a high level of CD39 and CD73 expression on their surface. These ectonucleotidases are involved in the production of immunosuppressive adenosine which induces T cell anergy (Deaglio et al., 2007). Finally, the regulatory T cells directly modulate the maturation and functions of APCs. Treg express on their surface inhibitory molecules such as LAG3 capable of binding to MHC-II and limiting activation of effector T cells (Huang et al., 2004). CTLA4 and its binding to its B7 ligand also play an important role in regulating immune responses (Paust et al., 2004; Read et al., 2006). In addition, CTLA4/B7 binding also induces indoleamine 2,3-dioxygenase activation by DCs, limiting tryptophan accessibility and thus effector T cell activation (Fallarino et al., 2003).

A defect in the development of Treg cells contributes to the development of autoimmune diseases, allergy, or rejection of transplants. After pathogen removal, Treg cells limit inflammatory responses and inhibit Th1, NK, and macrophage activity, thus limiting tissue damage caused by inflammation (Ouyang et al., 2011). The synergistic action of IL-10 associated with TGF-β reduces allergic reactions by suppressing IgE production (Jutel and Akdis, 2008). Treg cell activation is a target of choice to initiate tolerance in autoimmune diseases such as diabetes (Battaglia et al., 2006b) or multiple sclerosis (Pot et al., 2011).

Conversely, an aberrant development of these cells limits antiviral or antitumoral immune responses. Their immunosuppressive capacities tend to promote the growth of many tumor subtypes (Ladoire et al., 2011; Menetrier-Caux et al., 2012). However, their ability to suppress Th17-dependent inflammatory responses may be a good prognosis in some cancer models (Fridman et al., 2012).

5.5 TFh: B Lymphocyte Maturation

Tfh cells express the CXCR5 chemokine receptor, which contributes to their localization in B lymphoid follicles, where they promote the differentiation and maturation of B lymphocytes. The homing exerted by the CXCL13 chemokine (CXCR5 ligand) allows follicular migration. Through their secretion of IL-21, Tfh sparticipate in the expansion and differentiation of B lymphocytes. In addition, to their canonical cytokine IL-21, Tfh can secrete IL-4, IFNγ, or IL-10. This plasticity allows them to adapt the humoral response to the detected infectious agents by orienting the Ig isotypic switch (Reinhardt et al., 2009). Tfh lymphocytes also appear necessary for the implementation of an antitumor immune response, particularly through their participation in the development of tertiary lymphoid structures located on the periphery of cancer (Tangye et al., 2013).

Aberrant Tfh cell activity is associated with multiple pathologies such as autoimmune diseases, primary or acquired immunodeficiencies, and follicular and angioimmunoblastic T lymphomas (Gu-Trantien et al., 2013).

5.6 Th9: A New Player in Cancer and Autoimmune Disease

In humans and mice, Th9 cells are characterized by high secretion of IL-9 and IL-21. Only murine Th9 are capable of producing IL-10. IL-9 promotes the development of Th17 cells (Elyaman et al., 2009; Nowak et al., 2009) and induces the release of chemokines by resident cells causing the recruitment of immune cells leading to severe inflammatory responses (Dong et al., 1999; Lu et al., 2012). Th9 cells have important antitumor properties (Lu et al., 2012; Purwar et al., 2012; Vegran et al., 2014). IL-9 promotes the antitumor activity of mast cells (Abdul-Wahid et al., 2016), reflected in increased production of TNFα, IL-1β, and IL-6. On the other hand, IL-21 causes the secretion of IFNγ by NK and T $CD8^+$ cells, responsible for the elimination of tumor cells.

These Th9 lymphocytes induce strong proinflammatory responses and promote the development of autoimmune colitis (Dardalhon et al., 2008) or autoimmune disease (Jager et al., 2009). IL-9 is involved in the activation and development of mast cells and eosinophils, notably, strengthening their survival and secretory capacity (Gounni et al., 2000; Eklund et al., 1993; Lora et al., 2003; Wiener et al., 2004). Th9 cells thus contribute to the development of allergic responses and are particularly involved in asthma (Staudt et al., 2010).

5.7 T CD4 Cytotoxic Cells

A less well-known role of CD4 T cells is their ability to acquire cytotoxic properties similar to CD8 T cells. CD4 T cells with cytotoxic activity (ThCTL) have been observed in various immune responses. Identified in long-term cultures of CD4 T cells (Fleischer, 1984), ThCTL cells are characterized by their ability to secrete Granzyme B and perforin and kill the target cells in an MHC II–restricted manner by direct contact (Maimone et al., 1986; Lukacher et al., 1985). Although the ThCTL differentiation mechanisms remain unclear, it is believed that these cells are more closely related to IFNγ secreting Th1 cells and regulated by Eomes and/or T-bet transcription factors for their differentiation. Although ThCTL have been identified in other CD4 subsets, they are all characterized by the expression of the CRTAM cell surface marker. Their cytotoxic activity is further acquired after incubation of $CRTAM^{+}$ CD4 T cells with IL-2, and both CRTAM and IL-2 signals are required for optimal Eomes expression (Takeuchi et al., 2016).

ThCTL cells can apparently develop from Th0, Th1, Th2, Th17, and Treg effector subsets (Takeuchi and Saito, 2017). However, ThCTL derived from Th1 cells represent the majority of ThCTL, with production of IFNγ alone or with other cytokines such as TNFα and IL-2 (Appay et al., 2002). T-bet is the master regulator of Th1 differentiation and induces IFNγ production. In CD8 T cells, T-bet induces the expression of granzyme B and perforin which are required for their cytotoxic activity (Glimcher et al., 2004), these two observations suggest that T-bet can also promote ThCTL differentiation (Hua et al., 2013). It has been shown that the expression of Eomes but not that of T-bet is increased in the second phase of ThCTL differentiation, at least during viral infection (Brown et al., 2016). Thus ThCTLs could be regulated by both T-bet and Eomes, depending on their differentiation stage.

ThCTL have been observed mainly in virus infection models (Zaunders et al., 2004; Van Leeuwen et al., 2004), but contrary to Th1 cells that can differentiate upon exposure of IFNγ, IFNγ is not required for induction of the cytotoxic activity of ThCTL (Brown et al., 2012). IL-2–dependent signaling is essential for the induction of ThCTLs cells. IL-2 together with CRTAM induces Eomes expression, which in turn promotes the expression of IFNγ and cytotoxic granules and the degranulation marker CD107a (Pipkin et al., 2010; Takeuchi et al., 2016; Hua et al., 2013; Mucida et al., 2013). In addition, it should be noted that the TCR signal strength affects effector cell differentiation and CD4 T cell polarization (Van Panhuys et al., 2014) and probably that of ThCTL cells. Retinoic acid, TGF-β but also prolonged antigenic stimulation abrogate the expression of the ThPOK transcription factor and inversely upregulate the expression of both Runx3 and T-bet. These changes in transcription factor expression are required for reprogramming CD4 T cells into functional ThCTL cells (Sujino et al., 2016). It was also reported that the balance of Bcl6 and Blimp-1 transcription factors in ThCTL cells is opposite to that observed in TFh cells. Indeed, ThCTL cells express high levels of Blimp-1 and low levels of Bcl6 (Donnarumma et al., 2016).

ThCTL cells with cytotoxic activity are mainly localized in peripheral tissues. In mouse models, influenza virus infections are limited to the lungs. Under these conditions, not only APC but also the infected epithelial cells cancer can present an antigenic peptide on MHC-II. ThCTL cells can recognize these cells and contribute to virus clearance. ThCTL cells also have the potential to exacerbate autoimmune diseases. Indeed, ThCTL are critically involved in the induction of colitis (Takeuchi et al., 2016). In addition, ThCTL cells may be involved in late onset of EAE since Eomes-expressing $CD4^+$ T cells were identified during this disease development (Raveney et al., 2015).

6. CONCLUSION

The study of the functions of transcription factors in the CD4 T cell differentiation has long been limited to their isolated effect on the expression of a limited number of targets. The current accumulation of data reveals a complex world, where the polarization of CD4 T cells requires a transcriptional network involving many factors and cooperation between them. However, epigenetic processes affect the nuclear context in which these transcription factors operate. CD4 T cells are also defined by a dynamic epigenetic landscape, and it remains to determine how collaboration between epigenetic modifiers and transcriptional factors may contribute to the T cell development and functions. The microenvironment greatly influences the polarization of CD4 T cells. In addition, to the presence of cytokines, lymphocytes are particularly sensitive to metabolic change, and the influence of metabolism on CD4 T differentiation and immune responses remains a key issue. It is therefore a fundamental and clinical importance to clearly describe the molecular events involved in the programming of helper T cells. Such knowledge could be used to develop tools to target auxiliary T cells, either by promoting effective CD4 T responses or by limiting the development of harmful T cells in different physiopathological contexts.

REFERENCES

Abdul-Wahid, A., Cydzik, M., Prodeus, A., Alwash, M., Stanojcic, M., Thompson, M., Huang, E.H., Shively, J.E., Gray-Owen, S.D., Gariepy, J., 2016. Induction of antigen-specific TH 9 immunity accompanied by mast cell activation blocks tumor cell engraftment. Int. J. Cancer 139, 841–853.

Agarwal, S., Avni, O., Rao, A., 2000. Cell-type-restricted binding of the transcription factor NFAT to a distal IL-4 enhancer in vivo. Immunity 12, 643–652.

Apetoh, L., Quintana, F.J., Pot, C., Joller, N., Xiao, S., Kumar, D., Burns, E.J., Sherr, D.H., Weiner, H.L., Kuchroo, V.K., 2010. The aryl hydrocarbon receptor interacts with c-Maf to promote the differentiation of type 1 regulatory T cells induced by IL-27. Nat. Immunol. 11, 854–861.

Appay, V., Zaunders, J.J., Papagno, L., Sutton, J., Jaramillo, A., Waters, A., Easterbrook, P., Grey, P., Smith, D., Mcmichael, A.J., Cooper, D.A., Rowland-Jones, S.L., Kelleher, A.D., 2002. Characterization of CD4(+) CTLs ex vivo. J. Immunol. 168, 5954–5958.

Asseman, C., Mauze, S., Leach, M.W., Coffman, R.L., Powrie, F., 1999. An essential role for interleukin 10 in the function of regulatory T cells that inhibit intestinal inflammation. J. Exp. Med. 190, 995–1004.

Aujla, S.J., Chan, Y.R., Zheng, M., Fei, M., Askew, D.J., Pociask, D.A., Reinhart, T.A., Mcallister, F., Edeal, J., Gaus, K., Husain, S., Kreindler, J.L., Dubin, P.J., Pilewski, J.M., Myerburg, M.M., Mason, C.A., Iwakura, Y., Kolls, J.K., 2008. IL-22 mediates mucosal host defense against Gram-negative bacterial pneumonia. Nat. Med. 14, 275–281.

Awe, O., Hufford, M.M., Wu, H., Pham, D., Chang, H.C., Jabeen, R., Dent, A.L., Kaplan, M.H., 2015. PU.1 Expression in T Follicular helper cells limits CD40L-dependent germinal center B cell development. J. Immunol. 195, 3705–3715.

Bacchetta, R., Bigler, M., Touraine, J.L., Parkman, R., Tovo, P.A., Abrams, J., De waal Malefyt, R., DE Vries, J.E., Roncarolo, M.G., 1994. High levels of interleukin 10 production in vivo are associated with tolerance in SCID patients transplanted with HLA mismatched hematopoietic stem cells. J. Exp. Med. 179, 493–502.

Bacchetta, R., Passerini, L., Gambineri, E., Dai, M., Allan, S.E., Perroni, L., Dagna-Bricarelli, F., Sartirana, C., Matthes-Martin, S., Lawitschka, A., Azzari, C., Ziegler, S.F., Levings, M.K., Roncarolo, M.G., 2006. Defective regulatory and effector T cell functions in patients with FOXP3 mutations. J. Clin. Invest. 116, 1713–1722.

Bassil, R., Orent, W., Olah, M., Kurdi, A.T., Frangieh, M., Buttrick, T., Khoury, S.J., Elyaman, W., 2014. BCL6 controls Th9 cell development by repressing Il9 transcription. J. Immunol. 193, 198–207.

Battaglia, M., Stabilini, A., Draghici, E., Gregori, S., Mocchetti, C., Bonifacio, E., Roncarolo, M.G., 2006a. Rapamycin and interleukin-10 treatment induces T regulatory type 1 cells that mediate antigen-specific transplantation tolerance. Diabetes 55, 40–49.

Battaglia, M., Stabilini, A., Draghici, E., Migliavacca, B., Gregori, S., Bonifacio, E., Roncarolo, M.G., 2006b. Induction of tolerance in type 1 diabetes via both $CD4^{+}CD25^{+}$ T regulatory cells and T regulatory type 1 cells. Diabetes 55, 1571–1580.

Bauquet, A.T., Jin, H., Paterson, A.M., Mitsdoerffer, M., Ho, I.C., Sharpe, A.H., Kuchroo, V.K., 2009. The costimulatory molecule ICOS regulates the expression of c-Maf and IL-21 in the development of follicular T helper cells and TH-17 cells. Nat. Immunol. 10, 167–175.

Becher, B., Durell, B.G., Noelle, R.J., 2003. IL-23 produced by CNS-resident cells controls T cell encephalitogenicity during the effector phase of experimental autoimmune encephalomyelitis. J. Clin. Invest. 112, 1186–1191.

Bennett, C.L., Christie, J., Ramsdell, F., Brunkow, M.E., Ferguson, P.J., Whitesell, L., Kelly, T.E., Saulsbury, F.T., Chance, P.F., Ochs, H.D., 2001. The immune dysregulation, polyendocrinopathy, enteropathy, X-linked syndrome (IPEX) is caused by mutations of FOXP3. Nat. Genet. 27, 20–21.

Berger, H., Vegran, F., Chikh, M., Gilardi, F., Ladoire, S., Bugaut, H., Mignot, G., Chalmin, F., Bruchard, M., Derangere, V., Chevriaux, A., Rebe, C., Ryffel, B., Pot, C., Hichami, A., Desvergne, B., Ghiringhelli, F., Apetoh, L., 2013. SOCS3 transactivation by PPARgamma prevents IL-17-driven cancer growth. Cancer Res. 73, 3578–3590.

Bettelli, E., Carrier, Y., Gao, W., Korn, T., Strom, T.B., Oukka, M., Weiner, H.L., Kuchroo, V.K., 2006. Reciprocal developmental pathways for the generation of pathogenic effector TH17 and regulatory T cells. Nature 441, 235–238.

Bettini, M.L., Pan, F., Bettini, M., Finkelstein, D., Rehg, J.E., Floess, S., Bell, B.D., Ziegler, S.F., Huehn, J., Pardoll, D.M., Vignali, D.A., 2012. Loss of epigenetic modification driven by the Foxp3 transcription factor leads to regulatory T cell insufficiency. Immunity 36, 717–730.

Bi, E., Ma, X., Lu, Y., Yang, M., Wang, Q., Xue, G., Qian, J., Wang, S., Yi, Q., 2017. Foxo1 and Foxp1 play opposing roles in regulating the differentiation and antitumor activity of TH9 cells programmed by IL-7. Sci. Signal. 10.

Bollig, N., Brustle, A., Kellner, K., Ackermann, W., Abass, E., Raifer, H., Camara, B., Brendel, C., Giel, G., Bothur, E., Huber, M., Paul, C., Elli, A., Kroczek, R.A., Nurieva, R., Dong, C., Jacob, R., Mak, T.W., Lohoff, M., 2012. Transcription factor IRF4 determines germinal center formation through follicular T-helper cell differentiation. Proc. Natl. Acad. Sci. USA 109, 8664–8669.

Bourgeois, C., Rocha, B., Tanchot, C., 2002. A role for CD40 expression on $CD8^+$ T cells in the generation of $CD8^+$ T cell memory. Science 297, 2060–2063.

Breitfeld, D., Ohl, L., Kremmer, E., Ellwart, J., Sallusto, F., Lipp, M., Forster, R., 2000. Follicular B helper T cells express CXC chemokine receptor 5, localize to B cell follicles, and support immunoglobulin production. J. Exp. Med. 192, 1545–1552.

Brockmann, L., Gagliani, N., Steglich, B., Giannou, A.D., Kempski, J., Pelczar, P., Geffken, M., Mfarrej, B., Huber, F., Herkel, J., Wan, Y.Y., Esplugues, E., Battaglia, M., Krebs, C.F., Flavell, R.A., Huber, S., 2017. IL-10 receptor signaling is essential for TR1 cell function in vivo. J. Immunol. 198, 1130–1141.

Brogdon, J.L., Leitenberg, D., Bottomly, K., 2002. The potency of TCR signaling differentially regulates NFATc/p activity and early IL-4 transcription in naive $CD4^+$ T cells. J. Immunol. 168, 3825–3832.

Brown, D.M., Lampe, A.T., Workman, A.M., 2016. The differentiation and protective function of cytolytic CD4 T cells in influenza infection. Front. Immunol. 7, 93.

Brown, D.M., Lee, S., Garcia-Hernandez Mde, L., Swain, S.L., 2012. Multifunctional CD4 cells expressing gamma interferon and perforin mediate protection against lethal influenza virus infection. J. Virol. 86, 6792–6803.

Bruchard, M., Rebe, C., Derangere, V., Togbe, D., Ryffel, B., Boidot, R., Humblin, E., Hamman, A., Chalmin, F., Berger, H., Chevriaux, A., Limagne, E., Apetoh, L., Vegran, F., Ghiringhelli, F., 2015. The receptor NLRP3 is a transcriptional regulator of TH2 differentiation. Nat. Immunol. 16, 859–870.

Brunkow, M.E., Jeffery, E.W., Hjerrild, K.A., Paeper, B., Clark, L.B., Yasayko, S.A., Wilkinson, J.E., Galas, D., Ziegler, S.F., Ramsdell, F., 2001. Disruption of a new forkhead/winged-helix protein, scurfin, results in the fatal lymphoproliferative disorder of the scurfy mouse. Nat. Genet. 27, 68–73.

Bruno, L., Mazzarella, L., Hoogenkamp, M., Hertweck, A., Cobb, B.S., Sauer, S., Hadjur, S., Leleu, M., Naoe, Y., Telfer, J.C., Bonifer, C., Taniuchi, I., Fisher, A.G., Merkenschlager, M., 2009. Runx proteins regulate Foxp3 expression. J. Exp. Med. 206, 2329–2337.

Brustle, A., Heink, S., Huber, M., Rosenplanter, C., Stadelmann, C., Yu, P., Arpaia, E., Mak, T.W., Kamradt, T., Lohoff, M., 2007. The development of inflammatory T(H)-17 cells requires interferon-regulatory factor 4. Nat. Immunol. 8, 958–966.
Burchill, M.A., Yang, J., Vogtenhuber, C., Blazar, B.R., Farrar, M.A., 2007. IL-2 receptor beta-dependent STAT5 activation is required for the development of $Foxp3^+$ regulatory T cells. J. Immunol. 178, 280–290.
Burgler, S., Mantel, P.Y., Bassin, C., Ouaked, N., Akdis, C.A., Schmidt-Weber, C.B., 2010. RORC2 is involved in T cell polarization through interaction with the FOXP3 promoter. J. Immunol. 184, 6161–6169.
Buttrick, T.S., Wang, W., Yung, C., Trieu, K.G., Patel, K., Khoury, S.J., Ai, X., Elyaman, W., 2018. Foxo1 promotes Th9 cell differentiation and airway allergy. Sci. Rep. 8, 818.
Cao, X., Cai, S.F., Fehniger, T.A., Song, J., Collins, L.I., Piwnica-Worms, D.R., Ley, T.J., 2007. Granzyme B and perforin are important for regulatory T cell-mediated suppression of tumor clearance. Immunity 27, 635–646.
Chae, S.C., Shim, S.C., Chung, H.T., 2009. Association of TBX21 polymorphisms in a Korean population with rheumatoid arthritis. Exp. Mol. Med. 41, 33–41.
Chalmin, F., Mignot, G., Bruchard, M., Chevriaux, A., Vegran, F., Hichami, A., Ladoire, S., Derangere, V., Vincent, J., Masson, D., Robson, S.C., Eberl, G., Pallandre, J.R., Borg, C., Ryffel, B., Apetoh, L., Rebe, C., Ghiringhelli, F., 2012. Stat3 and Gfi-1 transcription factors control Th17 cell immunosuppressive activity via the regulation of ectonucleotidase expression. Immunity 36, 362–373.
Chang, H.C., Sehra, S., Goswami, R., Yao, W., Yu, Q., Stritesky, G.L., Jabeen, R., Mckinley, C., Ahyi, A.N., Han, L., Nguyen, E.T., Robertson, M.J., Perumal, N.B., Tepper, R.S., Nutt, S.L., Kaplan, M.H., 2010. The transcription factor PU.1 is required for the development of IL-9-producing T cells and allergic inflammation. Nat. Immunol. 11, 527–534.
Chang, H.C., Zhang, S., Thieu, V.T., Slee, R.B., Bruns, H.A., Laribee, R.N., Klemsz, M.J., Kaplan, M.H., 2005. PU.1 expression delineates heterogeneity in primary Th2 cells. Immunity 22, 693–703.
Chaudhry, A., Rudra, D., Treuting, P., Samstein, R.M., Liang, Y., Kas, A., Rudensky, A.Y., 2009. $CD4^+$ regulatory T cells control TH17 responses in a Stat3-dependent manner. Science 326, 986–991.
Chen, C., Rowell, E.A., Thomas, R.M., Hancock, W.W., Wells, A.D., 2006. Transcriptional regulation by Foxp3 is associated with direct promoter occupancy and modulation of histone acetylation. J. Biol. Chem. 281, 36828–36834.
Chen, H., Zhang, L., Wang, P., Su, H., Wang, W., Chu, Z., Zhang, L., Zhang, X., Zhao, Y., 2017. mTORC2 controls Th9 polarization and allergic airway inflammation. Allergy 72, 1510–1520.
Chen, J., Liu, X., 2009. The role of interferon gamma in regulation of $CD4^+$ T-cells and its clinical implications. Cell. Immunol. 254, 85–90.
Chen, W., Jin, W., Hardegen, N., Lei, K.J., Li, L., Marinos, N., Mcgrady, G., Wahl, S.M., 2003. Conversion of peripheral $CD4^+CD25^-$ naive T cells to $CD4^+CD25^+$ regulatory T cells by TGF-beta induction of transcription factor Foxp3. J. Exp. Med. 198, 1875–1886.
Choi, Y.S., Eto, D., Yang, J.A., Lao, C., Crotty, S., 2013. Cutting edge: STAT1 is required for IL-6-mediated Bcl6 induction for early follicular helper cell differentiation. J. Immunol. 190, 3049–3053.
Choi, Y.S., Gullicksrud, J.A., Xing, S., Zeng, Z., Shan, Q., Li, F., Love, P.E., Peng, W., Xue, H.H., Crotty, S., 2015. LEF-1 and TCF-1 orchestrate T(FH) differentiation by regulating differentiation circuits upstream of the transcriptional repressor Bcl6. Nat. Immunol. 16, 980–990.

Chornoguz, O., Hagan, R.S., Haile, A., Arwood, M.L., Gamper, C.J., Banerjee, A., Powell, J.D., 2017. mTORC1 promotes T-bet phosphorylation to regulate Th1 differentiation. J. Immunol. 198, 3939–3948.

Chtanova, T., Tangye, S.G., Newton, R., Frank, N., Hodge, M.R., Rolph, M.S., Mackay, C.R., 2004. T follicular helper cells express a distinctive transcriptional profile, reflecting their role as non-Th1/Th2 effector cells that provide help for B cells. J. Immunol. 173, 68–78.

Chung, A.S., Wu, X., Zhuang, G., Ngu, H., Kasman, I., Zhang, J., Vernes, J.M., Jiang, Z., Meng, Y.G., Peale, F.V., Ouyang, W., Ferrara, N., 2013. An interleukin-17-mediated paracrine network promotes tumor resistance to anti-angiogenic therapy. Nat. Med. 19, 1114–1123.

Chung, H.T., Kim, L.H., Park, B.L., Lee, J.H., Park, H.S., Choi, B.W., Hong, S.J., Chae, S.C., Kim, J.J., Park, C.S., Shin, H.D., 2003. Association analysis of novel TBX21 variants with asthma phenotypes. Hum. Mutat. 22, 257.

Cimmino, L., Martins, G.A., Liao, J., Magnusdottir, E., Grunig, G., Perez, R.K., Calame, K.L., 2008. Blimp-1 attenuates Th1 differentiation by repression of ifng, tbx21, and bcl6 gene expression. J. Immunol. 181, 2338–2347.

Ciofani, M., Madar, A., Galan, C., Sellars, M., Mace, K., Pauli, F., Agarwal, A., Huang, W., Parkhurst, C.N., Muratet, M., Newberry, K.M., Meadows, S., Greenfield, A., Yang, Y., Jain, P., Kirigin, F.K., Birchmeier, C., Wagner, E.F., Murphy, K.M., Myers, R.M., Bonneau, R., Littman, D.R., 2012. A validated regulatory network for Th17 cell specification. Cell 151, 289–303.

Collison, L.W., Workman, C.J., Kuo, T.T., Boyd, K., Wang, Y., Vignali, K.M., Cross, R., Sehy, D., Blumberg, R.S., Vignali, D.A., 2007. The inhibitory cytokine IL-35 contributes to regulatory T-cell function. Nature 450, 566–569.

Cretney, E., Xin, A., Shi, W., Minnich, M., Masson, F., Miasari, M., Belz, G.T., Smyth, G.K., Busslinger, M., Nutt, S.L., Kallies, A., 2011. The transcription factors Blimp-1 and IRF4 jointly control the differentiation and function of effector regulatory T cells. Nat. Immunol. 12, 304–311.

Crotty, S., Johnston, R.J., Schoenberger, S.P., 2010. Effectors and memories: Bcl-6 and Blimp-1 in T and B lymphocyte differentiation. Nat. Immunol. 11, 114–120.

Dang, E.V., Barbi, J., Yang, H.Y., Jinasena, D., Yu, H., Zheng, Y., Bordman, Z., Fu, J., Kim, Y., Yen, H.R., Luo, W., Zeller, K., Shimoda, L., Topalian, S.L., Semenza, G.L., Dang, C.V., Pardoll, D.M., Pan, F., 2011. Control of T(H)17/T(reg) balance by hypoxia-inducible factor 1. Cell 146, 772–784.

Dardalhon, V., Awasthi, A., Kwon, H., Galileos, G., Gao, W., Sobel, R.A., Mitsdoerffer, M., Strom, T.B., Elyaman, W., Ho, I.C., Khoury, S., Oukka, M., Kuchroo, V.K., 2008. IL-4 inhibits TGF-beta-induced Foxp3$^+$ T cells and, together with TGF-beta, generates IL-9$^+$ IL-10$^+$ Foxp3(−) effector T cells. Nat. Immunol. 9, 1347–1355.

Darnell JR., J.E., Kerr, I.M., Stark, G.R., 1994. Jak-STAT pathways and transcriptional activation in response to IFNs and other extracellular signaling proteins. Science 264, 1415–1421.

Das, J., Chen, C.H., Yang, L., Cohn, L., Ray, P., Ray, A., 2001. A critical role for NF-kappa B in GATA3 expression and TH2 differentiation in allergic airway inflammation. Nat. Immunol. 2, 45–50.

Deaglio, S., Dwyer, K.M., Gao, W., Friedman, D., Usheva, A., Erat, A., Chen, J.F., Enjyoji, K., Linden, J., Oukka, M., Kuchroo, V.K., Strom, T.B., Robson, S.C., 2007. Adenosine generation catalyzed by CD39 and CD73 expressed on regulatory T cells mediates immune suppression. J. Exp. Med. 204, 1257–1265.

Del Prete, G., 1992. Human Th1 and Th2 lymphocytes: their role in the pathophysiology of atopy. Allergy 47, 450–455.

Delgoffe, G.M., Kole, T.P., Zheng, Y., Zarek, P.E., Matthews, K.L., Xiao, B., Worley, P.F., Kozma, S.C., Powell, J.D., 2009. The mTOR kinase differentially regulates effector and regulatory T cell lineage commitment. Immunity 30, 832–844.

Delgoffe, G.M., Pollizzi, K.N., Waickman, A.T., Heikamp, E., Meyers, D.J., Horton, M.R., Xiao, B., Worley, P.F., Powell, J.D., 2011. The kinase mTOR regulates the differentiation of helper T cells through the selective activation of signaling by mTORC1 and mTORC2. Nat. Immunol. 12, 295–303.

Dent, A.L., Shaffer, A.L., Yu, X., Allman, D., Staudt, L.M., 1997. Control of inflammation, cytokine expression, and germinal center formation by BCL-6. Science 276, 589–592.

Djuretic, I.M., Levanon, D., Negreanu, V., Groner, Y., Rao, A., Ansel, K.M., 2007. Transcription factors T-bet and Runx3 cooperate to activate Ifng and silence Il4 in T helper type 1 cells. Nat. Immunol. 8, 145–153.

Dong, Q., Louahed, J., Vink, A., Sullivan, C.D., Messler, C.J., Zhou, Y., Haczku, A., Huaux, F., Arras, M., Holroyd, K.J., Renauld, J.C., Levitt, R.C., Nicolaides, N.C., 1999. IL-9 induces chemokine expression in lung epithelial cells and baseline airway eosinophilia in transgenic mice. Eur. J. Immunol. 29, 2130–2139.

Donnarumma, T., Young, G.R., Merkenschlager, J., Eksmond, U., Bongard, N., Nutt, S.L., Boyer, C., Dittmer, U., LE-Trilling, V.T., Trilling, M., Bayer, W., Kassiotis, G., 2016. Opposing development of cytotoxic and follicular helper CD4 T cells controlled by the TCF-1-bcl6 nexus. Cell Rep. 17, 1571–1583.

Du, J., Huang, C., Zhou, B., Ziegler, S.F., 2008. Isoform-specific inhibition of ROR alpha-mediated transcriptional activation by human FOXP3. J. Immunol. 180, 4785–4792.

Durant, L., Watford, W.T., Ramos, H.L., Laurence, A., Vahedi, G., Wei, L., Takahashi, H., Sun, H.W., Kanno, Y., Powrie, F., O'shea, J.J., 2010. Diverse targets of the transcription factor STAT3 contribute to T cell pathogenicity and homeostasis. Immunity 32, 605–615.

Durbin, J.E., Hackenmiller, R., Simon, M.C., Levy, D.E., 1996. Targeted disruption of the mouse Stat1 gene results in compromised innate immunity to viral disease. Cell 84, 443–450.

Eklund, K.K., Ghildyal, N., Austen, K.F., Stevens, R.L., 1993. Induction by IL-9 and suppression by IL-3 and IL-4 of the levels of chromosome 14-derived transcripts that encode late-expressed mouse mast cell proteases. J. Immunol. 151, 4266–4273.

Elyaman, W., Bradshaw, E.M., Uyttenhove, C., Dardalhon, V., Awasthi, A., Imitola, J., Bettelli, E., Oukka, M., Van Snick, J., Renauld, J.C., Kuchroo, V.K., Khoury, S.J., 2009. IL-9 induces differentiation of TH17 cells and enhances function of $FoxP3^+$ natural regulatory T cells. Proc. Natl. Acad. Sci. USA 106, 12885–12890.

Fallarino, F., Grohmann, U., Hwang, K.W., Orabona, C., Vacca, C., Bianchi, R., Belladonna, M.L., Fioretti, M.C., Alegre, M.L., Puccetti, P., 2003. Modulation of tryptophan catabolism by regulatory T cells. Nat. Immunol. 4, 1206–1212.

Feng, Y., Arvey, A., Chinen, T., Van Der Veeken, J., Gasteiger, G., Rudensky, A.Y., 2014. Control of the inheritance of regulatory T cell identity by a cis element in the Foxp3 locus. Cell 158, 749–763.

Fleischer, B., 1984. Acquisition of specific cytotoxic activity by human $T4^+$ T lymphocytes in culture. Nature 308, 365–367.

Fridman, W.H., Pagès, F., Sautès-Fridman, C., Galon, J., 2012. The immune contexture in human tumours: impact on clinical outcome. Nat. Rev. Cancer. 12 (4), 298–306. https://doi.org/10.1038/nrc3245.

Fujio, K., Okamura, T., Yamamoto, K., 2010. The Family of IL-10-secreting $CD4^+$ T cells. Adv. Immunol. 105, 99–130.

Gandhi, R., Kumar, D., Burns, E.J., Nadeau, M., Dake, B., Laroni, A., Kozoriz, D., Weiner, H.L., Quintana, F.J., 2010. Activation of the aryl hydrocarbon receptor induces human type 1 regulatory T cell-like and Foxp3(+) regulatory T cells. Nat. Immunol. 11, 846–853.

Glasmacher, E., Agrawal, S., Chang, A.B., Murphy, T.L., Zeng, W., Vander Lugt, B., Khan, A.A., Ciofani, M., Spooner, C.J., Rutz, S., Hackney, J., Nurieva, R., Escalante, C.R., Ouyang, W., Littman, D.R., Murphy, K.M., Singh, H., 2012. A genomic regulatory element that directs assembly and function of immune-specific AP-1-IRF complexes. Science 338, 975–980.

Glimcher, L.H., Townsend, M.J., Sullivan, B.M., Lord, G.M., 2004. Recent developments in the transcriptional regulation of cytolytic effector cells. Nat. Rev. Immunol. 4, 900–911.

Goswami, R., Jabeen, R., Yagi, R., Pham, D., Zhu, J., Goenka, S., Kaplan, M.H., 2012. STAT6-dependent regulation of Th9 development. J. Immunol. 188, 968–975.

Goswami, R., Kaplan, M.H., 2012. Gcn5 is required for PU.1-dependent IL-9 induction in Th9 cells. J. Immunol. 189, 3026–3033.

Gounni, A.S., Gregory, B., Nutku, E., Aris, F., Latifa, K., Minshall, E., North, J., Tavernier, J., Levit, R., Nicolaides, N., Robinson, D., Hamid, Q., 2000. Interleukin-9 enhances interleukin-5 receptor expression, differentiation, and survival of human eosinophils. Blood 96, 2163–2171.

Gourh, P., Agarwal, S.K., Divecha, D., Assassi, S., Paz, G., Arora-Singh, R.K., Reveille, J.D., Shete, S., Mayes, M.D., Arnett, F.C., Tan, F.K., 2009. Polymorphisms in TBX21 and STAT4 increase the risk of systemic sclerosis: evidence of possible gene-gene interaction and alterations in Th1/Th2 cytokines. Arthritis Rheum. 60, 3794–3806.

Groux, H., O'garra, A., Bigler, M., Rouleau, M., Antonenko, S., De Vries, J.E., Roncarolo, M.G., 1997. A $CD4^+$ T-cell subset inhibits antigen-specific T-cell responses and prevents colitis. Nature 389, 737–742.

Gu-Trantien, C., Loi, S., Garaud, S., Equeter, C., Libin, M., De Wind, A., Ravoet, M., LE Buanec, H., Sibille, C., Manfouo-Foutsop, G., Veys, I., Haibe-Kains, B., Singhal, S.K., Michiels, S., Rothe, F., Salgado, R., Duvillier, H., Ignatiadis, M., Desmedt, C., Bron, D., Larsimont, D., Piccart, M., Sotiriou, C., Willard-Gallo, K., 2013. CD4(+) follicular helper T cell infiltration predicts breast cancer survival. J. Clin. Invest. 123, 2873–2892.

Hallett, M.A., Venmar, K.T., Fingleton, B., 2012. Cytokine stimulation of epithelial cancer cells: the similar and divergent functions of IL-4 and IL-13. Cancer Res. 72, 6338–6343.

Harada, Y., Harada, Y., Elly, C., Ying, G., Paik, J.H., Depinho, R.A., Liu, Y.C., 2010. Transcription factors Foxo3a and Foxo1 couple the E3 ligase Cbl-b to the induction of Foxp3 expression in induced regulatory T cells. J. Exp. Med. 207, 1381–1391.

Harris, T.J., Grosso, J.F., Yen, H.R., Xin, H., Kortylewski, M., Albesiano, E., Hipkiss, E.L., Getnet, D., Goldberg, M.V., Maris, C.H., Housseau, F., Yu, H., Pardoll, D.M., Drake, C.G., 2007. Cutting edge: an in vivo requirement for STAT3 signaling in TH17 development and TH17-dependent autoimmunity. J. Immunol. 179, 4313–4317.

Higashi, T., Tsukada, J., Yoshida, Y., Mizobe, T., Mouri, F., Minami, Y., Morimoto, H., Tanaka, Y., 2005. Constitutive tyrosine and serine phosphorylation of STAT4 in T-cells transformed with HTLV-I. Gene Cell. 10, 1153–1162.

Ho, I.C., Lo, D., Glimcher, L.H., 1998. c-maf promotes T helper cell type 2 (Th2) and attenuates Th1 differentiation by both interleukin 4-dependent and -independent mechanisms. J. Exp. Med. 188, 1859–1866.

Hoey, T., Zhang, S., Schmidt, N., Yu, Q., Ramchandani, S., Xu, X., Naeger, L.K., Sun, Y.L., Kaplan, M.H., 2003. Distinct requirements for the naturally occurring splice forms Stat4alpha and Stat4beta in IL-12 responses. EMBO J. 22, 4237–4248.

Hori, S., 2012. The Foxp3 interactome: a network perspective of T(reg) cells. Nat. Immunol. 13, 943–945.

Hsieh, C.S., Heimberger, A.B., Gold, J.S., O'garra, A., Murphy, K.M., 1992. Differential regulation of T helper phenotype development by interleukins 4 and 10 in an alpha beta T-cell-receptor transgenic system. Proc. Natl. Acad. Sci. USA 89, 6065–6069.

Hua, L., Yao, S., Pham, D., Jiang, L., Wright, J., Sawant, D., Dent, A.L., Braciale, T.J., Kaplan, M.H., Sun, J., 2013. Cytokine-dependent induction of $CD4^+$ T cells with cytotoxic potential during influenza virus infection. J. Virol. 87, 11884–11893.

Huang, C.T., Workman, C.J., Flies, D., Pan, X., Marson, A.L., Zhou, G., Hipkiss, E.L., Ravi, S., Kowalski, J., Levitsky, H.I., Powell, J.D., Pardoll, D.M., Drake, C.G., Vignali, D.A., 2004. Role of LAG-3 in regulatory T cells. Immunity 21, 503–513.

Huang, W., Solouki, S., Koylass, N., Zheng, S.G., August, A., 2017. ITK signalling via the Ras/IRF4 pathway regulates the development and function of Tr1 cells. Nat. Commun. 8, 15871.

Huber, M., Steinwald, V., Guralnik, A., Brustle, A., Kleemann, P., Rosenplanter, C., Decker, T., Lohoff, M., 2008. IL-27 inhibits the development of regulatory T cells via STAT3. Int. Immunol. 20, 223–234.

Humblin, E., Thibaudin, M., Chalmin, F., Derangere, V., Limagne, E., Richard, C., Flavell, R.A., Chevrier, S., Ladoire, S., Berger, H., Boidot, R., Apetoh, L., Vegran, F., Ghiringhelli, F., 2017. IRF8-dependent molecular complexes control the Th9 transcriptional program. Nat. Commun. 8, 2085.

Hwang, E.S., Hong, J.H., Glimcher, L.H., 2005a. IL-2 production in developing Th1 cells is regulated by heterodimerization of RelA and T-bet and requires T-bet serine residue 508. J. Exp. Med. 202, 1289–1300.

Hwang, E.S., Szabo, S.J., Schwartzberg, P.L., Glimcher, L.H., 2005b. T helper cell fate specified by kinase-mediated interaction of T-bet with GATA-3. Science 307, 430–433.

Ise, W., Kohyama, M., Schraml, B.U., Zhang, T., Schwer, B., Basu, U., Alt, F.W., Tang, J., Oltz, E.M., Murphy, T.L., Murphy, K.M., 2011. The transcription factor BATF controls the global regulators of class-switch recombination in both B cells and T cells. Nat. Immunol. 12, 536–543.

Ivanov, I., Mckenzie, B.S., Zhou, L., Tadokoro, C.E., Lepelley, A., Lafaille, J.J., Cua, D.J., Littman, D.R., 2006. The orphan nuclear receptor RORgammat directs the differentiation program of proinflammatory $IL-17^+$ T helper cells. Cell 126, 1121–1133.

Jabeen, R., Goswami, R., Awe, O., Kulkarni, A., Nguyen, E.T., Attenasio, A., Walsh, D., Olson, M.R., Kim, M.H., Tepper, R.S., Sun, J., Kim, C.H., Taparowsky, E.J., Zhou, B., Kaplan, M.H., 2013. Th9 cell development requires a BATF-regulated transcriptional network. J. Clin. Invest. 123, 4641–4653.

Jager, A., Dardalhon, V., Sobel, R.A., Bettelli, E., Kuchroo, V.K., 2009. Th1, Th17, and Th9 effector cells induce experimental autoimmune encephalomyelitis with different pathological phenotypes. J. Immunol. 183, 7169–7177.

Janssen, E.M., Lemmens, E.E., Wolfe, T., Christen, U., Von Herrath, M.G., Schoenberger, S.P., 2003. $CD4^+$ T cells are required for secondary expansion and memory in $CD8^+$ T lymphocytes. Nature 421, 852–856.

Jash, A., Sahoo, A., Kim, G.C., Chae, C.S., Hwang, J.S., Kim, J.E., Im, S.H., 2012. Nuclear factor of activated T cells 1 (NFAT1)-induced permissive chromatin modification facilitates nuclear factor-kappaB (NF-kappaB)-mediated interleukin-9 (IL-9) transactivation. J. Biol. Chem. 287, 15445–15457.

Johnston, R.J., Choi, Y.S., Diamond, J.A., Yang, J.A., Crotty, S., 2012. STAT5 is a potent negative regulator of TFH cell differentiation. J. Exp. Med. 209, 243–250.

Johnston, R.J., Poholek, A.C., Ditoro, D., Yusuf, I., Eto, D., Barnett, B., Dent, A.L., Craft, J., Crotty, S., 2009. Bcl6 and Blimp-1 are reciprocal and antagonistic regulators of T follicular helper cell differentiation. Science 325, 1006–1010.

Jutel, M., Akdis, C.A., 2008. T-cell regulatory mechanisms in specific immunotherapy. Chem. Immunol. Allergy. 94, 158–177. https://doi.org/10.1159/000155000.

Kallies, A., Hawkins, E.D., Belz, G.T., Metcalf, D., Hommel, M., Corcoran, L.M., Hodgkin, P.D., Nutt, S.L., 2006. Transcriptional repressor Blimp-1 is essential for T cell homeostasis and self-tolerance. Nat. Immunol. 7, 466–474.

Kano, S., Sato, K., Morishita, Y., Vollstedt, S., Kim, S., Bishop, K., Honda, K., Kubo, M., Taniguchi, T., 2008. The contribution of transcription factor IRF1 to the interferon-gamma-interleukin 12 signaling axis and TH1 versus TH-17 differentiation of $CD4^+$ T cells. Nat. Immunol. 9, 34–41.

Kaplan, M.H., Schindler, U., Smiley, S.T., Grusby, M.J., 1996a. Stat6 is required for mediating responses to IL-4 and for development of Th2 cells. Immunity 4, 313–319.

Kaplan, M.H., Sun, Y.L., Hoey, T., Grusby, M.J., 1996b. Impaired IL-12 responses and enhanced development of Th2 cells in Stat4-deficient mice. Nature 382, 174–177.

Karwacz, K., Miraldi, E.R., Pokrovskii, M., Madi, A., Yosef, N., Wortman, I., Chen, X., Watters, A., Carriero, N., Awasthi, A., Regev, A., Bonneau, R., Littman, D., Kuchroo, V.K., 2017. Critical role of IRF1 and BATF in forming chromatin landscape during type 1 regulatory cell differentiation. Nat. Immunol. 18, 412–421.

Kerdiles, Y.M., Beisner, D.R., Tinoco, R., Dejean, A.S., Castrillon, D.H., Depinho, R.A., Hedrick, S.M., 2009. Foxo1 links homing and survival of naive T cells by regulating L-selectin, CCR7 and interleukin 7 receptor. Nat. Immunol. 10, 176–184.

Kerdiles, Y.M., Stone, E.L., Beisner, D.R., Mcgargill, M.A., CH'en, I.L., Stockmann, C., Katayama, C.D., Hedrick, S.M., 2010. Foxo transcription factors control regulatory T cell development and function. Immunity 33, 890–904.

Kiani, A., Garcia-Cozar, F.J., Habermann, I., Laforsch, S., Aebischer, T., Ehninger, G., Rao, A., 2001. Regulation of interferon-gamma gene expression by nuclear factor of activated T cells. Blood 98, 1480–1488.

Kiani, A., Viola, J.P., Lichtman, A.H., Rao, A., 1997. Down-regulation of IL-4 gene transcription and control of Th2 cell differentiation by a mechanism involving NFAT1. Immunity 7, 849–860.

Kim, J.I., Ho, I.C., Grusby, M.J., Glimcher, L.H., 1999. The transcription factor c-Maf controls the production of interleukin-4 but not other Th2 cytokines. Immunity 10, 745–751.

Kitoh, A., Ono, M., Naoe, Y., Ohkura, N., Yamaguchi, T., Yaguchi, H., Kitabayashi, I., Tsukada, T., Nomura, T., Miyachi, Y., Taniuchi, I., Sakaguchi, S., 2009. Indispensable role of the Runx1-Cbfbeta transcription complex for in vivo-suppressive function of $FoxP3^+$ regulatory T cells. Immunity 31, 609–620.

Klunker, S., Chong, M.M., Mantel, P.Y., Palomares, O., Bassin, C., Ziegler, M., Ruckert, B., Meiler, F., Akdis, M., Littman, D.R., Akdis, C.A., 2009. Transcription factors RUNX1 and RUNX3 in the induction and suppressive function of $Foxp3^+$ inducible regulatory T cells. J. Exp. Med. 206, 2701–2715.

Kobayashi, M., Kobayashi, H., Pollard, R.B., Suzuki, F., 1998. A pathogenic role of Th2 cells and their cytokine products on the pulmonary metastasis of murine B16 melanoma. J. Immunol. 160, 5869–5873.

Koch, S., Graser, A., Mirzakhani, H., Zimmermann, T., Melichar, V.O., Wolfel, M., Croteau-Chonka, D.C., Raby, B.A., Weiss, S.T., Finotto, S., 2016. Increased expression of nuclear factor of activated T cells 1 drives IL-9-mediated allergic asthma. J. Allergy Clin. Immunol. 137, 1898 e7–1902 e7.

Koh, B., Hufford, M.M., Pham, D., Olson, M.R., Wu, T., Jabeen, R., Sun, X., Kaplan, M.H., 2016. The ETS family transcription factors Etv5 and PU.1 function in parallel to promote Th9 cell development. J. Immunol. 197, 2465–2472.

Kohu, K., Ohmori, H., Wong, W.F., Onda, D., Wakoh, T., Kon, S., Yamashita, M., Nakayama, T., Kubo, M., Satake, M., 2009. The Runx3 transcription factor augments Th1 and down-modulates Th2 phenotypes by interacting with and attenuating GATA3. J. Immunol. 183, 7817–7824.

Korn, T., Bettelli, E., Gao, W., Awasthi, A., Jager, A., Strom, T.B., Oukka, M., Kuchroo, V.K., 2007. IL-21 initiates an alternative pathway to induce proinflammatory T(H)17 cells. Nature 448, 484–487.

Kretschmer, K., Apostolou, I., Hawiger, D., Khazaie, K., Nussenzweig, M.C., Von Boehmer, H., 2005. Inducing and expanding regulatory T cell populations by foreign antigen. Nat. Immunol. 6, 1219–1227.

Kryczek, I., Banerjee, M., Cheng, P., Vatan, L., Szeliga, W., Wei, S., Huang, E., Finlayson, E., Simeone, D., Welling, T.H., Chang, A., Coukos, G., Liu, R., Zou, W., 2009. Phenotype, distribution, generation, and functional and clinical relevance of Th17 cells in the human tumor environments. Blood 114, 1141–1149.

Kurebayashi, Y., Nagai, S., Ikejiri, A., Ohtani, M., Ichiyama, K., Baba, Y., Yamada, T., Egami, S., Hoshii, T., Hirao, A., Matsuda, S., Koyasu, S., 2012. PI3K-Akt-mTORC1-S6K1/2 axis controls Th17 differentiation by regulating Gfi1 expression and nuclear translocation of RORgamma. Cell Rep. 1, 360–373.

Ladoire, S, Martin, F, Ghiringhelli, F., 2011. Prognostic role of FOXP3+ regulatory T cells infiltrating human carcinomas: the paradox of colorectal cancer. Cancer Immunol Immunother 60 (7), 909–918. https://doi.org/10.1007/s00262-011-1046-y.

Lal, G., Zhang, N., Van Der Touw, W., Ding, Y., Ju, W., Bottinger, E.P., Reid, S.P., Levy, D.E., Bromberg, J.S., 2009. Epigenetic regulation of Foxp3 expression in regulatory T cells by DNA methylation. J. Immunol. 182, 259–273.

Laurence, A., Amarnath, S., Mariotti, J., Kim, Y.C., Foley, J., Eckhaus, M., O'shea, J.J., Fowler, D.H., 2012. STAT3 transcription factor promotes instability of nTreg cells and limits generation of iTreg cells during acute murine graft-versus-host disease. Immunity 37, 209–222.

Laurence, A., Tato, C.M., Davidson, T.S., Kanno, Y., Chen, Z., Yao, Z., Blank, R.B., Meylan, F., Siegel, R., Hennighausen, L., Shevach, E.M., O'shea, J.J., 2007. Interleukin-2 signaling via STAT5 constrains T helper 17 cell generation. Immunity 26, 371–381.

Lazarevic, V., Chen, X., Shim, J.H., Hwang, E.S., Jang, E., Bolm, A.N., Oukka, M., Kuchroo, V.K., Glimcher, L.H., 2011. T-bet represses T(H)17 differentiation by preventing Runx1-mediated activation of the gene encoding RORgammat. Nat. Immunol. 12, 96–104.

Le Gros, G., Ben-Sasson, S.Z., Seder, R., Finkelman, F.D., Paul, W.E., 1990. Generation of interleukin 4 (IL-4)-producing cells in vivo and in vitro: IL-2 and IL-4 are required for in vitro generation of IL-4-producing cells. J. Exp. Med. 172, 921–929.

Lee, K., Gudapati, P., Dragovic, S., Spencer, C., Joyce, S., Killeen, N., Magnuson, M.A., Boothby, M., 2010. Mammalian target of rapamycin protein complex 2 regulates differentiation of Th1 and Th2 cell subsets via distinct signaling pathways. Immunity 32, 743–753.

Lee, W., Kim, H.S., Baek, S.Y., Lee, G.R., 2016. Transcription factor IRF8 controls Th1-like regulatory T-cell function. Cell. Mol. Immunol. 13, 785–794.

Leonard, W.J., Spolski, R., 2005. Interleukin-21: a modulator of lymphoid proliferation, apoptosis and differentiation. Nat. Rev. Immunol. 5, 688–698.

Levy, D.E., Darnell JR., J.E., 2002. Stats: transcriptional control and biological impact. Nat. Rev. Mol. Cell Biol. 3, 651–662.

Li, B., Samanta, A., Song, X., Iacono, K.T., Bembas, K., Tao, R., Basu, S., Riley, J.L., Hancock, W.W., Shen, Y., Saouaf, S.J., Greene, M.I., 2007a. FOXP3 interactions with histone acetyltransferase and class II histone deacetylases are required for repression. Proc. Natl. Acad. Sci. USA 104, 4571–4576.

Li, M.O., Wan, Y.Y., Flavell, R.A., 2007b. T cell-produced transforming growth factor-beta1 controls T cell tolerance and regulates Th1- and Th17-cell differentiation. Immunity 26, 579–591.
Li, P., Spolski, R., Liao, W., Wang, L., Murphy, T.L., Murphy, K.M., Leonard, W.J., 2012. BATF-JUN is critical for IRF4-mediated transcription in T cells. Nature 490, 543–546.
Li, Z., Zhang, Y., Sun, B., 2011. Current understanding of Th2 cell differentiation and function. Protein Cell 2, 604–611.
Liao, W., Lin, J.X., Wang, L., Li, P., Leonard, W.J., 2011. Modulation of cytokine receptors by IL-2 broadly regulates differentiation into helper T cell lineages. Nat. Immunol. 12, 551–559.
Liao, W., Schones, D.E., Oh, J., Cui, Y., Cui, K., Roh, T.Y., Zhao, K., Leonard, W.J., 2008. Priming for T helper type 2 differentiation by interleukin 2-mediated induction of interleukin 4 receptor alpha-chain expression. Nat. Immunol. 9, 1288–1296.
Liao, W., Spolski, R., Li, P., Du, N., West, E.E., Ren, M., Mitra, S., Leonard, W.J., 2014. Opposing actions of IL-2 and IL-21 on Th9 differentiation correlate with their differential regulation of BCL6 expression. Proc. Natl. Acad. Sci. USA 111, 3508–3513.
Lighvani, A.A., Frucht, D.M., Jankovic, D., Yamane, H., Aliberti, J., Hissong, B.D., Nguyen, B.V., Gadina, M., Sher, A., Paul, W.E., O'shea, J.J., 2001. T-bet is rapidly induced by interferon-gamma in lymphoid and myeloid cells. Proc. Natl. Acad. Sci. USA 98, 15137–15142.
Lin, J.X., Mietz, J., Modi, W.S., John, S., Leonard, W.J., 1996. Cloning of human Stat5B. Reconstitution of interleukin-2-induced Stat5A and Stat5B DNA binding activity in COS-7 cells. J. Biol. Chem. 271, 10738–10744.
Linterman, M.A., Beaton, L., Yu, D., Ramiscal, R.R., Srivastava, M., Hogan, J.J., Verma, N.K., Smyth, M.J., Rigby, R.J., Vinuesa, C.G., 2010. IL-21 acts directly on B cells to regulate Bcl-6 expression and germinal center responses. J. Exp. Med. 207, 353–363.
Liu, X., Chen, X., Zhong, B., Wang, A., Wang, X., Chu, F., Nurieva, R.I., Yan, X., Chen, P., Van Der Flier, L.G., Nakatsukasa, H., Neelapu, S.S., Chen, W., Clevers, H., Tian, Q., Qi, H., Wei, L., Dong, C., 2014. Transcription factor achaete-scute homologue 2 initiates follicular T-helper-cell development. Nature 507, 513–518.
Liu, X., Nurieva, R.I., Dong, C., 2013. Transcriptional regulation of follicular T-helper (Tfh) cells. Immunol. Rev. 252, 139–145.
Liu, Z., Gerner, M.Y., Van Panhuys, N., Levine, A.G., Rudensky, A.Y., Germain, R.N., 2015. Immune homeostasis enforced by co-localized effector and regulatory T cells. Nature 528, 225–230.
Lohoff, M., Mittrucker, H.W., Prechtl, S., Bischof, S., Sommer, F., Kock, S., Ferrick, D.A., Duncan, G.S., Gessner, A., Mak, T.W., 2002. Dysregulated T helper cell differentiation in the absence of interferon regulatory factor 4. Proc. Natl. Acad. Sci. USA 99, 11808–11812.
Lora, J.M., Al-Garawi, A., Pickard, M.D., Price, K.S., Bagga, S., Sicoli, J., Hodge, M.R., Gutierrez-Ramos, J.C., Briskin, M.J., Boyce, J.A., 2003. FcepsilonRI-dependent gene expression in human mast cells is differentially controlled by T helper type 2 cytokines. J. Allergy Clin. Immunol. 112, 1119–1126.
Lovett-Racke, A.E., Rocchini, A.E., Choy, J., Northrop, S.C., Hussain, R.Z., Ratts, R.B., Sikder, D., Racke, M.K., 2004. Silencing T-bet defines a critical role in the differentiation of autoreactive T lymphocytes. Immunity 21, 719–731.
Lu, Y., Hong, S., Li, H., Park, J., Hong, B., Wang, L., Zheng, Y., Liu, Z., Xu, J., He, J., Yang, J., Qian, J., Yi, Q., 2012. Th9 cells promote antitumor immune responses in vivo. J. Clin. Invest. 122, 4160–4171.

Lugo-Villarino, G., Maldonado-Lopez, R., Possemato, R., Penaranda, C., Glimcher, L.H., 2003. T-bet is required for optimal production of IFN-gamma and antigen-specific T cell activation by dendritic cells. Proc. Natl. Acad. Sci. USA 100, 7749–7754.

Lukacher, A.E., Morrison, L.A., Braciale, V.L., Malissen, B., Braciale, T.J., 1985. Expression of specific cytolytic activity by H-2I region-restricted, influenza virus-specific T lymphocyte clones. J. Exp. Med. 162, 171–187.

Lund, R.J., Chen, Z., Scheinin, J., Lahesmaa, R., 2004. Early target genes of IL-12 and STAT4 signaling in th cells. J. Immunol. 172, 6775–6782.

Ma, C.S., Chew, G.Y., Simpson, N., Priyadarshi, A., Wong, M., Grimbacher, B., Fulcher, D.A., Tangye, S.G., Cook, M.C., 2008. Deficiency of Th17 cells in hyper IgE syndrome due to mutations in STAT3. J. Exp. Med. 205, 1551–1557.

Magnani, C.F., Alberigo, G., Bacchetta, R., Serafini, G., Andreani, M., Roncarolo, M.G., Gregori, S., 2011. Killing of myeloid APCs via HLA class I, Cd2 and CD226 defines a novel mechanism of suppression by human Tr1 cells. Eur. J. Immunol. 41, 1652–1662.

Maimone, M.M., Morrison, L.A., Braciale, V.L., Braciale, T.J., 1986. Features of target cell lysis by class I and class II MHC-restricted cytolytic T lymphocytes. J. Immunol. 137, 3639–3643.

Mangan, P.R., Harrington, L.E., O'quinn, D.B., Helms, W.S., Bullard, D.C., Elson, C.O., Hatton, R.D., Wahl, S.M., Schoeb, T.R., Weaver, C.T., 2006. Transforming growth factor-beta induces development of the T(H)17 lineage. Nature 441, 231–234.

Mantel, P.Y., Kuipers, H., Boyman, O., Rhyner, C., Ouaked, N., Ruckert, B., Karagiannidis, C., Lambrecht, B.N., Hendriks, R.W., Crameri, R., Akdis, C.A., Blaser, K., Schmidt-Weber, C.B., 2007. GATA3-driven Th2 responses inhibit TGF-beta1-induced FOXP3 expression and the formation of regulatory T cells. PLoS Biol. 5, e329.

Martin-Orozco, N., Muranski, P., Chung, Y., Yang, X.O., Yamazaki, T., Lu, S., Hwu, P., Restifo, N.P., Overwijk, W.W., Dong, C., 2009. T helper 17 cells promote cytotoxic T cell activation in tumor immunity. Immunity 31, 787–798.

Martinez, G.J., Zhang, Z., Chung, Y., Reynolds, J.M., Lin, X., Jetten, A.M., Feng, X.H., Dong, C., 2009. Smad3 differentially regulates the induction of regulatory and inflammatory T cell differentiation. J. Biol. Chem. 284, 35283–35286.

Martins, G.A., Cimmino, L., Shapiro-Shelef, M., Szabolcs, M., Herron, A., Magnusdottir, E., Calame, K., 2006. Transcriptional repressor Blimp-1 regulates T cell homeostasis and function. Nat. Immunol. 7, 457–465.

Maruyama, S., Kanoh, M., Matsumoto, A., Kuwahara, M., Yamashita, M., Asano, Y., 2015. A novel function of interferon regulatory factor-1: inhibition of Th2 cells by down-regulating the Il4 gene during Listeria infection. Int. Immunol. 27, 143–152.

Mascanfroni, I.D., Takenaka, M.C., Yeste, A., Patel, B., Wu, Y., Kenison, J.E., Siddiqui, S., Basso, A.S., Otterbein, L.E., Pardoll, D.M., Pan, F., Priel, A., Clish, C.B., Robson, S.C., Quintana, F.J., 2015. Metabolic control of type 1 regulatory T cell differentiation by AHR and HIF1-alpha. Nat. Med. 21, 638–646.

Mcgeachy, M.J., Bak-Jensen, K.S., Chen, Y., Tato, C.M., Blumenschein, W., Mcclanahan, T., Cua, D.J., 2007. TGF-beta and IL-6 drive the production of IL-17 and IL-10 by T cells and restrain T(H)-17 cell-mediated pathology. Nat. Immunol. 8, 1390–1397.

Ménétrier-Caux, C., Curiel, T., Faget, J., Manuel, M., Caux, C., Zou, W., 2012. Targeting regulatory T cells. Target Oncol. 7 (1), 15–28. https://doi.org/10.1007/s11523-012-0208-y. Epub 2012 Feb 12.

Meraz, M.A., White, J.M., Sheehan, K.C., Bach, E.A., Rodig, S.J., Dighe, A.S., Kaplan, D.H., Riley, J.K., Greenlund, A.C., Campbell, D., Carver-Moore, K., Dubois, R.N., Clark, R., Aguet, M., Schreiber, R.D., 1996. Targeted disruption of the Stat1 gene in mice reveals unexpected physiologic specificity in the JAK-STAT signaling pathway. Cell 84, 431–442.

Montes De Oca, M., Kumar, R., De Labastida Rivera, F., Amante, F.H., Sheel, M., Faleiro, R.J., Bunn, P.T., Best, S.E., Beattie, L., Ng, S.S., Edwards, C.L., Muller, W., Cretney, E., Nutt, S.L., Smyth, M.J., Haque, A., Hill, G.R., Sundar, S., Kallies, A., Engwerda, C.R., 2016. Blimp-1-Dependent IL-10 production by Tr1 cells regulates TNF-mediated tissue pathology. PLoS Pathog. 12, e1005398.

Mosmann, T.R., Coffman, R.L., 1989. TH1 and TH2 cells: different patterns of lymphokine secretion lead to different functional properties. Annu. Rev. Immunol. 7, 145–173.

Mucida, D., Husain, M.M., Muroi, S., Van Wijk, F., Shinnakasu, R., Naoe, Y., Reis, B.S., Huang, Y., Lambolez, F., Docherty, M., Attinger, A., Shui, J.W., Kim, G., Lena, C.J., Sakaguchi, S., Miyamoto, C., Wang, P., Atarashi, K., Park, Y., Nakayama, T., Honda, K., Ellmeier, W., Kronenberg, M., Taniuchi, I., Cheroutre, H., 2013. Transcriptional reprogramming of mature CD4(+) helper T cells generates distinct MHC class II-restricted cytotoxic T lymphocytes. Nat. Immunol. 14, 281–289.

Muranski, P., Boni, A., Antony, P.A., Cassard, L., Irvine, K.R., Kaiser, A., Paulos, C.M., Palmer, D.C., Touloukian, C.E., Ptak, K., Gattinoni, L., Wrzesinski, C., Hinrichs, C.S., Kerstann, K.W., Feigenbaum, L., Chan, C.C., Restifo, N.P., 2008. Tumor-specific Th17-polarized cells eradicate large established melanoma. Blood 112, 362–373.

Muranski, P., Borman, Z.A., Kerkar, S.P., Klebanoff, C.A., Ji, Y., Sanchez-Perez, L., Sukumar, M., Reger, R.N., Yu, Z., Kern, S.J., Roychoudhuri, R., Ferreyra, G.A., Shen, W., Durum, S.K., Feigenbaum, L., Palmer, D.C., Antony, P.A., Chan, C.C., Laurence, A., Danner, R.L., Gattinoni, L., Restifo, N.P., 2011. Th17 cells are long lived and retain a stem cell-like molecular signature. Immunity 35, 972–985.

Murugaiyan, G., Beynon, V., Pires Da Cunha, A., Joller, N., Weiner, H.L., 2012. IFN-gamma limits Th9-mediated autoimmune inflammation through dendritic cell modulation of IL-27. J. Immunol. 189, 5277–5283.

Nakayamada, S., Kanno, Y., Takahashi, H., Jankovic, D., Lu, K.T., Johnson, T.A., Sun, H.W., Vahedi, G., Hakim, O., Handon, R., Schwartzberg, P.L., Hager, G.L., O'shea, J.J., 2011. Early Th1 cell differentiation is marked by a Tfh cell-like transition. Immunity 35, 919–931.

Nishizuka, Y., Sakakura, T., 1969. Thymus and reproduction: sex-linked dysgenesia of the gonad after neonatal thymectomy in mice. Science 166, 753–755.

Nowak, E.C., Weaver, C.T., Turner, H., Begum-Haque, S., Becher, B., Schreiner, B., Coyle, A.J., Kasper, L.H., Noelle, R.J., 2009. IL-9 as a mediator of Th17-driven inflammatory disease. J. Exp. Med. 206, 1653–1660.

Nurieva, R., Yang, X.O., Martinez, G., Zhang, Y., Panopoulos, A.D., Ma, L., Schluns, K., Tian, Q., Watowich, S.S., Jetten, A.M., Dong, C., 2007. Essential autocrine regulation by IL-21 in the generation of inflammatory T cells. Nature 448, 480–483.

Nurieva, R.I., Chung, Y., Martinez, G.J., Yang, X.O., Tanaka, S., Matskevitch, T.D., Wang, Y.H., Dong, C., 2009. Bcl6 mediates the development of T follicular helper cells. Science 325, 1001–1005.

Nurieva, R.I., Podd, A., Chen, Y., Alekseev, A.M., Yu, M., Qi, X., Huang, H., Wen, R., Wang, J., Li, H.S., Watowich, S.S., Qi, H., Dong, C., Wang, D., 2012. STAT5 protein negatively regulates T follicular helper (Tfh) cell generation and function. J. Biol. Chem. 287, 11234–11239.

O'shea, J.J., Lahesmaa, R., Vahedi, G., Laurence, A., Kanno, Y., 2011. Genomic views of STAT function in CD4$^+$ T helper cell differentiation. Nat. Rev. Immunol. 11, 239–250.

O'shea, J.J., Plenge, R., 2012. JAK and STAT signaling molecules in immunoregulation and immune-mediated disease. Immunity 36, 542–550.

Oestreich, K.J., Huang, A.C., Weinmann, A.S., 2011. The lineage-defining factors T-bet and Bcl-6 collaborate to regulate Th1 gene expression patterns. J. Exp. Med. 208, 1001–1013.

Okada, S., Markle, J.G., Deenick, E.K., Mele, F., Averbuch, D., Lagos, M., Alzahrani, M., Al-Muhsen, S., Halwani, R., Ma, C.S., Wong, N., Soudais, C., Henderson, L.A., Marzouqa, H., Shamma, J., Gonzalez, M., Martinez-Barricarte, R., Okada, C., Avery, D.T., Latorre, D., Deswarte, C., Jabot-Hanin, F., Torrado, E., Fountain, J., Belkadi, A., Itan, Y., Boisson, B., Migaud, M., Arlehamn, C.S.L., Sette, A., Breton, S., Mccluskey, J., Rossjohn, J., De Villartay, J.P., Moshous, D., Hambleton, S., Latour, S., Arkwright, P.D., Picard, C., Lantz, O., Engelhard, D., Kobayashi, M., Abel, L., Cooper, A.M., Notarangelo, L.D., Boisson-Dupuis, S., Puel, A., Sallusto, F., Bustamante, J., Tangye, S.G., Casanova, J.L., 2015. Impairment of immunity to Candida and Mycobacterium in humans with bi-allelic RORC mutations. Science 349, 606–613.

Ono, M., Yaguchi, H., Ohkura, N., Kitabayashi, I., Nagamura, Y., Nomura, T., Miyachi, Y., Tsukada, T., Sakaguchi, S., 2007. Foxp3 controls regulatory T-cell function by interacting with AML1/Runx1. Nature 446, 685–689.

Oppmann, B., Lesley, R., Blom, B., Timans, J.C., Xu, Y., Hunte, B., Vega, F., Yu, N., Wang, J., Singh, K., Zonin, F., Vaisberg, E., Churakova, T., Liu, M., Gorman, D., Wagner, J., Zurawski, S., Liu, Y., Abrams, J.S., Moore, K.W., Rennick, D., De waal-Malefyt, R., Hannum, C., Bazan, J.F., Kastelein, R.A., 2000. Novel p19 protein engages IL-12p40 to form a cytokine, IL-23, with biological activities similar as well as distinct from IL-12. Immunity 13, 715–725.

Ouyang, W., Beckett, O., Flavell, R.A., Li, M.O., 2009. An essential role of the Forkhead-box transcription factor Foxo1 in control of T cell homeostasis and tolerance. Immunity 30, 358–371.

Ouyang, W., Beckett, O., Ma, Q., Paik, J.H., Depinho, R.A., Li, M.O., 2010. Foxo proteins cooperatively control the differentiation of Foxp3$^+$ regulatory T cells. Nat. Immunol. 11, 618–627.

Ouyang, W., Liao, W., Luo, C.T., Yin, N., Huse, M., Kim, M.V., Peng, M., Chan, P., Ma, Q., Mo, Y., Meijer, D., Zhao, K., Rudensky, A.Y., Atwal, G., Zhang, M.Q., Li, M.O., 2012. Novel Foxo1-dependent transcriptional programs control T(reg) cell function. Nature 491, 554–559.

Ouyang, W., Lohning, M., Gao, Z., Assenmacher, M., Ranganath, S., Radbruch, A., Murphy, K.M., 2000. Stat6-independent GATA-3 autoactivation directs IL-4-independent Th2 development and commitment. Immunity 12, 27–37.

Ouyang, W., Rutz, S., Crellin, N.K., Valdez, P.A., Hymowitz, S.G., 2011. Regulation and functions of the IL-10 family of cytokines in inflammation and disease. Annu. Rev. Immunol. 29, 71–109.

Pai, S.Y., Truitt, M.L., Ho, I.C., 2004. GATA-3 deficiency abrogates the development and maintenance of T helper type 2 cells. Proc. Natl. Acad. Sci. USA 101, 1993–1998.

Pan, J., Zhang, M., Wang, J., Wang, Q., Xia, D., Sun, W., Zhang, L., Yu, H., Liu, Y., Cao, X., 2004. Interferon-gamma is an autocrine mediator for dendritic cell maturation. Immunol. Lett. 94, 141–151.

Pandiyan, P., Zheng, L., Ishihara, S., Reed, J., Lenardo, M.J., 2007. CD4$^+$CD25$^+$Foxp3$^+$ regulatory T cells induce cytokine deprivation-mediated apoptosis of effector CD4$^+$ T cells. Nat. Immunol. 8, 1353–1362.

Park, C., Li, S., Cha, E., Schindler, C., 2000. Immune response in Stat2 knockout mice. Immunity 13, 795–804.

Paust, S., Lu, L., Mccarty, N., Cantor, H., 2004. Engagement of B7 on effector T cells by regulatory T cells prevents autoimmune disease. Proc. Natl. Acad. Sci. USA 101, 10398–10403.

Peters, A., Fowler, K.D., Chalmin, F., Merkler, D., Kuchroo, V.K., Pot, C., 2015. IL-27 induces Th17 differentiation in the absence of STAT1 signaling. J. Immunol. 195, 4144–4153.

Pipkin, M.E., Sacks, J.A., Cruz-Guilloty, F., Lichtenheld, M.G., Bevan, M.J., Rao, A., 2010. Interleukin-2 and inflammation induce distinct transcriptional programs that promote the differentiation of effector cytolytic T cells. Immunity 32, 79–90.

Pot, C., Apetoh, L., Kuchroo, V.K., 2011. Type 1 regulatory T cells (Tr1) in autoimmunity. Semin. Immunol. 23, 202–208.

Pot, C., Jin, H., Awasthi, A., Liu, S.M., Lai, C.Y., Madan, R., Sharpe, A.H., Karp, C.L., Miaw, S.C., Ho, I.C., Kuchroo, V.K., 2009. Cutting edge: IL-27 induces the transcription factor c-Maf, cytokine IL-21, and the costimulatory receptor ICOS that coordinately act together to promote differentiation of IL-10-producing Tr1 cells. J. Immunol. 183, 797–801.

Powrie, F., Carlino, J., Leach, M.W., Mauze, S., Coffman, R.L., 1996. A critical role for transforming growth factor-beta but not interleukin 4 in the suppression of T helper type 1-mediated colitis by CD45RB(low) $CD4^+$ T cells. J. Exp. Med. 183, 2669–2674.

Powrie, F., Menon, S., Coffman, R.L., 1993. Interleukin-4 and interleukin-10 synergize to inhibit cell-mediated immunity in vivo. Eur. J. Immunol. 23, 2223–2229.

Puel, A., Cypowyj, S., Bustamante, J., Wright, J.F., Liu, L., Lim, H.K., Migaud, M., Israel, L., Chrabieh, M., Audry, M., Gumbleton, M., Toulon, A., Bodemer, C., EL-Baghdadi, J., Whitters, M., Paradis, T., Brooks, J., Collins, M., Wolfman, N.M., Al-Muhsen, S., Galicchio, M., Abel, L., Picard, C., Casanova, J.L., 2011. Chronic mucocutaneous candidiasis in humans with inborn errors of interleukin-17 immunity. Science 332, 65–68.

Purwar, R., Schlapbach, C., Xiao, S., Kang, H.S., Elyaman, W., Jiang, X., Jetten, A.M., Khoury, S.J., Fuhlbrigge, R.C., Kuchroo, V.K., Clark, R.A., Kupper, T.S., 2012. Robust tumor immunity to melanoma mediated by interleukin-9-producing T cells. Nat. Med. 18, 1248–1253.

Qin, H., Wang, L., Feng, T., Elson, C.O., Niyongere, S.A., Lee, S.J., Reynolds, S.L., Weaver, C.T., Roarty, K., Serra, R., Benveniste, E.N., Cong, Y., 2009. TGF-beta promotes Th17 cell development through inhibition of SOCS3. J. Immunol. 183, 97–105.

Quintana, F.J., Basso, A.S., Iglesias, A.H., Korn, T., Farez, M.F., Bettelli, E., Caccamo, M., Oukka, M., Weiner, H.L., 2008. Control of T(reg) and T(H)17 cell differentiation by the aryl hydrocarbon receptor. Nature 453, 65–71.

Ramsdell, F., Ziegler, S.F., 2014. FOXP3 and scurfy: how it all began. Nat. Rev. Immunol. 14, 343–349.

Ranger, A.M., Oukka, M., Rengarajan, J., Glimcher, L.H., 1998. Inhibitory function of two NFAT family members in lymphoid homeostasis and Th2 development. Immunity 9, 627–635.

Raveney, B.J., Oki, S., Hohjoh, H., Nakamura, M., Sato, W., Murata, M., Yamamura, T., 2015. Eomesodermin-expressing T-helper cells are essential for chronic neuroinflammation. Nat. Commun. 6, 8437.

Ray, J.P., Marshall, H.D., Laidlaw, B.J., Staron, M.M., Kaech, S.M., Craft, J., 2014. Transcription factor STAT3 and type I interferons are corepressive insulators for differentiation of follicular helper and T helper 1 cells. Immunity 40, 367–377.

Ray, J.P., Staron, M.M., Shyer, J.A., Ho, P.C., Marshall, H.D., Gray, S.M., Laidlaw, B.J., Araki, K., Ahmed, R., Kaech, S.M., Craft, J., 2015. The interleukin-2-mTORc1 kinase Axis defines the signaling, differentiation, and Metabolism of T Helper 1 and follicular B helper T cells. Immunity 43, 690–702.

Read, S., Greenwald, R., Izcue, A., Robinson, N., Mandelbrot, D., Francisco, L., Sharpe, A.H., Powrie, F., 2006. Blockade of CTLA-4 on $CD4^+CD25^+$ regulatory T cells abrogates their function in vivo. J. Immunol. 177, 4376–4383.

Reinhardt, R.L., Liang, H.E., Locksley, R.M., 2009. Cytokine-secreting follicular T cells shape the antibody repertoire. Nat. Immunol. 10, 385–393.

Rengarajan, J., Mowen, K.A., Mcbride, K.D., Smith, E.D., Singh, H., Glimcher, L.H., 2002a. Interferon regulatory factor 4 (IRF4) interacts with NFATc2 to modulate interleukin 4 gene expression. J. Exp. Med. 195, 1003–1012.

Rengarajan, J., Tang, B., Glimcher, L.H., 2002b. NFATc2 and NFATc3 regulate T(H)2 differentiation and modulate TCR-responsiveness of naive T(H)cells. Nat. Immunol. 3, 48–54.

Rivera Vargas, T., Cai, Z., Shen, Y., Dosset, M., Benoit-Lizon, I., Martin, T., Roussey, A., Flavell, R.A., Ghiringhelli, F., Apetoh, L., 2017. Selective degradation of PU.1 during autophagy represses the differentiation and antitumour activity of TH9 cells. Nat. Commun. 8, 559.

Rudra, D., Deroos, P., Chaudhry, A., Niec, R.E., Arvey, A., Samstein, R.M., Leslie, C., Shaffer, S.A., Goodlett, D.R., Rudensky, A.Y., 2012. Transcription factor Foxp3 and its protein partners form a complex regulatory network. Nat. Immunol. 13, 1010–1019.

Rudra, D., Egawa, T., Chong, M.M., Treuting, P., Littman, D.R., Rudensky, A.Y., 2009. Runx-CBFbeta complexes control expression of the transcription factor Foxp3 in regulatory T cells. Nat. Immunol. 10, 1170–1177.

Sahoo, A., Alekseev, A., Tanaka, K., Obertas, L., Lerman, B., Haymaker, C., Clise-Dwyer, K., Mcmurray, J.S., Nurieva, R., 2015. Batf is important for IL-4 expression in T follicular helper cells. Nat. Commun. 6, 7997.

Sakaguchi, S., Sakaguchi, N., Asano, M., Itoh, M., Toda, M., 1995. Immunologic self-tolerance maintained by activated T cells expressing IL-2 receptor alpha-chains (CD25). Breakdown of a single mechanism of self-tolerance causes various autoimmune diseases. J. Immunol. 155, 1151–1164.

Sato, K., Miyoshi, F., Yokota, K., Araki, Y., Asanuma, Y., Akiyama, Y., Yoh, K., Takahashi, S., Aburatani, H., Mimura, T., 2011. Marked induction of c-Maf protein during Th17 cell differentiation and its implication in memory Th cell development. J. Biol. Chem. 286, 14963–14971.

Sauer, S., Bruno, L., Hertweck, A., Finlay, D., Leleu, M., Spivakov, M., Knight, Z.A., Cobb, B.S., Cantrell, D., O'connor, E., Shokat, K.M., Fisher, A.G., Merkenschlager, M., 2008. T cell receptor signaling controls Foxp3 expression via PI3k, Akt, and mTOR. Proc. Natl. Acad. Sci. USA 105, 7797–7802.

Savitsky, D., Tamura, T., Yanai, H., Taniguchi, T., 2010. Regulation of immunity and oncogenesis by the IRF transcription factor family. Cancer Immunol. Immunother. 59, 489–510.

Schaerli, P., Willimann, K., Lang, A.B., Lipp, M., Loetscher, P., Moser, B., 2000. CXC chemokine receptor 5 expression defines follicular homing T cells with B cell helper function. J. Exp. Med. 192, 1553–1562.

Schmitt, E., Beuscher, H.U., Huels, C., Monteyne, P., Van Brandwijk, R., Van Snick, J., Ruede, E., 1991. IL-1 serves as a secondary signal for IL-9 expression. J. Immunol. 147, 3848–3854.

Schmitt, E., Germann, T., Goedert, S., Hoehn, P., Huels, C., Koelsch, S., Kuhn, R., Muller, W., Palm, N., Rude, E., 1994. IL-9 production of naive $CD4^+$ T cells depends on IL-2, is synergistically enhanced by a combination of TGF-beta and IL-4, and is inhibited by IFN-gamma. J. Immunol. 153, 3989–3996.

Schraml, B.U., Hildner, K., Ise, W., Lee, W.L., Smith, W.A., Solomon, B., Sahota, G., Sim, J., Mukasa, R., Cemerski, S., Hatton, R.D., Stormo, G.D., Weaver, C.T., Russell, J.H., Murphy, T.L., Murphy, K.M., 2009. The AP-1 transcription factor Batf controls T(H)17 differentiation. Nature 460, 405–409.

Sekiya, T., Kashiwagi, I., Inoue, N., Morita, R., Hori, S., Waldmann, H., Rudensky, A.Y., Ichinose, H., Metzger, D., Chambon, P., Yoshimura, A., 2011. The nuclear orphan receptor Nr4a2 induces Foxp3 and regulates differentiation of $CD4^+$ T cells. Nat. Commun. 2, 269.

Seliger, B., Ruiz-Cabello, F., Garrido, F., 2008. IFN inducibility of major histocompatibility antigens in tumors. Adv. Cancer Res. 101, 249–276.

Shankaran, V., Ikeda, H., Bruce, A.T., White, J.M., Swanson, P.E., Old, L.J., Schreiber, R.D., 2001. IFNgamma and lymphocytes prevent primary tumour development and shape tumour immunogenicity. Nature 410, 1107–1111.

Sherman, M.A., Powell, D.R., Brown, M.A., 2002. IL-4 induces the proteolytic processing of mast cell STAT6. J. Immunol. 169, 3811–3818.

Shi, L.Z., Wang, R., Huang, G., Vogel, P., Neale, G., Green, D.R., Chi, H., 2011. HIF1alpha-dependent glycolytic pathway orchestrates a metabolic checkpoint for the differentiation of TH17 and Treg cells. J. Exp. Med. 208, 1367–1376.

Shi, M., Lin, T.H., Appell, K.C., Berg, L.J., 2008. Janus-kinase-3-dependent signals induce chromatin remodeling at the Ifng locus during T helper 1 cell differentiation. Immunity 28, 763–773.

Shimoda, K., Van Deursen, J., Sangster, M.Y., Sarawar, S.R., Carson, R.T., Tripp, R.A., Chu, C., Quelle, F.W., Nosaka, T., Vignali, D.A., Doherty, P.C., Grosveld, G., Paul, W.E., Ihle, J.N., 1996. Lack of IL-4-induced Th2 response and IgE class switching in mice with disrupted Stat6 gene. Nature 380, 630–633.

Shinnakasu, R., Yamashita, M., Kuwahara, M., Hosokawa, H., Hasegawa, A., Motohashi, S., Nakayama, T., 2008. Gfi1-mediated stabilization of GATA3 protein is required for Th2 cell differentiation. J. Biol. Chem. 283, 28216–28225.

Sica, A., Dorman, L., Viggiano, V., Cippitelli, M., Ghosh, P., Rice, N., Young, H.A., 1997. Interaction of NF-kappaB and NFAT with the interferon-gamma promoter. J. Biol. Chem. 272, 30412–30420.

Sokol, C.L., Chu, N.Q., Yu, S., Nish, S.A., Laufer, T.M., Medzhitov, R., 2009. Basophils function as antigen-presenting cells for an allergen-induced T helper type 2 response. Nat. Immunol. 10, 713–720.

Staudt, V., Bothur, E., Klein, M., Lingnau, K., Reuter, S., Grebe, N., Gerlitzki, B., Hoffmann, M., Ulges, A., Taube, C., Dehzad, N., Becker, M., Stassen, M., Steinborn, A., Lohoff, M., Schild, H., Schmitt, E., Bopp, T., 2010. Interferon-regulatory factor 4 is essential for the developmental program of T helper 9 cells. Immunity 33, 192–202.

Steinke, J.W., Borish, L., 2001. Th2 cytokines and asthma. Interleukin-4: its role in the pathogenesis of asthma, and targeting it for asthma treatment with interleukin-4 receptor antagonists. Respir. Res. 2, 66–70.

Sujino, T., London, M., Hoytema Van Konijnenburg, D.P., Rendon, T., Buch, T., Silva, H.M., Lafaille, J.J., Reis, B.S., Mucida, D., 2016. Tissue adaptation of regulatory and intraepithelial CD4(+) T cells controls gut inflammation. Science 352, 1581–1586.

Summers, S.A., Phoon, R.K., Odobasic, D., Dewage, L., Kitching, A.R., Holdsworth, S.R., 2011. Signal transducer and activation of transcription 6 (STAT6) regulates T helper type 1 (Th1) and Th17 nephritogenic immunity in experimental crescentic glomerulonephritis. Clin. Exp. Immunol. 166, 227–234.

Sun, Z., Unutmaz, D., Zou, Y.R., Sunshine, M.J., Pierani, A., Brenner-Morton, S., Mebius, R.E., Littman, D.R., 2000. Requirement for RORgamma in thymocyte survival and lymphoid organ development. Science 288, 2369–2373.

Suzuki, Y., Orellana, M.A., Schreiber, R.D., Remington, J.S., 1988. Interferon-gamma: the major mediator of resistance against Toxoplasma gondii. Science 240, 516–518.

Swain, S.L., Weinberg, A.D., English, M., 1990a. $CD4^+$ T cell subsets. Lymphokine secretion of memory cells and of effector cells that develop from precursors in vitro. J. Immunol. 144, 1788–1799.

Swain, S.L., Weinberg, A.D., English, M., Huston, G., 1990b. IL-4 directs the development of Th2-like helper effectors. J. Immunol. 145, 3796–3806.

Szabo, S.J., Kim, S.T., Costa, G.L., Zhang, X., Fathman, C.G., Glimcher, L.H., 2000. A novel transcription factor, T-bet, directs Th1 lineage commitment. Cell 100, 655–669.

Szabo, S.J., Sullivan, B.M., Stemmann, C., Satoskar, A.R., Sleckman, B.P., Glimcher, L.H., 2002. Distinct effects of T-bet in TH1 lineage commitment and IFN-gamma production in CD4 and CD8 T cells. Science 295, 338–342.

Takeda, K., Noguchi, K., Shi, W., Tanaka, T., Matsumoto, M., Yoshida, N., Kishimoto, T., Akira, S., 1997. Targeted disruption of the mouse Stat3 gene leads to early embryonic lethality. Proc. Natl. Acad. Sci. USA 94, 3801–3804.

Takeuchi, A., Badr Mel, S., Miyauchi, K., Ishihara, C., Onishi, R., Guo, Z., Sasaki, Y., Ike, H., Takumi, A., Tsuji, N.M., Murakami, Y., Katakai, T., Kubo, M., Saito, T., 2016. CRTAM determines the $CD4^+$ cytotoxic T lymphocyte lineage. J. Exp. Med. 213, 123–138.

Takeuchi, A., Saito, T., 2017. CD4 CTL, a cytotoxic subset of CD4(+) T cells, their differentiation and function. Front. Immunol. 8, 194.

Tamiya, T., Ichiyama, K., Kotani, H., Fukaya, T., Sekiya, T., Shichita, T., Honma, K., Yui, K., Matsuyama, T., Nakao, T., Fukuyama, S., Inoue, H., Nomura, M., Yoshimura, A., 2013. Smad2/3 and IRF4 play a cooperative role in IL-9-producing T cell induction. J. Immunol. 191, 2360–2371.

Tanaka, S., Suto, A., Iwamoto, T., Kashiwakuma, D., Kagami, S., Suzuki, K., Takatori, H., Tamachi, T., Hirose, K., Onodera, A., Suzuki, J., Ohara, O., Yamashita, M., Nakayama, T., Nakajima, H., 2014. Sox5 and c-Maf cooperatively induce Th17 cell differentiation via RORgammat induction as downstream targets of Stat3. J. Exp. Med. 211, 1857–1874.

Tangye, S.G., Ma, C.S., Brink, R., Deenick, E.K., 2013. The good, the bad and the ugly - TFH cells in human health and disease. Nat. Rev. Immunol. 13, 412–426.

Taniguchi, T., Ogasawara, K., Takaoka, A., Tanaka, N., 2001. IRF family of transcription factors as regulators of host defense. Annu. Rev. Immunol. 19, 623–655.

Thierfelder, W.E., Van Deursen, J.M., Yamamoto, K., Tripp, R.A., Sarawar, S.R., Carson, R.T., Sangster, M.Y., Vignali, D.A., Doherty, P.C., Grosveld, G.C., Ihle, J.N., 1996. Requirement for Stat4 in interleukin-12-mediated responses of natural killer and T cells. Nature 382, 171–174.

Tousa, S., Semitekolou, M., Morianos, I., Banos, A., Trochoutsou, A.I., Brodie, T.M., Poulos, N., Samitas, K., Kapasa, M., Konstantopoulos, D., Paraskevopoulos, G., Gaga, M., Hawrylowicz, C.M., Sallusto, F., Xanthou, G., 2017. Activin-A co-opts IRF4 and AhR signaling to induce human regulatory T cells that restrain asthmatic responses. Proc. Natl. Acad. Sci. USA 114, E2891–E2900.

Trinchieri, G., Pflanz, S., Kastelein, R.A., 2003. The IL-12 family of heterodimeric cytokines: new players in the regulation of T cell responses. Immunity 19, 641–644.
Usui, T., Nishikomori, R., Kitani, A., Strober, W., 2003. GATA-3 suppresses Th1 development by downregulation of Stat4 and not through effects on IL-12Rbeta2 chain or T-bet. Immunity 18, 415–428.
Van Leeuwen, E.M., Remmerswaal, E.B., Vossen, M.T., Rowshani, A.T., Wertheim-Van Dillen, P.M., Van Lier, R.A., Ten Berge, I.J., 2004. Emergence of a $CD4^{+}CD28^{-}$ granzyme B+, cytomegalovirus-specific T cell subset after recovery of primary cytomegalovirus infection. J. Immunol. 173, 1834–1841.
Van Panhuys, N., Klauschen, F., Germain, R.N., 2014. T-cell-receptor-dependent signal intensity dominantly controls CD4(+) T cell polarization in Vivo. Immunity 41, 63–74.
Vegran, F., Berger, H., Boidot, R., Mignot, G., Bruchard, M., Dosset, M., Chalmin, F., Rebe, C., Derangere, V., Ryffel, B., Kato, M., Prevost-Blondel, A., Ghiringhelli, F., Apetoh, L., 2014. The transcription factor IRF1 dictates the IL-21-dependent anticancer functions of TH9 cells. Nat. Immunol. 15, 758–766.
Veldhoen, M., Hirota, K., Westendorf, A.M., Buer, J., Dumoutier, L., Renauld, J.C., Stockinger, B., 2008a. The aryl hydrocarbon receptor links TH17-cell-mediated autoimmunity to environmental toxins. Nature 453, 106–109.
Veldhoen, M., Hocking, R.J., Atkins, C.J., Locksley, R.M., Stockinger, B., 2006. TGFbeta in the context of an inflammatory cytokine milieu supports de novo differentiation of IL-17-producing T cells. Immunity 24, 179–189.
Veldhoen, M., Uyttenhove, C., Van Snick, J., Helmby, H., Westendorf, A., Buer, J., Martin, B., Wilhelm, C., Stockinger, B., 2008b. Transforming growth factor-beta 'reprograms' the differentiation of T helper 2 cells and promotes an interleukin 9-producing subset. Nat. Immunol. 9, 1341–1346.
Vigne, S., Chalmin, F., Duc, D., Clottu, A.S., Apetoh, L., Lobaccaro, J.A., Christen, I., Zhang, J., Pot, C., 2017. IL-27-Induced type 1 regulatory T-cells produce oxysterols that constrain IL-10 production. Front. Immunol. 8, 1184.
Vijayanand, P., Seumois, G., Simpson, L.J., Abdul-Wajid, S., Baumjohann, D., Panduro, M., Huang, X., Interlandi, J., Djuretic, I.M., Brown, D.R., Sharpe, A.H., Rao, A., Ansel, K.M., 2012. Interleukin-4 production by follicular helper T cells requires the conserved Il4 enhancer hypersensitivity site V. Immunity 36, 175–187.
Villarino, A.V., Gallo, E., Abbas, A.K., 2010. STAT1-activating cytokines limit Th17 responses through both T-bet-dependent and -independent mechanisms. J. Immunol. 185, 6461–6471.
Walker, M.R., Kasprowicz, D.J., Gersuk, V.H., Benard, A., Van Landeghen, M., Buckner, J.H., Ziegler, S.F., 2003. Induction of FoxP3 and acquisition of T regulatory activity by stimulated human $CD4^{+}CD25^{-}$ T cells. J. Clin. Invest. 112, 1437–1443.
Wang, A., Pan, D., Lee, Y.H., Martinez, G.J., Feng, X.H., Dong, C., 2013. Cutting edge: Smad2 and Smad4 regulate TGF-beta-mediated Il9 gene expression via EZH2 displacement. J. Immunol. 191, 4908–4912.
Wang, H., Geng, J., Wen, X., Bi, E., Kossenkov, A.V., Wolf, A.I., Tas, J., Choi, Y.S., Takata, H., Day, T.J., Chang, L.Y., Sprout, S.L., Becker, E.K., Willen, J., Tian, L., Wang, X., Xiao, C., Jiang, P., Crotty, S., Victora, G.D., Showe, L.C., Tucker, H.O., Erikson, J., Hu, H., 2014. The transcription factor Foxp1 is a critical negative regulator of the differentiation of follicular helper T cells. Nat. Immunol. 15, 667–675.
Wang, Y., Su, M.A., Wan, Y.Y., 2011. An essential role of the transcription factor GATA-3 for the function of regulatory T cells. Immunity 35, 337–348.
Weaver, C.T., Hatton, R.D., Mangan, P.R., Harrington, L.E., 2007. IL-17 family cytokines and the expanding diversity of effector T cell lineages. Annu. Rev. Immunol. 25, 821–852.

Wei, G., Abraham, B.J., Yagi, R., Jothi, R., Cui, K., Sharma, S., Narlikar, L., Northrup, D.L., Tang, Q., Paul, W.E., Zhu, J., Zhao, K., 2011. Genome-wide analyses of transcription factor GATA3-mediated gene regulation in distinct T cell types. Immunity 35, 299–311.

Wen, Z., Zhong, Z., Darnell JR., J.E., 1995. Maximal activation of transcription by Stat1 and Stat3 requires both tyrosine and serine phosphorylation. Cell 82, 241–250.

Wiener, Z., Falus, A., Toth, S., 2004. IL-9 increases the expression of several cytokines in activated mast cells, while the IL-9-induced IL-9 production is inhibited in mast cells of histamine-free transgenic mice. Cytokine 26, 122–130.

Wong, M.T., Ye, J.J., Alonso, M.N., Landrigan, A., Cheung, R.K., Engleman, E., Utz, P.J., 2010. Regulation of human Th9 differentiation by type I interferons and IL-21. Immunol. Cell Biol. 88, 624–631.

Wynn, T.A., 2003. IL-13 effector functions. Annu. Rev. Immunol. 21, 425–456.

Xiao, X., Balasubramanian, S., Liu, W., Chu, X., Wang, H., Taparowsky, E.J., Fu, Y.X., Choi, Y., Walsh, M.C., Li, X.C., 2012. OX40 signaling favors the induction of T(H) 9 cells and airway inflammation. Nat. Immunol. 13, 981–990.

Xu, L., Cao, Y., Xie, Z., Huang, Q., Bai, Q., Yang, X., He, R., Hao, Y., Wang, H., Zhao, T., Fan, Z., Qin, A., Ye, J., Zhou, X., Ye, L., Wu, Y., 2015. The transcription factor TCF-1 initiates the differentiation of T(FH) cells during acute viral infection. Nat. Immunol. 16, 991–999.

Xu, L., Kitani, A., Stuelten, C., Mcgrady, G., Fuss, I., Strober, W., 2010. Positive and negative transcriptional regulation of the Foxp3 gene is mediated by access and binding of the Smad3 protein to enhancer I. Immunity 33, 313–325.

Yagi, R., Junttila, I.S., Wei, G., Urban JR., J.F., Zhao, K., Paul, W.E., Zhu, J., 2010. The transcription factor GATA3 actively represses RUNX3 protein-regulated production of interferon-gamma. Immunity 32, 507–517.

Yang, X.O., Angkasekwinai, P., Zhu, J., Peng, J., Liu, Z., Nurieva, R., Liu, X., Chung, Y., Chang, S.H., Sun, B., Dong, C., 2009. Requirement for the basic helix-loop-helix transcription factor Dec2 in initial TH2 lineage commitment. Nat. Immunol. 10, 1260–1266.

Yang, X.O., Nurieva, R., Martinez, G.J., Kang, H.S., Chung, Y., Pappu, B.P., Shah, B., Chang, S.H., Schluns, K.S., Watowich, S.S., Feng, X.H., Jetten, A.M., Dong, C., 2008a. Molecular antagonism and plasticity of regulatory and inflammatory T cell programs. Immunity 29, 44–56.

Yang, X.O., Panopoulos, A.D., Nurieva, R., Chang, S.H., Wang, D., Watowich, S.S., Dong, C., 2007. STAT3 regulates cytokine-mediated generation of inflammatory helper T cells. J. Biol. Chem. 282, 9358–9363.

Yang, X.O., Pappu, B.P., Nurieva, R., Akimzhanov, A., Kang, H.S., Chung, Y., Ma, L., Shah, B., Panopoulos, A.D., Schluns, K.S., Watowich, S.S., Tian, Q., Jetten, A.M., Dong, C., 2008b. T helper 17 lineage differentiation is programmed by orphan nuclear receptors ROR alpha and ROR gamma. Immunity 28, 29–39.

Yao, Z., Cui, Y., Watford, W.T., Bream, J.H., Yamaoka, K., Hissong, B.D., Li, D., Durum, S.K., Jiang, Q., Bhandoola, A., Hennighausen, L., O'shea, J.J., 2006. Stat5a/b are essential for normal lymphoid development and differentiation. Proc. Natl. Acad. Sci. USA 103, 1000–1005.

Yashiro, T., Kubo, M., Ogawa, H., Okumura, K., Nishiyama, C., 2015. PU.1 suppresses Th2 cytokine expression via silencing of GATA3 transcription in dendritic cells. PLoS One 10, e0137699.

Ye, B.H., Cattoretti, G., Shen, Q., Zhang, J., Hawe, N., De Waard, R., Leung, C., Nouri-Shirazi, M., Orazi, A., Chaganti, R.S., Rothman, P., Stall, A.M., Pandolfi, P.P., Dalla-Favera, R., 1997. The BCL-6 proto-oncogene controls germinal-centre formation and Th2-type inflammation. Nat. Genet. 16, 161–170.

Yoshida, H., Nishina, H., Takimoto, H., Marengere, L.E., Wakeham, A.C., Bouchard, D., Kong, Y.Y., Ohteki, T., Shahinian, A., Bachmann, M., Ohashi, P.S., Penninger, J.M., Crabtree, G.R., Mak, T.W., 1998. The transcription factor NF-ATc1 regulates lymphocyte proliferation and Th2 cytokine production. Immunity 8, 115–124.

Yu, D., Rao, S., Tsai, L.M., Lee, S.K., He, Y., Sutcliffe, E.L., Srivastava, M., Linterman, M., Zheng, L., Simpson, N., Ellyard, J.I., Parish, I.A., Ma, C.S., Li, Q.J., Parish, C.R., Mackay, C.R., Vinuesa, C.G., 2009. The transcriptional repressor Bcl-6 directs T follicular helper cell lineage commitment. Immunity 31, 457–468.

Yusuf, I., Kageyama, R., Monticelli, L., Johnston, R.J., Ditoro, D., Hansen, K., Barnett, B., Crotty, S., 2010. Germinal center T follicular helper cell IL-4 production is dependent on signaling lymphocytic activation molecule receptor (CD150). J. Immunol. 185, 190–202.

Zaunders, J.J., Dyer, W.B., Wang, B., Munier, M.L., Miranda-Saksena, M., Newton, R., Moore, J., Mackay, C.R., Cooper, D.A., Saksena, N.K., Kelleher, A.D., 2004. Identification of circulating antigen-specific $CD4^+$ T lymphocytes with a $CCR5^+$, cytotoxic phenotype in an HIV-1 long-term nonprogressor and in CMV infection. Blood 103, 2238–2247.

Zeng, W.P., Chang, C., Lai, J.J., 2009. Immune suppressive activity and lack of T helper differentiation are differentially regulated in natural regulatory T cells. J. Immunol. 183, 3583–3590.

Zhang, D.H., Cohn, L., Ray, P., Bottomly, K., Ray, A., 1997. Transcription factor GATA-3 is differentially expressed in murine Th1 and Th2 cells and controls Th2-specific expression of the interleukin-5 gene. J. Biol. Chem. 272, 21597–21603.

Zhang, F., Meng, G., Strober, W., 2008. Interactions among the transcription factors Runx1, RORgammat and Foxp3 regulate the differentiation of interleukin 17-producing T cells. Nat. Immunol. 9, 1297–1306.

Zhang, P., Lee, J.S., Gartlan, K.H., Schuster, I.S., Comerford, I., Varelias, A., Ullah, M.A., Vuckovic, S., Koyama, M., Kuns, R.D., Locke, K.R., Beckett, K.J., Olver, S.D., Samson, L.D., Montes De Oca, M., De Labastida Rivera, F., Clouston, A.D., Belz, G.T., Blazar, B.R., Macdonald, K.P., Mccoll, S.R., Thomas, R., Engwerda, C.R., Degli-Esposti, M.A., Kallies, A., Tey, S.K., Hill, G.R., 2017. Eomesodermin promotes the development of type 1 regulatory T (TR1) cells. Sci. Immunol. 2.

Zheng, Y., Chaudhry, A., Kas, A., Deroos, P., Kim, J.M., Chu, T.T., Corcoran, L., Treuting, P., Klein, U., Rudensky, A.Y., 2009. Regulatory T-cell suppressor program co-opts transcription factor IRF4 to control T(H)2 responses. Nature 458, 351–356.

Zheng, Y., Josefowicz, S.Z., Kas, A., Chu, T.T., Gavin, M.A., Rudensky, A.Y., 2007. Genome-wide analysis of Foxp3 target genes in developing and mature regulatory T cells. Nature 445, 936–940.

Zhou, L., Ivanov, I., Spolski, R., Min, R., Shenderov, K., Egawa, T., Levy, D.E., Leonard, W.J., Littman, D.R., 2007. IL-6 programs T(H)-17 cell differentiation by promoting sequential engagement of the IL-21 and IL-23 pathways. Nat. Immunol. 8, 967–974.

Zhou, L., Lopes, J.E., Chong, M.M., Ivanov, I., Min, R., Victora, G.D., Shen, Y., Du, J., Rubtsov, Y.P., Rudensky, A.Y., Ziegler, S.F., Littman, D.R., 2008. TGF-beta-induced Foxp3 inhibits T(H)17 cell differentiation by antagonizing RORgammat function. Nature 453, 236–240.

Zhu, J., Guo, L., Min, B., Watson, C.J., Hu-Li, J., Young, H.A., Tsichlis, P.N., Paul, W.E., 2002. Growth factor independent-1 induced by IL-4 regulates Th2 cell proliferation. Immunity 16, 733–744.

Zhu, J., Guo, L., Watson, C.J., HU-Li, J., Paul, W.E., 2001. Stat6 is necessary and sufficient for IL-4's role in Th2 differentiation and cell expansion. J. Immunol. 166, 7276–7281.

Zhu, J., Jankovic, D., Grinberg, A., Guo, L., Paul, W.E., 2006a. Gfi-1 plays an important role in IL-2-mediated Th2 cell expansion. Proc. Natl. Acad. Sci. USA 103, 18214–18219.

Zhu, J., Jankovic, D., Oler, A.J., Wei, G., Sharma, S., Hu, G., Guo, L., Yagi, R., Yamane, H., Punkosdy, G., Feigenbaum, L., Zhao, K., Paul, W.E., 2012. The transcription factor T-bet is induced by multiple pathways and prevents an endogenous Th2 cell program during Th1 cell responses. Immunity 37, 660–673.

Zhu, J., Min, B., Hu-Li, J., Watson, C.J., Grinberg, A., Wang, Q., Killeen, N., Urban JR., J.F., Guo, L., Paul, W.E., 2004. Conditional deletion of Gata3 shows its essential function in T(H)1-T(H)2 responses. Nat. Immunol. 5, 1157–1165.

Zhu, J., Yamane, H., Cote-Sierra, J., Guo, L., Paul, W.E., 2006b. GATA-3 promotes Th2 responses through three different mechanisms: induction of Th2 cytokine production, selective growth of Th2 cells and inhibition of Th1 cell-specific factors. Cell Res. 16, 3–10.

Zhu, J., Yamane, H., Paul, W.E., 2010. Differentiation of effector CD4 T cell populations (*). Annu. Rev. Immunol. 28, 445–489.

CHAPTER TWO

The Single-Cell Phenotypic Identity of Human CD8$^+$ and CD4$^+$ T Cells

Jolanda Brummelman*[,a], Karolina Pilipow*[,a] and Enrico Lugli*[,§,1]
*Laboratory of Translational Immunology, Humanitas Clinical and Research Center, Milan, Italy
§Humanitas Flow Cytometry Core, Humanitas Clinical and Research Center, Milan, Italy
[1]Corresponding author: E-mail: enrico.lugli@humanitasresearch.it

Contents

[a] These authors contributed equally to this work.

International Review of Cell and Molecular Biology, Volume 341
ISSN 1937-6448
https://doi.org/10.1016/bs.ircmb.2018.05.007

Abstract

On antigen encounter, naïve $CD8^+$ and $CD4^+$ T cells differentiate into a large number of effector cells that migrate to inflamed tissues to fight infections or tumors. Following elimination of the target, a few cells remain in the long-term, the so-called memory T cells that are capable to reexpand and respond more vigorously on a second encounter with the cognate antigen. While the naïve T cell compartment is fairly homogenous, effector and memory T cells are largely diverse, comprising dozens of subsets with diverse functions, molecular characteristics, and localization in the body. In addition, $CD4^+$ and, to some extent, also $CD8^+$ T cells can differentiate into several effector subsets according to their pattern of cytokine expression, including the T helper 1 (Th1), Th2, and Th17 cells. It has become clear that specific subsets of T cells dominate different types of infections and pathological conditions and have different capacities to infiltrate and reject tumors. Their correct phenotypic identification is therefore of foremost importance for their live purification by magnetic or fluorescence-activated cell sorting and subsequent molecular characterization. Here, we present a comprehensive list of the main T cell subpopulations with a major focus on human cells along with their surface phenotypic properties.

1. INTRODUCTION

Single-positive $CD8^+$ and $CD4^+$ T cells develop within the thymus from double-positive precursor cells. Following positive and negative selection, T cells are released in the periphery as mature naïve T (T_N) cells that recirculate between secondary lymphoid organs (SLOs) such as the spleen and lymph nodes. Following encounter with antigen presenting cells (APCs) bearing their cognate antigen, T_N cells differentiate into a large number of effector cells that migrate to inflamed sites and kill infected cells. After antigen clearance, the majority (as much as 90%–95%) of these effector cells undergo apoptosis, whereas a small proportion survives, giving rise to memory cells that persist in the long-term and that are capable to respond more rapidly and more effectively after a subsequent encounter (Harty and Badovinac, 2008; Williams and Bevan, 2007; Zhang and Bevan, 2011).

The memory T cell pool is largely diverse. In fact, subsets of memory T cells are characterized by different capacities to persist in the long-term, to migrate in the body (SLOs vs. peripheral tissues), and to stably reside in such tissues (i.e., tissue residency). In addition to diversity according to these features, $CD4^+$ and, to some extent also, $CD8^+$ T cells can differentiate into several effector subsets, e.g., T helper 1 (Th1), Th2, Th17, or regulatory subsets, e.g., regulatory T cells (Tregs) and type 1 regulatory (Tr1) cells, depending on soluble mediators such as cytokines and metabolites that are present in the local environment (Tubo and Jenkins, 2014; Christie and Zhu, 2014; Sallusto and Lanzavecchia, 2009). Following stimulation, these Th subsets produce prototypical effector cytokines including IFNγ for Th1, interleukin-4 (IL-4) for Th2, and IL-17 for Th17 cells and are involved in protection to specific pathogens, such as intracellular viruses/bacteria, helminths, and fungi, respectively.

The diversity within the immune system is enormous. Here we present a comprehensive list of T cell subpopulations with important functions in the regulation of immune responses and their phenotypic properties. While functional differences among different subsets are reflected also at the gene expression, chromatin, metabolic and, ultimately, functional level, we will focus on antigens expressed on the surface of T cells. These are of fundamental need for their live purification by magnetic or fluorescence-activated cell sorting (FACS) and for their subsequent characterization. Historically, multiparameter flow cytometry has been instrumental in identifying new subsets of T cells with a defined set of markers and specialized functions (Lugli et al., 2010). The present review reflects knowledge obtained with this approach. However, the recent development of high-content technologies at the level of single cells, such as RNA sequencing (scRNA-seq) or proteomics, is currently redefining the phenotypic classification and functional landscape of T cells. While the existence of major distinct subsets of T cells is not put into question, it is now clear that some cells in the CD4$^+$ and CD8$^+$ T cell compartments exist in plastic states that can differentiate into multiple subpopulations. While here we offer a simplified (not necessarily meaning simple!) phenotypic classification of T cell subpopulations, we expect this to be improved or to change in light of new discoveries generated by advanced single-cell technologies.

2. NAÏVE AND MEMORY PHENOTYPES

The T cell compartment can be divided in different subsets based on their capability to maintain immunological memory in the long-term. Historically, naïve and memory T cells have been separated based on the combination of surface expression of different molecules, such as CD45RA, CD45RO, leukocytes function associate antigen 1 (LFA-1), and LFA-3 (CD58) (Sanders et al., 1988). CD45RA and CD45RO are both splicing isoforms of the tyrosine phosphatase CD45, with high and low molecular weight, respectively (Terry et al., 1988). CD45RA was originally thought to be expressed only on T_N cells, whereas it was lost during activation and differentiation to memory cells and replaced by CD45RO (Akbar et al., 1988; Cossarizza et al., 1996). In fact, CD45RO$^+$ T cells are rare in neonates, while their frequency increases with age (Yamada et al., 1992). In addition, both LFA-1 and LFA-3 are present on previously activated memory T cells (Lindestam Arlehamn et al., 2013; Li et al., 2009). LFA-1, composed by CD11a (αL) and CD18 (β2) subunits, is the receptor for intracellular adhesion molecule-1 (ICAM-1), ICAM-2, and ICAM-3 (Li et al., 2009). Its expression is required for T cell homing into SLOs and sites of inflammation, where it facilitates the adhesion of T cells to APCs (von Andrian and Mackay, 2000). Thus, LFA-1 regulates T cell activation and signal transduction through the immunological synapse (Perez et al., 2003; Li et al., 2009). LFA-3 expression is important to amplify responsiveness to T cell receptor (TCR) triggering by a process called costimulation, through binding of ligands expressed on APCs and subsequent intracellular signaling (Sanders et al., 1988). LFA-3 costimulation induces CD8$^+$ T cell proliferation and effector cytokine production and mediates cytotoxicity (Parra et al., 1997).

A diverse progeny of T cells exists following establishment of long-term memory. Despite it was previously demonstrated by 4-color flow cytometry that memory T cells are heterogeneous for the expression of tissue-homing markers (Picker et al., 1993a,b), classification of memory T cells into subsets according to functional capacity (Hamann et al., 1997; Sallusto et al., 1999a) and localization in the body (Sallusto et al., 1999a; Masopust et al., 2001), i.e., the central memory T (T_{CM}) and the effector memory T (T_{EM}) cells, was reported only later. While initially thought to be separate states of memory generated following resolution of the immune response, T_{CM} and T_{EM} cells were later demonstrated to retain a precursor—progeny relationship, according to which less-differentiated T_{CM} cells self-renew and give rise to more differentiated progeny, including the T_{EM} and the terminal effector/T_{EM} reexpressing CD45RA (T_{TE} or T_{EMRA}) cells, in a stem cell—like fashion (Sallusto et al., 2004). More recently, a small subpopulation of CD45RA$^+$ naïve-like cells but memory in function and capable to self-renew while retaining multipotency (i.e., capable to generate all memory subsets; for this reason defined as stem cell memory [T_{SCM}]) could be identified (Gattinoni et al., 2011; Lugli et al., 2013a). These data suggest the existence of a linear differentiation model of CD8$^+$ T cells with decreasing memory potential among subsets yet increasing effector functions (Lanzavecchia and Sallusto, 2002; Gattinoni et al., 2012; Crompton et al., 2016;

Restifo and Gattinoni, 2013). These subsets contribute differently to the circulating T cell compartments in the blood of healthy donors. The CD8$^+$ T cell compartment consists of around 40% T_N, 2% T_{SCM}, 10% T_{CM}, 30% T_{EM}, and 20% T_{TE} cells, whereas the CD4$^+$ T cell compartment comprises around 40% T_N, 2% T_{SCM}, 40% T_{CM}, 20% T_{EM}, and 2% T_{TE} cells (Lugli et al., 2013a; Sathaliyawala et al., 2013). It should be noted that these proportions are only indicative as they can change considerably according to age, during which naïve T cells decrease progressively in favor of memory cells or in the presence of chronic infections, such as cytomegalovirus (CMV) and Epstein-Barr virus (EBV), which cause the specific accumulation of T_{TE} cells.

Besides subsets that circulate between the blood and SLOs, T cells that do not recirculate but are maintained at virtually all nonlymphoid tissues were identified (Masopust et al., 2001; Jiang et al., 2012; Mueller and Mackay, 2016). These tissue-resident memory T (T_{RM}) cells constitute a population diverse from circulating memory T cells and are crucial in the protection against infections at peripheral sites (Schenkel et al., 2014; Masopust et al., 2010). The surface markers expressed by circulating and resident T cells are discussed in detail below and are summarized in Fig. 1 and Table 1.

2.1 Naïve T Cells

T_N cells are antigen-unprimed and quiescent T cells that egress from the thymus and recirculate between the bloodstream and SLOs. T_N cells are supposed to be fairly homogeneous at the phenotypic and functional level, except for CD4$^+$ T_N cells, where a subset of recent thymic emigrants (T_{RTE}) can be identified (see below) (Kimmig et al., 2002). T_N cells express high levels of CD45RA, chemokine receptor CCR7, and selectin CD62L (L-selectin). Both CCR7 and CD62L are widely used together with CD45RA and CD45RO to distinguish T_N cells from memory subsets (Sallusto et al., 1999a). CCR7 and CD62L expressions are relatively redundant on the cell surface and can thus be used interchangeably. It is important to note, however, that CD62L expression is lost with freeze-thaw procedures, thus making its measurement unreliable on frozen cells (Lugli, E., personal communication). While CD62L directs binding to endothelial cells of high endothelial venules (HEV) in lymph nodes via interaction with its ligands, sialylated, fucosylated, and sulfated N- and O-glycans on glycoproteins (Rosen, 2004), CCR7 is important for extravasation following binding to CCL19 and CCL21 and guides interaction of T cells with activated

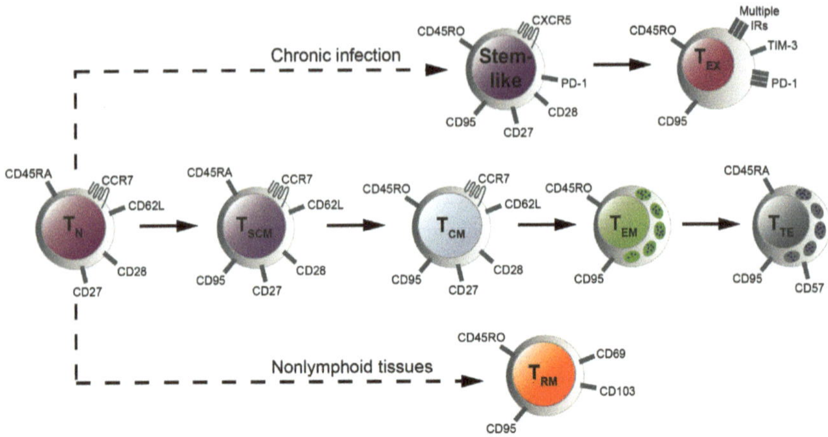

Figure 1 ***Surface markers defining the different memory T cell subsets.*** Naïve T cells differentiate into large numbers of effector cells on recognizing their cognate antigen presented by antigen-presenting cells. Although the majority of these effector cells die after antigen clearance, a small subset of activated cells gives rise to memory cells. Based on the surface marker expression, multiple memory subsets can be identified and are characterized by decreasing memory potential yet increasing effector functions. Tissue-resident memory T cells are retained within nonlymphoid tissues and their origin is at the moment not completely defined. Recently, a subset of $CD8^+$ T cells with stem cell–like properties, i.e., self-renewal and multipotency, but different from circulating T_{SCM} cells was identified in the exhausted pool of T cells of chronically infected mice. These cells can generate more differentiated yet exhausted T cells. T_{CM}, central memory T cells; T_{EM}, effector memory T cells; T_{EX}, exhausted T cells; T_N, naive T cells; T_{RM}, tissue-resident memory T cells; T_{SCM}, stem cell memory T cells; T_{TE}, terminal effector T cells.

dendritic cells (DCs) (Jorgensen et al., 2017). Although both CCL19 and CCL21 are secreted by stromal cells in lymph nodes and HEV cells, thereby orchestrating homing of T cells into the lymph nodes, CCL21 is also expressed by afferent lymphatic vessels, promoting DC homing into the lymph nodes. Therefore, CCL21–CCR7 interaction is thought to be essential in the intranodal positioning of both T cells as activated DCs. The secretion of CCL19 by activated DCs may be important for the short-lived interaction between DCs and T cells (Muthuswamy et al., 2010; Sallusto et al., 1999b). Homing of T_N cells to SLOs seems largely dependent on CD62L and CCR7, as demonstrated by experiments in mice using knockout models of these molecules. Differently, other C–X–C and C–C chemokine receptors that are responsible for T cell migration in vivo and are generally abundant on different memory T cell subtypes are absent in T_N cells. CXCR4 is, however, an exception as it is constitutively

Table 1 Expression of Surface Markers on Naïve and Memory T cells[a]

Subset	Surface Phenotype	Subset	Function	References
T_N	CD45RA$^+$		Regulation of T cell antigen signaling	Akbar et al. (1988), Cossarizza et al. (1996), and Sallusto et al. (1999a, 2004)
	CD45RO$^-$		Regulation of T cell antigen signaling	Akbar et al. (1988), Cossarizza et al. (1996), Sallusto et al. (1999a, 2004), and Hamann et al. (1997)
	CCR7$^+$		Homing to SLOs	Sallusto et al. (1999a, 2004)
	CD62L$^+$		Homing to SLOs	Sallusto et al. (1999a, 2004)
	CD27$^+$		Costimulatory receptor	Akbar et al. (1988) and Sallusto et al. (1999a)
	CD28$^+$		Costimulatory receptor	Azuma et al. (1992), Lenschow et al. (1996), Hamann et al. (1997), Sallusto et al. (2004), and Riley and June (2005)
	CD127$^+$		α-chain of IL-7 receptor; survival	Schluns et al. (2000), Tan et al. (2002), and Hassan and Reen (1998)
	CD95$^-$		Death receptor; costimulatory receptor	Klebanoff et al. (2016), Corsini et al. (2009), and Gattinoni et al. (2011)
	CXCR4$^+$		Homing to bone marrow/SLOs	Gattinoni et al. (2011) and Song et al. (2005)
	CXCR3$^{-/+}$	CD4$^+$	Migration into inflamed sites	Gattinoni et al. (2011), Song et al. (2005), and Brodie et al. (2013)
	CCR4$^{-/+}$	CD4$^+$	Homing to skin	Song et al. (2005) and Brodie et al. (2013)
	CCR6$^{-/+}$	CD4$^+$	Homing to gut, lung, and inflamed skin	Song et al. (2005), Brodie et al. (2013), and Kastirr et al. (2015)

(*Continued*)

Table 1 Expression of Surface Markers on Naïve and Memory T cells[a]—cont'd

Subset	Surface Phenotype	Subset	Function	References
T_{SCM}	$CD45RA^+$		Regulation of T cell activation	Akbar et al. (1988), Cossarizza et al. (1996), and Sallusto et al. (1999a, 2004)
	$CD45RO^-$		Regulation of T cell activation	Akbar et al. (1988), Cossarizza et al. (1996), Sallusto et al. (1999a, 2004), and Hamann et al. (1997)
	$CCR7^+$		Homing to SLOs	Sallusto et al. (1999a, 2004)
	$CD62L^+$		Homing to SLOs	Sallusto et al. (1999a, 2004)
	$CD27^+$		Costimulatory receptor	Akbar et al. (1988) and Sallusto et al. (1999a)
	$CD28^+$		Costimulatory receptor	Azuma et al. (1992), Lenschow et al. (1996), Hamann et al. (1997), Sallusto et al. (2004), and Riley and June (2005)
	$CD127^+$		α-chain of IL-7 receptor; survival	Schluns et al. (2000), Tan et al. (2002), and Hassan and Reen (1998)
	$CD122^+$		β-chain of IL-2 and IL-15 receptors; survival	Gattinoni et al. (2011) and Lugli et al. (2013a)
	$CD95^+$		Death receptor; costimulatory receptor	Klebanoff et al. (2016), Corsini et al. (2009), and Gattinoni et al. (2011)
	$LFA\text{-}1^+$		Homing to SLOs and inflamed sites; adhesion to APCs	Gattinoni et al. (2011) and Lugli et al. (2013a)
	$LFA\text{-}3^+$		Costimulatory receptor	Gattinoni et al. (2011) and Lugli et al. (2013a,b)
	$CXCR3^+$	$CD8^+$	Migration into inflamed sites	Gattinoni et al. (2011)
	$CXCR4^+$		Homing to bone marrow/SLOs	Gattinoni et al. (2011) and Brodie et al. (2013)

T_{CM}	CD45RA$^-$		Regulation of T cell antigen signaling	Akbar et al. (1988), Cossarizza et al. (1996), and Sallusto et al. (1999a, 2004)
	CD45RO$^+$		Regulation of T cell antigen signaling	Akbar et al. (1988), Cossarizza et al. (1996), Sallusto et al. (1999a, 2004), and Hamann et al. (1997)
	CCR7$^+$		Homing to SLOs	Sallusto et al. (1999a, 2004)
	CD62L$^+$		Homing to SLOs	Sallusto et al. (1999a, 2004)
	CD27$^+$		Costimulatory receptor	Akbar et al. (1988) and Sallusto et al. (1999a)
	CD28$^+$		Costimulatory receptor	Azuma et al. (1992), Lenschow et al. (1996), Hamann et al. (1997), Sallusto et al. (2004), and Riley and June (2005)
	CD127$^+$		α-chain of IL-7 receptor; survival	Schluns et al. (2000), Tan et al. (2002), and Hassan and Reen (1998)
	CD122$^+$		β-chain of IL-2 and IL-15 receptors; survival	Gattinoni et al. (2011) and Lugli et al. (2013a)
	CD95$^+$		Death receptor; costimulatory receptor	Klebanoff et al. (2016), Corsini et al. (2009), and Gattinoni et al. (2011)
	LFA-1$^+$		Homing to SLOs and inflamed sites; adhesion to APCs	Gattinoni et al. (2011) and Lugli et al. (2013a)
	LFA-3$^+$		Costimulatory receptor	Gattinoni et al. (2011) and Lugli et al. (2013a,b)
	CD161$^{-/+}$	CD4$^+$	Unknown	Takahashi et al. (2006)
	IL-18Rα^+		α-chain of IL-18 receptor; antigen-independent IFNγ production	Mahnke et al. (2013)
	CXCR3$^+$	Th1 T_{CM}/CD8$^+$	Migration into inflamed sites	Gattinoni et al. (2011), Rivino et al. (2004), and Sallusto et al. (2004)
	CXCR4$^+$		Homing to bone marrow/SLOs	Gattinoni et al. (2011) and Brodie et al. (2013)

(*Continued*)

Table 1 Expression of Surface Markers on Naïve and Memory T cells[a]—cont'd

Subset	Surface Phenotype	Subset	Function	References
	$CXCR5^{-/+}$	Tfh T_{CM}/$CD8^{+}$	Homing to B cell follicles	Chevalier et al. (2011), Sallusto et al. (2004), Im et al. (2016), Lin et al. (2016), He et al. (2016), and Quigley et al. (2007)
	$CCR4^{-/+}$	Th2 T_{CM}/Tc2	Homing to skin	Rivino et al. (2004), Sallusto et al. (2004), and Geginat et al. (2003)
	$CCR6^{-/+}$	Th17 T_{CM}	Homing to gut, lung, and inflamed skin	Rivino et al. (2004)
T_{EM}	$CD45RA^{-}$		Regulation T cell antigen signaling	Akbar et al. (1988), Cossarizza et al. (1996), and Sallusto et al. (1999a, 2004)
	$CD45RO^{+}$		Regulation of T cell antigen signaling	Akbar et al. (1988), Cossarizza et al. (1996), Sallusto et al. (1999a, 2004), and Hamann et al. (1997)
	$CCR7^{-}$		Homing to SLOs	Sallusto et al. (1999a, 2004)
	$CD62L^{-}$		Homing to SLOs	Sallusto et al. (1999a, 2004)
	$CD27^{-/+}$		Costimulatory receptor	Akbar et al. (1988), Sallusto et al. (1999a), and Romero et al. (2007)
	$CD28^{-/+}$		Costimulatory receptor	Azuma et al. (1992), Lenschow et al. (1996), Hamann et al. (1997), Sallusto et al. (2004), Riley and June (2005), and Romero et al. (2007)
	$CD127^{-/+}$		α-chain of IL-7 receptor; survival	Schluns et al. (2000), Tan et al. (2002), Hassan and Reen (1998), and Romero et al. (2007)
	$CD122^{+}$		β-chain of IL-2 and IL-15 receptors; survival	Gattinoni et al. (2011) and Lugli et al. (2013a)

CD95$^+$		Death receptor; costimulatory receptor	Klebanoff et al. (2016), Corsini et al. (2009), and Gattinoni et al. (2011)
LFA-1$^+$		Homing to SLOs and inflamed sites; adhesion to APCs	Gattinoni et al. (2011) and Lugli et al. (2013a)
LFA-3$^+$		Costimulatory receptor	Gattinoni et al. (2011) and Lugli et al. (2013a,b)
CD161$^{-/+}$	CD8$^+$	Unknown	Takahashi et al. (2006)
IL-18Rα^+		α-chain of IL-18 receptor; antigen-independent IFNγ production	Mahnke et al. (2013)
CD57$^+$		Unknown	Stubbe et al. (2008) and Verma et al. (2017)
KLRG1$^+$		Inhibitory receptor	Stubbe et al. (2008)
CXCR3$^+$	Th1 T_{EM}/CD8$^+$	Migration into inflamed sites	Gattinoni et al. (2011) and Sallusto et al. (2004)
CXCR4$^+$		Homing to bone marrow/SLOs	Gattinoni et al. (2011) and Brodie et al. (2013)
CCR4$^{-/+}$	Th2 T_{EM}/Th17 T_{EM}	Homing to skin	Sallusto et al. (2004) and Lintermans et al. (2017)
CRTh2$^{-/+}$	Th2 T_{EM}	Promotion of chemotaxis and proinflammatory cytokine production by Th2 cells	Lintermans et al. (2017) and Sallusto et al. (2004)
CCR5$^{-/+}$		Homing to inflamed sites	Sallusto et al. (2004)
CCR6$^{-/+}$	Th17 T_{EM}	Homing to gut, lung, and inflamed skin	Lintermans et al. (2017) and Sallusto et al. (2004)

(Continued)

Table 1 Expression of Surface Markers on Naïve and Memory T cells[a]—cont'd

Subset	Surface Phenotype	Subset	Function	References
T_{TE}/T_{EMRA}	$CD45RA^+$		Regulation of T cell antigen signaling	Akbar et al. (1988), Cossarizza et al. (1996), Sallusto et al. (1999a, 2004), and Hamann et al. (1997)
	$CD45RO^-$		Regulation of T cell antigen signaling	Akbar et al. (1988), Cossarizza et al. (1996), and Sallusto et al. (1999a, 2004)
	$CCR7^-$		Homing to SLOs	Sallusto et al. (1999a, 2004) and Faint et al. (2001)
	$CD62L^-$		Homing to SLOs	Sallusto et al. (1999a, 2004)
	$CD27^-$		Costimulatory receptor	Akbar et al. (1988) and Sallusto et al. (1999a)
	$CD28^-$		Costimulatory receptor	Azuma et al. (1992), Lenschow et al. (1996), Hamann et al. (1997), Sallusto et al. (2004), and Riley and June (2005)
	$CD127^-$		α-chain of IL-7 receptor; survival	Schluns et al. (2000), Tan et al. (2002), and Hassan and Reen (1998)
	$CD122^+$		β-chain of IL-2 and IL-15 receptors; survival	Gattinoni et al. (2011) and Lugli et al. (2013a)
	$CD95^+$		Death receptor; costimulatory receptor	Takahashi et al. (2006)
	$LFA\text{-}1^+$		Homing to SLOs and inflamed sites; adhesion to APCs	Gattinoni et al. (2011) and Lugli et al. (2013a)
	$LFA\text{-}3^+$		Costimulatory receptor	Gattinoni et al. (2011) and Lugli et al. (2013a,b)
	$IL\text{-}18R\alpha^+$		α-chain of IL-18 receptor; antigen-independent IFNγ production	Mahnke et al. (2013)
	$CD57^+$		Unknown	Faint et al. (2001)

$KLRG1^{+}$	Coinhibitory receptor	Weng et al. (2009)
KIR2DL2	Inhibitory receptor	Weng et al. (2009)
KIR3DL2	Inhibitory receptor	Weng et al. (2009)
KLRD1	Activating/inhibitory receptor	Weng et al. (2009)
KLRF1	Activating receptor	Weng et al. (2009)
KLRK1	Activating receptor	Weng et al. (2009)
NCR1	Activating receptor	Weng et al. (2009)
2B4	Activating receptor	Weng et al. (2009)
$CXCR3^{-/+}$	Migration into inflamed sites	Gattinoni et al. (2011), Rivino et al. (2004), and Sallusto et al. (2004)
$CXCR4^{-/+}$	Homing to bone marrow/SLOs	Gattinoni et al. (2011), Brodie et al. (2013), and Faint et al. (2001)
$CX3CR1^{+}$	Homing to inflamed sites	Hertoghs et al. (2010)
$CCR5^{+}$	Homing to inflamed sites	Faint et al. (2001) and Sallusto et al. (2004)

T_{CM}, central memory T cells; T_{EM}, effector memory T cells; T_{EX}, exhausted T cells; T_N, naive T cells; T_{RM}, tissue-resident memory T cells; T_{SCM}, stem cell memory T cells; T_{TE}, terminal effector T cells.

[a]−/+ indicates that a subset of the given population expresses the marker. Differences between $CD4^{+}$ and $CD8^{+}$ T cells are indicated when occurring.

expressed by T_N cells (Gattinoni et al., 2011) and mediates homing to the bone marrow (Liu et al., 2015), while facilitating cell migration into SLOs (von Andrian and Mackay, 2000; Miyasaka and Tanaka, 2004; Sallusto et al., 1998). Small subpopulations of phenotypically defined $CD4^+$ T_N cells seem to express low levels of CXCR3, CCR4, and CCR6 (Song et al., 2005; Brodie et al., 2013; Kastirr et al., 2015), and these $CD4^+$ T_N cells were, however, demonstrated to share features with conventional memory cells (Song et al., 2005; Brodie et al., 2013; Kastirr et al., 2015).

For their full activation, T_N cells are largely dependent on signals derived from costimulatory molecules such as CD27 and CD28 (Azuma et al., 1992; Fagnoni et al., 1996) that are in fact abundant on their cell surface. CD28 is a major costimulatory receptor, responsible for the optimal antigen-mediated T cell activation, proliferation, and survival (Lenschow et al., 1996; Riley and June, 2005). CD28 binding induces high levels of IL-2 production, IL-2Rα chain (CD25) expression, entry into the cell cycle, and upregulation of antiapoptotic transcription factor Bcl-xL (Lenschow et al., 1996; Riley and June, 2005). By contrast, CD27 promotes antigen-mediated T cell expansion independent of IL-2 production and cell cycle entry, but solely by enhancing T cell survival (Hendriks et al., 2003).

Cytokines belonging to the common γ (γ_c) family such as IL-2, IL-7, and IL-15 play a major role in the homeostasis of T cells, specifically by mediating their survival in the long-term and their slow turnover in the absence of antigen (Surh and Sprent, 2008). Homeostasis of T_N cells is mostly mediated by IL-7 signaling following binding to the IL-7 receptor α chain (IL-7Rα or CD127) that dimerizes with the γ_c receptor, also known as CD132, to form the IL-7R. IL-7R works in combination with tonic TCR signaling following contact with self-peptides presented by major histocompatibility complex (MHC) molecules (Murali-Krishna et al., 1999; Schluns et al., 2000; Tan et al., 2002; Hassan and Reen, 1998). Therefore, T_N cells are mostly defined in literature as $CD45RA^+$ $CD45RO^-CCR7^+CD62L^+CD27^+CD28^+CD127^+$ cells by flow cytometry. Absence of CXCR3, CCR4, and CCR6 may exclude memory contaminants (only in the $CD4^+$ population) and improves the identification of pure T_N cells.

In the past, two subsets of T_N cells were described in the $CD4^+$ but not in the $CD8^+$ compartment. CD31 is used to distinguish between $CD31^+$ recent thymic emigrants (T_{RTE}) and $CD31^-$ T_N cells that underwent homeostatic proliferation in the periphery, as revealed by their largely different T cell receptor excision circle (TREC) content (Kimmig et al., 2002).

$CD31^-$ T_N cells increase in proportion within the T_N pool with age, a process that is consistent with the drastic decrease in thymic activity and with the increased homeostatic proliferation of T_N cells (Kohler and Thiel, 2009).

2.2 Stem Cell Memory T Cells

T_{SCM} cells are the least-differentiated memory T cell subset identified so far in humans and nonhuman primates (Gattinoni et al., 2011; Lugli et al., 2013a; Hong et al., 2016; Vahidi et al., 2017). They constitute 2%–3% of the circulating $CD8^+$ and $CD4^+$ T cells (Gattinoni et al., 2011) and preferentially circulate through blood and lymph nodes (Lugli et al., 2013a). Antigen-specific T_{SCM} cells are both self-renewing and multipotent (Gattinoni et al., 2011; Lugli et al., 2013a; Ahmed et al., 2016), and, when induced by vaccination, e.g., for yellow fever (Fuertes Marraco et al., 2015), or following adoptive transfer in humans (Biasco et al., 2015), they can persist for decades. Interestingly, their frequency does not change with age and is not affected like other memory subsets by CMV serostatus (generally causing an accumulation of T cells with T_{EM}/T_{TE} phenotypes), thus suggesting that T_{SCM} cells constitute a stable memory phenotype (Di Benedetto et al., 2015). Due to their enhanced stemness and their capability to generate potent effectors following antigen recognition, T_{SCM} cells are of interest for immunotherapy of cancer, which requires long-term immune responses for antitumor surveillance (Klebanoff et al., 2012; Biasco et al., 2015). Due to their paucity in humans, many in vitro protocols to generate T_{SCM} cells from T_N precursors have been proposed. These utilize cytokine cocktails such as IL-7 and IL-15 (Cieri et al., 2013) or IL-7 and IL-21 (Sabatino et al., 2016), blunted TCR stimulation (Zanon et al., 2017), or small-molecule inhibitors of the Glycogen synthase kinase-3 beta (GSK-3β) (Gattinoni et al., 2009, 2011; Forget et al., 2012; Scholz et al., 2016), Phosphoinoside 3-kinase (PI-3K) (Majchrzak et al., 2017), Protein kinase B (AKT) (Crompton et al., 2015), and Bromodomain and Extra-Terminal Domain (BET) (Kagoya et al., 2016) pathways. T_{SCM} cells have a T_N-like surface phenotype, i.e., $CD45RA^+CD45RO^-CCR7^+$ $CD62L^+CD27^+CD28^+CD127^+$ (Lugli et al., 2013a). However, they can be distinguished from T_N cells by the expression of CD95 (Fas), interleukin-2 receptor β (IL-2Rβ, or CD122), LFA-1, and LFA-3 (Gattinoni et al., 2011; Lugli et al., 2013a). CD95 is a surface receptor triggered by members of the TNF family and functions as a death receptor because its signaling induces apoptosis via activation of caspase-8 (Peter et al., 2007). Recent data, however, indicate that CD95 may also function as a costimulatory rather than a death receptor (Klebanoff et al., 2016; Corsini et al., 2009). CD95 is upregulated by activation and is stably expressed by all memory cells (Aggarwal and Gupta, 1998; Fagnoni et al., 2000). Indeed, only the CD95$^+$ cells are able to proliferate vigorously and produce effector cytokines rapidly in response to IL-15, highlighting the memory characteristics of CD95$^+$ cells and the naïve characteristics of the CD95$^-$ cells (Gattinoni et al., 2011). CD122 is the β-chain of both the IL-2 and IL-15 receptors and is progressively upregulated with differentiation of memory T cells. While IL-2 signaling is involved in the secondary expansion of memory T cells, IL-15 signaling is important for the survival and homeostatic proliferation of all memory T cell subsets (Geginat et al., 2001, 2003). Therefore, T_{SCM} cells are generally defined as CD45RA$^+$CD45RO$^-$CCR7$^+$ CD62L$^+$CD27$^+$CD28$^+$CD127$^+$CD95$^+$CD122$^+$LFA-3$^+$ cells by flow cytometry. T_{SCM} cells have a similar homing profile to T_N cells, due to their shared expression of CCR7, CD62L, and CXCR4, although the latter is expressed to a higher extent than on T_N cells. Moreover, T_{SCM} cells also express high levels CXCR3 (Gattinoni et al., 2011), possibly allowing their migration into inflammatory sites guided by CXCR3 ligands CXCL9 (MIG), CXCL10 (IP-10), and CXCL11 (I-TAC) (Groom and Luster, 2011).

2.3 T_{CM} Cells

T_{CM} cells, like T_N and T_{SCM} cells, patrol SLOs, where on antigen encounter, they undergo fast and robust proliferation, as well as differentiation into effector cells (von Andrian and Mackay, 2000; Sallusto et al., 1999a). Similar to T_N and T_{SCM} cells, T_{CM} cells express high levels of CCR7, CD62L, CD27, CD28, and CD127, yet they can be discriminated from T_N and T_{SCM} cells by the lack of CD45RA and gain of CD45RO surface expression (Hamann et al., 1997; Sallusto et al., 1999a). These cells also express the memory markers CD95, CD122, LFA-1, and LFA-3 to a higher extent than T_{SCM} cells. Therefore, T_{CM} cells are defined as CD45RA$^-$ CD45RO$^+$CCR7$^+$CD62L$^+$CD27$^+$CD28$^+$CD127$^+$CD95$^+$CD122$^+$ cells by flow cytometry.

Besides CCR7, several other chemokine receptors have been identified on subsets of T_{CM} cells, yet with some differences between the CD4$^+$ and the CD8$^+$ compartments (von Andrian and Mackay, 2000; Miyasaka and Tanaka, 2004; Gattinoni et al., 2011). T_{CM} cells from both compartments express CXCR4 to a higher extent than T_{SCM} cells. In addition, CD4$^+$ T_{CM} cells express chemokine receptors that correspond to their Th phenotype (discussed in detail in the CD4$^+$ T cell–specific phenotypes section in this review), e.g., CXCR3 on Th1 T_{CM} cells, CCR4 on Th2 T_{CM} cells, CCR6 on Th17 T_{CM} cells, and CXCR5 on follicular T helper (Tfh) T_{CM} cells (Chevalier et al., 2011; Rivino et al., 2004). The majority of CD8$^+$ T_{CM} cells express CXCR3, whereas a subset expresses CCR4 or CXCR5. CCR4 defines CD8$^+$ T_{CM} cells that secrete Th2-associated cytokines (e.g., IL-5 and IL-13) and are the so-called type 2 polarized (Tc2) cells (Geginat et al., 2003). These cells were found to propagate Th2-mediated allergy and to contribute to arthritis (Tang et al., 2012). Interestingly, these Tc2 cells are undetectable in the T_{EM} and T_{TE} compartments, thus indicating that they confine solely to the T_{CM} compartment (Geginat et al., 2003). CXCR5 was shown, in recent studies in mice infected with a chronic virus, to be expressed by a subset of CD8$^+$ T cells that displayed high self-renewal capacity and that can be defined by the expression of programmed cell death 1 (PD-1) and T cell factor 1 (TCF1), while lacking T cell immunoglobulin and mucin-domain containing-3 (TIM-3) expression (Im et al., 2016; Lin et al., 2016; He et al., 2016). Because there is conflicting evidence regarding CCR7 and/or CD62L expression by this subset, it remains unclear if they truly have a T_{CM} phenotype (Im et al., 2016; Lin et al., 2016; He et al., 2016; Quigley et al., 2007).

In addition, 25% of circulating CD4$^+$ and CD8$^+$ T cells express CD161 (Takahashi et al., 2006), a natural killer cell receptor, though the function of this molecule on T cells remains elusive. Around 50% of the CD161$^+$ CD4$^+$ T cells display a T_{CM} phenotype, whereas no CD161$^+$ CD8$^+$ T cells have a T_{CM} phenotype (Takahashi et al., 2006). Interestingly, the CD161$^+$ CD4$^+$ T cells secrete more IFNγ, TNFα, and IL-4 than the CD161$^-$ CD4$^+$ T cells (Takahashi et al., 2006), suggesting that CD161 identifies CD4$^+$ T_{CM} cells with effector T cell commitment. Also IL-18Rα (CD218) is expressed on a subset of T_{CM} cells (Mahnke et al., 2013), which together with IL-18Rβ forms the receptor for IL-18. IL-18 together with IL-12 signaling has been shown to induce antigen-independent production of IFNγ by T cells (Bou Ghanem et al., 2011).

2.4 Effector Memory T Cells

T_{EM} cells form a memory subset that is ready for rapid effector responses, e.g., cytolysis of target cells, on antigen recognition. T_{EM} cells express high levels of cytolytic molecules, e.g., granzyme B and perforin. Compared with T_{CM} cells, these cells are enriched within the circulation and not in

SLOs, most likely due to the lack of the expression of SLO homing receptors CCR7 and CD62L (Sallusto et al., 1999a). Instead, T_{EM} cells express a vast array of chemokine receptors and tissue-homing molecules, which enables them to migrate to nonlymphoid tissues or to sites of inflammation (Butcher and Picker, 1996; Mackay et al., 1990). T_{EM} cells from both compartments express CXCR4 and to a higher extent than T_{CM} cells. In addition, the majority of CD4 T_{EM} and all CD8 T_{EM} express CCR5, which enables their migration into inflamed tissues, where CCR5 ligands RANTES or macrophage inflammatory proteins (MIPs) are highly expressed (Wills et al., 2002). Moreover, similar to $CD4^+$ T_{CM} cells, the $CD4^+$ T_{EM} cells express chemokine receptors that identify their Th differentiation pattern, e.g., CXCR3 on Th1 T_{EM} cells, CCR4 and CRTh2 on Th2 T_{EM} cells, CCR4 and CCR6 on Th17 T_{EM} cells, and CXCR5 on Tfh T_{EM} cells (Lintermans et al., 2017; Sallusto et al., 2004). The majority of $CD8^+$ T_{CM} cells express CXCR3, whereas a small subset expresses CCR4 or CXCR5. In addition, the expression of CD122, LFA-1, LFA-3, and IL-18Rα is increased on T_{EM} cells compared with T_{CM} cells (Gattinoni et al., 2011; Mahnke et al., 2013).

Generally, T_{EM} cells are regarded to have reduced or no expression of CD28, CD27, and CD127. Therefore, T_{EM} cells are usually defined as $CD45RA^-CD45RO^+CCR7^-CD62L^-CD27^-CD28^-CD127^-CD95^+$ $CD122^+$ cells. However, one study showed that four different T_{EM} ($CD45RA^-CCR7^-$) populations could be identified in human blood; $CD27^+CD28^+$, $CD27^+CD28^-$, $CD27^-CD28^+$, and $CD27^-CD28^-$ (Romero et al., 2007). The populations expressing CD28 also expressed high levels of CD127 and low levels of effector mediators such as granzyme B, suggesting that these $CD28^+CD45RA^-CCR7^-$ cells are more similar to T_{CM} cells. In contrast, the $CD28^-$ populations express low levels of CD127 and high levels of effector mediators. The loss of CD28 expression in $CD8^+$ T cells most likely occurs due to the repeated antigen and/or γ_c chain cytokine stimulations (Weng et al., 2009; Vallejo et al., 2000; Valenzuela and Effros, 2002; Borthwick et al., 2000; Chiu et al., 2006; Vallejo, 2005).

Subsets of T_{EM} cells also express the coinhibitory receptor killer cell lectin–like receptor G1 (KLRG1) and CD57 (Stubbe et al., 2008; Verma et al., 2017). KLRG1 signaling can inhibit both $CD4^+$ and $CD8^+$ T cell activation, yet only when KLRG1 ligands and the MHC/antigen are present on the same target cell (Rosshart et al., 2008). CD57 is a surface membrane carbohydrate epitope implicated in cell-to-cell and cell-to-extracellular matrix adhesion (Uusitalo and Kivela, 2001). Its presence on $CD4^+$ and $CD8^+$ T cells, but also natural killer (NK) cells, is used as a general marker of proliferative inability and short telomeres (Brenchley et al., 2003). CD161 is also expressed on a subset of T_{EM} cells. Most of the CD161$^+$ CD8$^+$ T cells have a T_{EM} phenotype, whereas a smaller percentage of the CD161$^+$ CD4$^+$ T cells have a T_{EM} compared to T_{CM} phenotype (Takahashi et al., 2006).

2.5 T_{TE}/T_{EMRA} Cells

T_{TE} or T_{EMRA} cells are the end-stage differentiated T cells. T_{TE} cells possess potent effector functions (Dunne et al., 2005; Henson et al., 2012) but are poorly proliferative (Dunne et al., 2005; Henson et al., 2012), due to the high levels of cell cycle inhibitors p16 and p21, have increased activation of senescence signaling pathways, and have a high propensity to undergo apoptosis in vitro (Verma et al., 2017; Geginat et al., 2003). This subset is abundant in the CD8$^+$ compartment but can be found in the CD4$^+$ compartment as well (Sallusto et al., 1999a). The frequency of T_{TE} cells increases with age and during persistent viral infections such as CMV (Effros, 1997; Khan et al., 2002; Appay et al., 2002a; Wills et al., 2002; Derhovanessian et al., 2011; Faint et al., 2001). Moreover, vaccination studies revealed that increased frequencies of CD28$^-$ CD8$^+$ T_{TE} cells correlate with defective humoral responses to influenza vaccine (Saurwein-Teissl et al., 2002).

Phenotypically, T_{TE} cells are a subset of T_{EM} cells that reexpress CD45RA (Sallusto et al., 1999a; Henson et al., 2012). Yet, in contrast to T_N and T_{SCM} cells that also express CD45RA, T_{TE} cells lack the expression of CCR7, CD62L, CD27, CD28, and CD127 (Romero et al., 2007). Therefore, T_{TE} cells are defined as CD45RA$^+$CD45RO$^-$CCR7$^-$CD62L$^-$CD27$^-$CD28$^-$CD127$^-$CD95$^+$CD122$^+$ cells by flow cytometry. In addition, the expression levels of CD122, LFA-1, LFA-3, and IL-18Rα on T_{TE} cells are similar to those on T_{EM} cells (Gattinoni et al., 2011; Mahnke et al., 2013).

The absence of CCR7 and CD62L on the surface of T_{TE} cells suggests that they are excluded from SLOs. This is confirmed by the fact that CMV-specific CD8$^+$ T_{TE} cells are absent from the lymph nodes but are found in high frequencies in the peripheral blood of humans (Remmerswaal et al., 2012). Moreover, under physiological condition, T_{TE} cells preferentially home to the spleen and lung (Sathaliyawala et al., 2013; Thome et al., 2014), suggesting that receptors other than CCR7 and CD62L might be required for T_{TE} cells to home to the spleen. In addition, T_{TE} cells express several homing receptors that allow them to migrate to sites of inflammation, including LFA-1, CX_3CR1, and CCR5 (Faint et al., 2001). High levels of CX_3CR1, which binds fractalkine expressed on stressed endothelium, allow the homing of T_{TE} cells to inflamed vascular endothelium (Hertoghs et al., 2010), whereas CCR5 expression allows migration into inflamed tissues, where CCR5 ligands RANTES or MIPs are highly expressed (Wills et al., 2002). T_{TE} cells also express CXCR3 and CXCR4, although the level of expression is lower than on T_{CM} and T_{EM} cells (Gattinoni et al., 2011). T_{TE} cells also express some of the markers of senescence such as KLRG1 and CD57 at a higher level than T_{EM} cells (Blaser et al., 1998; Voehringer et al., 2002; Appay et al., 2008; Romero et al., 2007; Koch et al., 2008; Brenchley et al., 2003). The increase of T_{TE} cells with age is in line with the increase in percentage of KLRG1$^+$ T cells from 40% of CD8$^+$ T cells (20% of CD4$^+$ T cells) in young adults (Voehringer et al., 2002) to 90% in individuals over 65 years of age (Ouyang et al., 2003; Henson et al., 2009).

An unusual feature of the T_{TE} subpopulation is the acquisition of several receptors associated with NK cells. These include KIR2DL2 (CD158B1), KIR3DL2 (CD158K), KLRD1 (CD94), KLRF1, KLRK1 (NKG2D), NCR1 (CD355), and 2B4 (CD244) (Weng et al., 2009). KLRD1 forms dimers with different molecules. Together with KLRC1 (NKG2A) it binds to HLA-E and inhibits cytotoxicity, whereas together with KLRD1 or KLRK1 (NKG2D) it has a stimulatory function (Braud et al., 1998; Guma et al., 2005). The presence of the activating receptor 2B4 has been shown to induce effector functions of CMV-specific effector CD8$^+$ T cells (Speiser et al., 2001).

3. TISSUE-RESIDENT MEMORY T CELL PHENOTYPES

T_{RM} cells are T cells that do not recirculate but are maintained at virtually all nonlymphoid tissues, including skin, lung, intestine, and brain (Masopust et al., 2001; Jiang et al., 2012; Mueller and Mackay, 2016). T_{RM} cells have shown to be crucial in the protection against infections at those sites (Schenkel et al., 2014; Masopust et al., 2010). The low expression of CD62L on these cells suggests that they share features with T_{EM} cells. However, it has been demonstrated that T_{RM} and circulating T_{EM} cells have a clearly distinct transcriptional program (Hombrink et al., 2016). Moreover, it was shown by several migration studies in mice that only effector cells, and not memory cells, are competent to enter the nonlymphoid tissues, suggesting that T_{RM} cells do not derive from T_{EM} cells (Hickey et al., 1991; Hofmann and Pircher, 2011; Wakim et al., 2010;

Masopust et al., 2010). The expression of surface markers, such as integrins, differs between T_{RM} cells within different nonlymphoid tissues, yet CD69 and CD103 (α_E integrin) expression is considered a common feature of T_{RM} cells. CD69, a member of the NK receptor family, is regarded as an early T cell activation marker; however, T_{RM} cells do not express other markers of activation, such as CD25, CD38, and HLA-DR, suggesting a different role of CD69 in T_{RM} cells (Kumar et al., 2017). Mouse studies revealed that CD69 is not only involved in the retention of recently activated T cells in lymph nodes (Shiow et al., 2006) but also of resident T cells in nonlymphoid tissues because CD69$^{-/-}$ T cells are not retained in those tissues (Skon et al., 2013). This is mediated by CD69-dependent downregulation of sphingosine-1-phosphate receptor 1 (S1PR1), normally critical for the recirculation of T_N cells (Skon et al., 2013). CD103 dimerizes with the β7 integrin to form the complete $\alpha_E\beta7$ integrin that binds E-cadherin present on epithelial cells (Agace et al., 2000), necessary for the retention of cells within tissues. Unfortunately, CD103 is not a good marker for CD4$^+$ T_{RM} cells because it is only expressed by a subset of skin CD4$^+$ T_{RM} cells (Gebhardt et al., 2011; Turner et al., 2014). The markers expressed on T_{RM} cells from the most studied tissues are discussed below.

In the skin, two different CD8$^+$ T_{RM} subsets can be identified based on CD103 expression both in humans and mice (Li et al., 2016; Watanabe et al., 2015). Mouse studies revealed that CD103$^+$ T_{RM} cells are mainly found in the epidermis, whereas CD103$^-$ T_{RM} cells are found in the dermis (Watanabe et al., 2015). In addition, the vast majority of skin T_{RM} cells express high levels of CD69, cutaneous lymphocyte-associated antigen (CLA), CCR4, and CCR6. The expression of CLA, CCR4, and CCR6 promotes homing to the skin. CLA is the ligand for E-selectin expressed by vascular endothelial cells at sites of inflamed skin (Rossiter et al., 1994; Clark et al., 2006), whereas skin homing via CCR4 is guided by binding to CCL17 and CCL22 (Imai et al., 1999). CCR6 is a gut, lung, and inflamed skin homing receptor, facilitated by the expression of CCL20 in those tissues (Schutyser et al., 2003). In addition, around 50% of T_{RM} express CCR8 and CXCR6 (Clark et al., 2006). Both chemokine receptors have been implicated in trafficking of T cells toward the skin (Schaerli et al., 2004). Moreover, a subset of T_{RM} cells in inflamed skin expresses CCR5 and CXCR3 (Kunkel et al., 2002), both involved in the trafficking to inflammatory sites. CD49a (VLA-1, $\alpha1$ integrin) expression, although mainly regarded as a marker for lung homing T_{RM} cells, also identifies a small subset of $CD8^+$ T_{RM} cells with cytotoxic activity in human skin (Cheuk et al., 2017; Purwar et al., 2011).

In the gut, T_{RM} cells can be found both in the lamina propria and in the intraepithelial compartment. In mice, around 50% of the $CD8^+$ T_{RM} cells within the lamina propria express CD103, whereas in the intraepithelial lymphocyte compartment, all $CD8^+$ T_{RM} cells express this integrin (Kim et al., 1999). Similarly, though on the mRNA level, *ITGAE* (encoding CD103) is expressed in human $CD4^+$ and $CD8^+$ T_{RM} cells within both the lamina propria and the intraepithelial lymphocyte compartment, although to a higher extent in the latter (Raine et al., 2015). Furthermore, gut T_{RM} cells are characterized by the expression of CD69, CCR6, CCR9, and CD49d (α4 integrin) (Kunkel et al., 2000; Purwar et al., 2011). CCR6, CCR9, and CD49d are involved in the homing of T cells into the gut mucosa, which is facilitated by binding to CCL20, CCL25, and mucosal addressin cell adhesion molecule-1 (MAdCAM-1), respectively (Eksteen et al., 2009). In addition, both CCR5 and CXCR3 are expressed on gut T_{RM} cells, guiding their migration toward inflammatory sites (Kunkel et al., 2002; Campbell et al., 2001).

In the lung, both $CD103^+$ and $CD103^-$ $CD8^+$ T_{RM} cells can be identified in humans, where all $CD103^+$ T_{RM}, but only 70%–95% of the $CD103^-$ T_{RM} cells express CD69 (Hombrink et al., 2016). In addition, lung $CD8^+$ and $CD4^+$ T_{RM} cells uniformly express CD49a and P-selectin glycoprotein ligand-1 (PSGL-1) (Purwar et al., 2011). Both CD49a, the receptor for type IV collagen, and PSGL-1, binding E-selectin, are considered to be essential in the homing of the T cells to the lungs. These cells also express CCR5 and CXCR3 chemokine receptors, similar to skin and gut T_{RM} cells, and additionally express CCR6 (mainly on $CD4^+$ T_{RM} cells) and CXCR4 (Hombrink et al., 2016; Purwar et al., 2011).

4. PHENOTYPES OF T CELL DYSFUNCTIONALITY

Under specific circumstances, e.g., chronic viral infection or cancer, T cells can become dysfunctional. T cell dysfunction is found in both the $CD8^+$ and $CD4^+$ T cell compartments, yet T cell dysfunction within the $CD8^+$ compartment has been more extensively studied. Moreover, there are differences between the two compartments regarding surface marker expression. Two different states of T cell dysfunction are described: exhaustion and senescence. In addition, in some instances, T cell dysfunction may be defined as "anergy", a biological state induced to prevent autoimmunity. Cells that are in each of these states can be identified by the expression of specific surface markers, as described below.

4.1 Exhaustion

T cell exhaustion was first described in chronic infection, although later a similar phenotype was found in tumor-infiltrating lymphocytes. Features of T cell exhaustion include (1) loss of proliferative capacity, (2) loss of effector cytokine production, (3) loss of cytotoxicity as compared to effector T cells, (4) gradually increase in number and level of inhibitory receptors, and (5) poor responsiveness to IL-7 and IL-15 mediating self-renewal and survival of T cells (Wherry et al., 2003; Fourcade et al., 2012; Shin et al., 2007). T cell exhaustion has been extensively studied in chronic viral infections, such as human immunodeficiency virus (HIV), simian immunodeficiency virus (SIV), hepatitis B and C virus (HBV and HCV), as well as in cancer (Fourcade et al., 2012; Blackburn et al., 2009). Different studies with exhausted virus-specific CD8$^+$ T cells show that high antigen load, persistence of infection, and absence of CD4$^+$ T cell help intensify the severity of T cell exhaustion. Regarding the tumor-infiltrating CD8$^+$ T cells, not only a suppressive tumor microenvironment but also high TCR affinity, chronic encounter with tumor antigens, and T cell terminal differentiation status contribute to their generalized hyporesponsiveness.

Exhausted CD4$^+$ and CD8$^+$ T cells are generally identified by common immunophenotypes characterized by the expression of inhibitory receptors, such as PD-1, TIM-3, lymphocyte activation gene 3 (LAG-3), cytotoxic T lymphocyte—associated protein 4 (CTLA-4), CD160, T cell immunoreceptor with Ig and ITIM domains (TIGIT), and B and T lymphocyte attenuator (BTLA), and their increasing expression levels on a per-cell basis (Blackburn et al., 2009; Wherry, 2011). However, lineage-specific differences can be identified. For example, during a chronic infection, exhausted CD4$^+$ T cells preferentially express more CTLA-4, CD200 (OX2), and BTLA, whereas exhausted CD8$^+$ T cells express more TIM-3 and CD160 (Kaufmann et al., 2007; Porichis et al., 2011; Crawford et al., 2014; Morou et al., 2014). Importantly, surface expression of single inhibitory receptors is not per se limited to exhausted T cells and as a consequence cannot be used as a hallmark of impaired T cell function. The presence of inhibitory receptors depends on differentiation and activation status of T cells. For example, PD-1 is a major inhibitory receptor described on exhausted T cells (Trautmann et al., 2006; Day et al., 2006; Barber et al., 2006; Petrovas et al., 2006).

However, stable expression of PD-1 is found on up to 60% of human memory $CD8^+$ T cells, and PD-1^{hi} cells are able to secrete effector cytokines on rechallenge (Duraiswamy et al., 2011). Moreover, its transient expression during T cell activation is necessary to prevent hyperactivation and/or autoimmunity (Keir et al., 2008; Nishimura et al., 1999). During chronic $CD8^+$ T cell stimulation and in the presence of persistent inflammatory signals, the *PDCD1* gene locus undergoes epigenetic alterations leading to high, long-term PD-1 expression (Youngblood et al., 2011, 2013). This indicates that expression of PD-1 or other inhibitory receptors alone does not identify exhausted T cells. Hence, coexpression of multiple inhibitory receptors should be used for identification of exhausted T cells. Many different combinations of inhibitory receptors can be found on T cells, yet the relationship between such phenotypes and the grade of dysfunction still remains to be elucidated in full. However, the general assumption is that progressive accumulation of inhibitory receptors correlates with increased exhaustion.

Importantly, T cell exhaustion can partly be reversed by blocking the signaling via the inhibitory receptors on the T cells. Therefore, blockade of inhibitory receptors (checkpoint blockade) using antibodies has become a promising therapy for cancer and chronic viral infections (Hodi et al., 2010; Topalian et al., 2012b; Brahmer et al., 2012; Velu et al., 2009). Ipilimumab, a CTLA-4-blocking monoclonal antibody for the treatment of melanoma, was the first FDA-approved drug of this class (Hodi et al., 2010). Based on several studies, it became clear that PD-1 signaling is essential in T cell exhaustion because the blockade of the PD-1—PD-1 ligand 1 (PD-L1) interaction by anti-PD-L1 antibodies alone or in combination with anti-CTLA-4 or anti-TIM-3 antibodies restored $CD8^+$ T cell function in advanced cancer and during chronic viral infections (Barber et al., 2006; Blackburn et al., 2009; Nakamoto et al., 2009; Jin et al., 2010; Topalian et al., 2012a; Pauken and Wherry, 2015). Recent phase I clinical trials with melanoma patients reported that combining PD-1 and CTLA-4 blockade results in synergistic effects in improving adaptive immune responses; however, a higher incidence of toxicity was observed compared to monotherapy (Wolchok et al., 2013; Hamid et al., 2013). Novel studies investigating the mechanism of PD-1 blockade have shed light on the existence of two distinct exhausted $CD8^+$ T cell subpopulations expressing different levels of PD-1 molecule: PD-1^{int} and PD-1^{hi} (Paley et al., 2012). Intermediate expression of PD-1 characterizes so-called "progenitor" cells, which are able to proliferate in response to persisting antigen stimulation yet ultimately differentiate into the more mature "terminal progeny"

PD-1^{hi} cells. These PD-1^{hi} cells exhibit higher expression of other inhibitory receptors also, do not replicate, and display low cytolytic activity (Blackburn et al., 2008). Interestingly, these two subsets respond differently to PD-1 blockade treatment because mainly the PD-1^{int} subset responds to the treatment (Wherry and Kurachi, 2015). Therefore, the ratio between these distinct exhausted phenotypes has impact on the effectiveness of checkpoint blockade therapy (Blackburn et al., 2008).

4.2 Senescence

Senescent T cells undergo irreversible permanent cell cycle arrest, commonly reflected by telomere shortening [the so-called Hayflick limit (Hayflick, 1961)] and expression of cell cycle inhibitors p16 and p21 (Campisi, 2005). In addition, senescent T cells are resistant to apoptosis, in contrast to exhausted T cells (Spaulding et al., 1999). Early-senescent $CD8^+$ T cells accumulate with age but are viable, maintain metabolic activity, and are capable to produce cytokines (Onyema et al., 2015), whereas late-senescent T cells tend to lose their effector functions. Senescent $CD8^+$ T cells have similar characteristics as the T_{TE} cells because senescent T cells have similar high expression of KLRG1 and CD57 and loss of CD27 and CD28 expression (Zelle-Rieser et al., 2016; Mondal et al., 2013; Brenchley et al., 2003; Effros, 2004). In addition, late-senescent cells lose the surface expression of CD95 (Onyema et al., 2015). Besides the difference in apoptosis resistance, senescent T cells can be distinguished from exhausted T cells based on the expression of specific surface markers: high levels of KLRG1 and low levels of PD-1 (Akbar and Henson, 2011).

Senescent T cells lack expression of SLO homing receptors CCR7 and CD62L, indicating their inability to migrate into SLOs, similar to T_{TE} cells. In addition, $CD57^+CD28^-$ $CD8^+$ T cells display low expression level of chemokine receptor CCR5 important for migration into nonlymphoid tissues (Onyema et al., 2015; Fukada et al., 2002). In contrast, senescent cells express high levels of CXCR2, a chemokine receptor involved in the migration into tumor sites and is associated with senescence reinforcement (Onyema et al., 2015; Peng et al., 2010; Acosta et al., 2008). Moreover, it has been demonstrated that the frequency of senescent cells increases in the circulation during strenuous physical exercise. This suggests that in resting conditions, these cells are likely localized outside the bloodstream, possibly in extralymphatic tissues (such as liver or bone marrow) (Simpson et al., 2007, 2008).

4.3 Anergy

When T cells encounter their cognate antigen presented by APCs, but costimulation is absent, T cells can become anergic (Schwartz, 2003). Anergic T cells are intrinsically functionally inactivated and thereby unable to produce IL-2 or to proliferate on a secondary encounter with their cognate antigen, even in optimal conditions where costimulation is present (Jenkins and Schwartz, 1987). Tregs were found to be responsible for the generation of anergic T cells in vitro, likely by controlling the expression of costimulatory molecules (CD80 and CD86) on the surface of APCs, facilitating the lack of optimal costimulation for T cells. Under physiological conditions, T cell anergy is thought to play a role in the regulation of self-reactive T cells (Maeda et al., 2014). The role of T cell anergy has also been studied in the context of cancer; however, there is only indirect evidence for the involvement of T cell anergy in the tumor microenvironment [reviewed in the study by Crespo et al. (2013)]. Clear markers to identify anergic T cells in vivo are still lacking; however, a recent study suggests that expression of inhibitory makers CTLA-4, PD-1, LAG-3, and TIM-3 on naïve ($CCR7^+CD45RA^+$) self-antigen Melan-A-specific $CD8^+$ T cells might identify anergic T cells within healthy donors (Maeda et al., 2014).

5. $CD4^+$ T CELL–SPECIFIC PHENOTYPES

In contrast to $CD8^+$ T cells, $CD4^+$ T_N cells can differentiate into several effector subsets, e.g., Th1, Th2, Th17, or regulatory subsets, e.g., Tregs and Tr1 cells, depending on factors in the local environment, such as cytokines and TCR signaling strength (Tubo and Jenkins, 2014; Christie and Zhu, 2014; Sallusto and Lanzavecchia, 2009). Each subset has its own phenotypical characteristics and function. Th1, Th2, Th17 cells are involved in the protection against intracellular pathogens, extracellular parasites, and extracellular bacteria and fungi, respectively. Tfh cells are essential for B cell help, whereas Tregs are essential for immune homeostasis. Each $CD4^+$ T cell subset is generally identified by the expression of subset-specific master transcription factors (TFs), which promote lineage-specific gene expression, and cytokine secretion. It should be noted, however, that opposite master TFs can also be induced in already differentiated T cells, thereby leading to reprogramming (Kanno et al., 2012). This suggests notable plasticity within the $CD4^+$ T cell subsets [reviewed in the studies by Zhou et al. (2009) and Sallusto (2016)]. Because purification of cells by flow cytometry

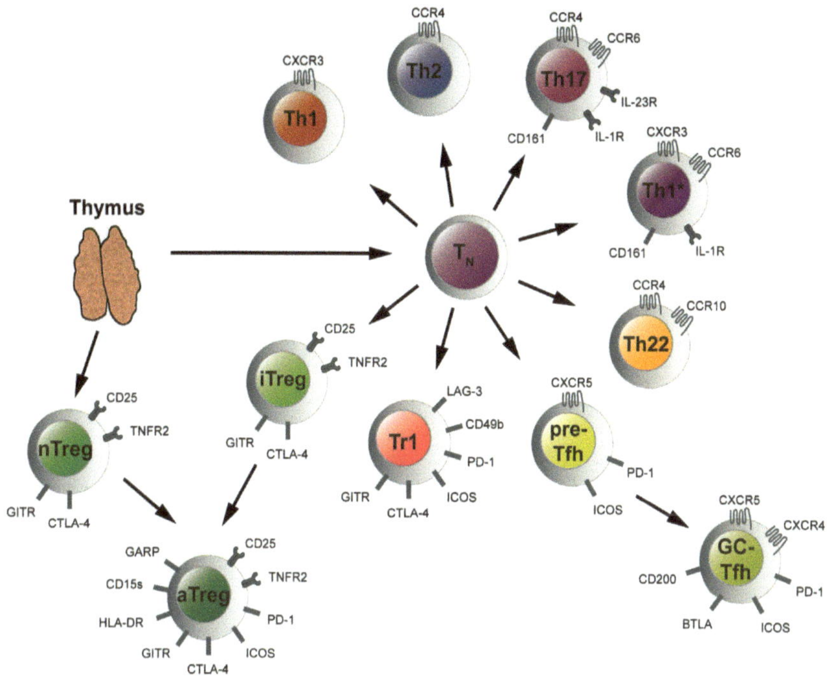

Figure 2 ***Surface markers defining the different CD4$^+$ T cell subsets.*** Naive T cells, egressing from the thymus after maturation, can differentiate into multiple T helper or regulatory subsets. Each population expresses (a combination of) specific surface markers. *aTreg*, activated regulatory T cells; *GC*, germinal center; *iTreg*, induced regulatory T cells; *nTreg*, natural regulatory T cells; *Tfh*, T follicular helper cells; *Th*, T helper cells; *T_N*, naive T cells; *Tr1*, type 1 regulatory cells.

based on the expression of TFs would preclude downstream functional analysis, scientists have devised strategies based on the combinatorial expression of surface chemokine receptors and/or other surface markers to enable the identification of the different CD4$^+$ T cell subsets. The surface markers expressed by these subsets are discussed below and are summarized in Fig. 2 and Table 2.

5.1 Th1 Cells

Th1 cells are generated when CD4$^+$ T_N cells are activated in the presence of IL-12. Th1 cells play an important role in the protection against intracellular pathogens, including bacteria and viruses (Fietta and Delsante, 2009). They are characterized by the expression of master transcription factor T-bet and the secretion of IL-2 and IFNγ (Szabo et al., 2015). The latter activates

Table 2 Expression of Surface Markers on $CD4^+$ T Cells[a]

Subset	Surface Phenotype	Function	References
Th1	$CXCR3^+$	Migration into inflamed sites	Lord et al. (2005), Rivino et al. (2004), Sallusto et al. (1998), and Sallusto (2016)
	$CCR4^-$	Homing to skin	Sallusto et al. (1998), Sallusto (2016), and Rivino et al. (2004)
	$CCR6^-$	Homing to gut, lung, and inflamed skin	Sallusto (2016) and Rivino et al. (2004)
Th2	$CXCR3^-$	Migration into inflamed sites	Sallusto (2016) and Rivino et al. (2004)
	$CCR4^+$	Homing to skin	Imai et al. (1999), Rivino et al. (2004), and Sallusto (2016)
	$CCR6^-$	Homing to gut, lung, and inflamed skin	Sallusto (2016) and Rivino et al. (2004)
	$CRTh2^{-/+}$	Promotion of chemotaxis and proinflammatory cytokine production by Th2 cells	Cosmi et al. (2000)
Th17	$CXCR3^-$	Migration into inflamed sites	Acosta-Rodriguez et al. (2007) and Sallusto (2016)
	$CCR4^+$	Homing to skin	Acosta-Rodriguez et al. (2007) and Sallusto (2016)
	$CCR6^+$	Homing to gut, lung, and inflamed skin	Acosta-Rodriguez et al. (2007) and Sallusto (2016)
	$IL\text{-}23R^+$	IL-23 receptor; induction IL-17A production and proliferation	Acosta-Rodriguez et al. (2007), Wilson et al. (2007), and Sallusto (2016)
	$CD161^+$	Unknown	Maggi et al. (2010)
	$IL\text{-}1R^+$	IL-1 receptor	Lee et al. (2010)
Th1*	$CXCR3^+$	Migration into inflamed sites	Acosta-Rodriguez et al. (2007), Maggi et al. (2012), and Sallusto (2016)
	$CCR4^-$	Homing to skin	Acosta-Rodriguez et al. (2007) and Sallusto (2016)

	CCR6$^+$	Homing to gut, lung, and inflamed skin	Acosta-Rodriguez et al. (2007), Maggi et al. (2012), and Sallusto (2016)
	CD161$^+$	Unknown	Maggi et al. (2012)
	IL-1R$^+$	IL-1 receptor	Maggi et al. (2012)
Th22	CXCR3$^-$	Migration into inflamed sites	Trifari et al. (2009)
	CCR4$^+$	Homing to skin	Trifari et al. (2009)
	CCR6$^-$	Homing to gut, lung, and inflamed skin	Trifari et al. (2009)
	CCR10$^+$	Homing to skin	Trifari et al. (2009)
Th9	Unknown		
Pre-Tfh	CXCR5$^+$	Homing to B cell follicles	Haynes et al. (2007), Suh (2015), and Kroenke et al. (2012)
	ICOS$^+$	Costimulatory receptor	Suh (2015) and Kroenke et al. (2012)
	PD-1$^+$	Inhibition of effector function	Haynes et al. (2007), Suh (2015), and Kroenke et al. (2012)
	Ly9$^+$	Interaction with B cells and DCs	Ma et al. (2009)
	CD84$^+$	Interaction with B cells and DCs	Ma et al. (2009)
	SLAMF6$^+$	Interaction with B cells and DCs	Ma et al. (2009)
	CCR7$^-$	Homing to SLOs	Haynes et al. (2007)
	PSGL-1^{lo}	Homing to SLOs	Suh (2015)
GC-Tfh	CXCR5^{hi}	Homing to B cell follicles	Schaerli et al. (2000), Ma et al. (2009), Haynes et al. (2007), Kroenke et al. (2012), Wong et al. (2015), and Rasheed et al. (2006)
	ICOS^{hi}	Costimulatory receptor	Ma et al. (2009), Kroenke et al. (2012), Wong et al. (2015), and Rasheed et al. (2006)
	PD-1^{hi}	Inhibition of effector function	Schaerli et al. (2000), Ma et al. (2009), Kroenke et al. (2012), and Rasheed et al. (2006)
	Ly9$^+$	Interaction with B cells and DCs	Ma et al. (2009) and Suh (2015)

(Continued)

Table 2 Expression of Surface Markers on $CD4^+$ T Cells[a]—cont'd

Subset	Surface Phenotype	Function	References
	$CD84^+$	Interaction with B cells and DCs	Ma et al. (2009) and Suh (2015)
	$SLAMF6^+$	Interaction with B cells and DCs	Ma et al. (2009) and Suh (2015)
	$CCR7^-$	Homing to SLOs	Schaerli et al. (2000), Haynes et al. (2007), and Lim and Kim (2007)
	$PSGL1^{lo}$	Homing to SLOs	Suh (2015)
	$CD69^+$	Early activation marker; retention in tissues	Schaerli et al. (2000)
	$BTLA^{hi}$	Inhibitory receptor	Ma et al. (2009)
	$CD200^+$	Inhibitory receptor	Ma et al. (2009)
	$CD127^{lo}$	α-chain of IL-7 receptor	Ma et al. (2009) and Lim and Kim (2007)
	$CCR6^{lo}$	Homing to gut, lung and inflamed skin	Ma et al. (2009)
	$CXCR3^{lo}$	Migration into inflamed sites	Ma et al. (2009)
	$CXCR4^+$	Homing to bone marrow/SLOs	Ma et al. (2009), Kroenke et al. (2012), and Lim and Kim (2007)
	$CD57^{-/+}$	Unknown	Ma et al. (2009), Lim and Kim (2007), Kim et al. (2001), Wong et al. (2015), and Rasheed et al. (2006)
cTfh	$CXCR5^+$	Homing to B cell follicles	Chevalier et al. (2011), Locci et al. (2013), He et al. (2013), Schmitt et al. (2014), and Morita et al. (2011)
	$ICOS^{-/+}$	Costimulatory receptor	Chevalier et al. (2011), Locci et al. (2013), He et al. (2013), Schmitt et al. (2014), and Morita et al. (2011)

	PD-1$^{-/+}$	Inhibition of effector function	Chevalier et al. (2011), Locci et al. (2013), He et al. (2013), Schmitt et al. (2014), and Morita et al. (2011)
	CCR7$^+$	Homing to SLOs	Chevalier et al. (2011), Locci et al. (2013), He et al. (2013), Schmitt et al. (2014), and Morita et al. (2011)
	CD69$^-$	Early activation marker; retention in tissues	Ma et al. (2009) and Kim et al. (2001)
	CXCR3$^{-/+}$	Migration into inflamed sites	Morita et al. (2011) and Schmitt et al. (2014)
	CCR4$^{-/+}$	Homing to skin	Morita et al. (2011) and Schmitt et al. (2014)
	CCR6$^{-/+}$	Homing to gut, lung, and inflamed skin	Morita et al. (2011) and Schmitt et al. (2014)
rTreg	CD25$^+$	α-chain of IL-2 receptor	Sakaguchi et al. (2010), Miyara et al. (2009), Dong et al. (2013), and Jonuleit et al. (2001)
	CD127^{lo}	α-chain of IL-7 receptor	Sakaguchi et al. (2010), Miyara et al. (2009), Dong et al. (2013), and Jonuleit et al. (2001)
	CD45RA$^+$	Regulation of T cell antigen signaling	Miyara et al. (2009), Dong et al. (2013), and Jonuleit et al. (2001)
	TNFR2$^+$	TNF receptor; stabilizing Treg phenotype	van Mierlo et al. (2008) and Chen et al. (2010)
	CTLA-4$^+$	Inhibition of effector function	Dong et al. (2013)
	GITR$^+$	Inhibition of effector function	Ronchetti et al. (2015)
aTreg	CD25^{br}	α-chain of IL-2 receptor	Sakaguchi et al. (2010)
	CD127^{lo}	α-chain of IL-7 receptor	Sakaguchi et al. (2010)
	CD45RA$^-$	Regulation of T cell antigen signaling	Miyara et al. (2009), Dong et al. (2013), and Jonuleit et al. (2001)
	TNFR2$^+$	TNF receptor; stabilizing Treg phenotype	van Mierlo et al. (2008) and Chen et al. (2010)
	CTLA-4^{hi}	Inhibition of effector function	Wang et al. (2008)
	GITR$^+$	Inhibition of effector function	Sakaguchi et al. (2010) and Ronchetti et al. (2015)

(*Continued*)

Table 2 Expression of Surface Markers on CD4^{+} T Cells[a]—cont'd

Subset	Surface Phenotype	Function	References
	HLA-DR^{hi}	Peptide presentation	Wang et al. (2008)
	PD-1^{+}	Inhibition of effector function	Sakaguchi et al. (2010)
	CD15s	Migration into tissues	Miyara et al. (2015)
	ICOS^{hi}	Costimulatory receptor	Wang et al. (2008)
	GARP^{+}	Controlling FoxP3 expression	Wang et al. (2008) and Probst-Kepper et al. (2009)
	CCR6$^{-/+}$	Homing to gut, lung, and inflamed skin	Lee et al. (2017)
	CD39$^{-/+}$	Ectonucleotidases	Borsellino et al. (2007)
	LAG-3$^{-/+}$	Inhibition of effector function	Camisaschi et al. (2010)
Tr1	LAG-3^{+}	Inhibition of effector function	Gagliani et al. (2013)
	CD49b^{+}	Homing to gut	Gagliani et al. (2013)
	CD25^{lo}	α-chain of IL-2 receptor	Vieira et al. (2004) and Facciotti et al. (2016)
	CD127^{lo}	α-chain of IL-7 receptor	Facciotti et al. (2016)
	CTLA-4^{hi}	Inhibition of effector function	Gregori et al. (2012)
	PD-1^{hi}	Inhibition of effector function	Gregori et al. (2012)
	GITR^{+}	Inhibition of effector function	Ronchetti et al. (2015)
	ICOS^{hi}	Costimulatory receptor	Gregori et al. (2012)
	LFA-1^{+}	Homing to SLOs and inflamed sites; adhesion to APCs	Magnani et al. (2011)
	CD2^{+}	Adhesion to myeloid APC; activation	Magnani et al. (2011)
	CD226^{+}	Adhesion to myeloid APC; activation	Magnani et al. (2011)
	CD39$^{-/+}$	Ectonucleotidases	Mandapathil et al. (2010)

	CD73$^{-/+}$	Ectonucleotidases	Mandapathil et al. (2010)
CD4^{+} CTL	CD107a^{+}	Lysosome-associated membrane glycoproteins	Lin et al. (2014) and Nemes et al. (2010)
	CD107b^{+}	Lysosome-associated membrane glycoproteins	Lin et al. (2014) and Nemes et al. (2010)
	KLRC1^{+}	Activator/inhibitory receptor	Saez-Borderias et al. (2006) and Graham et al. (2007)
	KLRK1^{+}	Activator receptor	Saez-Borderias et al. (2006) and Graham et al. (2007)
	CRTAM^{+}	Induction of effector molecule expression	Takeuchi et al. (2016)
	CD27^{lo}	Costimulatory receptor	Appay et al. (2002b) and van Leeuwen et al. (2004)
	CD28^{lo}	Costimulatory receptor	Appay et al. (2002b) and van Leeuwen et al. (2004)

aTreg, activated regulatory T cells; *GC*, germinal center; *iTreg*, induced regulatory T cells; *nTreg*, natural regulatory T cells; *Tfh*, T follicular helper cells; *Th*, T helper cells; *Tr1*, type 1 regulatory cells.

[a] −/+ indicates that a subset of the given population expresses the marker.

mononuclear phagocytes such as macrophages and increases their phagocytic activity. Th1 cells can be characterized by the expression of chemokine receptor CXCR3 (Lord et al., 2005; Rivino et al., 2004) and the absence of chemokine receptors associated with the other Th subsets such as CCR4 (Th2) and CCR6 (Th17), as discussed below. CXCR3 is essential for the entry of Th1 cells in peripheral tissues and inflammatory sites, which is facilitated by the presence of CXCL9, CXCL10, and CXCL11 at those sites (Groom and Luster, 2011).

5.2 Th2 Cells

Th2 cells are generated when $CD4^+$ T_N cells are primed in the presence of IL-4. This cytokine induces the Th2 master transcription factor GATA-3, which on its turn induces production of Th2-cytokines IL-4, IL-5, and IL-13 (Zheng and Flavell, 1997; Lantelme et al., 2001). Th2 cells play an important role in the protection against extracellular parasites. However, they are also involved in allergic reactions (Romagnani, 1994). Surface markers that can identify Th2 cells are prostaglandin D2 receptor CRTh2 and chemokine receptor CCR4 (Cosmi et al., 2000; Imai et al., 1999). CRTh2 promotes chemotaxis and proinflammatory cytokine production by Th2 cells, whereas it also prevents apoptosis (Xue et al., 2009). CRTh2 has been found to identify highly lineage committed and activated Th2 cells, yet it might not identify resting Th2 cells in vivo (Nagata et al., 1999). CCR4 is necessary for the recruitment of Th2 cells from the blood toward the skin, guided by its ligands CCL17 and CCL22 expressed by APCs (Imai et al., 1999). Besides on Th2 cells, CCR4 is also expressed on a subset of Tregs and on Th17 and Th22 cells (Acosta-Rodriguez et al., 2007; Gobert et al., 2009). Therefore, Th2 cells can purely be identified by the expression of CCR4 and the absence of CCR6 (Th17) and CCR10 (Th22).

5.3 Th17 Cells

Although multiple cytokines such as TGFβ, IL-6, IL-1β, IL-21, and/or IL-23 are involved, the requirements for Th17 differentiation from $CD4^+$ T_N cells in humans are not completely clear (Veldhoen et al., 2006; Shainheit et al., 2008; Zhou et al., 2007; Korn et al., 2007a). Th17 cells are important in the immune responses against extracellular bacteria and fungi but have been also implicated in autoimmunity development (Korn et al., 2007b). The Th17 master transcription factor is RORγt, which induces the production of Th17-associated cytokines IL-17A and IL-17F.

Both cytokines promote the production of neutrophil chemoattractants and β-defensins by epithelial cells, whereas only IL-17A can induce proinflammatory cytokine production by macrophages and T cells (Ishigame et al., 2009). Th17 cells can be characterized by the expression of chemokine receptor CCR6 and IL-23 receptor (IL-23R) (Acosta-Rodriguez et al., 2007; Wilson et al., 2007). CCR6 orchestrates the homing of Th17 cells to different tissues, such as gut, lung, and inflamed skin, via binding to CCL20 (Schutyser et al., 2003), whereas signaling via IL-23R induces IL-17A production by and proliferation of Th17 cells (Wilson et al., 2007). As mentioned previously, Th17 cells also express CCR4, the chemokine receptor highly expressed on Th2 cells, which facilitates their homing to the skin. Other surface markers expressed by Th17 cells are CD161 and IL-1 receptor (IL-1R). CD161 expression is found on all IL-17–producing and RORγt-expressing CD4$^+$ T cells in human blood, although the function of this marker on this subset remains unknown (Maggi et al., 2010). Of note, also small numbers of Th1 and Th2 cells are CD161$^+$ cells; therefore, identification of Th17 cells based on solely CD161 expression is not possible. IL-1R is also expressed by IL-17–producing CD4$^+$ T cells, yet it does not define all IL-17 producing CD4$^+$ T cells (Lee et al., 2010). Signaling through IL-1R by binding its ligand IL-1β induces an increase in IL-17 production, indicating a role in the effector functions of Th17 cells. In addition, the IL-1R$^+$ cells have increased levels of CCR4, CCR6, IL-23R, and CD161 than IL-1R$^-$ cells (Lee et al., 2010).

5.4 Th1* Cells

Th1* cells (also referred to as Th1/Th17 or nonclassical Th1 cells) are cells that have characteristics of both Th1 and Th17 cells; they express both Th1 and Th17 master transcription factors T-bet and RORγt, respectively (Maggi et al., 2012). Despite the expression of RORγt, Th1* cells do not produce IL-17. The origin of Th1* cells still need to be defined, although there are several possibilities suggested (Sallusto, 2016). One possibility is that these cells derive from Th17 cells that convert to Th1* cells under the influence of IL-12, TNFα, and/or IL-1β. Another possibility is that Th1* cells differentiate directly from CD4$^+$ T_N cells in a RORγt-dependent fashion. Interestingly, the contribution of Th1* and classical Th1 cells in the immune response is dependent on the pathogen. For example, memory cells specific for influenza were found in both the Th1* and Th1 compartments, whereas those specific for *Mycobacterium tuberculosis* mainly have a Th1* phenotype (Lindestam Arlehamn et al., 2013). Th1* cells are characterized by coexpression of the

Th1-associated molecule CXCR3 and the Th17-associated CCR6, CD161, and IL-1R (Maggi et al., 2012).

5.5 Th22 Cells

Th22 cell development depends on the expression of master transcription factor aryl hydrocarbon receptor (AHR), although the cytokines responsible for this development are currently unknown. Th22 cells are characterized by high production of IL-22, without IFNγ, IL-4, or IL-17 production (Trifari et al., 2009; Eyerich et al., 2009). Similar to Th17 cells, Th22 cells express chemokine receptors CCR4 and CCR6; however, they can be distinguished from Th17 cells by the expression of CCR10 (Trifari et al., 2009). Both CCR4 and CCR10 are homing receptors for the skin; therefore, unsurprisingly, Th22 cells are mainly found in the skin where they play a role in wound healing (Eyerich et al., 2009). In addition, in many inflammatory skin diseases, Th22 cells have been identified within the inflamed skin (Fujita, 2013).

5.6 Th9 Cells

One of the most recently recognized Th subset is the Th9 subset. The differentiation of $CD4^+$ T_N cells into Th9 cells requires the presence of IL-4 and TGF-β, while also IL-2, IL-1, IL-33, and IL-25 can induce IL-9 production by T cells [reviewed in the study by Jia and Wu (2014)]. The Th9 subset is dependent on master transcription factor PU.1 and is characterized by expression of high levels of IL-9 (Chang et al., 2010). Besides PU.1, other transcription factors, such as interferon regulatory factor 4 (IRF4), IRF8, and basic leucine zipper ATF-like transcription factor (BATF), are crucial for Th9 development (Humblin et al., 2017). Th9 cells have been shown to facilitate antitumor immune responses (Purwar et al., 2012) and immunity against intestinal worms (Licona-Limon et al., 2013). Conversely, Th9 cells also contribute to immunopathology in autoimmunity and allergy (Chang et al., 2010). Plasticity of the Th9 cells has been found; in the presence of Th2 polarizing cytokines, a large portion of Th9 cells acquired the Th2 phenotype, although in the presence of Th1 or Th17 polarizing cytokines they kept their Th9 phenotype (Sakaguchi et al., 2010). Unfortunately, no surface markers to define Th9 cells have been identified.

5.7 Follicular T Helper Cells

Tfh cells are T cells specialized in providing help to B cells. They are crucial for germinal center (GC) formation, antibody class switching, affinity maturation, and development of high affinity antibodies and memory B cells (Hams et al., 2011; Good-Jacobson et al., 2010; Kaji et al., 2012; Victora et al., 2010). The master transcription factor of Tfh cells is BCL6, and they secrete high levels of IL-4 and IL-21. On the surface marker level, Tfh cells are generally characterized by the expression of CXCR5, chemokine receptor for CXCL13 that is highly expressed in B cell follicles, inducible T cell costimulator (ICOS), and PD-1 (Schaerli et al., 2000; Ma et al., 2009; Haynes et al., 2007; Kroenke et al., 2012; Wong et al., 2015; Rasheed et al., 2006). Moreover, Tfh cells express members of the signaling lymphocyte activation molecule (SLAM) family of surface receptors, Ly9 (SLAMF3 or CD229), CD84 (SLAMF5), and SLAMF6 (NTB-A), which signal through transcription factor SLAM-associated protein (SAP) (Ma et al., 2009) and are involved in the interaction between Tfh cells and B cells (Schwartzberg et al., 2009).

The differentiation of Tfh cells occurs in several steps [reviewed in the studies by Suh (2015) and Schmitt et al. (2014)], and during this differentiation, the cells express different surface markers. First, $CD4^+$ T_N cells are activated by DCs within the T cell zone and differentiate toward a precursor Tfh (pre-Tfh) phenotype; they upregulate CXCR5, PD-1, and ICOS and downregulate CCR7 and PSGL-1 (Haynes et al., 2007; Suh, 2015; Kroenke et al., 2012). Second, the pre-Tfh cells migrate toward the T—B border orchestrated by their simultaneous increased expression of CXCR5, a homing receptor for the B cell zone, and decreased expression of CCR7 and PSGL-1, respectively, a homing receptor for the T cell zone and an anchor for T cells to the T cell zone extracellular matrix. Third, in the T—B border, the pre-Tfh cells make a stable conjugation with B cells that share the same specificity. This stable conjugation, which is promoted by the SLAM family receptors expressed on the pre-Tfh cells, leads to the migration of the pre-Tfh into the GC. Furthermore, it leads to the differentiation of the pre-Tfh into mature germinal center Tfh (GC-Tfh) cells. GC-Tfh cells express increased levels of CXCR5, PD-1, and ICOS compared with pre-Tfh cells. In addition, GC-Tfh cells express high levels of CXCR4, CD69, and inhibitory receptors BTLA and CD200 (OX2), while they express low levels of CD127, CXCR3, and CCR6 (Ma et al., 2009; Kroenke et al., 2012; Lim and Kim, 2007; Schaerli et al., 2000), as was shown in human tonsils. A subset of GC-Tfh cells expresses also CD57 (Kim et al., 2001; Ma et al., 2009; Rasheed et al., 2006; Lim and Kim, 2007). This was confirmed by a recent study that applied high-dimensional mass cytometry analysis to identify heterogeneity of human Tfh cells (Wong et al., 2015). In this study,

CD57 expressing Tfh cells were identified as the last stage of GC-Tfh cell progression (Wong et al., 2015).

Besides in SLOs, Tfh cells can also be identified in blood (circulating Tfh [cTfh]) (Forster et al., 1994). The relationship between the cTfh and GC-Tfh cells is still a subject of discussion, which is mainly due to the difference in phenotype between cTfh and GC-Tfh cells. For example, in contrast to GC-Tfh cells, cTfh cells express low or no expression of ICOS, PD-1, and CD69 (Chevalier et al., 2011; Locci et al., 2013; He et al., 2013; Schmitt et al., 2014; Morita et al., 2011; Ma et al., 2009). Three major subsets of cTfh cells can be identified based on ICOS, PD-1, and CCR7 expression (Chevalier et al., 2011; Locci et al., 2013; He et al., 2013). Almost 70% of the cTfh cells are $ICOS^{-}PD\text{-}1^{-}CCR7^{hi}$ (although the level of CCR7 expression is still lower than on $CXCR5^{-}$ $CD4^{+}$ T cells), around 30% are $ICOS^{-}PD\text{-}1^{+}CCR7^{int}$, and less than 1% is $ICOS^{+}PD\text{-}1^{hi}CCR7^{lo}$. The study of He et al. showed that the $PD\text{-}1^{-}CCR7^{hi}$ and $PD\text{-}1^{hi}CCR7^{lo}$ cTfh subsets are primarily generated before T cells differentiate into mature GC-Tfh cells because patients diagnosed with C-linked lymphoproliferative disease (lacking SAP expression which is essential in the maturation of GC-Tfh from pre-Tfh cells) had unaltered levels of the two cTfh subsets (He et al., 2013). Moreover, this study showed using a mouse model that an increase of $PD\text{-}1^{hi}CCR7^{lo}$ cTfh cells reflects active Tfh cell differentiation in SLOs (He et al., 2013).

Within each of these three cTfh subsets defined by ICOS, PD-1, and CCR7 expression, other subsets can be identified based on CXCR3 and CCR6 expression: $CXCR3^{+}CCR6^{-}$ (Tfh1) cells, $CXCR3^{-}CCR6^{-}$ (Tfh2) cells, and $CXCR3^{-}CCR6^{+}$ (Tfh17) cells (Morita et al., 2011). These cells express respectively T-bet and IFNγ (hallmarks Th1 cells), GATA-3 and IL-4, IL-5 and IL-13 (hallmarks Th2 cells), and RORγT and IL-17A and IL-22 (hallmarks Th17 cells). Functional differences have been described between these subsets. Within the Tfh2 and Tfh17, the $ICOS^{-}PD\text{-}1^{+}CCR7^{int}$ cells, in contrast to the $ICOS^{-}PD\text{-}1^{-}CCR7^{hi}$ cells and both subsets within Tfh1, are efficient in inducing differentiation of memory B cells into IgG-secreting plasma cells (Locci et al., 2013). Moreover, these cells have a gene expression profile that resembles the tonsillar GC-Tfh cells (Locci et al., 2013).

5.8 Regulatory T Cells

Tregs are essential in maintaining immune homeostasis, preventing autoimmunity, immunopathology, and allergy and can exert their immunosuppressive functions via multiple mechanisms [reviewed in the studies by Vignali et al. (2008) and Sakaguchi et al. (2010)]. These include secretion of immunosuppressive soluble factors, cytolysis of the target cells, metabolic disruption, and modulation of DCs. Tregs can either originate from the thymus, indicated as natural Tregs (nTregs), or are generated in the periphery, named induced Tregs (iTregs). The nTregs and iTregs are quite similar in their surface marker expression, making it difficult to distinguish between them. Recent studies in mice, however, have shown that Nrp1 (CD304) could be a reliable surface marker for nTregs because the Nrp1^{+} population also expresses transcription factor Helios that is considered an nTreg marker (Weiss et al., 2012). Unfortunately, there are conflicting data on whether Nrp1 is expressed by human Tregs (Milpied et al., 2009; Battaglia et al., 2008); therefore, a reliable surface marker to distinguish between human nTregs and iTregs remains unknown.

Master transcription factor FoxP3 characterizes nTregs but not iTregs generated after activation of T_N in the presence of TGFβ. Both Treg subsets require IL-2 but not IL-7 for their survival, which is in line with the constitutively high expression of the high-affinity receptor for IL-2 (IL-2Rα or CD25) and low expression of the IL-7 receptor α chain (IL-7Rα or CD127). The combination of CD25^{hi} and CD127^{lo} is commonly used to define human Tregs (Sakaguchi et al., 2010). However, CD127 is also downregulated during activation of FoxP3^{-} CD4^{+} T cells (Aerts et al., 2008), which could interfere with identifying Tregs purely especially in inflammatory conditions. An additional marker for defining Tregs is tumor necrosis factor receptor 2 (TNFR2 or CD120B), a receptor for TNF, which is constitutively expressed by Tregs (van Mierlo et al., 2008; Chen et al., 2010). TNFR2 has been shown in mice to be essential in stabilizing the Treg phenotype in inflammatory conditions by sustaining FoxP3 expression (Chen et al., 2010). In addition, Tregs express the inhibitory receptor CTLA-4 (Dong et al., 2013), which plays a role in their suppressive functions, and GITR (Ronchetti et al., 2015), which, at least in mice, is involved in Treg expansion.

Within the FoxP3^{+} population in human peripheral blood, at least two distinct Treg subsets can be identified, i.e., FoxP3^{lo}CD45RA^{+} resting Tregs (rTregs) and FoxP3^{br}CD45RA^{-} activated Tregs (aTregs) (Miyara et al., 2009; Dong et al., 2013; Jonuleit et al., 2001). Although the rTregs possess suppressive activity of effector T cell responses, the aTregs are the most potent in this respect (Miyara et al., 2009; Dong et al., 2013; Lin et al., 2013; Jonuleit et al., 2001). Specifically the aTregs have been found to be enriched in several cancers, including colorectal cancer (Lin et al., 2013) and lung cancer (Kurose et al., 2015). In addition to these two Treg subsets, another FoxP3-expressing subset could be identified in the peripheral blood, i.e., $FoxP3^{lo}CD45RA^{-}$ cells; these cells, however, do not possess suppressive activity, and a subset of these cells produces proinflammatory cytokines, e.g., IL-17, IFNγ, and IL-4 (Aerts et al., 2008; Dong et al., 2013). Besides the difference in CD45RA expression by the rTreg and aTreg subsets, there are more phenotypical differences. The aTregs express an increased level of CTLA-4 and express HLA-DR, ICOS, PD-1, CD15s (Sialyl Lewis x),and GARP (Wang et al., 2008; Probst-Kepper et al., 2009; Sakaguchi et al., 2010; Miyara et al., 2015). CD15s is a ligand for selectins that promotes binding to endothelial cells and subsequent migration into the tissues (Polley et al., 1991), whereas GARP is important in the maintenance of Tregs (Sun et al., 2016). Other surface markers are expressed by subpopulations of the aTregs. CCR6 is expressed on >50% of aTregs, and these $CCR6^{+}$ Tregs are more potent in suppression than $CCR6^{-}$ Tregs (Lee et al., 2017). CD39, an ectonucleotidase that degrades ATP to adenosine, is also expressed on a subset of Tregs with a highly suppressive capacity (Borsellino et al., 2007). Interestingly, $CD39^{+}$ Tregs also coexpress CCR6 and HLA-DR. LAG-3, an inhibitory receptor, is expressed by about 3% of Tregs, yet the abundance of LAG-3^{+} Tregs increases to around 10% in lymph nodes and within tumor-infiltrating lymphocytes of cancer patients (Camisaschi et al., 2010).

An additional subset of $CD4^{+}$ T cells originating from Tregs but sharing characteristics with Tfh cells, namely, CXCR5 and ICOS expression, has been recently identified and has been named T follicular regulatory (Tfr) (Sage and Sharpe, 2016). Specifically, these Tfr cells suppress Tfh-mediated B cell activation and antibody production. Similar as conventional Tfh cells, there are multiple phases of Tfr differentiation, including first the interaction with DCs, imprinting the Tfr program, followed by maturation on interaction with B cells in the follicle/GC. They can be distinguished from conventional Tfh cells by the expression of transcription factor FoxP3.

5.9 Type 1 Regulatory Cells

Tr1 cells define another $CD4^{+}$ T cell subset with potent immunosuppressive capacities. These cells are distinguishable from Tregs by the lack of constitutive FoxP3 and CD25 expression (Vieira et al., 2004; Brun et al., 2009). Tr1 cells are defined by the secretion of high levels of IL-10 with clear distinct secretion kinetics from all other IL-10 producing cells (Cavani et al., 2000). Tr1 cells express inhibitory receptor LAG-3 and integrin

CD49b, which individually are also expressed on other CD4$^+$ T cell subsets (Boisvert et al., 2010; Camisaschi et al., 2010; Gagliani et al., 2013), but the combination is unique for Tr1 cells (Gagliani et al., 2013). In addition, low expression of both CD25 and CD127 identifies the CD4$^+$ T cell subset enriched for IL-10–producing cells (Tr1 like) (Facciotti et al., 2016).

Tr1 cells suppress immune responses mainly via the secretion of IL-10 and TGF-β. Similar to Tregs, Tr1 cells express high levels of inhibitory receptors CTLA-4, PD-1, and GITR, and costimulatory receptor ICOS (Gregori et al., 2012; Ronchetti et al., 2015), which can directly modulate T cell functions. In addition, Tr1 cells secrete high levels of granzyme B, causing lysis of specifically cells of myeloid origin (Magnani et al., 2011). This lysis is MHC class I dependent and is mediated by the binding of adhesion molecules LFA-1, CD2, and CD266 on Tr1 cells with their respective receptors CD54, CD58, and CD155 on myeloid cells (Magnani et al., 2011). A subset of Tr1 cells also expresses ectonucleotidases CD39 and CD73, which degrade ATP to adenosine, another mechanism by which immune responses can be modulated (Mandapathil et al., 2010).

5.10 CD4$^+$ Cytotoxic T Lymphocytes

CD4$^+$ cytotoxic T lymphocytes (CD4$^+$CTLs) are able to kill target cells in a similar fashion as CD8$^+$ CTLs, yet in an MHC class II restricted manner (Haigh et al., 2008; Klucar et al., 2008). CD4$^+$ CTLs express master transcription factor ThPOX and high levels of granzyme B and perforin (Workman et al., 2014; Lin et al., 2014). Their MHC class II restriction suggests that CD4$^+$ CTLs can only function in the immunosurveillance of APCs. Nonetheless, CD4$^+$ CTLs have been found to play a role in both antiviral immunity, highlighted by their increased frequency in peripheral blood after viral infection, such as CMV and HIV (van Leeuwen et al., 2004; Zaunders et al., 2004; Nemes et al., 2010) and antitumor immunity (Matsuzaki et al., 2015). This could be due to the induction of MHC class II expression on cells that usually do not express MHC class II molecules, such as tumor cells or infected epithelial cells. For example, irradiation or IFNγ treatment induces MHC class II expression on epithelial or tumor cells (Xie et al., 2010; Quezada et al., 2010). Also, lung epithelial cells have been found to express MHC class II molecules after infection with parainfluenza or *Mycobacterium tuberculosis* (Gao et al., 1999; Debbabi et al., 2005).

CD4$^+$ CTLs are mainly confined within the T_{TE} compartment, identified by CD45RA and KLRG1 expression and lack of CCR7 expression (Patil et al., 2018). CD4$^+$ CTLs share multiple surface markers with

$CD8^+$ CTLs or NK cells. These markers include lysosome-associated membrane glycoproteins, LAMP-1 (CD107a) and LAMP-2 (CD107b), which are expressed on the surface after degranulation (Lin et al., 2014; Nemes et al., 2010). Moreover, $CD4^+$ CTLs express natural killer group 2 members KLRC1 (NKG2A) and KLRK1 (NKG2D) (Saez-Borderias et al., 2006), and both $KLRC1^+$ and $KLRK1^+$ $CD4^+$ CTLs have been identified at effector sites (Graham et al., 2007; Saez-Borderias et al., 2006). The ligand of KLRC1, which forms dimers with KLRD1 (CD94), is the nonclassical MHC class I molecule HLA-E. Interestingly, for $CD8^+$ T cells, KLRC1 signaling has been shown to contribute to T cell dysfunction during chronic infections, most likely due to its suppression of activation signaling processes (Moser et al., 2002). Signaling via KLRK1, which forms homodimers, on NK cells is essential for cell polarization and degranulation (Belting et al., 2015), and its presence on $CD4^+$ CTLs suggests similar function in this subset. Another surface marker expressed on both $CD8^+$ and $CD4^+$ CTLs is MHC class I–restricted T cell-associated molecule (CRTAM) (Takeuchi et al., 2016). CTRAM induces Eomes, IFNγ, granzyme B, and perforin expression by $CD4^+$ CTLs (Takeuchi et al., 2016). However, CRTAM is only upregulated on activation making it, together with its dim expression, a limited marker for detection of $CD4^+$ CTLs in vivo or ex vivo. Low expression of both CD27 and CD28 is also used to identify $CD4^+$ CTLs (Appay et al., 2002b; van Leeuwen et al., 2004); however, $CD4^+$ CTLs have also been found within the $CD27^+$ population (Raveney et al., 2015). Therefore, although gating for low CD27 and CD28 expression allows for enrichment, it will likely not identify all $CD4^+$ CTLs. Currently, the best options for identification of $CD4^+$ CTLs are to stain for granzyme B and perforin intracellular or to stimulate the cells with peptides and stain for degranulation markers (CD107a and CD107b).

Recently, precursors of $CD4^+$ CTLs have been identified in blood of healthy controls within the T_{TE} compartment. These cells display lower expression levels of the surface and effector molecules described previously. Moreover, these precursors can be distinguished from the mature $CD4^+$ CTLs by high expression of CD127 (Patil et al., 2018).

6. CONCLUSION

In the recent years, increasing numbers of human T cell subsets have been identified, and it has become clear that some cells in the $CD4^+$ and

CD8+ T cell compartments exist in plastic states and can differentiate into multiple subsets. This shows high heterogeneity within both the CD4+ and CD8+ T cell compartments, although still there is no complete understanding of the complexity. Advances in the development of high-content technologies at the single-cell level such as RNA sequencing (scRNA-seq) or proteomics is essential in unraveling this complexity and will most likely redefine the phenotypic classification and functional landscape of T cells. Yet, the identification of markers expressed on the surface of the different T cell subsets, though differences are also reflected at the gene expression, chromatin, metabolic and functional level, remains crucial for live purification and subsequent (functional) characterization of the cells.

ACKNOWLEDGMENTS

The authors wish to thank the members of the Laboratory of Translational Immunology for helpful discussion. This work has been supported by grants from the Associazione Italiana per la Ricerca sul Cancro (AIRC; Investigator Grant No. 20607), the Italian Ministry of Health (Bando Giovani Ricercatori GR-2011-02347324), and Humanitas Clinical and Research Center to E.L. J.B. is a recipient of the "Fondo di beneficenza Intesa San Paolo" fellowship from AIRC. K.P. is a recipient of the 2018 Fondazione Umberto Veronesi postdoctoral fellowship. E.L. is an International Society for the Advancement of Cytometry (ISAC) Marylou Ingram scholar.

REFERENCES

Acosta, J.C., O'loghlen, A., Banito, A., Guijarro, M.V., Augert, A., Raguz, S., Fumagalli, M., Da Costa, M., Brown, C., Popov, N., Takatsu, Y., Melamed, J., D'adda Di Fagagna, F., Bernard, D., Hernando, E., Gil, J., 2008. Chemokine signaling via the CXCR2 receptor reinforces senescence. Cell 133, 1006–1018.

Acosta-Rodriguez, E.V., Rivino, L., Geginat, J., Jarrossay, D., Gattorno, M., Lanzavecchia, A., Sallusto, F., Napolitani, G., 2007. Surface phenotype and antigenic specificity of human interleukin 17-producing T helper memory cells. Nat. Immunol. 8, 639–646.

Aerts, N.E., Dombrecht, E.J., Ebo, D.G., Bridts, C.H., Stevens, W.J., De Clerck, L.S., 2008. Activated T cells complicate the identification of regulatory T cells in rheumatoid arthritis. Cell. Immunol. 251, 109–115.

Agace, W.W., Higgins, J.M., Sadasivan, B., Brenner, M.B., Parker, C.M., 2000. T-lymphocyte-epithelial-cell interactions: integrin alpha(E)(CD103)beta(7), LEEP-CAM and chemokines. Curr. Opin. Cell Biol. 12, 563–568.

Aggarwal, S., Gupta, S., 1998. Increased apoptosis of T cell subsets in aging humans: altered expression of Fas (CD95), Fas ligand, Bcl-2, and Bax. J. Immunol. 160, 1627–1637.

Ahmed, R., Roger, L., Costa Del Amo, P., Miners, K.L., Jones, R.E., Boelen, L., Fali, T., Elemans, M., Zhang, Y., Appay, V., Baird, D.M., Asquith, B., Price, D.A., Macallan, D.C., Ladell, K., 2016. Human stem cell-like memory T cells are maintained in a state of dynamic flux. Cell Rep. 17, 2811–2818.

Akbar, A.N., Henson, S.M., 2011. Are senescence and exhaustion intertwined or unrelated processes that compromise immunity? Nat. Rev. Immunol. 11, 289–295.

Akbar, A.N., Terry, L., Timms, A., Beverley, P.C., Janossy, G., 1988. Loss of CD45R and gain of UCHL1 reactivity is a feature of primed T cells. J. Immunol. 140, 2171–2178.

Appay, V., Dunbar, P.R., Callan, M., Klenerman, P., Gillespie, G.M., Papagno, L., Ogg, G.S., King, A., Lechner, F., Spina, C.A., Little, S., Havlir, D.V., Richman, D.D., Gruener, N., Pape, G., Waters, A., Easterbrook, P., Salio, M., Cerundolo, V., Mcmichael, A.J., Rowland-Jones, S.L., 2002a. Memory CD8+ T cells vary in differentiation phenotype in different persistent virus infections. Nat. Med. 8, 379–385.

Appay, V., Van Lier, R.A., Sallusto, F., Roederer, M., 2008. Phenotype and function of human T lymphocyte subsets: consensus and issues. Cytometry 73, 975–983.

Appay, V., Zaunders, J.J., Papagno, L., Sutton, J., Jaramillo, A., Waters, A., Easterbrook, P., Grey, P., Smith, D., Mcmichael, A.J., Cooper, D.A., Rowland-Jones, S.L., Kelleher, A.D., 2002b. Characterization of CD4(+) CTLs ex vivo. J. Immunol. 168, 5954–5958.

Azuma, M., Phillips, J.H., Lanier, L.L., 1992. CD28 co-stimulation of T-cell-mediated cytotoxicity. Int. J. Cancer Suppl. 7, 33–35.

Barber, D.L., Wherry, E.J., Masopust, D., Zhu, B., Allison, J.P., Sharpe, A.H., Freeman, G.J., Ahmed, R., 2006. Restoring function in exhausted CD8 T cells during chronic viral infection. Nature 439, 682–687.

Battaglia, A., Buzzonetti, A., Monego, G., Peri, L., Ferrandina, G., Fanfani, F., Scambia, G., Fattorossi, A., 2008. Neuropilin-1 expression identifies a subset of regulatory T cells in human lymph nodes that is modulated by preoperative chemoradiation therapy in cervical cancer. Immunology 123, 129–138.

Belting, L., Homberg, N., Przewoznik, M., Brenner, C., Riedel, T., Flatley, A., Polic, B., Busch, D.H., Rocken, M., Mocikat, R., 2015. Critical role of the NKG2D receptor for NK cell-mediated control and immune escape of B-cell lymphoma. Eur. J. Immunol. 45, 2593–2601.

Biasco, L., Scala, S., Basso Ricci, L., Dionisio, F., Baricordi, C., Calabria, A., Giannelli, S., Cieri, N., Barzaghi, F., Pajno, R., Al-Mousa, H., Scarselli, A., Cancrini, C., Bordignon, C., Roncarolo, M.G., Montini, E., Bonini, C., Aiuti, A., 2015. In vivo tracking of T cells in humans unveils decade-long survival and activity of genetically modified T memory stem cells. Sci. Transl. Med. 7, 273ra13.

Blackburn, S.D., Shin, H., Freeman, G.J., Wherry, E.J., 2008. Selective expansion of a subset of exhausted CD8 T cells by alphaPD-L1 blockade. Proc. Natl. Acad. Sci. U.S.A. 105, 15016–15021.

Blackburn, S.D., Shin, H., Haining, W.N., Zou, T., Workman, C.J., Polley, A., Betts, M.R., Freeman, G.J., Vignali, D.A., Wherry, E.J., 2009. Coregulation of CD8+ T cell exhaustion by multiple inhibitory receptors during chronic viral infection. Nat. Immunol. 10, 29–37.

Blaser, C., Kaufmann, M., Pircher, H., 1998. Virus-activated CD8 T cells and lymphokine-activated NK cells express the mast cell function-associated antigen, an inhibitory C-type lectin. J. Immunol. 161, 6451–6454.

Boisvert, M., Chetoui, N., Gendron, S., Aoudjit, F., 2010. Alpha2beta1 integrin is the major collagen-binding integrin expressed on human Th17 cells. Eur. J. Immunol. 40, 2710–2719.

Borsellino, G., Kleinewietfeld, M., Di Mitri, D., Sternjak, A., Diamantini, A., Giometto, R., Hopner, S., Centonze, D., Bernardi, G., Dell'acqua, M.L., Rossini, P.M., Battistini, L., Rotzschke, O., Falk, K., 2007. Expression of ectonucleotidase CD39 by Foxp3+ Treg cells: hydrolysis of extracellular ATP and immune suppression. Blood 110, 1225–1232.

Borthwick, N.J., Lowdell, M., Salmon, M., Akbar, A.N., 2000. Loss of CD28 expression on CD8(+) T cells is induced by IL-2 receptor gamma chain signalling cytokines and type I IFN, and increases susceptibility to activation-induced apoptosis. Int. Immunol. 12, 1005–1013.

Bou Ghanem, E.N., Nelson, C.C., D'orazio, S.E., 2011. T cell-intrinsic factors contribute to the differential ability of CD8+ T cells to rapidly secrete IFN-gamma in the absence of antigen. J. Immunol. 186, 1703–1712.
Brahmer, J.R., Tykodi, S.S., Chow, L.Q., Hwu, W.J., Topalian, S.L., Hwu, P., Drake, C.G., Camacho, L.H., Kauh, J., Odunsi, K., Pitot, H.C., Hamid, O., Bhatia, S., Martins, R., Eaton, K., Chen, S., Salay, T.M., Alaparthy, S., Grosso, J.F., Korman, A.J., Parker, S.M., Agrawal, S., Goldberg, S.M., Pardoll, D.M., Gupta, A., Wigginton, J.M., 2012. Safety and activity of anti-PD-L1 antibody in patients with advanced cancer. N. Engl. J. Med. 366, 2455–2465.
Braud, V.M., Allan, D.S., O'callaghan, C.A., Soderstrom, K., D'andrea, A., Ogg, G.S., Lazetic, S., Young, N.T., Bell, J.I., Phillips, J.H., Lanier, L.L., Mcmichael, A.J., 1998. HLA-E binds to natural killer cell receptors CD94/NKG2A, B and C. Nature 391, 795–799.
Brenchley, J.M., Karandikar, N.J., Betts, M.R., Ambrozak, D.R., Hill, B.J., Crotty, L.E., Casazza, J.P., Kuruppu, J., Migueles, S.A., Connors, M., Roederer, M., Douek, D.C., Koup, R.A., 2003. Expression of CD57 defines replicative senescence and antigen-induced apoptotic death of CD8+ T cells. Blood 101, 2711–2720.
Brodie, T., Brenna, E., Sallusto, F., 2013. OMIP-018: chemokine receptor expression on human T helper cells. Cytometry 83, 530–532.
Brun, V., Bastian, H., Neveu, V., Foussat, A., 2009. Clinical grade production of IL-10 producing regulatory Tr1 lymphocytes for cell therapy of chronic inflammatory diseases. Int. Immunopharmacol. 9, 609–613.
Butcher, E.C., Picker, L.J., 1996. Lymphocyte homing and homeostasis. Science 272, 60–66.
Camisaschi, C., Casati, C., Rini, F., Perego, M., De Filippo, A., Triebel, F., Parmiani, G., Belli, F., Rivoltini, L., Castelli, C., 2010. LAG-3 expression defines a subset of CD4(+)CD25(high)Foxp3(+) regulatory T cells that are expanded at tumor sites. J. Immunol. 184, 6545–6551.
Campbell, J.J., Brightling, C.E., Symon, F.A., Qin, S., Murphy, K.E., Hodge, M., Andrew, D.P., Wu, L., Butcher, E.C., Wardlaw, A.J., 2001. Expression of chemokine receptors by lung T cells from normal and asthmatic subjects. J. Immunol. 166, 2842–2848.
Campisi, J., 2005. Senescent cells, tumor suppression, and organismal aging: good citizens, bad neighbors. Cell 120, 513–522.
Cavani, A., Nasorri, F., Prezzi, C., Sebastiani, S., Albanesi, C., Girolomoni, G., 2000. Human CD4+ T lymphocytes with remarkable regulatory functions on dendritic cells and nickel-specific Th1 immune responses. J. Invest. Dermatol. 114, 295–302.
Chang, H.C., Sehra, S., Goswami, R., Yao, W., Yu, Q., Stritesky, G.L., Jabeen, R., Mckinley, C., Ahyi, A.N., Han, L., Nguyen, E.T., Robertson, M.J., Perumal, N.B., Tepper, R.S., Nutt, S.L., Kaplan, M.H., 2010. The transcription factor PU.1 is required for the development of IL-9-producing T cells and allergic inflammation. Nat. Immunol. 11, 527–534.
Chen, X., Subleski, J.J., Hamano, R., Howard, O.M., Wiltrout, R.H., Oppenheim, J.J., 2010. Co-expression of TNFR2 and CD25 identifies more of the functional CD4+FOXP3+ regulatory T cells in human peripheral blood. Eur. J. Immunol. 40, 1099–1106.
Cheuk, S., Schlums, H., Gallais Serezal, I., Martini, E., Chiang, S.C., Marquardt, N., Gibbs, A., Detlofsson, E., Introini, A., Forkel, M., Hoog, C., Tjernlund, A., Michaelsson, J., Folkersen, L., Mjosberg, J., Blomqvist, L., Ehrstrom, M., Stahle, M., Bryceson, Y.T., Eidsmo, L., 2017. CD49a expression defines tissue-resident CD8(+) T cells poised for cytotoxic function in human skin. Immunity 46, 287–300.
Chevalier, N., Jarrossay, D., Ho, E., Avery, D.T., Ma, C.S., Yu, D., Sallusto, F., Tangye, S.G., Mackay, C.R., 2011. CXCR5 expressing human central memory CD4 T cells and their relevance for humoral immune responses. J. Immunol. 186, 5556–5568.

Chiu, W.K., Fann, M., Weng, N.P., 2006. Generation and growth of CD28nullCD8+ memory T cells mediated by IL-15 and its induced cytokines. J. Immunol. 177, 7802–7810.
Christie, D., Zhu, J., 2014. Transcriptional regulatory networks for CD4 T cell differentiation. Curr. Top. Microbiol. Immunol. 381, 125–172.
Cieri, N., Camisa, B., Cocchiarella, F., Forcato, M., Oliveira, G., Provasi, E., Bondanza, A., Bordignon, C., Peccatori, J., Ciceri, F., Lupo-Stanghellini, M.T., Mavilio, F., Mondino, A., Bicciato, S., Recchia, A., Bonini, C., 2013. IL-7 and IL-15 instruct the generation of human memory stem T cells from naive precursors. Blood 121, 573–584.
Clark, R.A., Chong, B., Mirchandani, N., Brinster, N.K., Yamanaka, K., Dowgiert, R.K., Kupper, T.S., 2006. The vast majority of CLA+ T cells are resident in normal skin. J. Immunol. 176, 4431–4439.
Corsini, N.S., Sancho-Martinez, I., Laudenklos, S., Glagow, D., Kumar, S., Letellier, E., Koch, P., Teodorczyk, M., Kleber, S., Klussmann, S., Wiestler, B., Brustle, O., Mueller, W., Gieffers, C., Hill, O., Thiemann, M., Seedorf, M., Gretz, N., Sprengel, R., Celikel, T., Martin-Villalba, A., 2009. The death receptor CD95 activates adult neural stem cells for working memory formation and brain repair. Cell Stem Cell 5, 178–190.
Cosmi, L., Annunziato, F., Galli, M.I.G., Maggi, R.M.E., Nagata, K., Romagnani, S., 2000. CRTH2 is the most reliable marker for the detection of circulating human type 2 Th and type 2 T cytotoxic cells in health and disease. Eur. J. Immunol. 30, 2972–2979.
Cossarizza, A., Ortolani, C., Paganelli, R., Barbieri, D., Monti, D., Sansoni, P., Fagiolo, U., Castellani, G., Bersani, F., Londei, M., Franceschi, C., 1996. CD45 isoforms expression on CD4+ and CD8+ T cells throughout life, from newborns to centenarians: implications for T cell memory. Mech. Ageing Dev. 86, 173–195.
Crawford, A., Angelosanto, J.M., Kao, C., Doering, T.A., Odorizzi, P.M., Barnett, B.E., Wherry, E.J., 2014. Molecular and transcriptional basis of CD4(+) T cell dysfunction during chronic infection. Immunity 40, 289–302.
Crespo, J., Sun, H., Welling, T.H., Tian, Z., Zou, W., 2013. T cell anergy, exhaustion, senescence, and stemness in the tumor microenvironment. Curr. Opin. Immunol. 25, 214–221.
Crompton, J.G., Narayanan, M., Cuddapah, S., Roychoudhuri, R., Ji, Y., Yang, W., Patel, S.J., Sukumar, M., Palmer, D.C., Peng, W., Wang, E., Marincola, F.M., Klebanoff, C.A., Zhao, K., Tsang, J.S., Gattinoni, L., Restifo, N.P., 2016. Lineage relationship of CD8(+) T cell subsets is revealed by progressive changes in the epigenetic landscape. Cell. Mol. Immunol. 13, 502–513.
Crompton, J.G., Sukumar, M., Roychoudhuri, R., Clever, D., Gros, A., Eil, R.L., Tran, E., Hanada, K., Yu, Z., Palmer, D.C., Kerkar, S.P., Michalek, R.D., Upham, T., Leonardi, A., Acquavella, N., Wang, E., Marincola, F.M., Gattinoni, L., Muranski, P., Sundrud, M.S., Klebanoff, C.A., Rosenberg, S.A., Fearon, D.T., Restifo, N.P., 2015. Akt inhibition enhances expansion of potent tumor-specific lymphocytes with memory cell characteristics. Cancer Res. 75, 296–305.
Day, C.L., Kaufmann, D.E., Kiepiela, P., Brown, J.A., Moodley, E.S., Reddy, S., Mackey, E.W., Miller, J.D., Leslie, A.J., Depierres, C., Mncube, Z., Duraiswamy, J., Zhu, B., Eichbaum, Q., Altfeld, M., Wherry, E.J., Coovadia, H.M., Goulder, P.J., Klenerman, P., Ahmed, R., Freeman, G.J., Walker, B.D., 2006. PD-1 expression on HIV-specific T cells is associated with T-cell exhaustion and disease progression. Nature 443, 350–354.
Debbabi, H., Ghosh, S., Kamath, A.B., Alt, J., Demello, D.E., Dunsmore, S., Behar, S.M., 2005. Primary type II alveolar epithelial cells present microbial antigens to antigen-specific CD4+ T cells. Am. J. Physiol. Lung Cell Mol. Physiol. 289, L274–L279.
Derhovanessian, E., Maier, A.B., Hahnel, K., Beck, R., De Craen, A.J., Slagboom, E.P., Westendorp, R.G., Pawelec, G., 2011. Infection with cytomegalovirus but not herpes simplex virus induces the accumulation of late-differentiated CD4+ and CD8+ T-cells in humans. J. Gen. Virol. 92, 2746–2756.

Di Benedetto, S., Derhovanessian, E., Steinhagen-Thiessen, E., Goldeck, D., Muller, L., Pawelec, G., 2015. Impact of age, sex and CMV-infection on peripheral T cell phenotypes: results from the Berlin BASE-II Study. Biogerontology 16, 631–643.

Dong, S., Maiella, S., Xhaard, A., Pang, Y., Wenandy, L., Larghero, J., Becavin, C., Benecke, A., Bianchi, E., Socie, G., Rogge, L., 2013. Multiparameter single-cell profiling of human CD4+FOXP3+ regulatory T-cell populations in homeostatic conditions and during graft-versus-host disease. Blood 122, 1802–1812.

Dunne, P.J., Belaramani, L., Fletcher, J.M., Fernandez De Mattos, S., Lawrenz, M., Soares, M.V., Rustin, M.H., Lam, E.W., Salmon, M., Akbar, A.N., 2005. Quiescence and functional reprogramming of Epstein-Barr virus (EBV)-specific CD8+ T cells during persistent infection. Blood 106, 558–565.

Duraiswamy, J., Ibegbu, C.C., Masopust, D., Miller, J.D., Araki, K., Doho, G.H., Tata, P., Gupta, S., Zilliox, M.J., Nakaya, H.I., Pulendran, B., Haining, W.N., Freeman, G.J., Ahmed, R., 2011. Phenotype, function, and gene expression profiles of programmed death-1(hi) CD8 T cells in healthy human adults. J. Immunol. 186, 4200–4212.

Effros, R.B., 1997. Loss of CD28 expression on T lymphocytes: a marker of replicative senescence. Dev. Comp. Immunol. 21, 471–478.

Effros, R.B., 2004. Replicative senescence of CD8 T cells: effect on human ageing. Exp. Gerontol. 39, 517–524.

Eksteen, B., Mora, J.R., Haughton, E.L., Henderson, N.C., Lee-Turner, L., Villablanca, E.J., Curbishley, S.M., Aspinall, A.I., Von Andrian, U.H., Adams, D.H., 2009. Gut homing receptors on CD8 T cells are retinoic acid dependent and not maintained by liver dendritic or stellate cells. Gastroenterology 137, 320–329.

Eyerich, S., Eyerich, K., Pennino, D., Carbone, T., Nasorri, F., Pallotta, S., Cianfarani, F., Odorisio, T., Traidl-Hoffmann, C., Behrendt, H., Durham, S.R., Schmidt-Weber, C.B., Cavani, A., 2009. Th22 cells represent a distinct human T cell subset involved in epidermal immunity and remodeling. J. Clin. Invest. 119, 3573–3585.

Facciotti, F., Gagliani, N., Haringer, B., Alfen, J.S., Penatti, A., Maglie, S., Paroni, M., Iseppon, A., Moro, M., Crosti, M.C., Stolzel, K., Romagnani, C., Moroni, G., Ingegnoli, F., Torretta, S., Pignataro, L., Annoni, A., Russo, F., Pagani, M., Abrignani, S., Meroni, P., Flavell, R., Geginat, J., 2016. IL-10-producing forkhead box protein 3-negative regulatory T cells inhibit B-cell responses and are involved in systemic lupus erythematosus. J. Allergy Clin. Immunol. 137, 318–321 e5.

Fagnoni, F.F., Vescovini, R., Mazzola, M., Bologna, G., Nigro, E., Lavagetto, G., Franceschi, C., Passeri, M., Sansoni, P., 1996. Expansion of cytotoxic CD8+ CD28- T cells in healthy ageing people, including centenarians. Immunology 88, 501–507.

Fagnoni, F.F., Vescovini, R., Passeri, G., Bologna, G., Pedrazzoni, M., Lavagetto, G., Casti, A., Franceschi, C., Passeri, M., Sansoni, P., 2000. Shortage of circulating naive CD8(+) T cells provides new insights on immunodeficiency in aging. Blood 95, 2860–2868.

Faint, J.M., Annels, N.E., Curnow, S.J., Shields, P., Pilling, D., Hislop, A.D., Wu, L., Akbar, A.N., Buckley, C.D., Moss, P.A., Adams, D.H., Rickinson, A.B., Salmon, M., 2001. Memory T cells constitute a subset of the human CD8+CD45RA+ pool with distinct phenotypic and migratory characteristics. J. Immunol. 167, 212–220.

Fietta, P., Delsante, G., 2009. The effector T helper cell triade. Riv. Biol. 102, 61–74.

Forget, M.A., Huon, Y., Reuben, A., Grange, C., Liberman, M., Martin, J., Mes-Masson, A.M., Arbour, N., Lapointe, R., 2012. Stimulation of Wnt/ss-catenin pathway in human CD8+ T lymphocytes from blood and lung tumors leads to a shared young/memory phenotype. PLoS One 7, e41074.

Forster, R., Emrich, T., Kremmer, E., Lipp, M., 1994. Expression of the G-protein–coupled receptor BLR1 defines mature, recirculating B cells and a subset of T-helper memory cells. Blood 84, 830–840.

Fourcade, J., Sun, Z., Pagliano, O., Guillaume, P., Luescher, I.F., Sander, C., Kirkwood, J.M., Olive, D., Kuchroo, V., Zarour, H.M., 2012. CD8(+) T cells specific for tumor antigens can be rendered dysfunctional by the tumor microenvironment through upregulation of the inhibitory receptors BTLA and PD-1. Cancer Res 72, 887–896.

Fuertes Marraco, S.A., Soneson, C., Cagnon, L., Gannon, P.O., Allard, M., Abed Maillard, S., Montandon, N., Rufer, N., Waldvogel, S., Delorenzi, M., Speiser, D.E., 2015. Long-lasting stem cell-like memory CD8+ T cells with a naive-like profile upon yellow fever vaccination. Sci. Transl. Med. 7, 282ra48.

Fujita, H., 2013. The role of IL-22 and Th22 cells in human skin diseases. J. Dermatol. Sci. 72, 3–8.

Fukada, K., Sobao, Y., Tomiyama, H., Oka, S., Takiguchi, M., 2002. Functional expression of the chemokine receptor CCR5 on virus epitope-specific memory and effector CD8+ T cells. J. Immunol. 168, 2225–2232.

Gagliani, N., Magnani, C.F., Huber, S., Gianolini, M.E., Pala, M., Licona-Limon, P., Guo, B., Herbert, D.R., Bulfone, A., Trentini, F., Di Serio, C., Bacchetta, R., Andreani, M., Brockmann, L., Gregori, S., Flavell, R.A., Roncarolo, M.G., 2013. Coexpression of CD49b and LAG-3 identifies human and mouse T regulatory type 1 cells. Nat. Med. 19, 739–746.

Gao, J., De, B.P., Banerjee, A.K., 1999. Human parainfluenza virus type 3 up-regulates major histocompatibility complex class I and II expression on respiratory epithelial cells: involvement of a STAT1- and CIITA-independent pathway. J. Virol. 73, 1411–1418.

Gattinoni, L., Klebanoff, C.A., Restifo, N.P., 2012. Paths to stemness: building the ultimate antitumour T cell. Nat Rev Cancer 12, 671–684.

Gattinoni, L., Lugli, E., Ji, Y., Pos, Z., Paulos, C.M., Quigley, M.F., Almeida, J.R., Gostick, E., Yu, Z., Carpenito, C., Wang, E., Douek, D.C., Price, D.A., June, C.H., Marincola, F.M., Roederer, M., Restifo, N.P., 2011. A human memory T cell subset with stem cell-like properties. Nat. Med. 17, 1290–1297.

Gattinoni, L., Zhong, X.S., Palmer, D.C., Ji, Y., Hinrichs, C.S., Yu, Z., Wrzesinski, C., Boni, A., Cassard, L., Garvin, L.M., Paulos, C.M., Muranski, P., Restifo, N.P., 2009. Wnt signaling arrests effector T cell differentiation and generates CD8+ memory stem cells. Nat. Med. 15, 808–813.

Gebhardt, T., Whitney, P.G., Zaid, A., Mackay, L.K., Brooks, A.G., Heath, W.R., Carbone, F.R., Mueller, S.N., 2011. Different patterns of peripheral migration by memory CD4+ and CD8+ T cells. Nature 477, 216–219.

Geginat, J., Lanzavecchia, A., Sallusto, F., 2003. Proliferation and differentiation potential of human CD8+ memory T-cell subsets in response to antigen or homeostatic cytokines. Blood 101, 4260–4266.

Geginat, J., Sallusto, F., Lanzavecchia, A., 2001. Cytokine-driven proliferation and differentiation of human naive, central memory, and effector memory CD4(+) T cells. J. Exp. Med. 194, 1711–1719.

Gobert, M., Treilleux, I., Bendriss-Vermare, N., Bachelot, T., Goddard-Leon, S., Arfi, V., Biota, C., Doffin, A.C., Durand, I., Olive, D., Perez, S., Pasqual, N., Faure, C., Ray-Coquard, I., Puisieux, A., Caux, C., Blay, J.Y., Menetrier-Caux, C., 2009. Regulatory T cells recruited through CCL22/CCR4 are selectively activated in lymphoid infiltrates surrounding primary breast tumors and lead to an adverse clinical outcome. Cancer Res 69, 2000–2009.

Good-Jacobson, K.L., Szumilas, C.G., Chen, L., Sharpe, A.H., Tomayko, M.M., Shlomchik, M.J., 2010. PD-1 regulates germinal center B cell survival and the formation and affinity of long-lived plasma cells. Nat. Immunol. 11, 535–542.

Graham, C.M., Christensen, J.R., Thomas, D.B., 2007. Differential induction of CD94 and NKG2 in CD4 helper T cells. A consequence of influenza virus infection and interferon-gamma? Immunology 121, 238–247.

Gregori, S., Goudy, K.S., Roncarolo, M.G., 2012. The cellular and molecular mechanisms of immuno-suppression by human type 1 regulatory T cells. Front. Immunol. 3, 30.
Groom, J.R., Luster, A.D., 2011. CXCR3 in T cell function. Exp. Cell Res. 317, 620–631.
Guma, M., Busch, L.K., Salazar-Fontana, L.I., Bellosillo, B., Morte, C., Garcia, P., Lopez-Botet, M., 2005. The CD94/NKG2C killer lectin-like receptor constitutes an alternative activation pathway for a subset of CD8+ T cells. Eur. J. Immunol. 35, 2071–2080.
Haigh, T.A., Lin, X., Jia, H., Hui, E.P., Chan, A.T., Rickinson, A.B., Taylor, G.S., 2008. EBV latent membrane proteins (LMPs) 1 and 2 as immunotherapeutic targets: LMP-specific CD4+ cytotoxic T cell recognition of EBV-transformed B cell lines. J. Immunol. 180, 1643–1654.
Hamann, D., Baars, P.A., Rep, M.H., Hooibrink, B., Kerkhof-Garde, S.R., Klein, M.R., Van Lier, R.A., 1997. Phenotypic and functional separation of memory and effector human CD8+ T cells. J. Exp. Med. 186, 1407–1418.
Hamid, O., Robert, C., Daud, A., Hodi, F.S., Hwu, W.J., Kefford, R., Wolchok, J.D., Hersey, P., Joseph, R.W., Weber, J.S., Dronca, R., Gangadhar, T.C., Patnaik, A., Zarour, H., Joshua, A.M., Gergich, K., Elassaiss-Schaap, J., Algazi, A., Mateus, C., Boasberg, P., Tumeh, P.C., Chmielowski, B., Ebbinghaus, S.W., Li, X.N., Kang, S.P., Ribas, A., 2013. Safety and tumor responses with lambrolizumab (anti-PD-1) in melanoma. N. Engl. J. Med. 369, 134–144.
Hams, E., Mccarron, M.J., Amu, S., Yagita, H., Azuma, M., Chen, L., Fallon, P.G., 2011. Blockade of B7-H1 (programmed death ligand 1) enhances humoral immunity by positively regulating the generation of T follicular helper cells. J. Immunol. 186, 5648–5655.
Harty, J.T., Badovinac, V.P., 2008. Shaping and reshaping CD8+ T-cell memory. Nat. Rev. Immunol. 8, 107–119.
Hassan, J., Reen, D.J., 1998. IL-7 promotes the survival and maturation but not differentiation of human post-thymic CD4+ T cells. Eur. J. Immunol. 28, 3057–3065.
Hayflick, L., 1961. Moorhead PS: The serial cultivation of human diploid cell strains. Exp Cell Res 25, 585–621. PMID:13905658.
Haynes, N.M., Allen, C.D., Lesley, R., Ansel, K.M., Killeen, N., Cyster, J.G., 2007. Role of CXCR5 and CCR7 in follicular Th cell positioning and appearance of a programmed cell death gene-1high germinal center-associated subpopulation. J. Immunol. 179, 5099–5108.
He, J., Tsai, L.M., Leong, Y.A., Hu, X., Ma, C.S., Chevalier, N., Sun, X., Vandenberg, K., Rockman, S., Ding, Y., Zhu, L., Wei, W., Wang, C., Karnowski, A., Belz, G.T., Ghali, J.R., Cook, M.C., Riminton, D.S., Veillette, A., Schwartzberg, P.L., Mackay, F., Brink, R., Tangye, S.G., Vinuesa, C.G., Mackay, C.R., Li, Z., Yu, D., 2013. Circulating precursor CCR7(lo)PD-1(hi) CXCR5(+) CD4(+) T cells indicate Tfh cell activity and promote antibody responses upon antigen reexposure. Immunity 39, 770–781.
He, R., Hou, S., Liu, C., Zhang, A., Bai, Q., Han, M., Yang, Y., Wei, G., Shen, T., Yang, X., Xu, L., Chen, X., Hao, Y., Wang, P., Zhu, C., Ou, J., Liang, H., Ni, T., Zhang, X., Zhou, X., Deng, K., Chen, Y., Luo, Y., Xu, J., Qi, H., Wu, Y., Ye, L., 2016. Follicular CXCR5- expressing CD8(+) T cells curtail chronic viral infection. Nature 537, 412–428.
Hendriks, J., Xiao, Y., Borst, J., 2003. CD27 promotes survival of activated T cells and complements CD28 in generation and establishment of the effector T cell pool. J. Exp. Med. 198, 1369–1380.
Henson, S.M., Franzese, O., Macaulay, R., Libri, V., Azevedo, R.I., Kiani-Alikhan, S., Plunkett, F.J., Masters, J.E., Jackson, S., Griffiths, S.J., Pircher, H.P., Soares, M.V., Akbar, A.N., 2009. KLRG1 signaling induces defective Akt (ser473) phosphorylation and proliferative dysfunction of highly differentiated CD8+ T cells. Blood 113, 6619–6628.

Henson, S.M., Riddell, N.E., Akbar, A.N., 2012. Properties of end-stage human T cells defined by CD45RA re-expression. Curr. Opin. Immunol. 24, 476–481.

Hertoghs, K.M., Moerland, P.D., Van Stijn, A., Remmerswaal, E.B., Yong, S.L., Van De Berg, P.J., Van Ham, S.M., Baas, F., Ten Berge, I.J., Van Lier, R.A., 2010. Molecular profiling of cytomegalovirus-induced human CD8+ T cell differentiation. J. Clin. Invest. 120, 4077–4090.

Hickey, W.F., Hsu, B.L., Kimura, H., 1991. T-lymphocyte entry into the central nervous system. J. Neurosci. Res. 28, 254–260.

Hodi, F.S., O'day, S.J., Mcdermott, D.F., Weber, R.W., Sosman, J.A., Haanen, J.B., Gonzalez, R., Robert, C., Schadendorf, D., Hassel, J.C., Akerley, W., Van Den Eertwegh, A.J., Lutzky, J., Lorigan, P., Vaubel, J.M., Linette, G.P., Hogg, D., Ottensmeier, C.H., Lebbe, C., Peschel, C., Quirt, I., Clark, J.I., Wolchok, J.D., Weber, J.S., Tian, J., Yellin, M.J., Nichol, G.M., Hoos, A., Urba, W.J., 2010. Improved survival with ipilimumab in patients with metastatic melanoma. N. Engl. J. Med. 363, 711–723.

Hofmann, M., Pircher, H., 2011. E-cadherin promotes accumulation of a unique memory CD8 T-cell population in murine salivary glands. Proc. Natl. Acad. Sci. U. S. A. 108, 16741–16746.

Hombrink, P., Helbig, C., Backer, R.A., Piet, B., Oja, A.E., Stark, R., Brasser, G., Jongejan, A., Jonkers, R.E., Nota, B., Basak, O., Clevers, H.C., Moerland, P.D., Amsen, D., Van Lier, R.A., 2016. Programs for the persistence, vigilance and control of human CD8(+) lung-resident memory T cells. Nat. Immunol. 17, 1467–1478.

Hong, H., Gu, Y., Sheng, S.Y., Lu, C.G., Zou, J.Y., Wu, C.Y., 2016. The distribution of human stem cell-like memory T cell in lung cancer. J. Immunother. 39, 233–240.

Humblin, E., Thibaudin, M., Chalmin, F., Derangere, V., Limagne, E., Richard, C., Flavell, R.A., Chevrier, S., Ladoire, S., Berger, H., Boidot, R., Apetoh, L., Vegran, F., Ghiringhelli, F., 2017. IRF8-dependent molecular complexes control the Th9 transcriptional program. Nat. Commun. 8, 2085.

Im, S.J., Hashimoto, M., Gerner, M.Y., Lee, J., Kissick, H.T., Burger, M.C., Shan, Q., Hale, J.S., Lee, J., Nasti, T.H., Sharpe, A.H., Freeman, G.J., Germain, R.N., Nakaya, H.I., Xue, H.H., Ahmed, R., 2016. Defining CD8+ T cells that provide the proliferative burst after PD-1 therapy. Nature 537, 417–421.

Imai, T., Nagira, M., Takagi, S., Kakizaki, M., Nishimura, M., Wang, J., Gray, P.W., Matsushima, K., Yoshie, O., 1999. Selective recruitment of CCR4-bearing Th2 cells toward antigen-presenting cells by the CC chemokines thymus and activation-regulated chemokine and macrophage-derived chemokine. Int. Immunol. 11, 81–88.

Ishigame, H., Kakuta, S., Nagai, T., Kadoki, M., Nambu, A., Komiyama, Y., Fujikado, N., Tanahashi, Y., Akitsu, A., Kotaki, H., Sudo, K., Nakae, S., Sasakawa, C., Iwakura, Y., 2009. Differential roles of interleukin-17A and -17F in host defense against mucoepithelial bacterial infection and allergic responses. Immunity 30, 108–119.

Jenkins, M.K., Schwartz, R.H., 1987. Antigen presentation by chemically modified splenocytes induces antigen-specific T cell unresponsiveness in vitro and in vivo. J. Exp. Med. 165, 302–319.

Jia, L., Wu, C., 2014. Differentiation, regulation and function of Th9 cells. Adv. Exp. Med. Biol. 841, 181–207.

Jiang, X., Clark, R.A., Liu, L., Wagers, A.J., Fuhlbrigge, R.C., Kupper, T.S., 2012. Skin infection generates non-migratory memory CD8+ T(RM) cells providing global skin immunity. Nature 483, 227–231.

Jin, H.T., Anderson, A.C., Tan, W.G., West, E.E., Ha, S.J., Araki, K., Freeman, G.J., Kuchroo, V.K., Ahmed, R., 2010. Cooperation of Tim-3 and PD-1 in CD8 T-cell exhaustion during chronic viral infection. Proc. Natl. Acad. Sci. U. S. A. 107, 14733–14738.

Jonuleit, H., Schmitt, E., Stassen, M., Tuettenberg, A., Knop, J., Enk, A.H., 2001. Identification and functional characterization of human CD4(+)CD25(+) T cells with regulatory properties isolated from peripheral blood. J. Exp. Med. 193, 1285–1294.

Jorgensen, A.S., Rosenkilde, M.M., Hjorto, G.M., 2017. Biased signaling of G protein-coupled receptors - from a chemokine receptor CCR7 perspective. Gen. Comp. Endocrinol.

Kagoya, Y., Nakatsugawa, M., Yamashita, Y., Ochi, T., Guo, T., Anczurowski, M., Saso, K., Butler, M.O., Arrowsmith, C.H., Hirano, N., 2016. BET bromodomain inhibition enhances T cell persistence and function in adoptive immunotherapy models. J. Clin. Invest. 126, 3479–3494.

Kaji, T., Ishige, A., Hikida, M., Taka, J., Hijikata, A., Kubo, M., Nagashima, T., Takahashi, Y., Kurosaki, T., Okada, M., Ohara, O., Rajewsky, K., Takemori, T., 2012. Distinct cellular pathways select germline-encoded and somatically mutated antibodies into immunological memory. J. Exp. Med. 209, 2079–2097.

Kanno, Y., Vahedi, G., Hirahara, K., Singleton, K., O'shea, J.J., 2012. Transcriptional and epigenetic control of T helper cell specification: molecular mechanisms underlying commitment and plasticity. Annu. Rev. Immunol. 30, 707–731.

Kastirr, I., Crosti, M., Maglie, S., Paroni, M., Steckel, B., Moro, M., Pagani, M., Abrignani, S., Geginat, J., 2015. Signal strength and metabolic requirements control cytokine-induced Th17 differentiation of Uncommitted human T cells. J. Immunol. 195, 3617–3627.

Kaufmann, D.E., Kavanagh, D.G., Pereyra, F., Zaunders, J.J., Mackey, E.W., Miura, T., Palmer, S., Brockman, M., Rathod, A., Piechocka-Trocha, A., Baker, B., Zhu, B., Le Gall, S., Waring, M.T., Ahern, R., Moss, K., Kelleher, A.D., Coffin, J.M., Freeman, G.J., Rosenberg, E.S., Walker, B.D., 2007. Upregulation of CTLA-4 by HIV-specific CD4+ T cells correlates with disease progression and defines a reversible immune dysfunction. Nat. Immunol. 8, 1246–1254.

Keir, M.E., Butte, M.J., Freeman, G.J., Sharpe, A.H., 2008. PD-1 and its ligands in tolerance and immunity. Annu. Rev. Immunol. 26, 677–704.

Khan, N., Shariff, N., Cobbold, M., Bruton, R., Ainsworth, J.A., Sinclair, A.J., Nayak, L., Moss, P.A., 2002. Cytomegalovirus seropositivity drives the CD8 T cell repertoire toward greater clonality in healthy elderly individuals. J. Immunol. 169, 1984–1992.

Kim, C.H., Rott, L.S., Clark-Lewis, I., Campbell, D.J., Wu, L., Butcher, E.C., 2001. Subspecialization of CXCR5+ T cells: B helper activity is focused in a germinal center-localized subset of CXCR5+ T cells. J. Exp. Med. 193, 1373–1381.

Kim, S.K., Schluns, K.S., Lefrancois, L., 1999. Induction and visualization of mucosal memory CD8 T cells following systemic virus infection. J. Immunol. 163, 4125–4132.

Kimmig, S., Przybylski, G.K., Schmidt, C.A., Laurisch, K., Mowes, B., Radbruch, A., Thiel, A., 2002. Two subsets of naive T helper cells with distinct T cell receptor excision circle content in human adult peripheral blood. J. Exp. Med. 195, 789–794.

Klebanoff, C.A., Gattinoni, L., Restifo, N.P., 2012. Sorting through subsets: which T-cell populations mediate highly effective adoptive immunotherapy? J. Immunother. 35, 651–660.

Klebanoff, C.A., Scott, C.D., Leonardi, A.J., Yamamoto, T.N., Cruz, A.C., Ouyang, C., Ramaswamy, M., Roychoudhuri, R., Ji, Y., Eil, R.L., Sukumar, M., Crompton, J.G., Palmer, D.C., Borman, Z.A., Clever, D., Thomas, S.K., Patel, S., Yu, Z., Muranski, P., Liu, H., Wang, E., Marincola, F.M., Gros, A., Gattinoni, L., Rosenberg, S.A., Siegel, R.M., Restifo, N.P., 2016. Memory T cell-driven differentiation of naive cells impairs adoptive immunotherapy. J. Clin. Invest. 126, 318–334.

Klucar, P., Barnes, P.F., Kong, Y., Samten, B., Tvinnereim, A., Spallek, R., Nepom, G.T., Singh, M., Shams, H., 2008. Characterization of effector functions of human peptide-specific CD4+ T-cell clones for an intracellular pathogen. Hum. Immunol. 69, 475–483.

Koch, S., Larbi, A., Derhovanessian, E., Ozcelik, D., Naumova, E., Pawelec, G., 2008. Multiparameter flow cytometric analysis of CD4 and CD8 T cell subsets in young and old people. Immun. Ageing 5, 6.

Kohler, S., Thiel, A., 2009. Life after the thymus: CD31+ and CD31- human naive CD4+ T-cell subsets. Blood 113, 769–774.

Korn, T., Bettelli, E., Gao, W., Awasthi, A., Jager, A., Strom, T.B., Oukka, M., Kuchroo, V.K., 2007a. IL-21 initiates an alternative pathway to induce proinflammatory T(H)17 cells. Nature 448, 484–487.

Korn, T., Oukka, M., Kuchroo, V., Bettelli, E., 2007b. Th17 cells: effector T cells with inflammatory properties. Semin. Immunol. 19, 362–371.

Kroenke, M.A., Eto, D., Locci, M., Cho, M., Davidson, T., Haddad, E.K., Crotty, S., 2012. Bcl6 and Maf cooperate to instruct human follicular helper CD4 T cell differentiation. J. Immunol. 188, 3734–3744.

Kumar, B.V., Ma, W., Miron, M., Granot, T., Guyer, R.S., Carpenter, D.J., Senda, T., Sun, X., Ho, S.H., Lerner, H., Friedman, A.L., Shen, Y., Farber, D.L., 2017. Human tissue-resident memory T cells are defined by Core transcriptional and functional Signatures in lymphoid and mucosal sites. Cell Rep. 20, 2921–2934.

Kunkel, E.J., Boisvert, J., Murphy, K., Vierra, M.A., Genovese, M.C., Wardlaw, A.J., Greenberg, H.B., Hodge, M.R., Wu, L., Butcher, E.C., Campbell, J.J., 2002. Expression of the chemokine receptors CCR4, CCR5, and CXCR3 by human tissue-infiltrating lymphocytes. Am. J. Pathol. 160, 347–355.

Kunkel, E.J., Campbell, J.J., Haraldsen, G., Pan, J., Boisvert, J., Roberts, A.I., Ebert, E.C., Vierra, M.A., Goodman, S.B., Genovese, M.C., Wardlaw, A.J., Greenberg, H.B., Parker, C.M., Butcher, E.C., Andrew, D.P., Agace, W.W., 2000. Lymphocyte CC chemokine receptor 9 and epithelial thymus-expressed chemokine (TECK) expression distinguish the small intestinal immune compartment: epithelial expression of tissue-specific chemokines as an organizing principle in regional immunity. J. Exp. Med. 192, 761–768.

Kurose, K., Ohue, Y., Sato, E., Yamauchi, A., Eikawa, S., Isobe, M., Nishio, Y., Uenaka, A., Oka, M., Nakayama, E., 2015. Increase in activated Treg in TIL in lung cancer and in vitro depletion of Treg by ADCC using an antihuman CCR4 mAb (KM2760). J. Thorac. Oncol. 10, 74–83.

Lantelme, E., Mantovani, S., Palermo, B., Campanelli, R., Sallusto, F., Giachino, C., 2001. Kinetics of GATA-3 gene expression in early polarizing and committed human T cells. Immunology 102, 123–130.

Lanzavecchia, A., Sallusto, F., 2002. Progressive differentiation and selection of the fittest in the immune response. Nat. Rev. Immunol. 2, 982–987.

Lee, J.J., Kao, K.C., Chiu, Y.L., Jung, C.J., Liu, C.J., Cheng, S.J., Chang, Y.L., Ko, J.Y., Chia, J.S., 2017. Enrichment of human CCR6(+) regulatory T cells with Superior suppressive activity in oral cancer. J. Immunol. 199, 467–476.

Lee, W.W., Kang, S.W., Choi, J., Lee, S.H., Shah, K., Eynon, E.E., Flavell, R.A., Kang, I., 2010. Regulating human Th17 cells via differential expression of IL-1 receptor. Blood 115, 530–540.

Lenschow, D.J., Walunas, T.L., Bluestone, J.A., 1996. CD28/B7 system of T cell costimulation. Annu. Rev. Immunol. 14, 233–258.

Li, D., Molldrem, J.J., Ma, Q., 2009. LFA-1 regulates CD8+ T cell activation via T cell receptor-mediated and LFA-1-mediated Erk1/2 signal pathways. J. Biol. Chem. 284, 21001–21010.

Li, J., Olshansky, M., Carbone, F.R., Ma, J.Z., 2016. Transcriptional Analysis of T Cells resident in human skin. PLoS One 11, e0148351.

Licona-Limon, P., Henao-Mejia, J., Temann, A.U., Gagliani, N., Licona-Limon, I., Ishigame, H., Hao, L., Herbert, D.R., Flavell, R.A., 2013. Th9 cells drive host immunity against Gastrointestinal worm infection. Immunity 39, 744–757.

Lim, H.W., Kim, C.H., 2007. Loss of IL-7 receptor alpha on CD4+ T cells defines terminally differentiated B cell-helping effector T cells in a B cell-rich lymphoid tissue. J. Immunol. 179, 7448–7456.

Lin, L., Couturier, J., Yu, X., Medina, M.A., Kozinetz, C.A., Lewis, D.E., 2014. Granzyme B secretion by human memory CD4 T cells is less strictly regulated compared to memory CD8 T cells. BMC Immunol. 15, 36.

Lin, W.W., Nish, S.A., Yen, B., Chen, Y.H., Adams, W.C., Kratchmarov, R., Rothman, N.J., Bhandoola, A., Xue, H.H., Reiner, S.L., 2016. CD8(+) T lymphocyte self-renewal during effector cell Determination. Cell Rep. 17, 1773–1782.

Lin, Y.C., Mahalingam, J., Chiang, J.M., Su, P.J., Chu, Y.Y., Lai, H.Y., Fang, J.H., Huang, C.T., Chiu, C.T., Lin, C.Y., 2013. Activated but not resting regulatory T cells accumulated in tumor microenvironment and correlated with tumor progression in patients with colorectal cancer. Int J Cancer 132, 1341–1350.

Lindestam Arlehamn, C.S., Gerasimova, A., Mele, F., Henderson, R., Swann, J., Greenbaum, J.A., Kim, Y., Sidney, J., James, E.A., Taplitz, R., Mckinney, D.M., Kwok, W.W., Grey, H., Sallusto, F., Peters, B., Sette, A., 2013. Memory T cells in latent Mycobacterium tuberculosis infection are directed against three antigenic islands and largely contained in a CXCR3+CCR6+ Th1 subset. PLoS Pathog. 9, e1003130.

Lintermans, L.L., Rutgers, A., Stegeman, C.A., Heeringa, P., Abdulahad, W.H., 2017. Chemokine receptor co-expression reveals aberrantly distributed TH effector memory cells in GPA patients. Arthritis Res. Ther. 19, 136.

Liu, Q., Li, Z., Gao, J.L., Wan, W., Ganesan, S., Mcdermott, D.H., Murphy, P.M., 2015. CXCR4 antagonist AMD3100 redistributes leukocytes from primary immune organs to secondary immune organs, lung, and blood in mice. Eur. J. Immunol. 45, 1855–1867.

Locci, M., Havenar-Daughton, C., Landais, E., Wu, J., Kroenke, M.A., Arlehamn, C.L., Su, L.F., Cubas, R., Davis, M.M., Sette, A., Haddad, E.K., International, A.V.I.P.C.P.I., Poignard, P., Crotty, S., 2013. Human circulating PD-1+CXCR3-CXCR5+ memory Tfh cells are highly functional and correlate with broadly neutralizing HIV antibody responses. Immunity 39, 758–769.

Lord, G.M., Rao, R.M., Choe, H., Sullivan, B.M., Lichtman, A.H., Luscinskas, F.W., Glimcher, L.H., 2005. T-bet is required for optimal proinflammatory CD4+ T-cell trafficking. Blood 106, 3432–3439.

Lugli, E., Dominguez, M.H., Gattinoni, L., Chattopadhyay, P.K., Bolton, D.L., Song, K., Klatt, N.R., Brenchley, J.M., Vaccari, M., Gostick, E., Price, D.A., Waldmann, T.A., Restifo, N.P., Franchini, G., Roederer, M., 2013a. Superior T memory stem cell persistence supports long-lived T cell memory. J. Clin. Invest. 123, 594–599.

Lugli, E., Gattinoni, L., Roberto, A., Mavilio, D., Price, D.A., Restifo, N.P., Roederer, M., 2013b. Identification, isolation and in vitro expansion of human and nonhuman primate T stem cell memory cells. Nat. Protoc. 8, 33–42.

Lugli, E., Roederer, M., Cossarizza, A., 2010. Data analysis in flow cytometry: the future just started. Cytometry 77, 705–713.

Ma, C.S., Suryani, S., Avery, D.T., Chan, A., Nanan, R., Santner-Nanan, B., Deenick, E.K., Tangye, S.G., 2009. Early commitment of naive human CD4(+) T cells to the T follicular helper (T(FH)) cell lineage is induced by IL-12. Immunol. Cell Biol. 87, 590–600.

Mackay, C.R., Marston, W.L., Dudler, L., 1990. Naive and memory T cells show distinct pathways of lymphocyte recirculation. J. Exp. Med. 171, 801–817.

Maeda, Y., Nishikawa, H., Sugiyama, D., Ha, D., Hamaguchi, M., Saito, T., Nishioka, M., Wing, J.B., Adeegbe, D., Katayama, I., Sakaguchi, S., 2014. Detection of self-reactive CD8(+) T cells with an anergic phenotype in healthy individuals. Science 346, 1536–1540.

Maggi, L., Santarlasci, V., Capone, M., Peired, A., Frosali, F., Crome, S.Q., Querci, V., Fambrini, M., Liotta, F., Levings, M.K., Maggi, E., Cosmi, L., Romagnani, S., Annunziato, F., 2010. CD161 is a marker of all human IL-17-producing T-cell subsets and is induced by RORC. Eur. J. Immunol. 40, 2174–2181.

Maggi, L., Santarlasci, V., Capone, M., Rossi, M.C., Querci, V., Mazzoni, A., Cimaz, R., De Palma, R., Liotta, F., Maggi, E., Romagnani, S., Cosmi, L., Annunziato, F., 2012. Distinctive features of classic and nonclassic (Th17 derived) human Th1 cells. Eur. J. Immunol. 42, 3180–3188.

Magnani, C.F., Alberigo, G., Bacchetta, R., Serafini, G., Andreani, M., Roncarolo, M.G., Gregori, S., 2011. Killing of myeloid APCs via HLA class I, CD2 and CD226 defines a novel mechanism of suppression by human Tr1 cells. Eur. J. Immunol. 41, 1652–1662.

Mahnke, Y.D., Brodie, T.M., Sallusto, F., Roederer, M., Lugli, E., 2013. The who's who of T-cell differentiation: human memory T-cell subsets. Eur. J. Immunol. 43, 2797–2809.

Majchrzak, K., Nelson, M.H., Bowers, J.S., Bailey, S.R., Wyatt, M.M., Wrangle, J.M., Rubinstein, M.P., Varela, J.C., Li, Z., Himes, R.A., Chan, S.S.L., Paulos, C.M., 2017. beta-catenin and PI3Kdelta inhibition expands precursor Th17 cells with heightened stemness and antitumor activity. JCI Insight 2.

Mandapathil, M., Szczepanski, M.J., Szajnik, M., Ren, J., Jackson, E.K., Johnson, J.T., Gorelik, E., Lang, S., Whiteside, T.L., 2010. Adenosine and prostaglandin E2 cooperate in the suppression of immune responses mediated by adaptive regulatory T cells. J. Biol. Chem. 285, 27571–27580.

Masopust, D., Choo, D., Vezys, V., Wherry, E.J., Duraiswamy, J., Akondy, R., Wang, J., Casey, K.A., Barber, D.L., Kawamura, K.S., Fraser, K.A., Webby, R.J., Brinkmann, V., Butcher, E.C., Newell, K.A., Ahmed, R., 2010. Dynamic T cell migration program provides resident memory within intestinal epithelium. J. Exp. Med. 207, 553–564.

Masopust, D., Vezys, V., Marzo, A.L., Lefrancois, L., 2001. Preferential localization of effector memory cells in nonlymphoid tissue. Science 291, 2413–2417.

Matsuzaki, J., Tsuji, T., Luescher, I.F., Shiku, H., Mineno, J., Okamoto, S., Old, L.J., Shrikant, P., Gnjatic, S., Odunsi, K., 2015. Direct tumor recognition by a human CD4(+) T-cell subset potently mediates tumor growth inhibition and orchestrates anti-tumor immune responses. Sci. Rep. 5, 14896.

Milpied, P., Renand, A., Bruneau, J., Mendes-Da-Cruz, D.A., Jacquelin, S., Asnafi, V., Rubio, M.T., Macintyre, E., Lepelletier, Y., Hermine, O., 2009. Neuropilin-1 is not a marker of human Foxp3+ Treg. Eur. J. Immunol. 39, 1466–1471.

Miyara, M., Chader, D., Sage, E., Sugiyama, D., Nishikawa, H., Bouvry, D., Claer, L., Hingorani, R., Balderas, R., Rohrer, J., Warner, N., Chapelier, A., Valeyre, D., Kannagi, R., Sakaguchi, S., Amoura, Z., Gorochov, G., 2015. Sialyl Lewis x (CD15s) identifies highly differentiated and most suppressive FOXP3high regulatory T cells in humans. Proc. Natl. Acad. Sci. U. S. A. 112, 7225–7230.

Miyara, M., Yoshioka, Y., Kitoh, A., Shima, T., Wing, K., Niwa, A., Parizot, C., Taflin, C., Heike, T., Valeyre, D., Mathian, A., Nakahata, T., Yamaguchi, T., Nomura, T., Ono, M., Amoura, Z., Gorochov, G., Sakaguchi, S., 2009. Functional delineation and differentiation dynamics of human CD4+ T cells expressing the FoxP3 transcription factor. Immunity 30, 899–911.

Miyasaka, M., Tanaka, T., 2004. Lymphocyte trafficking across high endothelial venules: dogmas and enigmas. Nat. Rev. Immunol. 4, 360–370.

Mondal, A.M., Horikawa, I., Pine, S.R., Fujita, K., Morgan, K.M., Vera, E., Mazur, S.J., Appella, E., Vojtesek, B., Blasco, M.A., Lane, D.P., Harris, C.C., 2013. p53 isoforms regulate aging- and tumor-associated replicative senescence in T lymphocytes. J. Clin. Invest. 123, 5247–5257.

Morita, R., Schmitt, N., Bentebibel, S.E., Ranganathan, R., Bourdery, L., Zurawski, G., Foucat, E., Dullaers, M., Oh, S., Sabzghabaei, N., Lavecchio, E.M., Punaro, M., Pascual, V., Banchereau, J., Ueno, H., 2011. Human blood CXCR5(+)CD4(+) T cells are counterparts of T follicular cells and contain specific subsets that differentially support antibody secretion. Immunity 34, 108–121.

Morou, A., Palmer, B.E., Kaufmann, D.E., 2014. Distinctive features of CD4+ T cell dysfunction in chronic viral infections. Curr. Opin. HIV AIDS 9, 446–451.

Moser, J.M., Gibbs, J., Jensen, P.E., Lukacher, A.E., 2002. CD94-NKG2A receptors regulate antiviral CD8(+) T cell responses. Nat. Immunol. 3, 189–195.

Mueller, S.N., Mackay, L.K., 2016. Tissue-resident memory T cells: local specialists in immune defence. Nat. Rev. Immunol. 16, 79–89.

Murali-Krishna, K., Lau, L.L., Sambhara, S., Lemonnier, F., Altman, J., Ahmed, R., 1999. Persistence of memory CD8 T cells in MHC class I-deficient mice. Science 286, 1377–1381.

Muthuswamy, R., Mueller-Berghaus, J., Haberkorn, U., Reinhart, T.A., Schadendorf, D., Kalinski, P., 2010. PGE(2) transiently enhances DC expression of CCR7 but inhibits the ability of DCs to produce CCL19 and attract naive T cells. Blood 116, 1454–1459.

Nagata, K., Tanaka, K., Ogawa, K., Kemmotsu, K., Imai, T., Yoshie, O., Abe, H., Tada, K., Nakamura, M., Sugamura, K., Takano, S., 1999. Selective expression of a novel surface molecule by human Th2 cells in vivo. J. Immunol. 162, 1278–1286.

Nakamoto, N., Cho, H., Shaked, A., Olthoff, K., Valiga, M.E., Kaminski, M., Gostick, E., Price, D.A., Freeman, G.J., Wherry, E.J., Chang, K.M., 2009. Synergistic reversal of intrahepatic HCV-specific CD8 T cell exhaustion by combined PD-1/CTLA-4 blockade. PLoS Pathog. 5, e1000313.

Nemes, E., Bertoncelli, L., Lugli, E., Pinti, M., Nasi, M., Manzini, L., Manzini, S., Prati, F., Borghi, V., Cossarizza, A., Mussini, C., 2010. Cytotoxic granule release dominates gag-specific CD4+ T-cell response in different phases of HIV infection. AIDS 24, 947–957.

Nishimura, H., Nose, M., Hiai, H., Minato, N., Honjo, T., 1999. Development of lupus-like autoimmune diseases by disruption of the PD-1 gene encoding an ITIM motif-carrying immunoreceptor. Immunity 11, 141–151.

Onyema, O.O., Njemini, R., Forti, L.N., Bautmans, I., Aerts, J.L., De Waele, M., Mets, T., 2015. Aging-associated subpopulations of human CD8+ T-lymphocytes identified by their CD28 and CD57 phenotypes. Arch. Gerontol. Geriatr. 61, 494–502.

Ouyang, Q., Wagner, W.M., Voehringer, D., Wikby, A., Klatt, T., Walter, S., Muller, C.A., Pircher, H., Pawelec, G., 2003. Age-associated accumulation of CMV-specific CD8+ T cells expressing the inhibitory killer cell lectin-like receptor G1 (KLRG1). Exp. Gerontol. 38, 911–920.

Paley, M.A., Kroy, D.C., Odorizzi, P.M., Johnnidis, J.B., Dolfi, D.V., Barnett, B.E., Bikoff, E.K., Robertson, E.J., Lauer, G.M., Reiner, S.L., Wherry, E.J., 2012. Progenitor and terminal subsets of CD8+ T cells cooperate to contain chronic viral infection. Science 338, 1220–1225.

Parra, E., Wingren, A.G., Hedlund, G., Kalland, T., Dohlsten, M., 1997. The role of B7-1 and LFA-3 in costimulation of CD8+ T cells. J. Immunol. 158, 637–642.

Patil, V.S., Madrigal, A., Schmiedel, B.J., Clarke, J., O'rourke, P., De Silva, A.D., Harris, E., Peters, B., Seumois, G., Weiskopf, D., Sette, A., Vijayanand, P., 2018. Precursors of human CD4(+) cytotoxic T lymphocytes identified by single-cell transcriptome analysis. Sci Immunol 3.

Pauken, K.E., Wherry, E.J., 2015. Overcoming T cell exhaustion in infection and cancer. Trends Immunol. 36, 265–276.

Peng, W., Ye, Y., Rabinovich, B.A., Liu, C., Lou, Y., Zhang, M., Whittington, M., Yang, Y., Overwijk, W.W., Lizee, G., Hwu, P., 2010. Transduction of tumor-specific T cells with CXCR2 chemokine receptor improves migration to tumor and antitumor immune responses. Clin Cancer Res 16, 5458–5468.

Perez, O.D., Mitchell, D., Jager, G.C., South, S., Murriel, C., Mcbride, J., Herzenberg, L.A., Kinoshita, S., Nolan, G.P., 2003. Leukocyte functional antigen 1 lowers T cell activation thresholds and signaling through cytohesin-1 and Jun-activating binding protein 1. Nat. Immunol. 4, 1083–1092.

Peter, M.E., Budd, R.C., Desbarats, J., Hedrick, S.M., Hueber, A.O., Newell, M.K., Owen, L.B., Pope, R.M., Tschopp, J., Wajant, H., Wallach, D., Wiltrout, R.H., Zornig, M., Lynch, D.H., 2007. The CD95 receptor: apoptosis revisited. Cell 129, 447–450.

Petrovas, C., Casazza, J.P., Brenchley, J.M., Price, D.A., Gostick, E., Adams, W.C., Precopio, M.L., Schacker, T., Roederer, M., Douek, D.C., Koup, R.A., 2006. PD-1 is a regulator of virus-specific CD8+ T cell survival in HIV infection. J. Exp. Med. 203, 2281–2292.

Picker, L.J., Treer, J.R., Ferguson-Darnell, B., Collins, P.A., Bergstresser, P.R., Terstappen, L.W., 1993a. Control of lymphocyte recirculation in man. II. Differential regulation of the cutaneous lymphocyte-associated antigen, a tissue-selective homing receptor for skin-homing T cells. J. Immunol. 150, 1122–1136.

Picker, L.J., Treer, J.R., Ferguson-Darnell, B., Collins, P.A., Buck, D., Terstappen, L.W., 1993b. Control of lymphocyte recirculation in man. I. Differential regulation of the peripheral lymph node homing receptor L-selectin on T cells during the virgin to memory cell transition. J. Immunol. 150, 1105–1121.

Polley, M.J., Phillips, M.L., Wayner, E., Nudelman, E., Singhal, A.K., Hakomori, S., Paulson, J.C., 1991. CD62 and endothelial cell-leukocyte adhesion molecule 1 (ELAM-1) recognize the same carbohydrate ligand, sialyl-Lewis x. Proc. Natl. Acad. Sci. U. S. A. 88, 6224–6228.

Porichis, F., Kwon, D.S., Zupkosky, J., Tighe, D.P., Mcmullen, A., Brockman, M.A., Pavlik, D.F., Rodriguez-Garcia, M., Pereyra, F., Freeman, G.J., Kavanagh, D.G., Kaufmann, D.E., 2011. Responsiveness of HIV-specific CD4 T cells to PD-1 blockade. Blood 118, 965–974.

Probst-Kepper, M., Geffers, R., Kroger, A., Viegas, N., Erck, C., Hecht, H.J., Lunsdorf, H., Roubin, R., Moharregh-Khiabani, D., Wagner, K., Ocklenburg, F., Jeron, A., Garritsen, H., Arstila, T.P., Kekalainen, E., Balling, R., Hauser, H., Buer, J., Weiss, S., 2009. GARP: a key receptor controlling FOXP3 in human regulatory T cells. J. Cell Mol. Med. 13, 3343–3357.

Purwar, R., Campbell, J., Murphy, G., Richards, W.G., Clark, R.A., Kupper, T.S., 2011. Resident memory T cells (T(RM)) are abundant in human lung: diversity, function, and antigen specificity. PLoS One 6, e16245.

Purwar, R., Schlapbach, C., Xiao, S., Kang, H.S., Elyaman, W., Jiang, X., Jetten, A.M., Khoury, S.J., Fuhlbrigge, R.C., Kuchroo, V.K., Clark, R.A., Kupper, T.S., 2012. Robust tumor immunity to melanoma mediated by interleukin-9-producing T cells. Nat. Med. 18, 1248–1253.

Quezada, S.A., Simpson, T.R., Peggs, K.S., Merghoub, T., Vider, J., Fan, X., Blasberg, R., Yagita, H., Muranski, P., Antony, P.A., Restifo, N.P., Allison, J.P., 2010. Tumor-reactive CD4(+) T cells develop cytotoxic activity and eradicate large established melanoma after transfer into lymphopenic hosts. J. Exp. Med. 207, 637–650.

Quigley, M.F., Gonzalez, V.D., Granath, A., Andersson, J., Sandberg, J.K., 2007. CXCR5+ CCR7- CD8 T cells are early effector memory cells that infiltrate tonsil B cell follicles. Eur. J. Immunol. 37, 3352–3362.

Raine, T., Liu, J.Z., Anderson, C.A., Parkes, M., Kaser, A., 2015. Generation of primary human intestinal T cell transcriptomes reveals differential expression at genetic risk loci for immune-mediated disease. Gut 64, 250–259.
Rasheed, A.U., Rahn, H.P., Sallusto, F., Lipp, M., Muller, G., 2006. Follicular B helper T cell activity is confined to CXCR5(hi)ICOS(hi) CD4 T cells and is independent of CD57 expression. Eur. J. Immunol. 36, 1892–1903.
Raveney, B.J., Oki, S., Hohjoh, H., Nakamura, M., Sato, W., Murata, M., Yamamura, T., 2015. Eomesodermin-expressing T-helper cells are essential for chronic neuroinflammation. Nat. Commun. 6, 8437.
Remmerswaal, E.B., Havenith, S.H., Idu, M.M., Van Leeuwen, E.M., Van Donselaar, K.A., Ten Brinke, A., Van Der Bom-Baylon, N., Bemelman, F.J., Van Lier, R.A., Ten Berge, I.J., 2012. Human virus-specific effector-type T cells accumulate in blood but not in lymph nodes. Blood 119, 1702–1712.
Restifo, N.P., Gattinoni, L., 2013. Lineage relationship of effector and memory T cells. Curr. Opin. Immunol. 25, 556–563.
Riley, J.L., June, C.H., 2005. The CD28 family: a T-cell rheostat for therapeutic control of T-cell activation. Blood 105, 13–21.
Rivino, L., Messi, M., Jarrossay, D., Lanzavecchia, A., Sallusto, F., Geginat, J., 2004. Chemokine receptor expression identifies Pre-T helper (Th)1, Pre-Th2, and nonpolarized cells among human CD4+ central memory T cells. J. Exp. Med. 200, 725–735.
Romagnani, S., 1994. Regulation of the development of type 2 T-helper cells in allergy. Curr. Opin. Immunol. 6, 838–846.
Romero, P., Zippelius, A., Kurth, I., Pittet, M.J., Touvrey, C., Iancu, E.M., Corthesy, P., Devevre, E., Speiser, D.E., Rufer, N., 2007. Four functionally distinct populations of human effector-memory CD8+ T lymphocytes. J. Immunol. 178, 4112–4119.
Ronchetti, S., Ricci, E., Petrillo, M.G., Cari, L., Migliorati, G., Nocentini, G., Riccardi, C., 2015. Glucocorticoid-induced tumour necrosis factor receptor-related protein: a key marker of functional regulatory T cells. J Immunol Res 2015, 171520.
Rosen, S.D., 2004. Ligands for L-selectin: homing, inflammation, and beyond. Annu. Rev. Immunol. 22, 129–156.
Rosshart, S., Hofmann, M., Schweier, O., Pfaff, A.K., Yoshimoto, K., Takeuchi, T., Molnar, E., Schamel, W.W., Pircher, H., 2008. Interaction of KLRG1 with E-cadherin: new functional and structural insights. Eur. J. Immunol. 38, 3354–3364.
Rossiter, H., Van Reijsen, F., Mudde, G.C., Kalthoff, F., Bruijnzeel-Koomen, C.A., Picker, L.J., Kupper, T.S., 1994. Skin disease-related T cells bind to endothelial selectins: expression of cutaneous lymphocyte antigen (CLA) predicts E-selectin but not P-selectin binding. Eur. J. Immunol. 24, 205–210.
Sabatino, M., Hu, J., Sommariva, M., Gautam, S., Fellowes, V., Hocker, J.D., Dougherty, S., Qin, H., Klebanoff, C.A., Fry, T.J., Gress, R.E., Kochenderfer, J.N., Stroncek, D.F., Ji, Y., Gattinoni, L., 2016. Generation of clinical-grade CD19-specific CAR-modified CD8+ memory stem cells for the treatment of human B-cell malignancies. Blood 128, 519–528.
Saez-Borderias, A., Guma, M., Angulo, A., Bellosillo, B., Pende, D., Lopez-Botet, M., 2006. Expression and function of NKG2D in CD4+ T cells specific for human cytomegalovirus. Eur. J. Immunol. 36, 3198–3206.
Sage, P.T., Sharpe, A.H., 2016. T follicular regulatory cells. Immunol. Rev. 271, 246–259.
Sakaguchi, S., Miyara, M., Costantino, C.M., Hafler, D.A., 2010. FOXP3+ regulatory T cells in the human immune system. Nat. Rev. Immunol. 10, 490–500.
Sallusto, F., 2016. Heterogeneity of human CD4(+) T cells against Microbes. Annu. Rev. Immunol. 34, 317–334.
Sallusto, F., Geginat, J., Lanzavecchia, A., 2004. Central memory and effector memory T cell subsets: function, generation, and maintenance. Annu. Rev. Immunol. 22, 745–763.

Sallusto, F., Lanzavecchia, A., 2009. Heterogeneity of CD4+ memory T cells: functional modules for tailored immunity. Eur. J. Immunol. 39, 2076–2082.

Sallusto, F., Lenig, D., Forster, R., Lipp, M., Lanzavecchia, A., 1999a. Two subsets of memory T lymphocytes with distinct homing potentials and effector functions. Nature 401, 708–712.

Sallusto, F., Lenig, D., Mackay, C.R., Lanzavecchia, A., 1998. Flexible programs of chemokine receptor expression on human polarized T helper 1 and 2 lymphocytes. J. Exp. Med. 187, 875–883.

Sallusto, F., Palermo, B., Lenig, D., Miettinen, M., Matikainen, S., Julkunen, I., Forster, R., Burgstahler, R., Lipp, M., Lanzavecchia, A., 1999b. Distinct patterns and kinetics of chemokine production regulate dendritic cell function. Eur. J. Immunol. 29, 1617–1625.

Sanders, M.E., Makgoba, M.W., Shaw, S., 1988. Human naive and memory T cells: reinterpretation of helper-inducer and suppressor-inducer subsets. Immunol. Today 9, 195–199.

Sathaliyawala, T., Kubota, M., Yudanin, N., Turner, D., Camp, P., Thome, J.J., Bickham, K.L., Lerner, H., Goldstein, M., Sykes, M., Kato, T., Farber, D.L., 2013. Distribution and compartmentalization of human circulating and tissue-resident memory T cell subsets. Immunity 38, 187–197.

Saurwein-Teissl, M., Lung, T.L., Marx, F., Gschosser, C., Asch, E., Blasko, I., Parson, W., Bock, G., Schonitzer, D., Trannoy, E., Grubeck-Loebenstein, B., 2002. Lack of antibody production following immunization in old age: association with CD8(+)CD28(-) T cell clonal expansions and an imbalance in the production of Th1 and Th2 cytokines. J. Immunol. 168, 5893–5899.

Schaerli, P., Ebert, L., Willimann, K., Blaser, A., Roos, R.S., Loetscher, P., Moser, B., 2004. A skin-selective homing mechanism for human immune surveillance T cells. J. Exp. Med. 199, 1265–1275.

Schaerli, P., Willimann, K., Lang, A.B., Lipp, M., Loetscher, P., Moser, B., 2000. CXC chemokine receptor 5 expression defines follicular homing T cells with B cell helper function. J. Exp. Med. 192, 1553–1562.

Schenkel, J.M., Fraser, K.A., Beura, L.K., Pauken, K.E., Vezys, V., Masopust, D., 2014. T cell memory. Resident memory CD8 T cells trigger protective innate and adaptive immune responses. Science 346, 98–101.

Schluns, K.S., Kieper, W.C., Jameson, S.C., Lefrancois, L., 2000. Interleukin-7 mediates the homeostasis of naive and memory CD8 T cells in vivo. Nat. Immunol. 1, 426–432.

Schmitt, N., Bentebibel, S.E., Ueno, H., 2014. Phenotype and functions of memory Tfh cells in human blood. Trends Immunol. 35, 436–442.

Scholz, G., Jandus, C., Zhang, L., Grandclement, C., Lopez-Mejia, I.C., Soneson, C., Delorenzi, M., Fajas, L., Held, W., Dormond, O., Romero, P., 2016. Modulation of mTOR signalling Triggers the formation of stem cell-like memory T cells. EBioMedicine 4, 50–61.

Schutyser, E., Struyf, S., Van Damme, J., 2003. The CC chemokine CCL20 and its receptor CCR6. Cytokine Growth Factor Rev. 14, 409–426.

Schwartz, R.H., 2003. T cell anergy. Annu. Rev. Immunol. 21, 305–334.

Schwartzberg, P.L., Mueller, K.L., Qi, H., Cannons, J.L., 2009. SLAM receptors and SAP influence lymphocyte interactions, development and function. Nat. Rev. Immunol. 9, 39–46.

Shainheit, M.G., Smith, P.M., Bazzone, L.E., Wang, A.C., Rutitzky, L.I., Stadecker, M.J., 2008. Dendritic cell IL-23 and IL-1 production in response to schistosome eggs induces Th17 cells in a mouse strain prone to severe immunopathology. J. Immunol. 181, 8559–8567.

Shin, H., Blackburn, S.D., Blattman, J.N., Wherry, E.J., 2007. Viral antigen and extensive division maintain virus-specific CD8 T cells during chronic infection. J. Exp. Med. 204, 941–949.

Shiow, L.R., Rosen, D.B., Brdickova, N., Xu, Y., An, J., Lanier, L.L., Cyster, J.G., Matloubian, M., 2006. CD69 acts downstream of interferon-alpha/beta to inhibit S1P1 and lymphocyte egress from lymphoid organs. Nature 440, 540–544.

Simpson, R.J., Cosgrove, C., Ingram, L.A., Florida-James, G.D., Whyte, G.P., Pircher, H., Guy, K., 2008. Senescent T-lymphocytes are mobilised into the peripheral blood compartment in young and older humans after exhaustive exercise. Brain Behav. Immun. 22, 544–551.

Simpson, R.J., Florida-James, G.D., Cosgrove, C., Whyte, G.P., Macrae, S., Pircher, H., Guy, K., 2007 Jul. J Appl Physiol (1985) 103 (1), 396–401. Epub 2007 Mar 22. PMID:17379755.

Skon, C.N., Lee, J.Y., Anderson, K.G., Masopust, D., Hogquist, K.A., Jameson, S.C., 2013. Transcriptional downregulation of S1pr1 is required for the establishment of resident memory CD8+ T cells. Nat. Immunol. 14, 1285–1293.

Song, K., Rabin, R.L., Hill, B.J., De Rosa, S.C., Perfetto, S.P., Zhang, H.H., Foley, J.F., Reiner, J.S., Liu, J., Mattapallil, J.J., Douek, D.C., Roederer, M., Farber, J.M., 2005. Characterization of subsets of CD4+ memory T cells reveals early branched pathways of T cell differentiation in humans. Proc. Natl. Acad. Sci. U. S. A. 102, 7916–7921.

Spaulding, C., Guo, W., Effros, R.B., 1999. Resistance to apoptosis in human CD8+ T cells that reach replicative senescence after multiple rounds of antigen-specific proliferation. Exp. Gerontol. 34, 633–644.

Speiser, D.E., Colonna, M., Ayyoub, M., Cella, M., Pittet, M.J., Batard, P., Valmori, D., Guillaume, P., Lienard, D., Cerottini, J.C., Romero, P., 2001. The activatory receptor 2B4 is expressed in vivo by human CD8+ effector alpha beta T cells. J. Immunol. 167, 6165–6170.

Stubbe, M., Vanderheyde, N., Pircher, H., Goldman, M., Marchant, A., 2008. Characterization of a subset of antigen-specific human central memory CD4+ T lymphocytes producing effector cytokines. Eur. J. Immunol. 38, 273–282.

Suh, W.K., 2015. Life of T follicular helper cells. Mol Cells 38, 195–201.

Sun, L., Jin, H., Li, H., 2016. GARP: a surface molecule of regulatory T cells that is involved in the regulatory function and TGF-beta releasing. Oncotarget 7, 42826–42836.

Surh, C.D., Sprent, J., 2008. Homeostasis of naive and memory T cells. Immunity 29, 848–862.

Szabo, S.J., Kim, S.T., Costa, G.L., Zhang, X., Fathman, C.G., Glimcher, L.H., 2015. Pillars article: a novel transcription factor, T-bet, directs Th1 lineage commitment. Cell 2000 (100), 655–669. *J Immunol*, 194, 2961-75.

Takahashi, T., Dejbakhsh-Jones, S., Strober, S., 2006. Expression of CD161 (NKR-P1A) defines subsets of human CD4 and CD8 T cells with different functional activities. J. Immunol. 176, 211–216.

Takeuchi, A., Badr Mel, S., Miyauchi, K., Ishihara, C., Onishi, R., Guo, Z., Sasaki, Y., Ike, H., Takumi, A., Tsuji, N.M., Murakami, Y., Katakai, T., Kubo, M., Saito, T., 2016. CRTAM determines the CD4+ cytotoxic T lymphocyte lineage. J. Exp. Med. 213, 123–138.

Tan, J.T., Ernst, B., Kieper, W.C., Leroy, E., Sprent, J., Surh, C.D., 2002. Interleukin (IL)-15 and IL-7 jointly regulate homeostatic proliferation of memory phenotype CD8+ cells but are not required for memory phenotype CD4+ cells. J. Exp. Med. 195, 1523–1532.

Tang, Y., Guan, S.P., Chua, B.Y., Zhou, Q., Ho, A.W., Wong, K.H., Wong, K.L., Wong, W.S., Kemeny, D.M., 2012. Antigen-specific effector CD8 T cells regulate allergic responses via IFN-gamma and dendritic cell function. J. Allergy Clin. Immunol. 129, 1611–1620 e4.

Terry, L.A., Brown, M.H., Beverley, P.C., 1988. The monoclonal antibody, UCHL1, recognizes a 180,000 MW component of the human leucocyte-common antigen, CD45. Immunology 64, 331–336.

Thome, J.J., Yudanin, N., Ohmura, Y., Kubota, M., Grinshpun, B., Sathaliyawala, T., Kato, T., Lerner, H., Shen, Y., Farber, D.L., 2014. Spatial map of human T cell compartmentalization and maintenance over decades of life. Cell 159, 814–828.

Topalian, S.L., Drake, C.G., Pardoll, D.M., 2012a. Targeting the PD-1/B7-H1(PD-L1) pathway to activate anti-tumor immunity. Curr. Opin. Immunol. 24, 207–212.

Topalian, S.L., Hodi, F.S., Brahmer, J.R., Gettinger, S.N., Smith, D.C., Mcdermott, D.F., Powderly, J.D., Carvajal, R.D., Sosman, J.A., Atkins, M.B., Leming, P.D., Spigel, D.R., Antonia, S.J., Horn, L., Drake, C.G., Pardoll, D.M., Chen, L., Sharfman, W.H., Anders, R.A., Taube, J.M., Mcmiller, T.L., Xu, H., Korman, A.J., Jure-Kunkel, M., Agrawal, S., Mcdonald, D., Kollia, G.D., Gupta, A., Wigginton, J.M., Sznol, M., 2012b. Safety, activity, and immune correlates of anti-PD-1 antibody in cancer. N. Engl. J. Med. 366, 2443–2454.

Trautmann, L., Janbazian, L., Chomont, N., Said, E.A., Gimmig, S., Bessette, B., Boulassel, M.R., Delwart, E., Sepulveda, H., Balderas, R.S., Routy, J.P., Haddad, E.K., Sekaly, R.P., 2006. Upregulation of PD-1 expression on HIV-specific CD8+ T cells leads to reversible immune dysfunction. Nat. Med. 12, 1198–1202.

Trifari, S., Kaplan, C.D., Tran, E.H., Crellin, N.K., Spits, H., 2009. Identification of a human helper T cell population that has abundant production of interleukin 22 and is distinct from T(H)-17, T(H)1 and T(H)2 cells. Nat. Immunol. 10, 864–871.

Tubo, N.J., Jenkins, M.K., 2014. TCR signal quantity and quality in CD4(+) T cell differentiation. Trends Immunol. 35, 591–596.

Turner, D.L., Bickham, K.L., Thome, J.J., Kim, C.Y., D'ovidio, F., Wherry, E.J., Farber, D.L., 2014. Lung niches for the generation and maintenance of tissue-resident memory T cells. Mucosal Immunol. 7, 501–510.

Uusitalo, M., Kivela, T., 2001. The HNK-1 carbohydrate epitope in the eye: basic science and functional implications. Prog. Retin. Eye Res. 20, 1–28.

Vahidi, Y., Faghih, Z., Talei, A.R., Doroudchi, M., Ghaderi, A., 2017. Memory CD4(+) T cell subsets in tumor draining lymph nodes of breast cancer patients: a focus on T stem cell memory cells. Cell. Oncol.

Valenzuela, H.F., Effros, R.B., 2002. Divergent telomerase and CD28 expression patterns in human CD4 and CD8 T cells following repeated encounters with the same antigenic stimulus. Clin. Immunol. 105, 117–125.

Vallejo, A.N., 2005. CD28 extinction in human T cells: altered functions and the program of T-cell senescence. Immunol. Rev. 205, 158–169.

Vallejo, A.N., Schirmer, M., Weyand, C.M., Goronzy, J.J., 2000. Clonality and longevity of CD4+CD28null T cells are associated with defects in apoptotic pathways. J. Immunol. 165, 6301–6307.

Van Leeuwen, E.M., Remmerswaal, E.B., Vossen, M.T., Rowshani, A.T., Wertheim-Van Dillen, P.M., Van Lier, R.A., Ten Berge, I.J., 2004. Emergence of a CD4+CD28- granzyme B+, cytomegalovirus-specific T cell subset after recovery of primary cytomegalovirus infection. J. Immunol. 173, 1834–1841.

Van Mierlo, G.J., Scherer, H.U., Hameetman, M., Morgan, M.E., Flierman, R., Huizinga, T.W., Toes, R.E., 2008. Cutting edge: TNFR-shedding by CD4+CD25+ regulatory T cells inhibits the induction of inflammatory mediators. J. Immunol. 180, 2747–2751.

Veldhoen, M., Hocking, R.J., Atkins, C.J., Locksley, R.M., Stockinger, B., 2006. TGFbeta in the context of an inflammatory cytokine milieu supports de novo differentiation of IL-17-producing T cells. Immunity 24, 179–189.

Velu, V., Titanji, K., Zhu, B., Husain, S., Pladevega, A., Lai, L., Vanderford, T.H., Chennareddi, L., Silvestri, G., Freeman, G.J., Ahmed, R., Amara, R.R., 2009. Enhancing SIV-specific immunity in vivo by PD-1 blockade. Nature 458, 206–210.

Verma, K., Ogonek, J., Varanasi, P.R., Luther, S., Bunting, I., Thomay, K., Behrens, Y.L., Mischak-Weissinger, E., Hambach, L., 2017. Human CD8+ CD57- TEMRA cells: Too young to be called "old". PLoS One 12, e0177405.

Victora, G.D., Schwickert, T.A., Fooksman, D.R., Kamphorst, A.O., Meyer-Hermann, M., Dustin, M.L., Nussenzweig, M.C., 2010. Germinal center dynamics revealed by multi-photon microscopy with a photoactivatable fluorescent reporter. Cell 143, 592–605.

Vieira, P.L., Christensen, J.R., Minaee, S., O'neill, E.J., Barrat, F.J., Boonstra, A., Barthlott, T., Stockinger, B., Wraith, D.C., O'garra, A., 2004. IL-10-secreting regulatory T cells do not express Foxp3 but have comparable regulatory function to naturally occurring CD4+CD25+ regulatory T cells. J. Immunol. 172, 5986–5993.

Vignali, D.A., Collison, L.W., Workman, C.J., 2008. How regulatory T cells work. Nat. Rev. Immunol. 8, 523–532.

Voehringer, D., Koschella, M., Pircher, H., 2002. Lack of proliferative capacity of human effector and memory T cells expressing killer cell lectinlike receptor G1 (KLRG1). Blood 100, 3698–3702.

Von Andrian, U.H., Mackay, C.R., 2000. T-cell function and migration. Two sides of the same coin. N. Engl. J. Med. 343, 1020–1034.

Wakim, L.M., Woodward-Davis, A., Bevan, M.J., 2010. Memory T cells persisting within the brain after local infection show functional adaptations to their tissue of residence. Proc. Natl. Acad. Sci. U. S. A. 107, 17872–17879.

Wang, R., Wan, Q., Kozhaya, L., Fujii, H., Unutmaz, D., 2008. Identification of a regulatory T cell specific cell surface molecule that mediates suppressive signals and induces Foxp3 expression. PLoS One 3, e2705.

Watanabe, R., Gehad, A., Yang, C., Scott, L.L., Teague, J.E., Schlapbach, C., Elco, C.P., Huang, V., Matos, T.R., Kupper, T.S., Clark, R.A., 2015. Human skin is protected by four functionally and phenotypically discrete populations of resident and recirculating memory T cells. Sci. Transl. Med. 7, 279ra39.

Weiss, J.M., Bilate, A.M., Gobert, M., Ding, Y., Curotto De Lafaille, M.A., Parkhurst, C.N., Xiong, H., Dolpady, J., Frey, A.B., Ruocco, M.G., Yang, Y., Floess, S., Huehn, J., Oh, S., Li, M.O., Niec, R.E., Rudensky, A.Y., Dustin, M.L., Littman, D.R., Lafaille, J.J., 2012. Neuropilin 1 is expressed on thymus-derived natural regulatory T cells, but not mucosa-generated induced Foxp3+ T reg cells. J. Exp. Med. 209, 1723–1742. S1.

Weng, N.P., Akbar, A.N., Goronzy, J., 2009. CD28(-) T cells: their role in the age-associated decline of immune function. Trends Immunol. 30, 306–312.

Wherry, E.J., 2011. T cell exhaustion. Nat. Immunol. 12, 492–499.

Wherry, E.J., Blattman, J.N., Murali-Krishna, K., Van Der Most, R., Ahmed, R., 2003. Viral persistence alters CD8 T-cell immunodominance and tissue distribution and results in distinct stages of functional impairment. J. Virol. 77, 4911–4927.

Wherry, E.J., Kurachi, M., 2015. Molecular and cellular insights into T cell exhaustion. Nat. Rev. Immunol. 15, 486–499.

Williams, M.A., Bevan, M.J., 2007. Effector and memory CTL differentiation. Annu. Rev. Immunol. 25, 171–192.

Wills, M.R., Okecha, G., Weekes, M.P., Gandhi, M.K., Sissons, P.J., Carmichael, A.J., 2002. Identification of naive or antigen-experienced human CD8(+) T cells by expression of costimulation and chemokine receptors: analysis of the human cytomegalovirus-specific CD8(+) T cell response. J. Immunol. 168, 5455–5464.

Wilson, N.J., Boniface, K., Chan, J.R., Mckenzie, B.S., Blumenschein, W.M., Mattson, J.D., Basham, B., Smith, K., Chen, T., Morel, F., Lecron, J.C., Kastelein, R.A., Cua, D.J., Mcclanahan, T.K., Bowman, E.P., De Waal Malefyt, R., 2007. Development, cytokine profile and function of human interleukin 17-producing helper T cells. Nat. Immunol. 8, 950–957.

Wolchok, J.D., Kluger, H., Callahan, M.K., Postow, M.A., Rizvi, N.A., Lesokhin, A.M., Segal, N.H., Ariyan, C.E., Gordon, R.A., Reed, K., Burke, M.M., Caldwell, A., Kronenberg, S.A., Agunwamba, B.U., Zhang, X., Lowy, I., Inzunza, H.D., Feely, W., Horak, C.E., Hong, Q., Korman, A.J., Wigginton, J.M., Gupta, A., Sznol, M., 2013. Nivolumab plus ipilimumab in advanced melanoma. N. Engl. J. Med. 369, 122–133.

Wong, M.T., Chen, J., Narayanan, S., Lin, W., Anicete, R., Kiaang, H.T., De Lafaille, M.A., Poidinger, M., Newell, E.W., 2015. Mapping the diversity of follicular helper T cells in human blood and tonsils using high-dimensional mass cytometry analysis. Cell Rep. 11, 1822–1833.

Workman, A.M., Jacobs, A.K., Vogel, A.J., Condon, S., Brown, D.M., 2014. Inflammation enhances IL-2 driven differentiation of cytolytic CD4 T cells. PLoS One 9, e89010.

Xie, Y., Akpinarli, A., Maris, C., Hipkiss, E.L., Lane, M., Kwon, E.K., Muranski, P., Restifo, N.P., Antony, P.A., 2010. Naive tumor-specific CD4(+) T cells differentiated in vivo eradicate established melanoma. J. Exp. Med. 207, 651–667.

Xue, L., Barrow, A., Pettipher, R., 2009. Novel function of CRTH2 in preventing apoptosis of human Th2 cells through activation of the phosphatidylinositol 3-kinase pathway. J. Immunol. 182, 7580–7586.

Yamada, A., Kaneyuki, T., Hara, A., Rothstein, D.M., Yokoyama, M.M., 1992. CD45 isoform expression on human neonatal T cells: expression and turnover of CD45 isoforms on neonatal versus adult T cells after activation. Cell. Immunol. 142, 114–124.

Youngblood, B., Noto, A., Porichis, F., Akondy, R.S., Ndhlovu, Z.M., Austin, J.W., Bordi, R., Procopio, F.A., Miura, T., Allen, T.M., Sidney, J., Sette, A., Walker, B.D., Ahmed, R., Boss, J.M., Sekaly, R.P., Kaufmann, D.E., 2013. Cutting edge: Prolonged exposure to HIV reinforces a poised epigenetic program for PD-1 expression in virus-specific CD8 T cells. J. Immunol. 191, 540–544.

Youngblood, B., Oestreich, K.J., Ha, S.J., Duraiswamy, J., Akondy, R.S., West, E.E., Wei, Z., Lu, P., Austin, J.W., Riley, J.L., Boss, J.M., Ahmed, R., 2011. Chronic virus infection enforces demethylation of the locus that encodes PD-1 in antigen-specific CD8(+) T cells. Immunity 35, 400–412.

Zanon, V., Pilipow, K., Scamardella, E., De Paoli, F., De Simone, G., Price, D.A., Martinez Usatorre, A., Romero, P., Mavilio, D., Roberto, A., Lugli, E., 2017. Curtailed T-cell activation curbs effector differentiation and generates CD8(+) T cells with a naturally-occurring memory stem cell phenotype. Eur. J. Immunol. 47, 1468–1476.

Zaunders, J.J., Dyer, W.B., Wang, B., Munier, M.L., Miranda-Saksena, M., Newton, R., Moore, J., Mackay, C.R., Cooper, D.A., Saksena, N.K., Kelleher, A.D., 2004. Identification of circulating antigen-specific CD4+ T lymphocytes with a CCR5+, cytotoxic phenotype in an HIV-1 long-term nonprogressor and in CMV infection. Blood 103, 2238–2247.

Zelle-Rieser, C., Thangavadivel, S., Biedermann, R., Brunner, A., Stoitzner, P., Willenbacher, E., Greil, R., Johrer, K., 2016. T cells in multiple myeloma display features of exhaustion and senescence at the tumor site. J. Hematol. Oncol. 9, 116.

Zhang, N., Bevan, M.J., 2011. CD8(+) T cells: foot soldiers of the immune system. Immunity 35, 161–168.

Zheng, W., Flavell, R.A., 1997. The transcription factor GATA-3 is necessary and sufficient for Th2 cytokine gene expression in CD4 T cells. Cell 89, 587–596.

Zhou, L., Chong, M.M., Littman, D.R., 2009. Plasticity of CD4+ T cell lineage differentiation. Immunity 30, 646–655.

Zhou, L., Ivanov, I., Spolski, R., Min, R., Shenderov, K., Egawa, T., Levy, D.E., Leonard, W.J., Littman, D.R., 2007. IL-6 programs T(H)-17 cell differentiation by promoting sequential engagement of the IL-21 and IL-23 pathways. Nat. Immunol. 8, 967–974.

CHAPTER THREE

T Lymphocytes and Autoimmunity

Uqba Khan[1] and Hareem Ghazanfar
St. John Hospital and Medical Centre, Detroit, MI, United States
[1]Corresponding author: E-mail: uqba.md@gmail.com

Contents

International Review of Cell and Molecular Biology, Volume 342
ISSN 1937-6448
https://doi.org/10.1016/bs.ircmb.2018.05.008

Abstract

T lymphocytes play a central role in regulation of the immune system. Both effector and regulatory T cells work in equilibrium to provide optimal immune response against foreign pathogens. Normally, T cells do not react against self-antigens because of the presence of central and peripheral immunogenic tolerance. Central tolerance eliminates autoreactive naive T cells that develop in thymus by presenting them with self-antigens on the thymic cells. The autoreactive T cells that escape thymus are subjected to additional mechanisms such as clonal anergy, ignorance, and deletion. Moreover, the regulatory T cells (Tregs), specifically $CD4^{+}CD25^{+}Foxp3^{+}$ Tregs, exert a tight control over autoreactive B and T cell responses in the periphery. Failure of any one of these checkpoints can cause uncontrolled expansion of these self-reactive T cells leading to the development of autoimmune diseases. In this chapter we discuss the key role of T cells in the underlying pathogenesis of autoimmune responses. We review the role of T-cell receptor and signaling pathways including costimulatory pathways. We also review T cell inhibitory receptors such as programmed cell death protein 1 and cytotoxic T lymphocyte—associated antigen 4 and T cell—related important cytokines (interleukin [IL]-2, IL-6, IL-17, IL-7, and IL-33) involved in autoimmunity. Defects in genes responsible for T-cell regulation and function are also discussed in detail which forms the basis of many important autoimmune disorders. Various therapeutic measures that target T cells in the management of autoimmune diseases are also highlighted.

1. INTRODUCTION

Autoimmunity is defined as an immune response directed against an antigen within the body of the host. Autoimmune diseases are chronic conditions initiated by the loss of immunological tolerance to self-antigens and represent a heterogeneous group of disorders that can be organ specific or involve multiple organ systems within a host (Anaya, 2012). Autoimmunity is relatively a common phenomenon and develops in up to 5% of the population (Goodnow, 2007). It has significant impact on health-care delivery across the world. There are a number of autoimmune diseases that can affect human beings and can be broadly divided into organ-specific or systemic autoimmune diseases. Organ-specific autoimmune disease includes Graves disease, type 1 diabetes mellitus, and autoimmune hemolytic anemia, whereas examples of systemic autoimmune diseases include systemic lupus erythematosus (SLE) and rheumatoid arthritis (RA). These diseases are complex and heterogeneous because of their etiology, pathogenesis, clinical presentation, and treatment modalities. Humans have intricate immune system that works in series of various molecular checks and balances to ensure measured and appropriate response to foreign threats. However, in case of autoimmune disorders, these regulatory mechanisms are dysfunctional and can serve as potential avenues for therapeutic intervention (Zhang and Vignali, 2016). Several factors play an important role in development of autoimmunity including genetic susceptibility, environmental stimuli, and host factors. However, T cells remain one of the most important factors in pathogenesis of autoimmunity. This chapter will focus on the role of T cells in pathogenesis and regulation of autoimmunity.

2. MECHANISMS OF AUTOIMMUNITY

Autoimmunity develops because of loss of self-tolerance to an individual's own antigens. It is crucial to understand the mechanisms by which the body keeps immunologic tolerance before explaining detailed pathogenesis of autoimmunity itself. Following is a brief overview of the way immune response and self-tolerance work to prevent development of autoimmune diseases.

2.1 Immunologic Tolerance

Immune system has to work constantly to generate specific responses against a variety of harmful molecules that enter the human body. It has the ability to adapt itself to create enormous diversity in T-cell and B-cell receptors (Kambayashi, 2018). The precise and specific immune response is developed because of immunologic specificity of the antigen receptors of T cells and B cells as a result of random shuffling of many genes (V [Variable], D [Diversity], and J [Joining] genes) that form the DNA code for the antigen-binding site of these receptors (Delves and Roitt, 2000). Theoretically this process could generate 10^9 different T-cell receptors (TCRs) (Kamradt and Mitchison, 2001). As a result, the system inevitably generates many receptors that recognize self-antigens. To overcome this issue, immune system has developed mechanisms to stop self-destruction (Schwartz, 2012). These mechanisms are called self-tolerance or immunologic tolerance mechanisms. These self-tolerance mechanisms eliminate or inactivate lymphocytes with receptors capable of recognizing self-antigens. Immunologic tolerance mechanisms are divided into two groups: central tolerance and peripheral tolerance.

2.2 Central Tolerance

During development and maturation of lymphocytes in central lymphoid organs (thymus for T cells and bone marrow for B cells), self-reactive lymphocytes are destroyed by central tolerance mechanisms. When immature T cells migrate from the bone marrow to the thymus, they are presented with peptides derived from endogenous proteins bound to major histocompatibility complex (MHC) molecules (Kamradt and Mitchison, 2001). Initially the double-positive lymphocytes ($CD4^+$ $CD8^+$ lymphocytes) are presented with self-peptides bound to class I or II MHC molecules present on the cortical epithelial cells. The lymphocytes that bind with intermediate affinity get a survival signal and differentiate into mature single-positive lymphocytes. This is called positive selection that is responsible for commitment of lymphocytes to either $CD4^+$ or $CD8^+$ lineage (Sha et al., 1988). These lymphocytes then migrate to the corticomedullary junction of the thymus where they interact with MHC—self-peptide complex present on bone marrow—derived macrophages and dendritic cells. At this stage the lymphocytes that bind with high affinity to MHC—self-peptide complex undergo negative selection and die by apoptosis. In this way autoreactive T cells get eliminated from the body (Klein et al., 2014).

> *Consequently, positive and negative selection in the thymus generates a diverse collection of T cells containing a variety of TCRs that recognize self-MHC but are not strong enough to cause self-reactivity and hence autoimmunity (Kambayashi, 2018). It is important to mention the role of medullary thymic epithelial cells (mTECs). These cells express genes specific for different tissues from the whole body and present these tissue-restricted antigens (TRAs) to the developing T cells and help in negative selection of the T cells that bind strongly with these TRAs. Another function of mTECs is that they skew autoreactive T cells into regulatory T cells (Treg) that migrate to the periphery and help in suppressing the autoreactive T cells that may have escaped central tolerance.*
>
> ***Shi and Zhu (2013)***

mTECs also possess specific transcription factors i.e., autoimmune regulator transcription factors (AIRE) that play an important role in central tolerance. These proteins upregulate the expression of TRAs and thus help in efficient negative selection of autoreactive T cells. Mutation of AIRE causes breakdown of central tolerance due to decreased expression of TRAs (Siggs et al., 2006; Kyewski and Klein, 2006; Anderson et al., 2002). This may lead to the development of multisystem autoimmune diseases such as autoimmune polyglandular syndrome-type I (APS-1, also known as APECED). Mice with mutations in AIRE gene have been shown to develop pathological autoimmune features similar to APECED with multiorgan lymphocytic infiltration and autoantibody production (Peterson et al., 2008).

2.3 Peripheral Tolerance

Central tolerance mechanism remains imperfect and incomplete. Consequently, autoreactive T cells escape the central tolerance and enter into periphery. Fortunately, various mechanisms exist which can counteract these autoreactive T cells to prevent autoimmunity. These mechanisms are collectively called peripheral tolerance. A brief overview of these mechanisms is described in the following.

2.4 Anergy

Anergy is a phenomenon by which lymphocytes that recognize self-antigens become functionally unresponsive. Activation of T cells requires the recognition of the antigen by the TCR as well by the costimulator molecules present on the T cells. Anergy may be induced if the normal antigenic stimulus is received for a prolonged time and/or is received in the absence of costimulation or from an altered and/or chronic TCR stimulus (Choi and Schwartz, 2007). One of the most prominent costimulatory pathways is the CD28:B7 axis that promotes T-cell proliferation and survival through amplification of TCR signaling and interleukin-2 (IL-2) production (Lafferty and Cunningham, 1975; Mueller et al., 1989). Prolonged antigen recognition alone without the activation of these costimulatory pathways may lead to anergy. In this way T cells become unresponsive when they are continuously displayed to self-antigens in the absence of innate immunity and strong costimulation (Abbas et al., 2017).

Several mechanisms have been described in literature to explain how the state of anergy is maintained. T cells that recognize self-antigens receive an inhibitory signal from receptors that are structurally similar to CD28. Two well-described inhibitory receptors are cytotoxic T lymphocyte—associated protein 4 (CTLA-4) and programmed cell death protein 1 (PD-1) (Vinay Kumar et al., 2015). The role of this mechanism is described in detail later in the chapter. One less known mechanism is blockage of TCR-induced signal transduction by T cells. Other mechanisms include activation of ubiquitin ligases that ubiquinate TCR-associated proteins and cause their proteolytic degradation in lysosomes.

2.5 Clonal Ignorance

Clonal ignorance is a phenomenon in which autoreactive T cells simply ignore self-antigen through multiple proposed mechanisms. One explanation is that the amount of antigen is not enough to induce activation of T cells that can cause autoimmunity (Akkaraju et al., 1997; Ferber et al., 1994). In other instances, there is a physical barrier that can prevent autoreactive T cells from attacking self-antigens e.g., blood—brain barrier (Barker and Billingham, 1977) or lack of ability of lymphocytes for penetrance into normal tissue other than lymphoid organs (Butcher and Picker, 1996). Moreover, T cells express both Fas and its ligand; hence on activation, the engagement between the two molecules can induce apoptosis (Dhein et al., 1995).

2.6 Deletion

Autoreactive T cells that recognize self-antigens undergo apoptosis instead of activation; this process is called deletion. A proposed mechanism for this process is through Fas—Fas ligand system. Interaction of the Fas receptor with Fas ligand (FasL) promotes apoptosis in Fas-positive cells (Suda et al., 1993). Because both Fas and FasL are present on T lymphocytes, engagement of Fas with FasL induces apoptosis of T cells. This mechanism carries extreme importance in pathogenesis of autoimmune lymphoproliferative syndrome (ALPS) in which there is mutation in Fas gene that regulates apoptosis of lymphocytes. This syndrome is clinically characterized by development of autoimmunity and lymphoproliferative disorders (Kuehn et al., 2011).

2.7 Inhibition

As discussed earlier, T-cell activation requires stimulation of TCR as well as of costimulatory receptors (B7:CD28 complexes). T cells also possess inhibitory receptors, which keep a check on the magnitude of immune response. Many such inhibitory receptors are described in literature. Among them, CTLA-4 and PD-1 receptors remain the most studied and carry great therapeutic implications as well (Khan et al., 2017).

2.8 Cytotoxic T Lymphocyte—Associated Antigen 4

Cytotoxic T lymphocyte—associated antigen 4 (CTLA-4) is an inhibitory receptor, which is present on T cells. It is a homolog of CD28 and binds both to B7-1 and B7-2 with greater affinities than CD28 (Leach et al., 1996). CTLA-4 is induced after activation on $CD4^+$ and $CD8^+$ conventional T cells; however, it is constitutively expressed on Treg. Of note, CTLA-4 expression has been reported in other cells as well, including B cells, dendritic cells, monocytes, granulocytes, $CD34^+$ stem cells, placental fibroblasts, mouse embryonic cells, and embryoid bodies (Schildberg et al., 2016).

T cells tightly regulate CTLA-4 expression. The majority of CTLA-4 molecules are contained in the intracellular compartments of perinuclear Golgi vesicles, endosomes, and lysosomes (Linsley et al., 1996). After

TCR stimulation, CTLA-4 migrates to the cell surface (Schneider et al., 1999). CTLA-4 is an endocytic receptor, and its cytoplasmic tail contains a motif that connects it to clathrin, a protein involved in receptor-mediated endocytosis. Once CTLA-4 binds to B7, it causes removal and ingestion of the complex. In this way it lowers down the B7 levels and thus causes decreased costimulation via B7—CD28 interaction (Abbas et al., 2017).

2.9 Programmed Cell Death Protein 1

PD-1 is another very well-studied and important inhibitory receptor that puts break into immune response and hence prevents autoreactive T cells from causing damage to normal organs. PD-1 is expressed on various cells, including $CD4^+$ and $CD8^+$ T cells, natural killer cells, natural killer T cells, B cells, macrophages, and some dendritic subsets as well. PD-1 recognizes two ligands, PD-L1 and PD-L2. Both PD-L1 and PD-L2 are expressed on various cells of hematopoietic and nonhematopoietic origin.

After PD-1 binds with either of its ligand PD-L1 or PD-L2, it is phosphorylated on its Immunoreceptor Tyrosine Based Inhibitory Motif (ITIM) (Y223) and Immunoreceptor Tyrosine Based Switch Motif (ITSM) (Y248) tyrosine motifs, which results in recruitment of Protooncogene Tyrosine Protein Kinase C (Src) homology region 2 domain—containing phosphatases and downregulation of TCR signaling through dephosphorylation of signaling intermediates such as CD3ζ, ZAP-70, and PKCθ in T cells (Schildberg et al., 2016; Chemnitz et al., 2004; Okazaki et al., 2001; Sheppard et al., 2004).

Both CTLA-4 and PD-1 are in forefront of cancer immunotherapy approaches. There are number of CTLA-4 and PD-1/PD-l1 inhibitors (nivolumab, pembrolizumab, atezolizumab, and durvalumab) approved for treatment of various malignancies, including melanoma, non—small cell lung cancer, Hodgkin lymphoma, microsatellite instability—high (MSI) tumors, etc.

2.10 B and T Lymphocyte Attenuator

Another inhibitory receptor is B and T lymphocyte attenuator (BTLA) that is similar to CTLA-4. It has two immunoreceptor tyrosine—based inhibitory motifs.

It is expressed by T helper cells type 1. Crosslinking BTLA with antigen receptor leads to its tyrosine phosphorylation and recruitment of SH2-containing protein tyrosine phosphatases and attenuates the production of IL-2. BTLA deficient mice show increased proliferation of T lymphocytes and increases antibody responses (Watanabe et al., 2003). In future, BTLA might play a role in immunotherapy against malignancies. In vitro studies show that antibodies that block BTLA interaction with Herpesvirus entry mediator (HVEM) lead to increased proliferation of melanoma-specific $CD8^+$ T cells, resulting in efficient immune response against cancer cells (Spodzieja et al., 2017).

2.11 V-Domain Ig-Containing Suppressor of T-Cell Activation

V-domain Ig-containing suppressor of T-cell activation (VISTA) is another inhibitory receptor that is expressed by T cells. Its overexpression causes suppression of T-cell activation, proliferation, and cytokine production. In vivo blockade of VISTA has been found to enhance T-cell response and exacerbate the development of experimental autoimmune encephalomyelitis.

Experiments also demonstrate that blockade of VISTA can impair tumor growth in malignancies by augmenting T-cell response (Lines et al., 2014).

2.12 LAG3

Lymphocyte Activation Gene 3 (LAG3) (CD223) is a recently discovered cell surface inhibitory molecule present on activated T cells, NK cells, and dendritic cells. The interaction between LAG3 and its ligand MHC class II suppresses $CD8^+$ T lymphocytes. In vitro trials show that monoclonal anti-LAG3 antibodies that block LAG3/class II interaction may be effective in immunotherapy against cancers such as melanoma (Goldberg and Drake, 2011).

2.13 T Cell Immunoglobulin and Mucin Domain 3

T cell immunoglobulin and mucin domain 3 (TIM-3) is a member of TIM gene family that includes TIM-1, TIM-2, and TIM-4 in humans and TIM 1–8 in mice. It is involved in suppressing proximal T-cell signaling in $CD4^+$ and $CD8^+$ T cells. In TIM-3^+ $CD8^+$ effector T cells, Gal-9 is shown to induce colocalization of TIM-3 with receptor phosphatases CD45 and CD148, leading to inhibition of TCR signaling. In this way it is involved in mediating immune tolerance (Du et al., 2017). Treatment with TIM-3 mAb enhances the severity of experimental autoimmune encephalomyelitis (EAE) (Monney et al., 2002). Also, lower expression of TIM-3 on T cells and higher production of interferon (IFN)-γ can be detected in patients suffering from multiple sclerosis (MS) (Yang et al., 2008). Similar to other inhibitory molecules, blockade of TIM-3 pathway may prove to be effective in enhancing immune response against tumors as demonstrated in several preclinical trials (Du et al., 2017).

2.14 Role of Tregs

Tregs are a subset of helper T cells, main function if which is the regulation or suppression of immune response. There are several types of T cells that have regulatory activity, including subpopulations of γδ T cells, NKT cells, and $CD8^+$ and $CD4^+$ T cells (Bach, 2003). Treg cells can develop either in the thymus or in the periphery and are critical for maintaining tolerance. CD4 Tregs can be divided into two categories i.e., naturally occurring and induced. Naturally occurring Tregs are $CD25^+$ and $Foxp3^+$. FoxP3 is a member of the forkhead family of transcription factors and is important for the development and function of most Tregs. Naturally occurring Tregs are primarily generated in the thymus but can also arise from periphery by naïve CD4 T cells, whereas induced T cells differentiate from naïve T cells in the periphery (Shimon Sakaguchi, 2013).

In vitro experiments have shown that once $Foxp3^+CD25^+$Tregs are activated by TCR stimulation by specific antigen, they suppress other CD4 and CD8 T cells, irrespective of their antigen specificities (Piccirillo and Thornton, 2004). They can also suppress the function of already differentiated Th1 and Th2 cells in vitro and suppress B cells, NK cells, and NKT cells (Wing et al., 2005; Shimon Sakaguchi, 2013). There are several ways by which Tregs exert their suppressive effect. Some of these mechanisms include production of immunosuppressive cytokines such as IL-10 and Transforming Growth Factor (TGF)-β and consumption of IL-2 to deprive other immune cells of this growth factor. Moreover, they can also dampen the activity of antigen-presenting cells (APCs) to stimulate T cells especially through CTLA-4 interaction.

It is important to mention the importance of another subset of Tregs called Tr1 cells (Salvatore Albani, 2016). They are characterized by the production of IL-10 and the expression of CD49b and LAG-3. TR1 cells suppress APC responses as well as T-cell responses through the secretion of IL-10 and TGF-β. Tr1 cells inhibit T-cell responses by CTLA-4 or PD-1 and by disrupting the metabolic state of T effector cells via the production of the ectoenzymes CD39 and CD73. They also release granzyme B and perforin and kill myeloid cells. That is why it is suggested that Tr1 cells may have a role in underlying pathogenesis of infectious diseases, autoimmunity, and transplant rejection (Roncarolo et al., 2014).

3. WHAT GOES WRONG IN AUTOIMMUNITY WITH REGARD TO T CELLS?

Immune system constantly works to fight off the offending molecules in the human body. There is a fine balance between the process of self-tolerance and autoimmunity. Whenever the balance is shifted away from self-tolerance, autoimmunity develops, and that can lead to autoimmune diseases as shown in Fig. 1.

Autoimmunity is not a simple process to develop. There are several pathological mechanisms, which are still not clearly understood, that lead to the development of autoimmunity and autoimmune diseases. These mechanisms may include infections, environmental factors, and genetic alterations. Following are some of these mechanisms that play a role in development of autoimmunity.

3.1 Role of T Cell Receptor and Signaling Pathways

Once the T cell is stimulated by the attachment of the TCR with MHC molecule on an APC, a number of intricate signaling processes occur which lead to gene expression, cytokine production, and promotion of effector function. In autoimmune diseases, these signaling pathways have aberrations

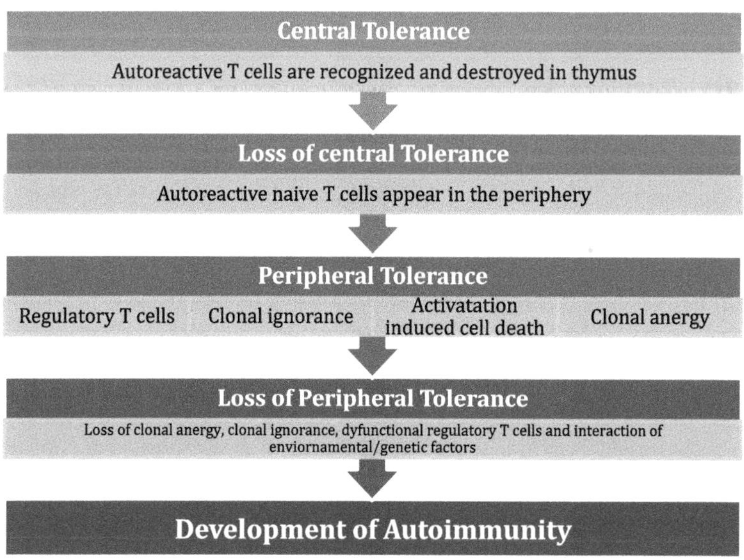

Figure 1 Tolerance and autoimmunity.

that contribute to development of autoimmunity. Apart from signaling dysfunction, autoreactive T cells also have unusual binding properties for their cognate MHC-peptide ligands (Bluestone et al., 2015).

T cells that have abnormal TCR topologies escape thymic deletion because of aberrant binding to MHC that is needed for apoptosis (Wucherpfennig et al., 2009). It is reported that TCRs from patients with MS and type 1 diabetes have altered structure that makes them autoreactive for disease-specific self-peptides (Hahn et al., 2005; Bulek et al., 2012). Similarly, defects in T-cell signaling also impacts the development of autoimmunity. In case of SLE, T cells isolated from the patients show increased phosphorylation of signaling proteins and an enhanced calcium influx (Moulton et al., 2015; Crispin et al., 2008). In SLE T cells the expression of the CD3ζ chain is decreased as compared with healthy individuals. This leads to defective signaling through the TCR, resulting in increased calcium influx. Restoration of the levels of CD3ζ in SLE T cells corrected the IL-2 production, hence emphasizing the importance of the CD3ζ chain in T-cell signaling (Nambiar et al., 2003). Defects in downstream signaling pathways, such as MAPK pathway, and extracellular signal—regulated kinase signaling pathway also have a role in autoimmunity, especially in SLE (Moulton et al., 2015).

3.2 Gene Expression and T Cells

There are a number of studies that have shown strong genetic basis of autoimmune diseases. For instance, in RA, the patient's genome clearly plays a role in susceptibility and severity of the disease. The most noticeable genes are the class II MHC (Okada et al., 2014). Similarly, T-cell signaling phosphatase encoded by PTPN11 can increase risk by about twofold in individuals with RA. Presence of Human Leukocyte Antigen (HLA)-B27 increases the chance of development of ankylosing spondylitis by 100 times compared with individuals who are B27 negative. The association of class II HLA-DR and HLA-DQ alleles with autoimmune diseases has received great attention, mainly because class II MHC molecules are involved in the selection and activation of $CD4^+$ T cells and $CD4^+$ T cells regulate both humoral and cell-mediated immune responses to protein antigens (Abbas et al., 2017). It has been established in certain cases of autoimmune diseases that the disease-associated MHC molecule may present a self-peptide, which leads to activation of pathological T cells. In certain cases it is hypothesized that certain genetic alleles are protective against autoimmunity probably because of induction of negative selection of autoreactive T cells. One of them is FasL gene that causes death receptor–mediated apoptosis through Fas. Others include Bcl-2 family member Bim and orphan steroid receptor nur77. Bim induces mitochondrial dysfunction leading to apoptosis of autoreactive T cells, whereas the molecular mechanism of nur77-mediated apoptosis remains unclear (Tischner et al., 2010).

Various non-HLA genes associated with autoimmunity have also been identified especially with the help of whole genome sequencing. Some of these genes, which are involved in T-cell function and regulation, are described in the following.

3.3 Protein Tyrosine Phosphatase PTPN22

Various autoimmune diseases including RA, type 1 diabetes, and autoimmune thyroiditis are associated with a variant of the protein tyrosine phosphatase PTPN22 (arginine at position 620 is replaced with a tryptophan). PTPN22 gene encodes a lymphoid-specific phosphatase (Lyp) and is present at chromosome 1p13.3–p13.1. Lyp is bound through a proline-rich motif to the SH3 domain of the Csk kinase. It has an important function of suppressing the kinases that mediate T-cell activation (Cohen et al., 1999; Lee et al., 2007). Defect of this gene causes the disruption of Lyp and Csk interaction, which ultimately causes hyperresponsiveness of T cells and consequently increases the chances of autoimmunity (Bottini et al., 2004; Begovich et al., 2004).

3.4 IL Receptor Polymorphism

There are a number of cytokines that are involved in activation, maintenance, and differentiation of various subtypes of T cells. Polymorphisms in the receptor of these interleukins (for example, IL-2, IL-23R, IL-7Rα, and IL-12B) may be involved in increased susceptibility for development of autoimmune diseases such as SLE, RA, and psoriasis.

3.5 Lymphocyte Regulation Genes

There are many genes that are involved in regulation of lymphocyte function, including CTLA-4 and FcγRIIB. These are specifically involved in Tregs function and T-cell inhibition. Studies have shown their involvement in development of various autoimmune diseases, such as SLE, type 1 diabetes, and RA.

4. ROLE OF COSTIMULATORY PATHWAYS

There are many costimulatory pathways involved in normal T-cell interaction with APCs which activate and enhance the function of T cells. These pathways may be altered in autoimmune diseases leading to development and progression of these diseases. Some of these pathways are briefly discussed in the following.

4.1 CD40 Ligand

CD40L is present on T cells, whereas CD40 is present on various cells including APCs and B cells. The interactions of CD40L with CD40 play an important role in autoimmunity. First, CD40 along with Tumor Necrosis Factor Receptor (TNFR) family member Receptor Activator of Nuclear Factor Kappa B (RANK) promotes the development of mTECs. The mTECs mediate negative selection against autoreactive T cells by expressing peripheral-restricted tissue antigens. Thus disruption of CD40–CD40L interactions in thymus could possibly contribute to failure of central tolerance by downregulating the mTECs, which ultimately increases the susceptibility toward autoimmunity (Akiyama et al., 2008). Second, CD40 signaling also increases IL-6 and other proinflammatory cytokines production, which enhances the T-cell differentiation into Th17 cells (Peters et al., 2009; Iezzi et al., 2009). Third, aberrant expression of CD40 on tissues in which it is normally undetectable may also be responsible in initiating autoimmunity. Aberrant CD40 expression on the thyroid tissue has been demonstrated to be involved in causing Graves' disease (GD) (Jacobson et al., 2007). The blockade of CD40L–CD40 interaction is a potential therapeutic target for new drug development as well.

4.2 ICOS

Another important costimulatory pathway is Inducible T Cell Costimulator (ICOS)–B7RP1. ICOS is heavily expressed on T cells of patients with autoimmune diseases, especially SLE, and B7RP1 is expressed on APCs and B cells. Increased activation of this pathway may result in development of autoimmunity. In vitro experiments established that ICOS enhances T-cell responses, including proliferation, cytokine production, upregulation of molecules facilitating cell—cell interaction, and production of antibodies by B cells (Hu et al., 2009).

4.3 CD27—CD70 Pathway

CD27 is a member of the TNF receptor family. It is also constitutively expressed on central memory T and naive T cells. CD27 interacts with CD70 that is present on B cells, dendritic cells, and activated T cells (Goronzy and Weyand, 2008). CD27 signaling enhances proliferation, survival, and differentiation of T cells (Hendriks et al., 2003; Lens et al., 1998). CD27—CD70 pathway may play an important role in autoimmunity which has been demonstrated in several murine models (Nakajima et al., 2000; Arens et al., 2001).

4.4 CD28-Mediated Costimulation

CD28 is expressed on all naive T cells and is stimulated by CD80 and CD86 (Goronzy and Weyand, 2008). CD80 and CD86 are present on dendritic cells, B cells, and other APCs, such as monocytes. Overactivation of this costimulatory pathway may contribute to autoimmunity and is an area of therapeutic approach as well to stop the interaction of CD28 with CD80/CD86 to control the autoimmune diseases. Abatacept is a drug that prevents CD28 from binding to CD80/CD86. This prevents increased T effector cell activity. It also prevents the suppression of Treg activity (Dilek et al., 2013).

4.5 Signaling Lymphocyte Activation Molecule Family Receptors

Signaling lymphocyte activation molecule family (SLAMF) has been identified as a group of receptors that regulate the activation and differentiation of a broad range of cell types involved in innate and adaptive immunity. There is an increasing evidence of their involvement in development of autoimmune diseases; for instance, increased expression of SLAM receptors SLAMF3 and SLAMF6 in SLE T lymphocytes promotes Th17 differentiation (Chatterjee et al., 2012).

Role of costimulatory pathways in autoimmunity is elaborated in Table 1.

Table 1 Costimulatory Pathways and Their Role in Autoimmunity

Costimulatory Pathways	Role in Pathogenesis of Autoimmunity	Autoimmune Diseases	References
CD40L-CD40 pathway	Disruption of this pathway in thymus causes downregulation of mTECs and loss of central tolerance. Excessive activation enhances T-cell differentiation into Th17 cells by increased production of IL-6 and other proinflammatory cytokines. Aberrant expression of CD40 on tissues where it is normally undetectable causes loss of peripheral tolerance.	Graves' disease and type 1 diabetes	Akiyama et al. (2008), Peters et al. (2009), Iezzi et al. (2009), and Jacobson et al. (2007)
ICOS–B7RP1 pathway	Excessive activation enhances T-cell responses, including proliferation, cytokine production, upregulation of molecules facilitating cell–cell interaction, and production of antibodies by B cells.	SLE	Hu et al. (2009)
CD27-CD70 pathway	Abnormal signaling enhances proliferation, survival, and differentiation of T cells, promoting autoimmunity.	Murine models	Goronzy and Weyand (2008), Hendriks et al. (2003), Lens et al. (1998), and Nakajima et al. (2000)
CD28-CD80/86 pathway	Overactivation of this pathway causes excessive proliferation and differentiation of naïve T cells.	Type I diabetes and rheumatoid arthritis	Goronzy and Weyand (2008)
SLAMF receptors	Overactivation causes enhanced differentiation of T cells into Th17 cells.	SLE	Chatterjee et al. (2012)

IL-6, interleukin-6; *mTECs*, medullary thymic epithelial cells; *SLAMF*, signaling lymphocyte activation molecule family; *SLE*, systemic lupus erythematosus.

5. ROLE OF CYTOKINES

5.1 Interleukin-2

IL-2 is a very important cytokine involved in T-cell function. It is a type 1 cytokine with four α-helical bundles produced mainly by $CD4^+$ T cells and also by $CD8^+$ cells and APCs (Liao et al., 2013). Receptors for IL-2 are expressed by a number of immune cells including activated T cells, B cells, and monocytes. IL-2 receptors can also be present in soluble form, which has low affinity for IL-2 and consequently limited ability to absorb excessive IL-2. IL-2 plays an important role in generation and sustained expansion of Treg cells and also regulates the clonal expansion of T cells. Tregs are key cells in prevention of autoimmunity and maintaining immune homeostasis. Animal experiments have shown that IL-2 treatment increases Treg cell numbers and improves allergic encephalitis and myasthenia gravis (Boyman and Sprent, 2012). It is also considered that IL-2 is deficient at functional level in SLE patients due to which Tregs may not be properly regulated. A recent study showed that low-dose IL-2 treatment can selectively correct Treg defects in patients with SLE (von Spee-Mayer et al., 2016).

Another important role of IL-2 is proliferation of cytotoxic T cells to help fight off infections especially in autoimmune diseases in which the patient immunity is already compromised either because of disease itself or secondary to immunosuppressive treatment (Tsokos, 2011).

5.2 Interleukin-6

IL-6 is secreted by a range of cells and facilitates activation and differentiation of T and B lymphocytes. IL-6 in combination with TGF-β favorably stimulates the differentiation of naive CD4-positive T cells into Th17 cells. IL-6 inhibits TGF-β—induced regulatory T-cell development as well. Owing to this imbalance of Th17/Treg, autoimmunity progresses (Tanaka and Kishimoto, 2012). A number of studies have shown its involvement in development of autoimmune diseases, especially SLE. Elevated levels of IL-6 in SLE patients correlate variably with severity of symptoms. In mice studies, IL-6—deficient mice showed less severe SLE nephritis (Richards et al., 1998).

5.3 Interleukin-17

IL-17A and IL-17F are produced mainly by activated T cells (Korn et al., 2009). They have important role in host defense against fungi and bacteria.

IL-17 is associated in the pathogenesis of several autoimmune diseases, including MS and SLE (Nalbandian et al., 2009). IL-17 levels are shown to be increased in synovial fluid of patients with RA. It is also present in organs affected by SLE such as lungs, skin, and kidney. This perhaps highlights its role in pathogenesis of autoimmune diseases.

5.4 Interleukin-7

IL-7 is expressed mainly by the stromal cells of primary lymphoid organs. It plays a critical role in development and expansion of T cells, which abundantly express IL-7 receptors. IL-7 induces Th1- and Th17-associated cytokine secretion and is responsible for T-cell–dependent activation of macrophages, dendritic cells, and B cells. Moreover, overexpression of IL-7 activates innate lymphoid tissue inducer cells, leading to ectopic lymphoid aggregate formation (Bikker et al., 2012). Thus it can act as a key mediator in T-cell–associated autoimmunity, chronic inflammation, and tissue destruction. Raised levels of IL-7 have been found to be associated with many autoimmune diseases such as RA, type I diabetes mellitus, MS, and SLE.

5.5 Interleukin-33

IL-33 is a dual function protein that acts as a cytokine and a nuclear factor. It interacts with its specific receptor ST2 receptor complex present on the surface of Th2 cells, $CD8^+$ T cells, B cells, mast cells, basophils, eosinophils, macrophages, and dendritic cells. It induces the production of IL-5, IL-13, TNF-α, IFN-γ, and IL-2 from these cells (Zhao and Chen, 2014). Its levels are raised in many autoimmune diseases such as SLE, MS, inflammatory bowel disease, and RA, indicating possible involvement in causing autoimmunity and inflammatory destruction in these diseases.

Role of cytokines in causing autoimmunity is summarized in Table 2.

6. AUTOIMMUNE DISEASES

There are dozens of autoimmune diseases described in literature; some are common, and others are very rare (Garg et al., 2017). Although it is beyond the scope of this chapter to discuss all autoimmune diseases, however, some of the common and important autoimmune diseases are described in the following with special emphasis on the role of T cells in causing autoimmunity.

Table 2 Cytokines and Their Role in Autoimmunity

S. No	Cytokine	Role in Pathogenesis of Autoimmunity	Examples of Autoimmune Disorders	References
1.	Interleukin-2	1. Functional defect/deficiency of IL-2 leads to decreased generation and expansion of regulatory T cells. 2. Deficiency of IL-2 also decreases proliferation of cytotoxic T cells making the patient more susceptible to infections.	SLE and myasthenia gravis	Liao et al. (2013), Boyman and Sprent (2012), von Spee-Mayer et al. (2016), and Tsokos (2011)
2.	Interleukin-6	IL-6 in combination with TGF-β stimulates the differentiation of naive $CD4^+$ T cells into Th17 cells. It also inhibits TGF-β —induced regulatory T cell development. Imbalance of Th17/Treg leads to autoimmunity.	SLE	Tanaka and Kishimoto (2012) and Richards et al. (1998)
3.	Interleukin-17	IL-17 A and F are produced by activated T cells. Levels are increased in autoimmune diseases leading to loss of self-tolerance.	SLE, multiple sclerosis, and rheumatoid arthritis	Nalbandian et al. (2009)
4.	Interleukin-7	1. IL-7 induces Th1- and Th17-associated cytokine secretion and is responsible for T cell—dependent activation of macrophages, dendritic cells, and B cells. 2. Its overexpression also causes ectopic lymphoid aggregate formation.	Rheumatoid arthritis, type I diabetes mellitus, multiple sclerosis, and SLE.	Bikker et al. (2012)
5.	Interleukin-33	Overexpression of IL-33 on T cells (especially Th2 cells) and B cells causes increased production of IL-5, IL-13, TNF-α, IFN-γ, and IL-2 leading to autoimmunity.	SLE, multiple sclerosis, inflammatory bowel disease, and rheumatoid arthritis.	Zhao and Chen (2014)

IFN-γ, interferom-γ; *IL*, interleukin; *SLE*, systemic lupus erythematosus.

7. TYPE 1 DIABETES MELLITUS

Type 1 diabetes mellitus is divided into two subtypes, Type 1A and Type 1B. Type 1A diabetes mellitus is immune mediated, whereas type 1B is nonimmune mediated. Type 1 diabetes mellitus usually presents clinically after a varying period of time during which beta cells of pancreatic islets are destroyed by autoimmune process to a level that body can no longer control the blood glucose (Concannon et al., 2009). Type 1 diabetes mellitus is the most common form of diabetes in childhood, and about half of the patients present before the age of 10 (Dabelea et al., 2007).

Type 1 diabetes mellitus usually presents with typical symptoms of polyuria, polydipsia, and weight loss. Many times these patients remain undiagnosed until they present with diabetic ketoacidosis (DKA), which is a medical emergency. DKA criteria as defined by a consensus statement from the International Society for Pediatric and Adolescent Diabetes (ISPAD) include combination of hyperglycemia (blood glucose > 200 mg/dL), metabolic acidosis (pH < 7.3 or plasma bicarbonate <15 mEq/L), and ketosis (Wolfsdorf et al., 2014).

It is the second most common form of clinical presentation of T1DM. It can become life threatening if not diagnosed immediately and promptly treated. Comprehensive clinical investigations should be performed at the time of suspicion of DKA, which includes complete blood counts, comprehensive metabolic panel, arterial blood gases, urine analysis, serum ketones levels, and other appropriate laboratory testing.

Both environmental and genetic factors are involved in the etiology of type 1 diabetes. Environmental factors include diet, viral infections, immunizations, vitamin D deficiency, etc. Genetic polymorphisms that are reported to be associated with the risk of T1DM include PTPN22 gene, IL-2 receptor, CTLA-4, BACH2, CTSH, C1QTNF6, HLA-DQ alpha/beta, and HLA-DR (Cooper et al., 2008). PTP gene polymorphism affects TCR signaling and is associated with many other autoimmune diseases as well (Kyogoku et al., 2004).

It is clear from animal studies that T cells play an important role in type 1 diabetes mellitus. Both Th1 and Th2 cells and their corresponding mediators contribute toward destruction of pancreatic islet beta cells and hence control onset and progression of type 1 diabetes mellitus (Almawi et al., 1999). One research demonstrates that interferon gamma–positive Th1 cells mediate insulitis in nonobese diabetic mice (NOD), and conversely anti-interferon gamma antibodies administration can slow down the destruction of pancreatic islet cells. IL-12 and IL-18 are important cytokines, which induce interferon gamma (Rothe et al., 1997). Moreover, it is reported that FOXP3-positive Tregs are also involved specifically in the development of neonatal type 1 diabetes and fatal autoimmune destruction (Wildin and Freitas, 2005).

Thus basic immune dysregulation in type 1 diabetes mellitus involves failure of self-tolerance mechanisms. T cells that escape thymic clonal deletion along with the defective Tregs result in loss of self tolerance. These auto-reactive cells then cause pancreatic beta cell injury. This involves different T cell subsets including Th1 and CD8-positive T cells that target autoantigens, including islet cell insulinoma—associated protein 2, autoantigen 512, zinc transporter ZnT8, and glutamic acid decarboxylase.

Treatment of type 1 diabetes typically involves strict glycemic control with insulin along with many other interventions including dietary modifications, exercise, regular physician visits, and assessment of psychosocial issues. Besides the standard of care, there are a number of areas being investigated to enhance the management of diabetes. Some of the investigational immunomodulatory interventions targeting T cells are described in the following.

7.1 Anti-CD3 Antibodies

There has been some experimental evidence that anti-CD3 monoclonal antibodies (OKT3) help in inducing anergy and possibly apoptosis of activated T cells while sparing Tregs, which fosters tolerance by reestablishing the balance between pathogenic and regulatory T cells (You et al., 2008; Smith et al., 1997; Penaranda et al., 2011). OKT3 bind to the epsilon chain of the CD3/TCR complex of T lymphocytes. It is possible that heterogeneity of TCR expression by different T-cell subsets results in variance of response of OKT3 on Tregs versus other effector T cells (Valle et al., 2015). OKT3 have many adverse events in humans; so, owing to this problem, humanized monoclonal antibodies with less cytokine-mediated adverse events of fever, headache, and hypotension have been developed.

7.2 Otelixizumab

Otelixizumab is a humanized monoclonal anti-CD3 antibody. This drug has been tested in various preclinical and clinical studies. In a phase II clinical trial at multiple centers in Europe, patients with new-onset type 1 diabetes were randomly assigned to receive placebo or otelixizumab for 6 consecutive days. This trial enrolled 80 patients, and they were followed up for 18 months, during which daily insulin needs and residual beta-cell function of patients were measured according to glucose-clamp—induced C-peptide release before and after the administration of glucagon (Keymeulen et al., 2005). The results of this trial showed that at 6, 12, and 18 months, residual beta-cell function was better maintained with otelixizumab than with placebo. The insulin dose increased in the placebo group but not in the otelixizumab group. In terms of adverse events, otelixizumab was associated with significant flu-like syndrome and transient symptoms of Epstein—Barr viral mononucleosis. The authors concluded that in patients with recent-onset type 1 diabetes, short-term treatment with CD3 antibody preserves residual beta-cell function for at least 18 months. However, a phase III DEFEND-1 (Durable Response Therapy Evaluation for Early or New-Onset Type 1 Diabetes) trial was conducted to test whether a lower dose of otelixizumab could preserve C-peptide secretion in new-onset type 1 diabetes patients. This trial showed that otelixizumab was well tolerated in patients with recent-onset type 1 diabetes at a total dose of 3.1 mg but did not achieve preservation of levels of C-peptide or other markers of metabolic control (Aronson et al., 2014).

7.3 Teplizumab

Teplizumab is an Fc receptor—nonbinding anti-CD3 monoclonal antibody. In a phase III multicontinental trial, 516 patients aged 8—35 years who had been diagnosed with type 1 diabetes mellitus for 12 weeks or less were randomized in 2:1:1:1 ratio to one of three regimens of teplizumab infusions (14-day full dose, 14-day low dose, or 6-day full dose) or placebo at baseline and at 26 weeks. The primary composite outcome of the study was the percentage of patients with insulin use of <0·5 U/kg per day and glycated hemoglobin A1C of <6·5% at 1 year. However, the findings of this study at 1 year showed no difference in primary outcomes between various groups (Sherry et al., 2011). In another randomized controlled trial it was shown that two courses of treatment with teplizumab reduced the loss of C-peptide 2 years after the first treatment in patients with new-onset type 1 diabetes. In this trial, 83 patients diagnosed with type 1 diabetes within 8 weeks of enrollment were randomly assigned to teplizumab or no therapy. Teplizumab therapy included 14-day course at baseline and 1 year. There was no significant difference in hemoglobin A1C levels; however, patients on teplizumab required less amount of insulin to achieve similar glycemic control (Herold et al., 2013).

7.4 Costimulation Modulation

Costimulation modulation of TCRs has been implied in investigational manner to manage autoimmune diseases. As discussed earlier, abatacept is a fusion protein of cytotoxic T lymphocyte–associated antigen 4 and the Fc portion of IgG1 that inhibits T-cell activation by binding to CD80 and CD86 on APCs, thus blocking their interaction with CD28 present on the T cells. It is currently approved for management of psoriatic arthritis, RA, and juvenile idiopathic arthritis. In a multicentre, double-blind, randomized controlled trial, patients of type 1 diabetes were randomly assigned to receive abatacept or placebo infusions intravenously on days 1, 14, and 28 and monthly for a total of 27 infusions over 2 years. Results of the trial suggested that costimulation modulation with abatacept slowed reduction in beta-cell function over 2 years. Obviously longer follow-up is needed to determine whether the beneficial effect continues after cessation of abatacept administrations (Orban et al., 2011).

7.5 Other Immunosuppressive Drugs

Immunosuppressive drugs with mechanisms to inhibit T-cell function have been also used though without any major success. Azathioprine is an immunosuppressive drug, which is an imidazolyl derivative of mercaptopurine. It prevents T-cell responses to antigens. Azathioprine was tested in a double-blind control trial of 49 newly diagnosed type I diabetes patients. Patients were randomly assigned to receive either azathioprine or placebo for 12 months. It was concluded that azathioprine alone does not induce the remission phase in newly diagnosed patients of type I diabetes (Cook et al., 1989). Mycophenolate mofetil and cyclosporine have also been used in type 1 diabetes but without any major success (Gottlieb et al., 2010; Christie et al., 2002).

8. SYSTEMIC LUPUS ERYTHEMATOSUS

SLE is a chronic multisystem autoimmune disease that results from immune-mediated tissue damage. The incidence rates of SLE vary from 1 to 25 per 100,000 in different parts of the world (Pons-Estel et al., 2010). In United States the incidence of SLE is about 5.5/100,000 per year, and prevalence is approximately 73 per 100,000. There is preponderance of female cases with female-to-male ratio of approximately 8:1 in adults. Majority of the cases have onset between the ages of 16 and 55 years.

The exact etiology of SLE is unknown, but a number of factors have been associated with SLE. Genetic susceptibility, environmental factors, drugs, microbial infections, hormones, and altered immune system function are some of the important elements of pathogenesis in SLE (Rullo and Tsao, 2013; Crow, 2014). With the help of genome-wide association studies, a number of genes have been found which are associated with SLE. The most common genetic predilection for SLE is found at the MHC locus. Some of the other genes include TNFAIP3, STAT4, ITGAM, IRAK1, BANK1, BLK, and LYN.

There is an extreme heterogeneity in the clinical presentation of SLE. This disease can involve virtually any organ of the body. The clinical severity also varies among patients and may be illustrated by periods of remissions and relapses. Common symptoms of SLE include skin lesions, fatigue, weight loss, fevers, arthralgias, Reynaud's phenomenon, and neuropsychiatric symptoms. Although there are clinical criteria such as the 1997 American College of Rheumatology (ACR) criteria and the 2012 Systemic Lupus International Collaborating Clinics (SLICC) criteria, which help in diagnosing SLE, however, diagnosis is sometimes challenging because of the heterogeneity of the disease.

There are several mechanisms that lead to loss of self-tolerance and organ dysfunction in SLE. Immune cells including T and B cells along with cytokines play an important role in pathogenesis of SLE. Specifically, early signaling events are amplified resulting in abnormal T cell antigen receptor—mediated activation (Tsokos, 2011). There are several processes involved in enhanced activation of T cells. First of all, in SLE CD3 complex is different because of the replacement of CD3ζ with FcRγ common chain (Crispin et al., 2008). Furthermore, in downstream intracellular signaling, spleen tyrosine kinase (Syk) is used instead of canonical ZAP-70 (Crispin et al., 2010). There are several mechanisms by which the expression of CD3ζ is reduced, including dysfunctional transcription, decreased mRNA stability, increased caspase 3—mediated protein degradation, and abnormal mRNA splicing (Crispin et al., 2010; Tsokos et al., 2003). Other reported signaling defects in SLE T cells include decreased activity of protein kinase C (PKC), increased calcium response, decreased CD45 phosphatase activity, increased phosphorylation of tyrosine residues, decreased MAP kinase activity, and increased lipid raft clustering (Crispin et al., 2010). These alterations along with the deficient production of IL-2 favors the development of Th17 cells, CD4-CD8 double-negative T cells, and T follicular helper (TFH) T cells, while inhibiting Treg differentiation leading to uncontrolled autoimmune response.

8.1 T-Cell Targeting Therapies in SLE

It is reported that in vitro forced expression of splicing protein ASF/SF2 increases the level of CD3ζ chain mRNA which restores IL-2 production by the SLE T cells (Moulton et al., 2013). Another drug tested in vitro is fostamatinib, which inhibits the kinase Syk leading to normalization of hyperactive SLE T cells (Deng et al., 2010).

Some important drugs with considerable evidence of clinical efficacy in SLE are described in the following.

8.2 Rapamycin and N Acetyl Cysteine

Rapamycin and other mammalian target of rapamycin (mTOR) inhibitors bind to mTOR on SLE T cells resulting in decreased calcium flux. Another drug N acetyl cysteine (NAC) also decreases calcium flux by reversing mitochondrial hyperpolarization. Both these drugs have proven efficacy against SLE and are reported to improve sign and symptoms of SLE and to decrease the use of corticosteroid (Fernandez et al., 2006; Lai et al., 2012).

8.3 Drugs Targeting CD154

SLE T cells exhibit increased and prolonged expression of CD154, which interacts with B cells and leads to the production of pathogenic autoantibodies.

Anti-CD154 antibodies have been found to be effective in treating SLE. When given to patients with SLE, it was seen that there is reduced production of anti-dsDNA along with raised C3 levels. Also, hematuria disappeared in all patients (Boumpas et al., 2003). Other drugs targeting CD154, which may benefit SLE patients, include tacrolimus and dypyrimodole.

8.4 Trichostatin A

This drug acts as a histone deacetylase (HDAC) inhibitor and alters the methylation of DNA and acetylation of histones. This leads to the reversal of abnormal SLE T-cell signaling phenotype including the expression of CD154 and interferon γ (Mishra et al., 2001).

8.5 Low-Dose IL-2

Low-dose IL-2 selectively activates Treg cells but does not cause widespread effector T-cell activation. It is suggested that controlled IL-2 infusion can be used in SLE patients to suppress autoimmune response (Konya, 2013).

9. GRAVES' DISEASE

GD is an organ-specific autoimmune disorder involving hyperactivity of the thyroid gland because of the presence of thyroid-stimulating antibodies. These antibodies mimic the action of thyroid-stimulating hormone (TSH) and activate the TSH receptors. This results in increased proliferation of the thyroid follicular cells leading to excessive production of thyroid hormones (T3 and T4). Clinical manifestations of GD include tachycardia, weight loss, anxiety, temperature intolerance, ophthalmopathy, and dermatopathy, among others.

GD causes 60%–90% of thyrotoxicosis in many regions of the world [103]. It is the most common cause of hyperthyroidism in the United States with an incidence of 30 per 100,000 people per year (Furszyfer et al., 1970).

The risk factors associated with GD include the presence of susceptibility genes such as HLA alleles, CTLA-4 gene, Thyroglobulin gene (Tg) and TSHR gene. Other risk factors are stress, pregnancy, influence of sex hormones, radiation, and infection (Morshed et al., 2012). The underlying pathogenesis results from loss of central as well as peripheral tolerance against the thyroid antigens. Both T and B cells along with their regulatory populations are involved in the progression and maintenance of GD. There is diffuse infiltration of thyroid gland by lymphocytes along with the production of autoantibodies by the B cells, which activates the TSH receptors and leads to hyperthyroidism. T cells play an important role in maintaining autoimmunity and inflammation in GD. Under normal conditions there is a balance between Th1 and Th2 immune responses which prevents autoimmunity. Th1 cells promote apoptotic pathways on the thyroid follicular cells, whereas Th2 cells facilitate thyroid growth and activity by inducing B lymphocytes to produce TSH receptor antibodies. In GD this balance is shifted away from Th1 toward Th2, leading to increased production of TSH receptor antibodies and ultimately thyroid hyperactivity. Moreover, it is found that the Tregs are also involved in promoting autoimmunity (Marazuela et al., 2006). These cells have CD25 and FoxP3 on their surface and normally suppress the effector T cells. But, in GD it is seen that there is increased number of Tregs in the thyroid gland as well as the periphery, which have defective immunosuppressive function that further augments the autoimmunity. Recent evidence suggests that Th17 cells may also be involved in progression of GD (Bohgaki et al., 2007). It was seen that the patients who underwent $CD34^+$ autologous stem cell transplantation had increased predisposition to develop GD. These patients had increased levels of IL-6 and TGF-β that caused priming of Th17 cells and subsequent development of GD.

Rituximab, which is an anti-CD20 monoclonal antibody, has been proved to be effective in treating GD. Although its main target is B cells, it indirectly inhibits $CD4^+$ cells which may also contribute to its efficacy. Further studies into understanding T-cell involvement in GD pathogenesis may provide basis to develop novel therapeutic drugs that can slow down the disease progression and improve clinical outcome (Hegedus et al., 2011).

10. CELIAC DISEASE

Celiac disease (CD)/sprue is a T-cell–mediated autoimmune disorder in which there is leukocyte infiltration of the small intestine, leading to altered tissue architecture. It is a fairly common genetic disorder with a prevalence of 1% in the general population. In the United States, its prevalence is estimated to be around 0.5%–1.26% (Dube et al., 2005).

In Celiac disease an inappropriate immune response occurs in genetically susceptible individuals when they are exposed to gluten proteins present in wheat, barley, and rye. Clinical manifestations include abdominal bloating and pain, chronic diarrhea, vomiting, foul-smelling stool, weight loss, and delayed growth and puberty (Sainsbury et al., 2013).

The alleles for HLA class II haplotypes DQ2 and DQ8 are found to be positive in most of the patients with CD [110]. The underlying pathogenesis involves activation of $CD4^+$ effector T cells in the lamina propria of the small intestine when exposed to gluten peptides bound to HLA-DQ2 and DQ8 (Du Pre and Sollid, 2015). The activated T cells produce proinflammatory cytokines that cause tissue injury. Apart from $CD4^+$ T cells, studies also suggest the involvement of $CD8^+$ cytotoxic T cells. Peptides bound to HLA class I molecules have been identified which activate $CD8^+$ T cells and cause lysis of the target cells by production of interferon γ (Mazzarella et al., 2008). Other studies have reported raised levels of IL-17 in the inflamed gut of patients with CD which indicates the involvement of Th17 cells in the progression of celiac disease (Stockinger et al., 2007). Moreover, the celiac intestinal mucosa also harbors Tregs (Tr1 and $Foxp^+$ Tregs) that produce antiinflammatory cytokines IL-10 and TGF-β. In this way these cells help to suppress the pathogenic response to gluten challenge (Gianfrani et al., 2006; Zanzi et al., 2011).

10.1 Development of Novel Therapies Targeting T-Cell Immunity

10.1.1 CCR9 and Integrin α4β7 Antagonists

Recruitment of T cells to the small intestine requires specific chemokine–chemokine receptor interaction. The small intestine releases the ligands CCL25 and integrin α4β7 that interact with their specific chemokine receptors (CCR9 and integrin α4β7, respectively) present on the T cells, enabling these cells to home to the small intestine (Schuppan et al., 2009). A study conducted on a mouse model of spontaneous ileitis shows that blockage of these pathways inhibits the histological damage caused by intestinal inflammation (Rivera-Nieves et al., 2006).

10.1.2 IL-15 Antagonists

IL-15 has been reported to be involved in the expansion of malignant T cells. It is suggested that development of IL-15 antagonist may prove to be helpful in slowing down the course of CD.

11. RHEUMATOID ARTHRITIS

RA is a chronic inflammatory autoimmune disorder mainly involving small joints of hand, feet, and cervical spine. Larger joints such as shoulder or knee joint may also be involved. The joints become painful, red, and swollen. The symptoms often worsen after rest. Extraarticular manifestations may also occur resulting in lung fibrosis and atherosclerosis, leading to increase risk of myocardial infarction, rheumatic nodules on skin, renal amyloidosis, anemia, and neuropathies. The incidence of RA is approximately three cases per 10,000 of the world's population, and the prevalence rate is approximately 1%. It is more common in Native Americans with the prevalence rate of 5%–6%. It predominantly affects the women (Ahlmen et al., 2010). Exact etiology is unknown, but various genetic and environmental factors are involved in increasing the risk of RA. It is strongly associated with inheritance of DRB1 genes for class II HLA molecules HLA-DR4 and HLA-DR1. Environmental triggers such as smoking, infection, stress, etc. are thought to be responsible for initiating the disease process in genetically susceptible individuals (Carlens et al., 2010). The autoimmune response generated is mainly T-cell mediated and leads to synovial hypertrophy and chronic joint inflammation. Ultimately uncontrolled inflammation and synovial proliferation lead to irreversible destruction of bone, cartilage, ligaments, and tendons.

The underlying pathogenesis involves abnormal activation of $CD4^+$ T cells, phagocytes, fibroblasts, osteoclasts, and neutrophils. There is increased production of inflammatory mediators such as TNF-α, IL-1, IL-6, IL-8, TGF-β, fibroblast growth factor (FGF), and platelet-derived growth factor (PDGF). There is also B-cell–induced production of autoantibodies (Arend and Firestein, 2012).

T cells have a central role in causing autoimmune response in patients with RA (Cope et al., 2007). The $CD4^+$ Th1 effector T cells infiltrate into the synovial membrane where they activate macrophages and synovial fibroblasts, resulting in tissue destruction (Weyand et al., 2000). Genetic analysis suggests the involvement of genes PTPN22 and CTLA-4 in causing uncontrolled $CD4^+$ T-cell activation. The pivotal role played by $CD4^+$ T cells in the pathogenesis of RA is indicated in the following phenomena (Skapenko et al., 2005):

1. Association of aggressive forms of RA with specific class II HLA alleles DR4 and DR1 (shared epitope).
2. Elevated levels of $CD4^+$ T cells in the synovial fluid and peripheral blood of patients with RA.
3. Initiation of disease process in animal models of inflammatory arthritis by $CD4^+$ T cells.
4. Improvement of the disease outcome after anti-T-cell–directed therapies.

Along with effector helper T cells, Tregs such as $CD25^+$ Tregs and Th2 cells are also impaired in patients with RA which leads to breakdown of peripheral tolerance (Skapenko et al., 2005). These further augment effector T cell–mediated immune response. Th17 cells have also been thought to be involved as indicated by elevated levels of IL-17 in many patients with RA (Miossec, 2007). Recent evidence suggests the involvement of $\gamma\delta$ T cells as well (Jacobs and Haynes, 1992). These cells bear an alternative TCR composed of $\gamma\delta$ chains instead of the usual $\alpha\beta$ chains. They are capable of recognizing self-peptides and do not require antigen processing or antigen presentation by MHC molecules. Normally they form a small part of the peripheral T cell pool, but in RA their levels are increased especially in the synovial infiltrates. One study demonstrates that the levels of γδ T cells in the synovium are directly related to the extent of tissue inflammation i.e., increased levels of γδ T cells causes increased tissue inflammation.

Detailed analysis of synovial T cells in patients with RA has shown that these cells undergo many changes to produce autoimmune response. This includes an intrinsic resistance to apoptosis and loss of expression of CD25, CD28, CD27, and CD40L along with increased expression of IFN-γ and TNF-α (Cope, 2008). Increased levels of $CD4^{+}CD28^{null}$ T cells are specifically associated with systemic manifestations and worse prognosis (Cope, 2008).

11.1 Antirheumatoid Therapies Targeting T Cells

Studies show that T cell—targeted therapies have clinical efficacy against RA. Some of them are described in the following.

11.2 IL-2 Antagonists

As we know that IL-2 has a central role in T-cell activation and proliferation, drugs that block IL-2 production such as cyclosporine suppress T-cell activity and have proven to be effective in reducing signs and symptoms as well as radiological progression in patients with RA (Isaacs, 2008).

11.3 Anti-TNF Drugs

High concentration of TNF suppresses the activity of Treg cells. In this context it is observed that treatment of RA patients with anti-TNF drugs slows down disease progression by enhancing Treg cell—induced suppression of inflammatory cytokines (INFγ and TNFα) produced by effector T cells. One such anti-TNF drug is infliximab that restores the Treg cell function and also improves sensitivity of effector T cells to Treg suppression (Ehrenstein et al., 2004).

11.4 IL-6 Antagonists

Drugs that block IL-6 receptor have also shown efficacy in treating RA. IL-6 blocks the activity of Tregs. By inhibiting IL-6, these drugs drive the $CD4^{+}$ cells differentiation toward Treg cells and away from Th17 cells. Tocilizumab is an IL-6 antagonist that has proven to be effective in improving the clinical outcome of RA (Emery et al., 2008).

11.5 CD28 Costimulation Blockers

These drugs bind to CD28 ligands (CD80 and CD86) with a higher affinity than CD28 receptor, leading to blockage of this costimulatory pathway and suppression of T cell—driven immune responses. Examples include abatacept (CTLA-4-Ig fusion protein) and belatacept. When used in optimal doses, these drugs selectively block the effector T cells and increase the levels of Tregs in the inflamed tissue, thereby suppressing T-cell immune responses and improving the course of disease in RA (Linsley and Nadler, 2009). In a metaanalysis it was concluded that abatacept seemed to be efficacious and safe in the treatment of RA (Maxwell and Singh, 2010). The FDA has approved Abatacept for use in RA, psoriatic arthritis, and juvenile idiopathic arthritis. Both intravenous (IV) and subcutaneous (SC) formulations are approved for administration. In a study comparing IV versus SC formulations, it was shown that SC abatacept has similar efficacy and safety profile as of IV abatacept, with low immunogenicity and high retention rates (Genovese et al., 2011). It is important to note here that abatacept should not be used concurrently with TNF inhibitors or with the IL-1 receptor antagonist anakinra.

11.6 Immunosuppressive Small-Molecule Drugs

Rapamycin and other mTOR inhibitors are small-molecule immunosuppressive drugs that promote proliferation of Tregs and block the effector T cells (Thomson et al., 2009). It is seen that patients treated with both rapamycin and methotrexate showed greater response rate than those treated with methotrexate alone.

11.7 Other Experimental Drugs

Drugs such as hydroxychloroquine (Toll-like receptor antagonist) and trichostatin (small-molecule inhibitor of histone deacetlyases) have shown to increase Tregs function in animal models and may prove to be efficacious in future to manage RA (Esensten et al., 2009).

11.8 Cellular Therapy

Human stem cell transplantation has proven to be curative in some complicated and refractory cases of RA. The drawback of this therapy is that prolonged immunosuppression predisposes the recipient to multiple infections. To combat this, one approach is to isolate only Treg cells form the patients, expand them, and then reinfuse them into the patients

(Brusko and Bluestone, 2008). This has proven efficacy in animal models and may be effective in selectively boosting Treg cell function in human subjects in future.

12. MULTIPLE SCLEROSIS

MS is a chronic autoimmune disorder in which there is demyelination and degeneration of the central nervous system i.e., brain and spinal cord. The immune system attacks myelin sheath of the axon, resulting in a wide array of signs and symptoms that occur in episodic manner over the years. Clinical manifestations include weakness, tingling, numbness, blurred vision, muscle stiffness, loss of bladder control, and cognitive impairment, among others.

Worldwide around 2.1% million people are affected by MS, predominantly women (Rosati, 2001). In the United States, prevalence rate is 58–95 per 10,000 populations. The average age of presentation is 29 years in women and 31 years in men. Multiple genetic, geographic, and environmental factors interplay to cause MS. There is strong evidence of association of HLA-DRB1 with MS susceptibility. Role of other genetic polymorphisms involving IL (IL)-2, IL-7, CXCR5, IL-12A, IL-12β, and IL-12Rβ1 pathways have also been implicated. The disease process initiates when environmental trigger such as viral infection or vitamin D deficiency activates self-reactive T cells against myelin sheath in genetically predisposed individuals (Høglund and Maghazachi, 2014).

The autoimmune response generated in MS is mainly T cell mediated. Animal studies have shown that injecting myelin-reactive T cells in the mice or other rodents leads to T cell–mediated autoimmune reaction against myelin with signs and symptoms similar to those of MS (Holmoy and Hestvik, 2008). $CD4^+$ T cells are the main cells involved in causing MS lesion by secreting INF γ. Th17 cells also have a key role (Hestvik, 2010). These cells can cross the blood–brain barrier via interaction between the chemokine receptor CCR-6 present on Th17 cells and CCL20/MIP-3α expressed on endothelium, resulting in increased permeability of these cells into the brain. They secrete proinflammatory cytokines IL-17 and IL 22 that not only mediate neural tissue damage but also increase permeability of the blood–brain barrier by disrupting endothelial tight junctions. This facilitates the entry of other immune cells into the CNS and exacerbation of the disease. Other T-cell subtypes are also involved. There is evidence of infiltration and clonal expansion of $CD8^+$ cytotoxic T cells around the lesion, resulting in cell-mediated cytotoxicity (Babbe et al., 2000). Clonal $\gamma\delta$ T cells are also seen around the lesion (Wucherpfennig et al., 1992). Moreover, there is evidence of reduced expression of FOXP3 and loss of suppressive function of $CD4^+$ $CD25^+$ Tregs (Venken et al., 2008).

12.1 T Cell—Targeted Therapeutic Modalities

12.1.1 Interferon β

Interferon β is considered first-line therapy for management of MS because of its immunomodulatory properties. It downregulates HLA class II molecules in APCs. It also upregulates the expression of PDL-2 inhibitory molecules, which interact with their respective receptors on the T cells and induce apoptosis. It also inhibits proliferation of macrophages and so activation of autoreactive T cells. It also decreases transmigration of T cells into the CNS by downregulating the VLA-4 molecule that interacts with the VCAM partner present on the endothelial cells. Furthermore, it enhances the activity of the Treg cells by downregulating the CTLA-4 molecule (Buc, 2013).

12.1.2 Glatiramer Acetate

This drug is also used as first-line therapy for MS. It induces the production of IL-1Ra, a natural inhibitor of IL-1. It also inhibits TNF and IL-12. Inhibition of these potent proinflammatory agents slows down the disease process. Moreover, it also causes increased expression of immunosuppressive IL-10 and reversal of impaired Treg cell function (Castro-Borrero et al., 2012).

12.1.3 Natalizumab

It is a second-line agent in the treatment of MS. It exerts its effect by blocking the VLA-4 molecule on T cells, thereby reducing their interaction with VCAM -1 on the endothelial cells leading to decreased transmigration of T cells into the CNS. However, it does not improve the Treg cell function (Skarica et al., 2011).

12.1.4 Alemtuzumab

It is a monoclonal antibody active against CD52 molecule present on T cells, B cells, NK cells, monocytes, granulocytes, and dendritic cells. CD52 mediates T-cell activation via tyrosine phosphorylation. Treatment with alemtuzumab causes rapid depletion of CD52-bearing cells. Among the effected cells, T cells take the longest time to recover. It is beneficial in treating the relapsing remitting forms of the disease. It also induces the production of neurotrophic factors from the autoreactive T cells, which have a neuroprotective effect on the CNS (Coles et al., 2006).

12.1.5 Daclizumab

It is a humanized monoclonal antibody against the alpha chain of IL-2 receptor (CD25). It binds to CD25 and prevents high-affinity IL-2 signaling pathway. It also causes expansion of NK cells that in turn kills the activated T cells. It has been shown to significantly reduce neurodegeneration and inflammation in MS (Bielekova, 2013).

Table 3 summarizes T cell–targeted therapies against autoimmune diseases.

Table 3 T Cell–Targeted Therapeutic Approaches for Autoimmune Diseases

S. No	Intervention	Example
1.	Reduction in T cell number or function	1. Total lymphoid irradiation
2.	Immunosuppressive drugs	1. Corticosteroids 2. Methotrexate 3. Cyclosporine 4. Tacrolimus 5. Sirolimus
3.	Biological agents	1. TCR vaccinations. 2. mAbs to surface receptors of T cells (e.g., Anti-CD3 otelixizumab, anti-CD52 alemtuzumab). 3. mAbs to surface receptors of cells interacting with T cells–like APCs. 4. Cytokines (e.g., Low-dose IL-2, INFβ). 5. mAb to cytokines (IL-6 antagonist tocilizumab, Anti-TNF drug infliximab, IL-15 antagonist). 6. Inhibitors of T cell/APC interactions such as costimulatory modulators (e.g., abatacept). 7. T cell recruitment inhibitors (e.g., CCR9 and integrin α4β7 antagonists, natalizumab).

APC, antigen-presenting cell; *IL*, interleukin; *INFβ*, interferon β; *mAB*, monoclonal antibody; *TCR*, T-cell receptor.

13. SUMMARY

T lymphocytes play a key role in causing autoimmune diseases. Both effectors and regulatory T cells are found to be defective in many important autoimmune diseases. Genetic susceptibility along with environmental triggers enables the T cells to escape central and/or peripheral tolerance and thus become autoreactive to self-antigens. The autoimmune response is further augmented by the defective function of Tregs that are unable to cause immunosuppression. As a result, there is uncontrolled expansion of T cells against self-antigens leading to organ damage. Drugs that inhibit effector T cells and/or enhance activity of Tregs have been developed which will considerably improve the clinical outcome of autoimmune disorders.

REFERENCES

Abbas, A.K., Lichtman, A.H.H., Pillai, S., 2017. Immunologic Tolerance and Autoimmunity.

Ahlmen, M., Svensson, B., Albertsson, K., Forslind, K., Hafstrom, I., 2010. Influence of gender on assessments of disease activity and function in early rheumatoid arthritis in relation to radiographic joint damage. Ann. Rheum. Dis. 69, 230–233.

Akiyama, T., Shimo, Y., Yanai, H., Qin, J., Ohshima, D., Maruyama, Y., Asaumi, Y., Kitazawa, J., Takayanagi, H., Penninger, J.M., Matsumoto, M., Nitta, T., Takahama, Y., Inoue, J., 2008. The tumor necrosis factor family receptors RANK and CD40 cooperatively establish the thymic medullary microenvironment and self-tolerance. Immunity 29, 423–437.

Akkaraju, S., Ho, W.Y., Leong, D., Canaan, K., Davis, M.M., Goodnow, C.C., 1997. A range of CD4 T cell tolerance: partial inactivation to organ-specific antigen allows nondestructive thyroiditis or insulitis. Immunity 7, 255–271.

Almawi, W.Y., Tamim, H., Azar, S.T., 1999. Clinical review 103: T helper type 1 and 2 cytokines mediate the onset and progression of type I (insulin-dependent) diabetes. J. Clin. Endocrinol. Metab. 84, 1497–1502.

Anaya, J.M., 2012. Common mechanisms of autoimmune diseases (the autoimmune tautology). Autoimmun. Rev. 11, 781–784.

Anderson, M.S., Venanzi, E.S., Klein, L., Chen, Z., Berzins, S.P., Turley, S.J., Von Boehmer, H., Bronson, R., Dierich, A., Benoist, C., Mathis, D., 2002. Projection of an immunological self shadow within the thymus by the aire protein. Science 298, 1395–1401.

Arend, W.P., Firestein, G.S., 2012. Pre-rheumatoid arthritis: predisposition and transition to clinical synovitis. Nat. Rev. Rheumatol. 8, 573–586.

Arens, R., Tesselaar, K., Baars, P.A., Van Schijndel, G.M., Hendriks, J., Pals, S.T., Krimpenfort, P., Borst, J., Van Oers, M.H., Van Lier, R.A., 2001. Constitutive CD27/CD70 interaction induces expansion of effector-type T cells and results in IFN-gamma-mediated B cell depletion. Immunity 15, 801–812.

Aronson, R., Gottlieb, P.A., Christiansen, J.S., Donner, T.W., Bosi, E., Bode, B.W., Pozzilli, P., 2014. Low-dose otelixizumab anti-CD3 monoclonal antibody DEFEND-1 study: results of the randomized phase III study in recent-onset human type 1 diabetes. Diabetes Care 37, 2746–2754.

Babbe, H., Roers, A., Waisman, A., Lassmann, H., Goebels, N., Hohlfeld, R., Friese, M., Schroder, R., Deckert, M., Schmidt, S., Ravid, R., Rajewsky, K., 2000. Clonal expansions of CD8(+) T cells dominate the T cell infiltrate in active multiple sclerosis lesions as shown by micromanipulation and single cell polymerase chain reaction. J. Exp. Med. 192, 393–404.

Bach, J.F., 2003. Regulatory T cells under scrutiny. Nat. Rev. Immunol. 3, 189–198.

Barker, C.F., Billingham, R.E., 1977. Immunologically privileged sites. Adv. Immunol. 25, 1–54.

Begovich, A.B., Carlton, V.E., Honigberg, L.A., Schrodi, S.J., Chokkalingam, A.P., Alexander, H.C., Ardlie, K.G., Huang, Q., Smith, A.M., Spoerke, J.M., Conn, M.T., Chang, M., Chang, S.Y., Saiki, R.K., Catanese, J.J., Leong, D.U., Garcia, V.E., Mcallister, L.B., Jeffery, D.A., Lee, A.T., Batliwalla, F., Remmers, E., Criswell, L.A., Seldin, M.F., Kastner, D.L., Amos, C.I., Sninsky, J.J., Gregersen, P.K., 2004. A missense single-nucleotide polymorphism in a gene encoding a protein tyrosine phosphatase (PTPN22) is associated with rheumatoid arthritis. Am. J. Hum. Genet. 75, 330–337.

Bielekova, B., 2013. Daclizumab therapy for multiple sclerosis. Neurotherapeutics 10, 55–67.

Bikker, A., Hack, C.E., Lafeber, F.P., Van Roon, J.A., 2012. Interleukin-7: a key mediator in T cell-driven autoimmunity, inflammation, and tissue destruction. Curr. Pharm. Des. 18, 2347–2356.

Bluestone, J.A., Bour-Jordan, H., Cheng, M., Anderson, M., 2015. T cells in the control of organ-specific autoimmunity. J. Clin. Invest. 125, 2250–2260.

Bohgaki, T., Atsumi, T., Koike, T., 2007. Multiple autoimmune diseases after autologous stem-cell transplantation. N. Engl. J. Med. 357, 2734–2736.

Bottini, N., Musumeci, L., Alonso, A., Rahmouni, S., Nika, K., Rostamkhani, M., Macmurray, J., Meloni, G.F., Lucarelli, P., Pellecchia, M., Eisenbarth, G.S., Comings, D., Mustelin, T., 2004. A functional variant of lymphoid tyrosine phosphatase is associated with type I diabetes. Nat. Genet. 36, 337–338.

Boumpas, D.T., Furie, R., Manzi, S., Illei, G.G., Wallace, D.J., Balow, J.E., Vaishnaw, A., 2003. A short course of BG9588 (anti-CD40 ligand antibody) improves serologic activity and decreases hematuria in patients with proliferative lupus glomerulonephritis. Arthritis Rheum. 48, 719–727.

Boyman, O., Sprent, J., 2012. The role of interleukin-2 during homeostasis and activation of the immune system. Nat. Rev. Immunol. 12, 180–190.

Brusko, T., Bluestone, J., 2008. Clinical application of regulatory T cells for treatment of type 1 diabetes and transplantation. Eur. J. Immunol. 38, 931–934.

Buc, M., 2013. Role of regulatory T cells in pathogenesis and biological therapy of multiple sclerosis. Mediators Inflamm. 2013, 963748.

Bulek, A.M., Cole, D.K., Skowera, A., Dolton, G., Gras, S., Madura, F., Fuller, A., Miles, J.J., Gostick, E., Price, D.A., Drijfhout, J.W., Knight, R.R., Huang, G.C., Lissin, N., Molloy, P.E., Wooldridge, L., Jakobsen, B.K., Rossjohn, J., Peakman, M., Rizkallah, P.J., Sewell, A.K., 2012. Structural basis for the killing of human beta cells by CD8(+) T cells in type 1 diabetes. Nat. Immunol. 13, 283–289.

Butcher, E.C., Picker, L.J., 1996. Lymphocyte homing and homeostasis. Science 272, 60–66.

Carlens, C., Hergens, M.P., Grunewald, J., Ekbom, A., Eklund, A., Hoglund, C.O., Askling, J., 2010. Smoking, use of moist snuff, and risk of chronic inflammatory diseases. Am. J. Respir. Crit. Care Med. 181, 1217–1222.

Castro-Borrero, W., Graves, D., Frohman, T.C., Flores, A.B., Hardeman, P., Logan, D., Orchard, M., Greenberg, B., Frohman, E.M., 2012. Current and emerging therapies in multiple sclerosis: a systematic review. Ther. Adv. Neurol. Disord. 5, 205–220.

Chatterjee, M., Rauen, T., Kis-Toth, K., Kyttaris, V.C., Hedrich, C.M., Terhorst, C., Tsokos, G.C., 2012. Increased expression of SLAM receptors SLAMF3 and SLAMF6 in systemic lupus erythematosus T lymphocytes promotes Th17 differentiation. J. Immunol. 188, 1206–1212.

Chemnitz, J.M., Parry, R.V., Nichols, K.E., June, C.H., Riley, J.L., 2004. SHP-1 and SHP-2 associate with immunoreceptor tyrosine-based switch motif of programmed death 1 upon primary human T cell stimulation, but only receptor ligation prevents T cell activation. J. Immunol. 173, 945–954.

Choi, S., Schwartz, R.H., 2007. Molecular mechanisms for adaptive tolerance and other T cell anergy models. Semin. Immunol. 19, 140–152.

Christie, M.R., Molvig, J., Hawkes, C.J., Carstensen, B., Mandrup-Poulsen, T., 2002. IA-2 antibody-negative status predicts remission and recovery of C-peptide levels in type 1 diabetic patients treated with cyclosporin. Diabetes Care 25, 1192–1197.

Cohen, S., Dadi, H., Shaoul, E., Sharfe, N., Roifman, C.M., 1999. Cloning and characterization of a lymphoid-specific, inducible human protein tyrosine phosphatase, Lyp. Blood 93, 2013–2024.

Coles, A.J., Cox, A., Le Page, E., Jones, J., Trip, S.A., Deans, J., Seaman, S., Miller, D.H., Hale, G., Waldmann, H., Compston, D.A., 2006. The window of therapeutic opportunity in multiple sclerosis: evidence from monoclonal antibody therapy. J. Neurol. 253, 98–108.

Concannon, P., Rich, S.S., Nepom, G.T., 2009. Genetics of type 1A diabetes. N. Engl. J. Med. 360, 1646–1654.

Cook, J.J., Hudson, I., Harrison, L.C., Dean, B., Colman, P.G., Werther, G.A., Warne, G.L., Court, J.M., 1989. Double-blind controlled trial of azathioprine in children with newly diagnosed type I diabetes. Diabetes 38, 779–783.

Cooper, J.D., Smyth, D.J., Smiles, A.M., Plagnol, V., Walker, N.M., Allen, J.E., Downes, K., Barrett, J.C., Healy, B.C., Mychaleckyj, J.C., Warram, J.H., Todd, J.A., 2008. Meta-analysis of genome-wide association study data identifies additional type 1 diabetes risk loci. Nat. Genet. 40, 1399–1401.

Cope, A.P., 2008. T cells in rheumatoid arthritis. Arthritis Res. Ther. 10, S1.

Cope, A.P., Schulze-Koops, H., Aringer, M., 2007. The central role of T cells in rheumatoid arthritis. Clin. Exp. Rheumatol. 25, S4–S11.

Crispin, J.C., Kyttaris, V.C., Juang, Y.T., Tsokos, G.C., 2008. How signaling and gene transcription aberrations dictate the systemic lupus erythematosus T cell phenotype. Trends Immunol. 29, 110–115.

Crispin, J.C., Liossis, S.N., Kis-Toth, K., Lieberman, L.A., Kyttaris, V.C., Juang, Y.T., Tsokos, G.C., 2010. Pathogenesis of human systemic lupus erythematosus: recent advances. Trends Mol. Med. 16, 47–57.

Crow, M.K., 2014. Type I interferon in the pathogenesis of lupus. J. Immunol. 192, 5459–5468.

Dabelea, D., Bell, R.A., D'agostino, R.B., Imperatore, G., Johansen, J.M., Linder, B., Liu, L.L., Loots, B., Marcovina, S., Mayer-Davis, E.J., Pettitt, D.J., Waitzfelder, B., 2007. Incidence of diabetes in youth in the United States. Jama 297, 2716–2724.

Delves, P.J., Roitt, I.M., 2000. The immune system. N. Engl. J. Med. 343, 37–49.

Deng, G.M., Liu, L., Bahjat, F.R., Pine, P.R., Tsokos, G.C., 2010. Suppression of skin and kidney disease by inhibition of spleen tyrosine kinase in lupus-prone mice. Arthritis Rheum. 62, 2086–2092.

Dhein, J., Walczak, H., Baumler, C., Debatin, K.M., Krammer, P.H., 1995. Autocrine T-cell suicide mediated by APO-1/(Fas/CD95). Nature 373, 438–441.

Dilek, N., Poirier, N., Hulin, P., Coulon, F., Mary, C., Ville, S., Vie, H., Clemenceau, B., Blancho, G., Vanhove, B., 2013. Targeting CD28, CTLA-4 and PD-L1 costimulation differentially controls immune synapses and function of human regulatory and conventional T-cells. PLoS One 8, e83139.

Du Pre, M.F., Sollid, L.M., 2015. T-cell and B-cell immunity in celiac disease. Best Pract. Res. Clin. Gastroenterol. 29, 413–423.

Du, W., Yang, M., Turner, A., Xu, C., Ferris, R.L., Huang, J., Kane, L.P., Lu, B., 2017. TIM-3 as a target for cancer immunotherapy and mechanisms of action. Int. J. Mol. Sci. 18, 645.

Dube, C., Rostom, A., Sy, R., Cranney, A., Saloojee, N., Garritty, C., Sampson, M., Zhang, L., Yazdi, F., Mamaladze, V., Pan, I., Macneil, J., Mack, D., Patel, D., Moher, D., 2005. The prevalence of celiac disease in average-risk and at-risk Western European populations: a systematic review. Gastroenterology 128, S57–S67.

Ehrenstein, M.R., Evans, J.G., Singh, A., Moore, S., Warnes, G., Isenberg, D.A., Mauri, C., 2004. Compromised function of regulatory T cells in rheumatoid arthritis and reversal by anti-TNFα therapy. J. Exp. Med. 200, 277–285.

Emery, P., Keystone, E., Tony, H.P., Cantagrel, A., Van Vollenhoven, R., Sanchez, A., Alecock, E., Lee, J., Kremer, J., 2008. IL-6 receptor inhibition with tocilizumab improves treatment outcomes in patients with rheumatoid arthritis refractory to anti-tumour necrosis factor biologicals: results from a 24-week multicentre randomised placebo-controlled trial. Ann. Rheum. Dis. 67, 1516–1523.

Esensten, J.H., Wofsy, D., Bluestone, J.A., 2009. Regulatory T cells as therapeutic targets in rheumatoid arthritis. Nat. Rev. Rheumatol. 5, 560–565.

Ferber, I., Schonrich, G., Schenkel, J., Mellor, A.L., Hammerling, G.J., Arnold, B., 1994. Levels of peripheral T cell tolerance induced by different doses of tolerogen. Science 263, 674–676.

Fernandez, D., Bonilla, E., Mirza, N., Niland, B., Perl, A., 2006. Rapamycin reduces disease activity and normalizes T cell activation-induced calcium fluxing in patients with systemic lupus erythematosus. Arthritis Rheum. 54, 2983–2988.

Furszyfer, J., Kurland, L.T., Mcconahey, W.M., Elveback, L.R., 1970. Graves' disease in Olmsted county, Minnesota, 1935 through 1967. Mayo Clin. Proc. 45, 636–644.

Garg, R., Khan, U., Alrajjal, A., Kafri, Z., 2017. Autoimmune hepatitis: a risk factor for cholangiocarcinoma. Case Rep. Gastroenterol. 11, 672–677.

Genovese, M.C., Covarrubias, A., Leon, G., Mysler, E., Keiserman, M., Valente, R., Nash, P., SIMON-Campos, J.A., Porawska, W., Box, J., Legerton 3rd, C., Nasonov, E., Durez, P., Aranda, R., Pappu, R., Delaet, I., Teng, J., Alten, R., 2011. Subcutaneous abatacept versus intravenous abatacept: a phase IIIb noninferiority study in patients with an inadequate response to methotrexate. Arthritis Rheum. 63, 2854–2864.

Gianfrani, C., Levings, M.K., Sartirana, C., Mazzarella, G., Barba, G., Zanzi, D., Camarca, A., Iaquinto, G., Giardullo, N., Auricchio, S., Troncone, R., Roncarolo, M.G., 2006. Gliadin-specific type 1 regulatory T cells from the intestinal mucosa of treated celiac patients inhibit pathogenic T cells. J. Immunol. 177, 4178–4186.

Goldberg, M.V., Drake, C.G., 2011. LAG-3 in cancer immunotherapy. Curr. Top. Microbiol. Immunol. 344, 269–278.

Goodnow, C.C., 2007. Multistep pathogenesis of autoimmune disease. Cell 130, 25–35.

Goronzy, J.J., Weyand, C.M., 2008. T-cell co-stimulatory pathways in autoimmunity. Arthritis Res. Ther. 10, S3.

Gottlieb, P.A., Quinlan, S., Krause-Steinrauf, H., Greenbaum, C.J., Wilson, D.M., Rodriguez, H., Schatz, D.A., Moran, A.M., Lachin, J.M., Skyler, J.S., 2010. Failure to preserve beta-cell function with mycophenolate mofetil and daclizumab combined therapy in patients with new- onset type 1 diabetes. Diabetes Care 33, 826–832.

Hahn, M., Nicholson, M.J., Pyrdol, J., Wucherpfennig, K.W., 2005. Unconventional topology of self peptide-major histocompatibility complex binding by a human autoimmune T cell receptor. Nat. Immunol. 6, 490–496.

Hegedus, L., Smith, T.J., Douglas, R.S., Nielsen, C.H., 2011. Targeted biological therapies for Graves' disease and thyroid-associated ophthalmopathy. Focus on B-cell depletion with Rituximab. Clin. Endocrinol. 74, 1–8.

Hendriks, J., Xiao, Y., Borst, J., 2003. CD27 promotes survival of activated T cells and complements CD28 in generation and establishment of the effector T cell pool. J. Exp. Med. 198, 1369–1380.

Herold, K.C., Gitelman, S.E., Ehlers, M.R., Gottlieb, P.A., Greenbaum, C.J., Hagopian, W., Boyle, K.D., Keyes-Elstein, L., Aggarwal, S., Phippard, D., Sayre, P.H., Mcnamara, J., Bluestone, J.A., 2013. Teplizumab (anti-CD3 mAb) treatment preserves C-peptide responses in patients with new-onset type 1 diabetes in a randomized controlled trial: metabolic and immunologic features at baseline identify a subgroup of responders. Diabetes 62, 3766–3774.

Hestvik, A.L.K., 2010. The double-edged sword of autoimmunity: lessons from multiple sclerosis. Toxins 2, 856.

Høglund, R.A., Maghazachi, A.A., 2014. Multiple sclerosis and the role of immune cells. World J. Exp. Med. 4, 27–37.

Holmoy, T., Hestvik, A.L., 2008. Multiple sclerosis: immunopathogenesis and controversies in defining the cause. Curr. Opin. Infect. Dis. 21, 271–278.

Hu, Y.L., Metz, D.P., Chung, J., Siu, G., Zhang, M., 2009. B7RP-1 blockade ameliorates autoimmunity through regulation of follicular helper T cells. J. Immunol. 182, 1421–1428.

Iezzi, G., Sonderegger, I., Ampenberger, F., Schmitz, N., Marsland, B.J., Kopf, M., 2009. CD40-CD40L cross-talk integrates strong antigenic signals and microbial stimuli to induce development of IL-17-producing $CD4^+$ T cells. Proc. Natl. Acad. Sci. U.S.A. 106, 876–881.

Isaacs, J.D., 2008. Therapeutic T-cell manipulation in rheumatoid arthritis: past, present and future. Rheumatology 47, 1461–1468.

Jacobs, M.R., Haynes, B.F., 1992. Increase in TCR gamma delta T lymphocytes in synovia from rheumatoid arthritis patients with active synovitis. J. Clin. Immunol. 12, 130–138.

Jacobson, E.M., Huber, A.K., Akeno, N., Sivak, M., Li, C.W., Concepcion, E., Ho, K., Tomer, Y., 2007. A CD40 Kozak sequence polymorphism and susceptibility to antibody-mediated autoimmune conditions: the role of CD40 tissue-specific expression. Genes Immun. 8, 205–214.

Kambayashi, T., 2018. Chapter 25-tolerance and autoimmunity A2-Hoffman, Ronald. In: Benz, E.J., Silberstein, L.E., Heslop, H.E., Weitz, J.I., Anastasi, J., Salama, M.E., Abutalib, S.A. (Eds.), Hematology, seventh ed. Elsevier.

Kamradt, T., Mitchison, N.A., 2001. Tolerance and autoimmunity. N. Engl. J. Med. 344, 655–664.

Keymeulen, B., Vandemeulebroucke, E., Ziegler, A.G., Mathieu, C., Kaufman, L., Hale, G., Gorus, F., Goldman, M., Walter, M., Candon, S., Schandene, L., Crenier, L., de Block, C., Seigneurin, J.M., de Pauw, P., Pierard, D., Weets, I., Rebello, P., Bird, P., Berrie, E., Frewin, M., Waldmann, H., Bach, J.F., Pipeleers, D., Chatenoud, L., 2005. Insulin needs after CD3-antibody therapy in new-onset type 1 diabetes. N. Engl. J. Med. 352, 2598–2608.

Khan, U., Ali, F., Khurram, M.S., Zaka, A., Hadid, T., 2017. Immunotherapy-associated autoimmune hemolytic anemia. J. Immunother. Cancer 5, 15.

Klein, L., Kyewski, B., Allen, P.M., Hogquist, K.A., 2014. Positive and negative selection of the T cell repertoire: what thymocytes see and don't see. Nat. Rev. Immunol. 14, 377–391.

Konya, C.K.V., 2013. T cells as treatment targets in systemic lupus erythematosus. Rheumatol. Curr. Res. 3 (2) https://doi.org/10.4172/2161-1149.1000120.

Korn, T., Bettelli, E., Oukka, M., Kuchroo, V.K., 2009. IL-17 and Th17 cells. Annu. Rev. Immunol. 27, 485–517.
Kuehn, H.S., Caminha, I., Niemela, J.E., Rao, V.K., Davis, J., Fleisher, T.A., Oliveira, J.B., 2011. FAS haploinsufficiency is a common disease mechanism in the human autoimmune lymphoproliferative syndrome. J. Immunol. 186, 6035–6043.
Kumar, V., Abbas, A.K., Fausto, N., Aster, J.C., 2015. Diseases of the immune system. In: Robbins and Cotran Pathologic Basis of Disease.
Kyewski, B., Klein, L., 2006. A central role for central tolerance. Annu. Rev. Immunol. 24, 571–606.
Kyogoku, C., Langefeld, C.D., Ortmann, W.A., Lee, A., Selby, S., Carlton, V.E., Chang, M., Ramos, P., Baechler, E.C., Batliwalla, F.M., Novitzke, J., Williams, A.H., Gillett, C., Rodine, P., Graham, R.R., Ardlie, K.G., Gaffney, P.M., Moser, K.L., Petri, M., Begovich, A.B., Gregersen, P.K., Behrens, T.W., 2004. Genetic association of the R620W polymorphism of protein tyrosine phosphatase PTPN22 with human SLE. Am. J. Hum. Genet. 75, 504–507.
Lafferty, K.J., Cunningham, A.J., 1975. A new analysis of allogeneic interactions. Aust. J. Exp. Biol. Med. Sci. 53, 27–42.
Lai, Z.W., Hanczko, R., Bonilla, E., Caza, T.N., Clair, B., Bartos, A., Miklossy, G., Jimah, J., Doherty, E., Tily, H., Francis, L., Garcia, R., Dawood, M., Yu, J., Ramos, I., Coman, I., Faraone, S.V., Phillips, P.E., Perl, A., 2012. N-acetylcysteine reduces disease activity by blocking mammalian target of rapamycin in T cells from systemic lupus erythematosus patients: a randomized, double-blind, placebo-controlled trial. Arthritis Rheum. 64, 2937–2946.
Leach, D.R., Krummel, M.F., Allison, J.P., 1996. Enhancement of antitumor immunity by CTLA-4 blockade. Science 271, 1734–1736.
Lee, Y.H., Rho, Y.H., Choi, S.J., Ji, J.D., Song, G.G., Nath, S.K., Harley, J.B., 2007. The PTPN22 C1858T functional polymorphism and autoimmune diseases—a meta-analysis. Rheumatology 46, 49–56.
Lens, S.M., Tesselaar, K., Van Oers, M.H., Van Lier, R.A., 1998. Control of lymphocyte function through CD27-CD70 interactions. Semin. Immunol. 10, 491–499.
Liao, W., Lin, J.X., Leonard, W.J., 2013. Interleukin-2 at the crossroads of effector responses, tolerance, and immunotherapy. Immunity 38, 13–25.
Lines, J.L., Sempere, L.F., Broughton, T., Wang, L., Noelle, R., 2014. VISTA is a novel broad-Spectrum negative checkpoint regulator for cancer immunotherapy. Cancer. Immunol. Res. 2, 510–517.
Linsley, P.S., Bradshaw, J., Greene, J., Peach, R., Bennett, K.L., Mittler, R.S., 1996. Intracellular trafficking of CTLA-4 and focal localization towards sites of TCR engagement. Immunity 4, 535–543.
Linsley, P.S., Nadler, S.G., 2009. The clinical utility of inhibiting CD28-mediated costimulation. Immunol. Rev. 229, 307–321.
Marazuela, M., Garcia-Lopez, M.A., Figueroa-Vega, N., De La Fuente, H., Alvarado-Sanchez, B., Monsivais-Urenda, A., Sanchez-Madrid, F., Gonzalez-Amaro, R., 2006. Regulatory T cells in human autoimmune thyroid disease. J. Clin. Endocrinol. Metab. 91, 3639–3646.
Maxwell, L.J., Singh, J.A., 2010. Abatacept for rheumatoid arthritis: a cochrane systematic review. J. Rheumatol. 37, 234–245.
Mazzarella, G., Stefanile, R., Camarca, A., Giliberti, P., Cosentini, E., Marano, C., Iaquinto, G., Giardullo, N., Auricchio, S., Sette, A., Troncone, R., Gianfrani, C., 2008. Gliadin activates HLA class I-restricted $CD8^+$ T cells in celiac disease intestinal mucosa and induces the enterocyte apoptosis. Gastroenterology 134, 1017–1027.

Miossec, P., 2007. Interleukin-17 in fashion, at last: ten years after its description, its cellular source has been identified. Arthritis Rheum. 56, 2111–2115.

Mishra, N., Brown, D.R., Olorenshaw, I.M., Kammer, G.M., 2001. Trichostatin A reverses skewed expression of CD154, interleukin-10, and interferon-gamma gene and protein expression in lupus T cells. Proc. Natl. Acad. Sci. U.S.A. 98, 2628–2633.

Monney, L., Sabatos, C.A., Gaglia, J.L., Ryu, A., Waldner, H., Chernova, T., Manning, S., Greenfield, E.A., Coyle, A.J., Sobel, R.A., Freeman, G.J., Kuchroo, V.K., 2002. Th1-specific cell surface protein Tim-3 regulates macrophage activation and severity of an autoimmune disease. Nature 415, 536–541.

Morshed, S.A., Latif, R., Davies, T.F., 2012. Delineating the autoimmune mechanisms in Graves' disease. Immunol. Res. 54, 191–203.

Moulton, V.R., Grammatikos, A.P., Fitzgerald, L.M., Tsokos, G.C., 2013. Splicing factor SF2/ASF rescues IL-2 production in T cells from systemic lupus erythematosus patients by activating IL-2 transcription. Proc. Natl. Acad. Sci. U.S.A. 110, 1845–1850.

Moulton, V.R., Nagpal, K., Tsokos, G.C., 2015. T cells and autoimmunity. In: Infection and Autoimmunity. Elsevier.

Mueller, D.L., Jenkins, M.K., Schwartz, R.H., 1989. Clonal expansion versus functional clonal inactivation: a costimulatory signalling pathway determines the outcome of T cell antigen receptor occupancy. Annu. Rev. Immunol. 7, 445–480.

Nakajima, A., Oshima, H., Nohara, C., Morimoto, S., Yoshino, S., Kobata, T., Yagita, H., Okumura, K., 2000. Involvement of CD70-CD27 interactions in the induction of experimental autoimmune encephalomyelitis. J. Neuroimmunol. 109, 188–196.

Nalbandian, A., Crispin, J.C., Tsokos, G.C., 2009. Interleukin-17 and systemic lupus erythematosus: current concepts. Clin. Exp. Immunol. 157, 209–215.

Nambiar, M.P., Fisher, C.U., Warke, V.G., Krishnan, S., Mitchell, J.P., Delaney, N., Tsokos, G.C., 2003. Reconstitution of deficient T cell receptor zeta chain restores T cell signaling and augments T cell receptor/CD3-induced interleukin-2 production in patients with systemic lupus erythematosus. Arthritis Rheum. 48, 1948–1955.

Okada, Y., Wu, D., Trynka, G., Raj, T., Terao, C., Ikari, K., Kochi, Y., Ohmura, K., Suzuki, A., Yoshida, S., Graham, R.R., Manoharan, A., Ortmann, W., Bhangale, T., Denny, J.C., Carroll, R.J., Eyler, A.E., Greenberg, J.D., Kremer, J.M., Pappas, D.A., Jiang, L., Yin, J., Ye, L., Su, D.F., Yang, J., Xie, G., Keystone, E., Westra, H.J., Esko, T., Metspalu, A., Zhou, X., Gupta, N., Mirel, D., Stahl, E.A., Diogo, D., Cui, J., Liao, K., Guo, M.H., Myouzen, K., Kawaguchi, T., Coenen, M.J., Van Riel, P.L., Van de Laar, M.A., Guchelaar, H.J., Huizinga, T.W., Dieude, P., Mariette, X., Bridges, S.L., Zhernakova, A., Toes, R.E., Tak, P.P., Miceli-Richard, C., Bang, S.Y., Lee, H.S., Martin, J., Gonzalez-Gay, M.A., Rodriguez-Rodriguez, L., Rantapaa-Dahlqvist, S., Arlestig, L., Choi, H.K., Kamatani, Y., Galan, P., Lathrop, M., Eyre, S., Bowes, J., Barton, A., de Vries, N., Moreland, L.W., Criswell, L.A., Karlson, E.W., Taniguchi, A., Yamada, R., Kubo, M., Liu, J.S., Bae, S.C., Worthington, J., Padyukov, L., Klareskog, L., Gregersen, P.K., Raychaudhuri, S., Stranger, B.E., de Jager, P.L., Franke, L., Visscher, P.M., Brown, M.A., Yamanaka, H., Mimori, T., Takahashi, A., Xu, H., Behrens, T.W., Siminovitch, K.A., Momohara, S., Matsuda, F., Yamamoto, K., Plenge, R.M., 2014. Genetics of rheumatoid arthritis contributes to biology and drug discovery. Nature 506, 376–381.

Okazaki, T., Maeda, A., Nishimura, H., Kurosaki, T., Honjo, T., 2001. PD-1 immunoreceptor inhibits B cell receptor-mediated signaling by recruiting src homology 2-domain-containing tyrosine phosphatase 2 to phosphotyrosine. Proc. Natl. Acad. Sci. U.S.A. 98, 13866–13871.

Orban, T., Bundy, B., Becker, D.J., Dimeglio, L.A., Gitelman, S.E., Goland, R., Gottlieb, P.A., Greenbaum, C.J., Marks, J.B., Monzavi, R., Moran, A., Raskin, P., Rodriguez, H., Russell, W.E., Schatz, D., Wherrett, D., Wilson, D.M., Krischer, J.P., Skyler, J.S., 2011. Co-stimulation modulation with abatacept in patients with recent-onset type 1 diabetes: a randomised, double-blind, placebo-controlled trial. Lancet 378, 412–419.

Penaranda, C., Tang, Q., Bluestone, J.A., 2011. Anti-CD3 therapy promotes tolerance by selectively depleting pathogenic cells while preserving regulatory T cells. J. Immunol. 187, 2015–2022.

Peters, A.L., Stunz, L.L., Bishop, G.A., 2009. CD40 and autoimmunity: the dark side of a great activator. Semin. Immunol. 21, 293–300.

Peterson, P., Org, T., Rebane, A., 2008. Transcriptional regulation by AIRE: molecular mechanisms of central tolerance. Nat. Rev. Immunol. 8, 948–957.

Piccirillo, C.A., Thornton, A.M., 2004. Cornerstone of peripheral tolerance: naturally occurring $CD4^+CD25^+$ regulatory T cells. Trends Immunol. 25, 374–380.

Pons-Estel, G.J., Alarcon, G.S., Scofield, L., Reinlib, L., Cooper, G.S., 2010. Understanding the epidemiology and progression of systemic lupus erythematosus. Semin. Arthritis Rheum. 39, 257–268.

Richards, H.B., Satoh, M., Shaw, M., Libert, C., Poli, V., Reeves, W.H., 1998. Interleukin 6 dependence of anti-DNA antibody production: evidence for two pathways of autoantibody formation in pristane-induced lupus. J. Exp. Med. 188, 985–990.

Rivera-Nieves, J., Ho, J., Bamias, G., Ivashkina, N., Ley, K., Oppermann, M., Cominelli, F., 2006. Antibody blockade of CCL25/CCR9 ameliorates early but not late chronic murine ileitis. Gastroenterology 131, 1518–1529.

Roncarolo, M.G., Gregori, S., Bacchetta, R., Battaglia, M., 2014. Tr1 cells and the counter-regulation of immunity: natural mechanisms and therapeutic applications. Curr. Top. Microbiol. Immunol. 380, 39–68.

Rosati, G., 2001. The prevalence of multiple sclerosis in the world: an update. Neurol. Sci. 22, 117–139.

Rothe, H., Jenkins, N.A., Copeland, N.G., Kolb, H., 1997. Active stage of autoimmune diabetes is associated with the expression of a novel cytokine, IGIF, which is located near Idd2. J. Clin. Invest. 99, 469–474.

Rullo, O.J., Tsao, B.P., 2013. Recent insights into the genetic basis of systemic lupus erythematosus. Ann. Rheum. Dis. 72 (Suppl. 2), 56–61.

Sainsbury, A., Sanders, D.S., Ford, A.C., 2013. Prevalence of irritable bowel syndrome-type symptoms in patients with celiac disease: a meta-analysis. Clin. Gastroenterol. Hepatol. 11, 359–365.e1.

Salvatore Albani, Lucy, R, Wedderburn, Berent Prakken, 2016. Textbook of Pediatric Rheumatology, Chapter 4, pp. 33–42.e3.

Schildberg, F.A., Klein, S.R., Freeman, G.J., Sharpe, A.H., 2016. Coinhibitory pathways in the B7-CD28 ligand-receptor family. Immunity 44, 955–972.

Schneider, H., Martin, M., Agarraberes, F.A., Yin, L., Rapoport, I., Kirchhausen, T., Rudd, C.E., 1999. Cytolytic T lymphocyte-associated antigen-4 and the TCR zeta/CD3 complex, but not CD28, interact with clathrin adaptor complexes AP-1 and AP-2. J. Immunol. 163, 1868–1879.

Schuppan, D., Junker, Y., Barisani, D., 2009. Celiac disease: from pathogenesis to novel therapies. Gastroenterology 137, 1912–1933.

Schwartz, R.H., 2012. Historical overview of immunological tolerance. Cold Spring Harb. Perspect. Biol. 4, a006908.

Sha, W.C., Nelson, C.A., Newberry, R.D., Kranz, D.M., Russell, J.H., Loh, D.Y., 1988. Positive and negative selection of an antigen receptor on T cells in transgenic mice. Nature 336, 73.

Sheppard, K.A., Fitz, L.J., Lee, J.M., Benander, C., George, J.A., Wooters, J., Qiu, Y., Jussif, J.M., Carter, L.L., Wood, C.R., Chaudhary, D., 2004. PD-1 inhibits T-cell receptor induced phosphorylation of the ZAP70/CD3zeta signalosome and downstream signaling to PKCtheta. FEBS Lett. 574, 37–41.

Sherry, N., Hagopian, W., Ludvigsson, J., Jain, S.M., Wahlen, J., Ferry, R.J., Bode, B., Aronoff, S., Holland, C., Carlin, D., King, K.L., Wilder, R.L., Pillemer, S., Bonvini, E., Johnson, S., Stein, K.E., Koenig, S., Herold, K.C., Daifotis, A.G., 2011. Teplizumab for treatment of type 1 diabetes (Protege study): 1-year results from a randomised, placebo-controlled trial. Lancet 378, 487–497.

Shi, Y., Zhu, M., 2013. Medullary thymic epithelial cells, the indispensable player in central tolerance. Sci. China Life Sci. 56, 392–398.

Shimon Sakaguchi, K.W.A.M.M., 2013. Regulatory T Cells. Clinical Immunology.

Siggs, O.M., Makaroff, L.E., Liston, A., 2006. The why and how of thymocyte negative selection. Curr. Opin. Immunol. 18, 175–183.

Skapenko, A., Leipe, J., Lipsky, P.E., Schulze-Koops, H., 2005. The role of the T cell in autoimmune inflammation. Arthritis Res. Ther. 7, S4.

Skarica, M., Eckstein, C., Whartenby, K.A., Calabresi, P.A., 2011. Novel mechanisms of immune modulation of natalizumab in multiple sclerosis patients. J. Neuroimmunol. 235, 70–76.

Smith, J.A., Tso, J.Y., Clark, M.R., Cole, M.S., Bluestone, J.A., 1997. Nonmitogenic anti-CD3 monoclonal antibodies deliver a partial T cell receptor signal and induce clonal anergy. J. Exp. Med. 185, 1413–1422.

Spodzieja, M., Lach, S., Iwaszkiewicz, J., Cesson, V., Kalejta, K., Olive, D., Michielin, O., Speiser, D.E., Zoete, V., Derré, L., Rodziewicz-Motowidło, S., 2017. Design of short peptides to block BTLA/HVEM interactions for promoting anticancer T-cell responses. PLoS One 12, e0179201.

Stockinger, B., Veldhoen, M., Martin, B., 2007. Th17 T cells: linking innate and adaptive immunity. Semin. Immunol. 19, 353–361.

Suda, T., Takahashi, T., Golstein, P., Nagata, S., 1993. Molecular cloning and expression of the Fas ligand, a novel member of the tumor necrosis factor family. Cell 75, 1169–1178.

Tanaka, T., Kishimoto, T., 2012. Targeting interleukin-6: all the way to treat autoimmune and inflammatory diseases. Int. J. Biol. Sci. 8, 1227–1236.

Thomson, A.W., Turnquist, H.R., Raimondi, G., 2009. Immunoregulatory functions of mTOR inhibition. Nat. Rev. Immunol. 9, 324–337.

Tischner, D., Woess, C., Ottina, E., Villunger, A., 2010. Bcl-2-regulated cell death signalling in the prevention of autoimmunity. Cell Death Dis. 1, e48.

Tsokos, G.C., 2011. Systemic lupus erythematosus. N. Engl. J. Med. 365, 2110–2121.

Tsokos, G.C., Nambiar, M.P., Tenbrock, K., Juang, Y.T., 2003. Rewiring the T-cell: signaling defects and novel prospects for the treatment of SLE. Trends Immunol. 24, 259–263.

Valle, A., Barbagiovanni, G., Jofra, T., Stabilini, A., Perol, L., Baeyens, A., Anand, S., Cagnard, N., Gagliani, N., Piaggio, E., Battaglia, M., 2015. Heterogeneous CD3 expression levels in differing T cell subsets correlate with the in vivo anti-CD3-mediated T cell modulation. J. Immunol. 194, 2117–2127.

Venken, K., Hellings, N., Thewissen, M., Somers, V., Hensen, K., Rummens, J.-L., Medaer, R., Hupperts, R., Stinissen, P., 2008. Compromised CD4(+) CD25(high) regulatory T-cell function in patients with relapsing-remitting multiple sclerosis is correlated with a reduced frequency of FOXP3-positive cells and reduced FOXP3 expression at the single-cell level. Immunology 123, 79–89.

Von Spee-Mayer, C., Siegert, E., Abdirama, D., Rose, A., Klaus, A., Alexander, T., Enghard, P., Sawitzki, B., Hiepe, F., Radbruch, A., Burmester, G.R., Riemekasten, G., Humrich, J.Y., 2016. Low-dose interleukin-2 selectively corrects regulatory T cell defects in patients with systemic lupus erythematosus. Ann. Rheum. Dis. 75, 1407–1415.

Watanabe, N., Gavrieli, M., Sedy, J.R., Yang, J., Fallarino, F., Loftin, S.K., Hurchla, M.A., Zimmerman, N., Sim, J., Zang, X., Murphy, T.L., Russell, J.H., Allison, J.P., Murphy, K.M., 2003. BTLA is a lymphocyte inhibitory receptor with similarities to CTLA-4 and PD-1. Nat. Immunol. 4, 670–679.

Weyand, C., Bryl, E., Goronzy, J., 2000. The Role of T Cells in Rheumatoid Arthritis.

Wildin, R.S., Freitas, A., 2005. IPEX and FOXP3: clinical and research perspectives. J. Autoimmun. 25 (Suppl), 56–62.

Wing, K., Suri-Payer, E., Rudin, A., 2005. $CD4^+CD25+$-regulatory T cells from mouse to man. Scand. J. Immunol. 62, 1–15.

Wolfsdorf, J.I., Allgrove, J., Craig, M.E., Edge, J., Glaser, N., Jain, V., Lee, W.W., Mungai, L.N., Rosenbloom, A.L., Sperling, M.A., Hanas, R., 2014. ISPAD clinical Practice consensus Guidelines 2014. Diabetic ketoacidosis and hyperglycemic hyperosmolar state. Pediatr. Diabetes 15 (Suppl. 20), 154–179.

Wucherpfennig, K.W., Call, M.J., Deng, L., Mariuzza, R., 2009. Structural alterations in peptide-MHC recognition by self-reactive T cell receptors. Curr. Opin. Immunol. 21, 590–595.

Wucherpfennig, K.W., Newcombe, J., Li, H., Keddy, C., Cuzner, M.L., Hafler, D.A., 1992. Gamma delta T-cell receptor repertoire in acute multiple sclerosis lesions. Proc. Natl. Acad. Sci. U.S.A. 89, 4588–4592.

Yang, L., Anderson, D.E., Kuchroo, J., Hafler, D.A., 2008. Lack of TIM-3 immunoregulation in multiple sclerosis. J. Immunol. 180, 4409–4414.

You, S., Candon, S., Kuhn, C., Bach, J.F., Chatenoud, L., 2008. CD3 antibodies as unique tools to restore self-tolerance in established autoimmunity their mode of action and clinical application in type 1 diabetes. Adv. Immunol. 100, 13–37.

Zanzi, D., Stefanile, R., Santagata, S., Iaffaldano, L., Iaquinto, G., Giardullo, N., Lania, G., Vigliano, I., Vera, A.R., Ferrara, K., Auricchio, S., Troncone, R., Mazzarella, G., 2011. IL-15 interferes with suppressive activity of intestinal regulatory T cells expanded in Celiac disease. Am. J. Gastroenterol. 106, 1308–1317.

Zhang, Q., Vignali, D.A.A., 2016. Co-stimulatory and co-inhibitory pathways in autoimmunity. Immunity 44, 1034–1051.

Zhao, Q., Chen, G., 2014. Role of IL-33 and its receptor in T cell-mediated autoimmune diseases. BioMed Res. Int. 2014, 587376.

CHAPTER FOUR

Impact of Immunosuppressive Drugs on the Metabolism of T Cells

Nicolas Pallet[*,§,¶,1], **Ana A. Fernández-Ramos**[*,§] **and Marie-Anne Loriot**[*,§,¶]
[*]Institut National de la Santé et de la recherche Médeicale (INSERM) U1147, Paris, France
[§]Paris Descartes University, Paris, France
[¶]Clinical Chemistry Department, Georges Pompidou European Hospital, APHP, Paris, France
[1]Corresponding author: E-mail: nicolas.pallet@parisdescartes.fr

Contents

International Review of Cell and Molecular Biology, Volume 342
ISSN 1937-6448
https://doi.org/10.1016/bs.ircmb.2018.05.009

Abstract

Energetic metabolism supports rapid cell growth and proliferation, differentiation, polarization, and effector functions of T cells. T lymphocytes have the remarkable plasticity that allows them to shape their metabolism to adapt to extracellular and intracellular cues, a process that involves molecular modules referred to as "metabolic checkpoints" that sense metabolic signals and transduce effector messages. These metabolic checkpoints may represent a novel therapeutic strategy for immune modulation. Chemical immunosuppressive drugs including mammalian target of rapamycin inhibitors (sirolimus and everolimus), calcineurin inhibitors (tacrolimus and cyclosporine), and purine and pyrimidine synthesis inhibitors (6-mercaptopurine, mycophenolic acid, and methotrexate) are widely prescribed for the treatment of autoimmune and inflammatory diseases and for controlling alloimmunity in interfering with the signals that activate and allow T cells to proliferate. Emerging evidence indicates that these drugs also target T-cell metabolism and metabolic checkpoints, which, as a consequence, could contribute to their immunosuppressive effects. These examples raise the issue of how the modulation of these metabolic checkpoints can regulate T-cell activation, differentiation, and function. In this review we highlight emerging concepts about the modulation of metabolic reprogramming in T-cell responses by immunosuppressive drugs and how potential therapeutic interventions influence T-cell fate and effector function.

1. INTRODUCTION

Cellular metabolism is a complex network of biochemical reactions necessary for the production of energy in the form of adenosine triphosphate (ATP) and the biosynthesis of macromolecules (Palsson-McDermott and O'Neill, 2013; Pearce, 2010). T lymphocytes metabolism and functions are closely related because T-cell activation is associated with variations in metabolic requirements to ensure effector functions and rapid cell proliferation (MacIver et al., 2013; Pearce and Pearce, 2013; Pearce et al., 2013). In turn, metabolic reprogramming allows the cell to adapt to extracellular and intracellular conditions that change over time and location. Activation of T cells is accompanied by a reorientation of metabolism to ensure massive clonal growth and expansion and also to allow lymphocyte differentiation and effector functions (Loftus and Finlay, 2016; Maciolek et al., 2014; Pearce and Pearce, 2013). Knowledge of the molecular bases regulating the metabolic reprogramming of the cells of the immune system, mainly T lymphocytes, has progressed considerably.

A critical aspect of research on immunometabolism is elucidating the impact of T-cell metabolism on immunity to identify new therapeutic targets for controlling or modulating lymphocytes responses (Macintyre and Rathmell, 2013; Pearce, 2010). Although the characterization of the metabolic circuitries that drive and support T-cell activation, proliferation, and polarization exponentially expands, our knowledge of the impact of immunosuppressive molecules (molecules that inhibit activating signals that follow T-cell receptor (TCR) activation by antigen presenting cells) is still in its infancy. However, a better understanding of the changes operated by immunosuppressive drugs in the T cells metabolic landscape is critical to tailor new therapeutic strategies and to optimize drugs combination.

2. IMMUNOSUPPRESSION AND IMMUNOLOGICAL DISEASES

Adaptive immune responses involve both naive and memory lymphocytes. In the peripheral organs, dendritic cells become activated and move to T-cell areas of secondary lymphoid organs. There, antigen-bearing dendritic cells engage with antigen-reactive naive T cells and central memory T cells that recirculate between lymphoid compartments. An antigen on the surface of dendritic cells which triggers T cells with cognate TCRs constitutes "signal 1," transduced through the CD3 complex. Dendritic cells provide costimulation or "signal 2," delivered when CD80 and CD86 on the surface of dendritic cells engage with CD28 on T cells (Halloran, 2004). Signals 1 and 2 activate three signal transduction pathways: the calcium–calcineurin pathway, the RAS–mitogen-activated protein kinase pathway, and the nuclear factor-κB pathway (Fig. 1). These pathways activate transcription factors that trigger the expression of many molecules, including interleukin (IL)-2, CD154, and CD25. IL-2 and IL-15 activate the mammalian target of rapamycin (mTOR) pathway to provide "signal 3," the trigger for cell proliferation (Halloran, 2004). Lymphocyte proliferation also requires nucleotide synthesis. Proliferation and differentiation lead to a large number of effector T cells. B cells are activated when antigen engages their antigen receptors, usually in lymphoid follicles or in extrafollicular sites.

Immunosuppression can be achieved by depleting lymphocytes, diverting lymphocyte traffic, or blocking lymphocyte response pathways. Immunosuppressive drugs include small-molecule drugs, depleting and nondepleting polyclonal and monoclonal antibodies, fusion proteins, intravenous immune globulin, and glucocorticoids (Halloran, 2004; Khan et al., 2016; Khan and Sewell, 2006; Lim et al., 2017). Because of

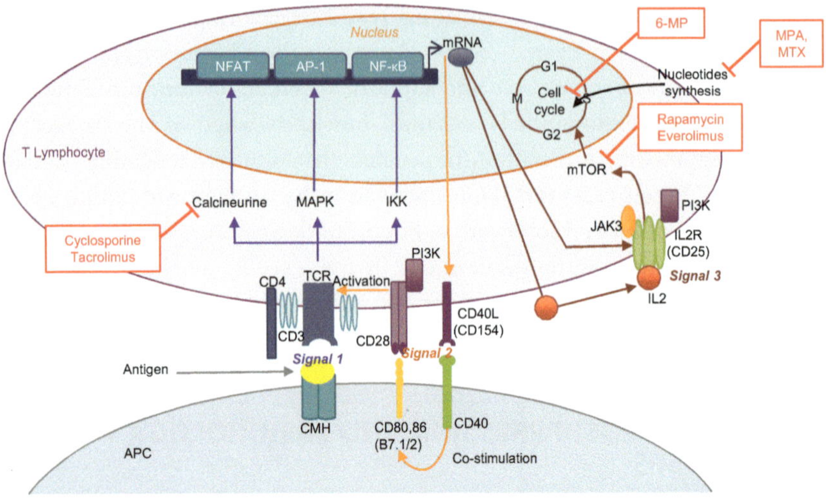

Figure 1 ***Activation pathways of T lymphocytes.*** After recognition of an antigen by the MHC presented by the APCs, the activation signal 1 by the TCR is triggered. Then three intracellular pathways can be activated: calcineurin with NFAT, MAPK with AP-1, and IKK with NF-κB. This allows the expression of costimulatory receptors such as CD40L necessary for complete activation of lymphocytes. Signals 1 and 2 allow synthesis of IL-2, which is secreted and binds to its IL-2R autocrine receptor to activate signal 3 for cell proliferation via mTOR and cell cycle progression. *6-MP*, 6-mercaptopurine; *AP-1*, activator protein 1; *APC*, antigen-presenting cell; *CD28*, differentiation cluster 28; *CD3*, differentiation cluster 3; *CD4*, differentiation cluster 4; *CD40*, differentiation cluster 40; *CD40L*, CD40 or CD154 ligand; *CD80/86*, 80/86 differentiation cluster; *IKK*, IKB kinase; *IL2*, interleukin 2; *IL2R*, IL2 receptor; *JAK3*, Janus kinase 3; *M*, mitosis; *MAPK*, mitogen-activated protein kinases; *MHC*, major histocompatibility complex; *mRNA*, messenger ribonucleic acid; *mTOR*, mammalian target of rapamycin; *MTX*, methotrexate; *NFAT*, nuclear factor of active T lymphocytes; *NF-κB*, nuclear factor-kappa B; *PI3K*, phosphatidylinositol 3-kinase.

space limitations, we will focus in this article on calcineurin inhibitors, mTOR inhibitors, and purine and pyrimydine synthesis inhibitors, which represent the great majority of chemical immunosuppressive molecules prescribed worldwide.

Immunosuppression is required in three major medical conditions: autoimmune diseases, autoinflammatory diseases including inflammatory bowel disease (IBD), and prevention and treatment of allograft rejection and graft versus host disease. These therapeutic strategies are intended to act selectively on the overactivated lymphocytes or better on auto/alloreactive clones. 6-Mercaptopurine (6-MP), mycophenolic acid (MPA), methotrexate (MTX), and rapamycin are immunosuppressive therapies intended to reduce the proliferation of lymphocytes, whereas calcineurin inhibitors inhibit cell activation (Fig. 1). These treatments are not devoid of side effects (drugs related and/or secondary to immunosuppression), which limit their use. Undesired effects of immunosuppressive drugs are related to the consequences of immunodeficiency and to the toxicity to other tissues. The intensity of immunosuppression, rather than the specific agent used, leads to infection (BK virus nephropathy, CMV or EBV reactivations) and cancers, such as posttransplantation lymphoproliferative disease or cutaneaous neoplasia. The other toxicity is drug specific and is related to a specific mechanism because each class of drugs targets molecules with physiologic roles in organs. For example, posttransplant diabetes induced by tacrolimus may reflect cytotoxicity of the drug for pancreatic beta cells, and nephrotoxicity of cyclosporine may reflect a role of calcineurin within the renal vasculature.

The treatment of autoimmune diseases, for example, systemic lupus erythematosus (SLE) and IBD, often uses an immunosuppression strategy to induce remission and to avoid relapses using maintenance immunosuppression, usually for years. Individuals with autoimmune diseases are not intended to receive long-term immunosuppressive drugs. Prevention of allograft rejection is based on the use of induction strategy followed by a long-term maintenance therapy. Induction with biological agents produces a state of immune nonresponse to the graft by depleting or blocking T-cell proliferation. Biological agents include polyclonal antilymphocyte immunoglobulins and anti-IL-2 (CD25) receptor antibodies. During the maintenance phase, azathioprine (AZA), MPA, rapamycin, tacrolimus, cyclosporine, corticosteroids (prednisone), and the belatacept fusion protein (which blocks the cosignal CD80/86 stimulation) are used in different combinations to target several signaling pathways and to prevent activation and proliferation of T cells (Halloran, 2004; Khan et al., 2016; Khan and Sewell, 2006; Lim et al., 2017). Among the calcineurin inhibitors, tacrolimus is widely used, whereas the use of cyclosporine is decreasing. Among the antimetabolite drugs, MPA is widely administered, whereas 6MP is no more prescribed except in some specific situation even though it was the first immunosuppressant used in renal transplantation. mTOR inhibitors are less potent immunosuppressive drugs during the initial phase of immunosuppression and are far less prescribed than tacrolimus (Lim et al., 2017).

3. IMMUNOSUPPRESSIVE DRUGS AND THEIR IMPACT ON CELL METABOLISM

Most of the immunosuppressive molecules currently used in the clinic to control the immune system have been discovered decades ago for purposes totally different to what they are actually used (in general for antibiotic development). The mechanisms supporting their immunomodulatory properties have been progressively discovered, and despite their longevity, the precise impact of these drugs on cell biology and how they interact with cells signaling and biochemistry are still an active field of research. Consistent with this, research on immunometabolism has progressively integrated the potential impact of immunosuppressive drugs, and our understanding of the metabolic reprogramming that would occur in cells exposed to immunosuppressive drugs is currently increasing. However, formal experimental evidence is limited, and emerging concepts derive from nonimmune models or models not adequately designed to mimic T-cell activation by antigen-presenting cells. Nevertheless, the available data clearly support the concept of metabolic reprogramming induced by immunosuppressive drugs. Advances made in this field will challenge established working models of immunosuppression in a mechanistic point of view and will drive the emerging concept of targeting energetic metabolism to dampen activated T cells, and current evidence supports the notion that immunosuppressive drugs interfere with "metabolic checkpoints" functions. Given their critical role in the fate of T cells, metabolic checkpoints constitute novel potential therapeutic targets (Pollizzi and Powell, 2014). Metabolic reprogramming is regulated at the transcriptional level by the transcription factors Myc and hypoxia-inducible factor 1α (HIF1α) and at the posttranscriptional level by 5′ adenosine monophosphate–activated protein kinase (AMPK) and mTOR. Theses nodes are referred to as "metabolic checkpoints" because they link immune signaling with metabolic activity (MacIver et al., 2013). After antigen recognition by T cells, these metabolic checkpoints control enzymatic activities that regulate catabolic and anabolic fluxes to promote T-cell activation, differentiation, proliferation, and functioning (Pearce and Pearce, 2013; Wang and Green, 2012; Green et al., 2014; Pollizzi and Powell, 2014).

3.1 Calcineurin Inhibitors

3.1.1 Pharmacology

Cyclosporine is an 11-amino acid cyclic peptide, which has been isolated from the *Tolypocladium inflatum* fungus (Allison, 2000; Tedesco and Haragsim, 2012). Tacrolimus is a lactone macrolide extracted from a culture filtrate of *Streptomyces tsukubaensis* (Allison, 2000; Kho et al., 2011; Tedesco and Haragsim, 2012) (Fig. 2). Both compounds are extremely lipophilic (Barbarino et al., 2013). Calcineurin inhibitors form complexes with immunophilins: cyclosporine binds to cyclophillin A (a peptidyl cis-trans isomerase) and tacrolimus binds to FKBP12 (12-kDa FK506-binding protein). The cyclosporine—cyclophilin or tacrolimus—FKBP12 complexes inhibit the activity of calcineurin. After activation of the TCRs and costimulation signals, three intracellular signaling pathways are activated (Fig. 3). One of these three pathways involves calcineurin, a protein phosphatase catalytic activity of which depends on calmodulin via a self-regulating pattern. Activated calcineurin dephosphorylates nuclear factor of activated T cells (NFAT), which can migrate to the nucleus. NFAT regulates the transcription of proinflammatory cytokines genes such as IL-2, IL-3, IL-4, interferon-γ, and tumor necrosis factor (TNF)-α (Allison, 2000; Barbarino et al., 2013; Naesens and Sarwal 2010a,b). Cyclosporine and tacrolimus have also been identified as inhibitors of c-jun-N-terminal kinase (JNK) and p38 mitogen-activated protein kinase (MAPK) signaling pathways, which promote transcription of IL-2 and other cytokines. Finally they induce the expression of transforming growth factor β1 (TGF-β1), a key cytokine involved in the regulation of immune cells. TGF-β1 blocks the proliferation of IL-2—dependent T cells and favors the polarization of Treg cells. Cyclosporine is indicated for the prevention of transplant rejection after solid organ transplantation and for the preventive or curative treatment of graft-versus-host disease. Cyclosporine is also used in inflammatory disorders such as uveitis, nephrotic syndromes, rheumatoid arthritis, psoriasis, and atypical dermatitis. Tacrolimus is indicated for the prevention of rejection of solid organ transplants.

3.1.2 Effects on Cellular Metabolism

Direct and indirect evidence suggests that calcineurin inhibitors may alter cellular metabolism in T cells and non-T cells. Cyclosporine and tacrolimus

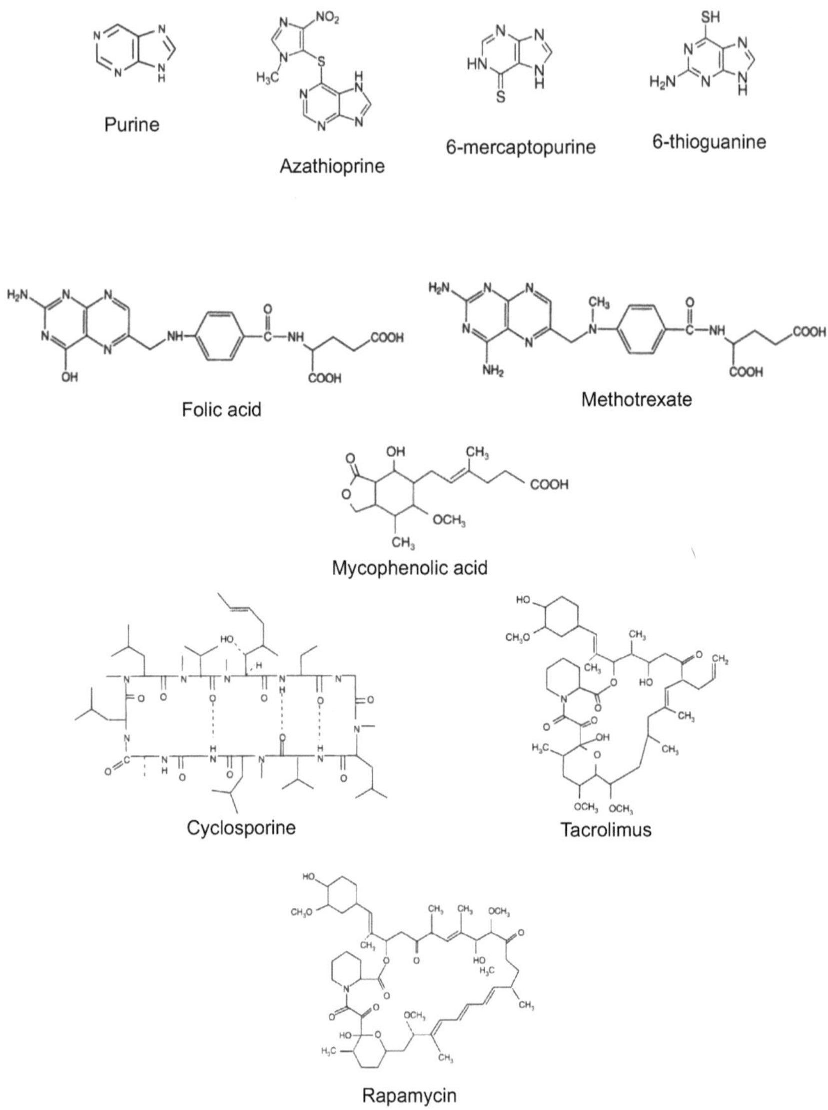

Figure 2 Chemical structures of immunosuppressive drugs.

inhibit the protein expression of HIF1α in human mast cells during hypoxia (Walczak-Drzewiecka et al., 2008). HIF1α promotes the transcription of several genes implicated in decreasing mitochondrial oxygen consumption, switching from oxidative phosphorylation (OXPHOS) to anaerobic glycolysis. Furthermore, it increases the rate of glycolytic flux and modulates glutaminolysis and glutamine uptake. Overall, HIF1α optimizes energy and

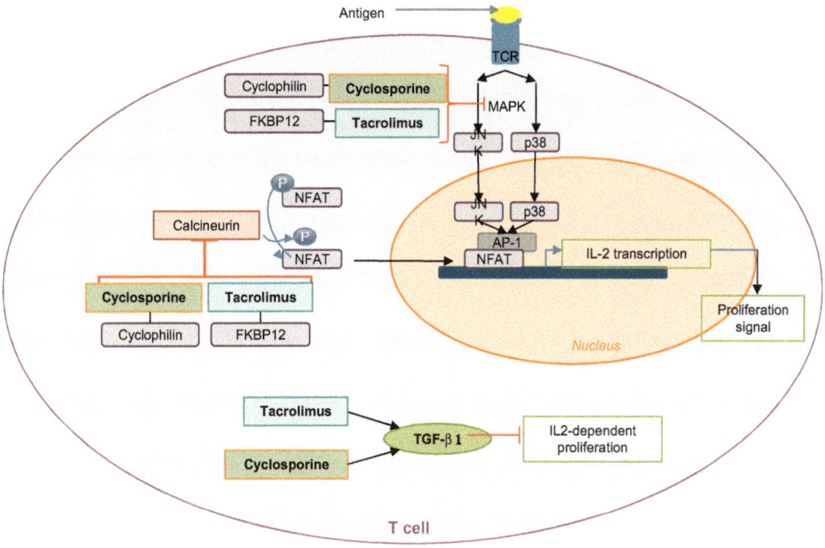

Figure 3 ***Mechanisms of action of tacrolimus and cyclosporine.*** Cyclosporine and tacrolimus have three major mechanisms of action: (1) The cyclosporine–cyclophiline and tacrolimus–FKBP12 complexes inhibit calcineurin, which will not be able to dephosphorylate NFAT so that it cannot migrate to the nucleus and promote IL-2 transcription. (2) The cyclosporin–cyclophiline and tacrolimus–FKBP12 complexes inhibit JNK and p38 MAPK pathways, preventing the transfer of AP-1 into the nucleus to play its role as NFAT cofactor. (3) Tacrolimus and cyclosporine also induce TGF-β1, which inhibits IL-2–dependent cell proliferation. *AP-1*, activator protein 1; *FKBP12*, 12-kDa FK506 binding protein; *IL-2*, interleukin 2; *JNK*, N-terminal c-Jun kinases; *MAPK*, mitogen-activated protein kinases; *NFAT*, nuclear factor of activated T cells; *TCR*, T-cell receptor; *TGF-β1*, transforming growth factor beta 1.

cellular homeostasis to promote cell survival (Nizet and Johnson, 2009; Iurlaro et al., 2014). Cyclosporine, by inhibition of calcineurin, which inhibits ubiquitination and proteosomal degradation of HIF1α, promotes its degradation in HEK293T cells (Liu et al., 2007).

The NFAT factor binds to a distal enhancer in the Myc promoter region to favor cell proliferation (Buchholz et al., 2006; Mognol et al., 2012; Singh et al., 2010; Wang et al., 2012). Myc is recognized as a key metabolic reprogramming checkpoint factor in T cells. When T cells are activated, Myc promotes the expression of genes implicated in enhancing glutamine influx, glutaminolysis, and glycolysis as well as those involved in regulating ribosomal and mitochondrial biogenesis (Pollizzi and Powell, 2015; Yang and Chi, 2015; MacIver et al., 2013; Pearce and Pearce, 2013; Wang et al., 2011; Karmaus and Chi, 2014; Maciolek et al., 2014;

Iurlaro et al., 2014). While NFAT controls the expression of genes involved in the activation and differentiation of T cells, cyclosporine reduces the expression of Myc in murine CD4+ T cells (Buchholz et al., 2006; Mognol et al., 2012). Hence modulation of cellular metabolism by the NFAT/Myc pathway and its inhibition by cyclosporine seem plausible but needs to be further demonstrated. Another example of the effect of calcineurin inhibitors on metabolic checkpoints comes from data indicating that the Akt-mTOR pathway is required for calcineurin inhibitors–induced tumor growth by removing the inhibition of mammalian target of rapamycin complex 1 (mTORC1) activity via PRAS40 phosphorylation in some cancer cells (Basu et al. 2011, 2012). mTOR is a metabolic checkpoint that regulates autophagy, cell growth, proliferation, and metabolic gene expression to increase glucose uptake, glycolysis, and nucleotide, protein, and lipid biosynthesis (Waickman and Powell, 2012; Pollizzi and Powell, 2014; Inoki et al., 2012; Chi, 2012; Wouters and Koritzinsky, 2008; Iurlaro et al., 2014). In the rat hippocampus, administration of cyclosporine increases phosphorylation of the metabolic checkpoint AMPK at the Thr-172 position and thus its activation (Park et al., 2011). AMPK enhances glucose uptake and catabolic processes, such as glycolysis, OXPHOS, and fatty acids oxidation (FAO), inhibits glucose, glycogen, and fatty acid synthesis, and is crucial for T cells survival and proliferation (Green et al., 2014; Pollizzi and Powell, 2015; MacIver et al., 2013; Pearce and Pearce, 2013; Gwinn et al., 2008).

Beyond metabolic checkpoints, calcineurin inhibitors target metabolic fluxes. For example, tacrolimus inhibits the activity of glucokinase, a glycolysis-limiting enzyme that allows phosphorylation of glucose into glucose 6-phosphate, resulting in a reduction in ATP production and energetic distress in pancreatic islets cells (Radu et al., 2004). In addition, calcineurin inhibitors, by inhibiting NFAT (which cooperates with the PI3K-AKT-mTOR pathway to regulate the expression of glucose transporters, glycolytic enzymes, and transcriptional regulators of glycolysis), promote the suppression of glucose-dependent metabolism, leading to T-cell anergy (Vaeth et al., 2017). These data suggest that the mechanism of action of calcineurin inhibitors includes the suppression of glucose-dependent metabolism and limiting the amount of IL-2 is not the primary cause of reduced proliferation in calcineurin-impaired T cells (Vaeth et al., 2017). In line with this and potentially explaining their metabolic side effects, calcineurin inhibitors impair the entry of glucose into muscles and adipose tissue. In human adipocytes and in the L6 myoblast line, these drugs inhibit glucose uptake by increasing endocytosis of the GLUT4 glucose transporter, thereby inducing insulin resistance (Pereira et al., 2014). Cyclosporine and tacrolimus induce mitochondrial dysfunction by affecting the respiratory chain II and III complexes (Sabine Illsinger et al., 2011a,b), decrease the intracellular transport of glucose, induce the expression of genes involved in lipogenesis (lipid synthesis), and reduce fatty acid oxidation in rat muscle and adipose tissues (Fuhrmann et al., 2014). In line with a potential impact of calcineurin inhibitors on OXPHOS, tacrolimus induces ATP depletion by a modification of the membrane structure of mitochondria in rat-isolated mitochondria (Simon et al., 2003). Finally cyclosporine blocks the expression of the leucine transporter SLC7A5 after TCR activation, causing a decrease in leucine entry into the cell, inducing anergic state, and inhibiting CD4+ T and CD8+ T cells differentiation (Sinclair et al., 2013). Overall, these data indicate that the spectrum of metabolic disturbances induced by calcineurin inhibitors is wide and a reduction in the activity of energetic metabolism, either at the level of glycolysis or OXPHOS, is causal in inducing T-cell anergy but may also be involved in some of the side effects of these drugs, for example, diabetes.

3.2 mTOR Inhibitors

3.2.1 Pharmacology

Rapamycin (sirolimus) and everolimus (2-hydroxyethyl-sirolimus) are macrocyclic and lipophilic lactones (Fig. 2). Sirolimus is naturally produced by the fermentation of the actinomycete *Streptomyces hygroscopicus*. Everolimus is more polar than rapamycin and has improved pharmacokinetic properties (Kirchner et al., 2004; Mahalati and Kahan, 2001; Sehgal, 2003; Taylor et al., 2005). Rapamycin and everolimus have similar modes of action. Thcy bind to FKBP12, forming an mTOR inhibitory complex (Sehgal, 2003). The mTOR protein is present in the mTORC1 and mammalian target of rapamycin complex 2 (mTORC2) complexes with different cellular functions. The activation of mTORC1 promotes cell growth by activating the synthesis of proteins, lipids, and nucleotides and the biogenesis of mRNAs. The mTORC2 complex is involved in the regulation of cell survival and metabolism. Rapamycin mainly inhibits the mTORC1 complex, and in the case of chronic exposure, it can also inhibit mTORC2 in certain cell types (Moes et al., 2015) (Fig. 4). As a result, treatment with rapamycin decreases translational activity, cell cycle progression, and cell proliferation (Dumont and Su, 1996; Halloran, 2004;

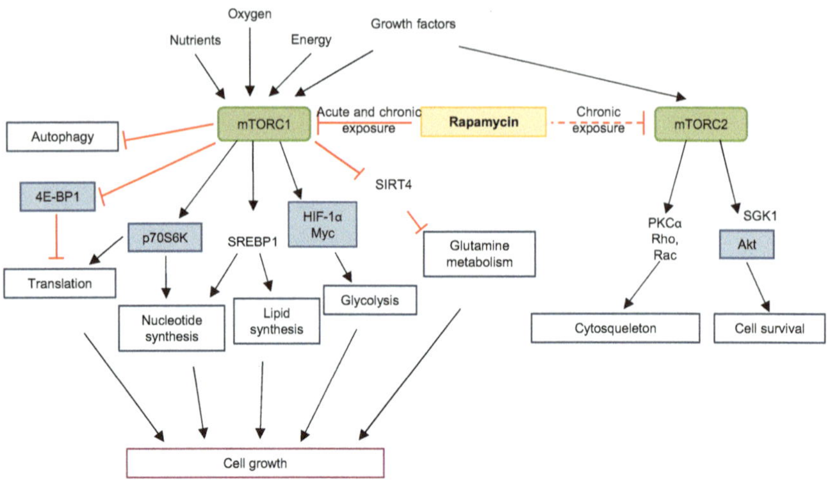

Figure 4 ***Cellular functions of mTORC1 and mTORC2 complexes and effects of rapamycin.*** The mTORC1 complex is activated by energy, oxygen, nutrients, and growth factors. It negatively regulates autophagy but activates the synthesis of proteins, nucleotides, and lipids, glutaminolytic metabolism, and glycolysis to promote cell growth and proliferation. The mTORC1 complex may be inhibited by rapamycin after acute or chronic exposure. The mTORC2 complex is upregulated by growth factors and downregulated by chronic exposure to rapamycin. It facilitates the organization of cytoskeletal actin and cell survival. *4E-BP1*, eukaryotic translation initiation factor 4E-binding protein 1; *Akt*, protein kinase B; *HIF-1α*, hypoxia inducible factor 1 α; *mTORC1*, mammalian target of rapamycin complex 1; *mTORC2*, mammalian target of rapamycin complex 2; *P70S6K*, ribosomal protein S6 beta-1 kinase; *PKCα*, protein kinase C alpha; *SGK1*, serine/threonine-protein kinase 1; *SIRT4*, sirtuin 4; *SREBP1*, sterol regulatory element–binding protein.

Moes et al., 2015; Taylor et al., 2005). During T-cell activation, signals 1 and 2 are triggered, resulting in IL-2 production and activation of its IL-2 receptor in an autocrine manner. Then lymphocytes enter into a phase of cell growth and proliferation regulated by mTOR (signal 3) (Fig. 1). Rapamycin-mediated inhibition of mTOR activity (signal 3) induces T-cell anergy (Powell and Delgoffe, 2010). Thus rapamycin inhibits the proliferation of T lymphocytes (and also B cells), causing immunosuppression. The PI3K/Akt/mTOR pathway is crucial for the proliferation of T cells but also for cancer cells in which inhibition of mTOR has a therapeutic interest (Li et al., 2014). Rapamycin (and everolimus) is indicated in adult patients with low-to-moderate immunologic risk receiving renal transplantation for prevention of organ rejection. For the effects of rapamycin being relatively modest as an immunosuppressant and anticancer drug, it is mainly prescribed in patients suffering from severe side effects related to calcineurin inhibitors (Li et al., 2014).

3.2.2 Effects on Cellular Metabolism

The central role of mTOR in orchestrating the relationship between T-cell metabolism and function, proliferation, and polarization has been extensively explored (Pollizzi and Powell, 2015). The mTOR pathway is critical for lymphocyte differentiation and polarization, and mTOR-deficient CD4+ T cells fail to differentiate into Th1 and Th17 subsets under activating condition and develop into forkhead box P3 (FoxP3+) regulatory cells (Pollizzi and Powell, 2015). By blocking mTOR activity, rapamycin inhibits effector $CD4^+$ cell differentiation, promotes the generation of Th2 CD4+ T cells, inhibits formation of resident memory cells in intestinal mucosa, increases the generation of Treg cells, and enhances quality and quantity of $CD8^+$ memory T cells (Powell and Delgoffe, 2010), and all of these phenotypic changes participate in the immunosuppressive effects of mTOR inhibitors (Pollizzi and Powell, 2014; Taylor et al., 2005; Zeiser and Negrin, 2008; Patel and Kobashigawa, 2006).

It is important to notice that mTOR inhibition in using genetic and chemical strategies has provided critical insights into how mTORC1 and mTORC2 regulate T-cell proliferation, function, and thereby metabolism. However, the specific effects of rapamycin as an immunosuppressive drug on T-cell metabolism cannot been simply extrapolated from these experiments. Indeed, rapamycin does not completely inhibit mTORC1, and its effects on downstream pathways are likely transient with feedback loops. In addition, rapamycin targets mTORC2 in some instances (Pollizzi and Powell, 2015). In addition, the role of mTOR in T cells has been deciphered on the basis of different in vitro and in vivo models targeting mTOR-specific proteins such as PDK1, mTOR, Rheb, RAPTOR, RIC TOR, TSC1, mSIN, or SGK1, which yield mTOR-specific effects, different effects related to the function of the targeted protein, and also nonspecific effects related to proliferation arrest. Consequently the immune effects of rapamycin in T-cell metabolism cannot be considered as the corollary of those produced by selective deletion of mTOR-specific proteins, leading to mTOR inhibition. In addition, the precise effects of rapamycin treatment on the mTOR signaling pathway and cell metabolism have not been systematically explored. An important (and often unresolved) issue in immunometabolism is to distinguish the causality links between metabolic reprogramming and T-cell phenotypic changes to differentiate the situations in which metabolic changes precede and dictate T-cell polarization from those in which metabolic plasticity helps cells to adapt to the new T-cell phenotype. However, there are experimental arguments supporting that rapamycin, in preventing effector T-cell energetic metabolism, favors conversion of effectors T cells into regulatory and/or memory cells (Powell and Delgoffe, 2010). For example, rapamycin, in reducing glycolytic flux in activated $CD8^+$ cells, promotes a shift toward $CD8^+$ memory T cells (Pollizzi, 2015; Powell and Delgoffe, 2010). In line with this, mTOR inhibition promotes fatty acid oxidation metabolism, which is associated with the generation of $CD8^+$ memory T cells (Pollizzi, 2015; Pollizzi and Powell, 2014; Patsoukis et al., 2015). In Jurkat T cells, rapamycin treatment is associated with a near-global shutdown of expression of genes implicated in glutaminolysis and nucleotide synthesis and several genes implicated in glycolysis (Fernandez-Ramos et al., 2017b). Rapamycin induces a progressive drop in intracellular ATP levels, reduces glucose and glutamine oxidation, and decreases glucose uptake, suggesting that the reduction of glucose oxidation could result from the reduction of glucose uptake, but without altering the expression of Glut1 (SLC2A1) and Glut3 (SLC2A3) (Fernandez-Ramos et al., 2017b) glucose transporters with a selective cell-intrinsic function in the metabolic reprogramming of T cells (Macintyre et al., 2014).

Rapamycin also targets mitochondrial metabolism and biogenesis, which are critically regulated by mTORC1 (Morita et al., 2013). Rapamycin promotes ATP depletion by acting on OXPHOS by a mechanism that involves modification of the membrane structure of isolated rat kidney mitochondria (Simon et al., 2003). Rapamycin also decreases mitochondrial membrane potential, oxygen uptake, and the ability of Jurkat T cells to synthesize ATP (Schieke et al., 2006). Rapamycin also decreases the transcription of genes involved in mitochondrial oxidation. In the Jurkat T cells line an early incubation time (30 min) with rapamycin decreases mitochondrial ATP production and fosters glycolysis and lactate production (Ramanathan and Schreiber, 2009). Together, these results indicate that rapamycin (and by extension everolimus, even if there is no direct experimental proof) profoundly alters proliferating T cell's metabolism and promotes energetic stress. Although most of the studies focused on glycolysis and mitochondrial metabolism, it is worth noting that given the wide spectrum of metabolic functions driven by mTOR, for example in fatty acids metabolism, the landscape of metabolic changes produced by mTOR in T cells is still largely uncovered.

3.3 Thiopurines

3.3.1 Pharmacology

Thiopurines, including 6-MP, and AZA, a prodrug metabolized into 6-MP by glutathione S-transferase, have a chemical structure derived from purines (Fig. 2). Thiopurines possess a sulfhydryl group attached to a carbon and are metabolized into thioguanines and thioguanine nucleotides. After intracellular uptake via nucleoside transporters, 6-MP is metabolized in the catabolic pathways of xanthine oxidase and thiopurine S-methyltransferase (TPMT) and in the anabolic pathway of hypoxanthine-guanine phosphoribosyltransferase (HPRT) (Fig. 5) (Bronk et al., 1988; Chouchana et al., 2012; Elion 1972, 1989; Fotoohi et al., 2006; Hartman and Buchanan, 1959; Zaza et al., 2005). HPRT catalyzes the formation of 6-thioinosine monophosphate (6-TIMP) nucleotide, resulting in the formation of 6-thioguanine nucleotides (6-TGN): thioguanine monophosphate (TGMP), diphosphate (TGDP), and triphosphate (TGTP) (Cara et al., 2004; Fong et al., 2015). 6-TGNs are the pharmacologically active metabolites that exert cytotoxic and immunosuppressive effects after their incorporation into RNA and DNA, resulting in the activation of the mismatch repair system and apoptosis (Cara et al., 2004). 6-TGNs also inhibit Rac 1: Ras-related C3 botulinum toxin substrate 1 and cause the apoptotic death of stimulated CD4+ T cells, supporting the immunosuppressive effects of thiopurines (Allison and Eugui, 1993; Fong et al., 2015; Taylor et al., 2005). 6-TIMP can also be metabolized by TPMT to generate methylated derivatives, 6-methyl mercaptopurines ribonucleotides (6MMP-R), which can inhibit de novo synthesis of purines and cause cell proliferation arrest. This pathway is crucial for purine synthesis because the salvage pathway is weakly active in lymphocytes (Chouchana et al., 2012). Therefore thiopurines promote immunosuppression by decreasing T-cell proliferation and activating apoptosis. AZA is used in IBD, rheumatoid arthritis, SLE, autoimmune hepatitis, autoimmune hemolytic anemia, and myasthenia gravis. It is currently a second-line drug after MPA in the maintenance phase for the prevention of rejection after kidney transplantation.

3.3.2 Effects on Cellular Metabolism

Numerous studies have highlighted the relationship between 6-MP and the pathways involved in metabolic reprogramming. 6-MP increases the amount of mRNA and stability of HIF1α in HEpG2 cells (Yoo et al., 2007).

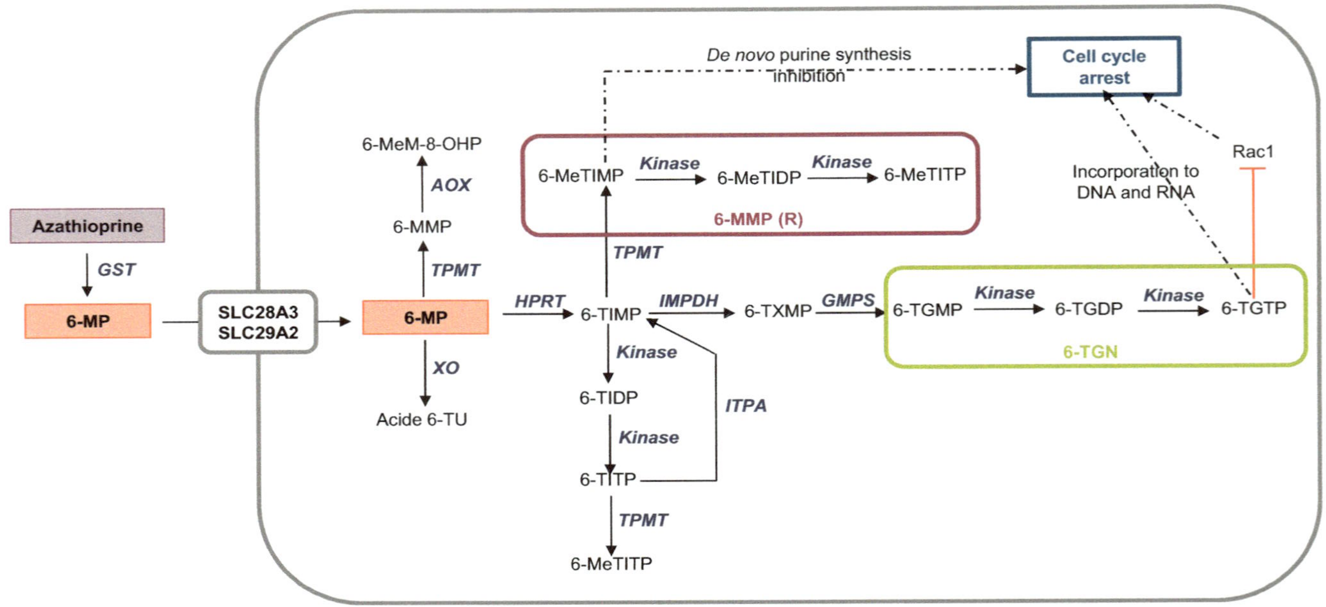

Figure 5 ***Metabolism and mechanism of action of thiopurines.*** Azathioprine is metabolized by TPMT6-MP, which enters cells through SLC28A3 or SLC29A2 transporters. Then 6-MP can follow three different routes: (1) oxidation by XO giving 6-TU acid; (2) methylation with TPMT, followed by AOX resulting in 6-MeM-8-OHP; or (3) metabolism by HPRT that produces 6-TIMP. Then 6-TIMP can be metabolized by three different pathways: (i) by successive kinases then again the TPMT to form the 6-MeTITP; (ii) by TPMT, followed by kinases to form 6-MMP (R); or (iii) by IMPDH, followed by GMPS and kinases to form 6-TGNs. 6-MeTIMP and 6-TGTP are responsible for cytotoxicity and cell death. *6-MeM-8-OHP*, 6-methylmercapto-8-hydroxypurine; *6-MeTIDP*, 6-methylthioinosine diphosphate; *6-MeTIMP*, 6-methylthioinosine monophosphate; *6-MeTITP*, 6-methylthioinosine triphosphate; *6-MMP (R)*, 6-methylmercaptopurine ribonucleotide; *6-MMP*, 6-methylmercaptopurine; *6-MP*, 6-mercaptopurine; *6-TGDP*, 6-thioguanine diphosphate; *6-TGMP*, 6-thioguanine monophosphate; *6-TGN*, 6-thioguanine nucleotides; *6-TGTP*, 6-thioguanine triphosphate; *6-TIDP*, 6-thioinosine diphosphate; *6-TIMP*, 6-thioinosine monophosphate; *6-TITP*, 6-thioinosine triphosphate; *6-TXMP*, 6-thioxanthosine monophosphate; *6-TU acid*, 6-thiouric acid; *AOX*, aldehyde oxidase; *GMPS*, guanosine monophosphate synthetase; *GST*, glutathione S-transferase; *HPRT*, hypoxanthine-guanine phosphoribosyltransferase; *IMPDH*, inosine monophosphate dehydrogenase; *ITPA*, inosine triphosphatase; *SLC28A3*, solute carrier family 28 member 3; *SLC29A2*, solute carrier family 29 member 2; *TPMT*, thiopurine S-methyltransferase; *XO*, xanthine oxidase.

Moreover, recent data indicate that 6-MP inhibits the PI3K/Akt/mTOR signaling pathway in a murine model of microglia, suggesting that it may interfere with metabolic checkpoints and modulate cellular metabolism. In this model, 6-MP modifies the activity of the family of nuclear NR: nuclear receptor group 4A orphan receptors, which regulate the metabolism of glucose and lipids in microglia (Huang et al., 2016). In addition, 6-MP inhibits the enzymes involved in glycolysis and ATP production, such as polyphosphate kinase and hexokinase, as well as in the release of lactate in the human-immortalized B lymphoblasts (Rodríguez-Enríquez et al., 2009). In a model of isolated mouse liver mitochondria, 6-MP has a direct effect on the decrease of mitochondrial respiration (Porceddu et al., 2012). Moreover, 6-MP induces oxidative stress, mitochondrial damage, and depletion of ATP in rat hepatocytes (Tapner et al., 2004). These data suggest a direct effect of 6-MP on oxidative metabolism, independent of metabolic checkpoint targeting.

6-MP profoundly impacts T cell's energetic metabolism by reducing ATP production and decreasing glycolytic and glutaminolytic fluxes in Jurkat T cells (Fernandez-Ramos et al., 2017a). 6-MP rapidly reduces intracellular ATP concentrations, leading to the activation of AMPK, whereas mTOR is inhibited, and the expression of HIF1α and Myc is reduced by 6-MP. Consequently, glucose and glutamine oxidation are strongly reduced. Notably, 6-MP has no impact on glucose uptake (Fernandez-Ramos et al., 2017a). In these cells, exposure to 6-MP induces mitochondrial dysfunction with loss of activity and increased production of reactive oxygen species, which can induce DNA damage (Zhang et al., 2013). Overall, these data indicate that the spectrum of metabolic disturbances induced by 6-MP and other thiopurines in T cells is only emerging but raise the possibility of an energetic failure promoted by 6 MP, which blocks de novo purine synthesis and therefore purine nucleotides synthesis, including ATP. In addition, if and how 6-MP impacts mitochondrial biogenesis and physiology, for example in interfering with mitochondrial DNA and RNA synthesis, is an appealing research avenue that could explain the disturbances in ATP production in T cells after 6-MP exposure.

3.4 Methotrexate

3.4.1 Pharmacology

MTX differs from folic acid only by the substitution of a hydroxyl radical by a 4-amino group in the pteridine ring and by addition of a methyl group in the N10 position (Bleyer, 1978; Grim et al., 2003; Inoue and Yuasa, 2014)

(Fig. 2). MTX is hydrophilic and penetrates into the cell mainly through active transport. The entry of MTX is mediated by the folate transporter 1 (SLC19A1), and its efflux requires the action of ATP-binding cassette transporters (Brown et al., 2016; Inoue and Yuasa, 2014). MTX is mainly converted into polyglutamine derivatives (MTX-PG) that are made of six glutamate residues that can be fixed on MTX by a folylpolyglutamate synthetase. The more the MTX is polyglutamine modified, the more it will be retained inside the cells. MTX-PGs have a potent inhibitory effect on pyrimidine and purine metabolism (Dogra and Mahajan 2013; Furst, 1997; Giverhaug et al., 1999; Grim et al., 2003) (Fig. 6). MTX is indicated in the active and severe forms of rheumatoid arthritis in adults and also in juvenile idiopathic arthritis and in psoriatic arthritis. MTX is also indicated for the treatment of severe and generalized psoriasis and for the treatment of IBD (Herfarth, 2016).

3.4.2 Effects on Cellular Metabolism

MTX could have an impact on metabolic checkpoints and cell metabolism in nonimmune cell lines. For example, in a human cell model of lung adenocarcinoma, MTX decreases Myc expression, thereby reducing cell growth (Serra et al., 2008). In human embryonic cells, MTX activates the AMPK signaling pathway to allow adaptation to hypoxic conditions (Thornton et al., 2016). In addition, in myotubes, HEK-293 cells, and isolated muscle cells, MTX activates AMPK with critical consequences on energetic metabolism, glucose entry, fatty acid oxidation, and mitochondrial biogenesis (Pirkmajer et al., 2015). In a model of breast cancer cells (MCF-7), MTX in combination with Aminoimidazole-4-carboxamide ribonucleotide (AICAR) blocks cell proliferation, but if administered separately, neither compound has an effect on cell growth. In addition, they activate AMPK and increase mitochondrial oxidation with reduction of glycolysis. These results indicate that the combination of AICAR and MTX influences cell metabolism and may be a promising therapeutic strategy for inducing cytostatic effects (Fodor et al., 2016). Notably, there are no relevant data available on the impact of MTX on the metabolism of T cells.

3.5 Mycophenolic Acid

3.5.1 Pharmacology

Mycophenolate mofetil (MMF) and mycophenolate sodium (MPS) are prodrugs of MPA (Fig. 2), a reversible inhibitor of eukaryotic inositol monophosphate dehydrogenase (IMPDH) type II (Allison and Eugui, 2000;

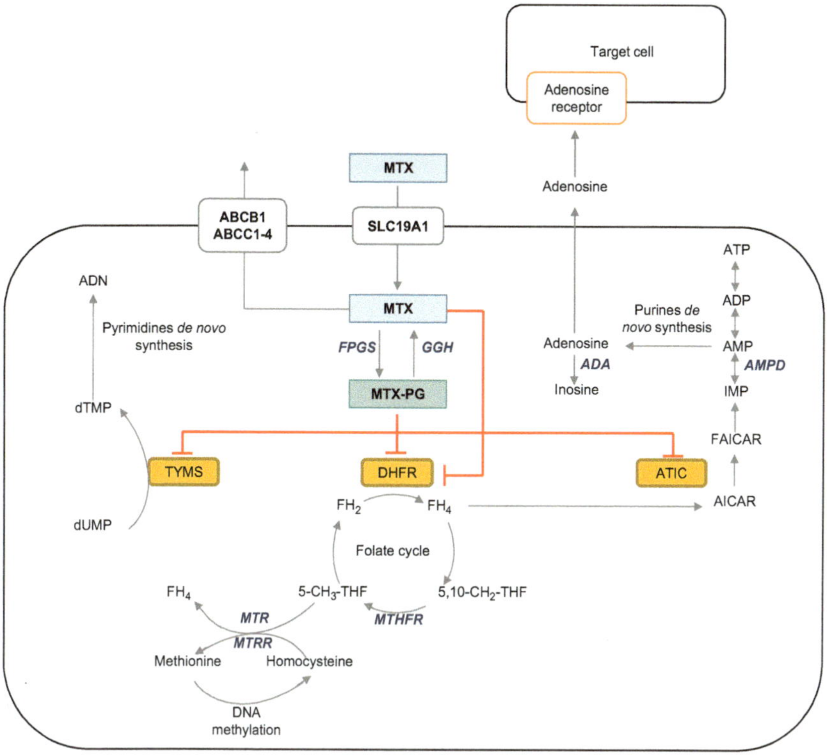

Figure 6 ***Metabolism of methotrexate (MTX) and mechanisms of action.*** MTX enters the cell via the SLC19A1 transporter. It is metabolized by FPGS to MTX-PG and by GGH to MTX. MTX can be transported out of the cell by ABCB1 and ABCC1-4 membrane efflux pumps. MTX-PG inhibits TYMS, which catalyzes the formation of dTMP from dUMP, involved in de novo synthesis of pyrimidines. MTX-PG and MTX can inhibit DHFR, the folate ring enzyme that converts FH2 to FH4. Then the ring intermediate, 5-CH3-THF, is converted to FH4, and homocysteine is recycled to methionine by the action of MTR and MTRR. The reverse reaction, from methionine to homocysteine, is necessary for the release of methyl groups and the methylation of DNA. Finally MTX-PG inhibits ATIC, an enzyme that transforms AICAR into FAICAR. Then FAICAR is transformed into IMP and AMP by AMPD. AMP can give ADP then ATP or be used in the de novo purine synthesis pathway to produce adenosine. Adenosine can be transformed into inosine by ADA or exit the cell and bind its receptor on a target cell to exert its antiinflammatory effects. *5,10-CH2-THF*, 5,10-methylene tetrahydrofolate; *5-CH3-THF*, 5-methyl tetrahydrofolate; *ABCB1*, ABCB1 transporter; *ABCC1-4*, adenosine triphosphate–binding cassette transporter C1-4; *ADA*, adenosine deaminase; *ADP*, adenosine diphosphate; *AICAR*, 5-aminoimidazole-4-carboxamide ribonucleotide; *AMP*, adenosine monophosphate; *AMPD*, adenosine monophosphate deaminase; *ATIC*, bifunctional protein of purine synthesis PURH; *ATP*, adenosine triphosphate; *DHFR*, dihydrofolate reductase; *dTMP*, deoxythymidine monophosphate; *dUMP*, deoxyuridine monophosphate; *FAICAR*, 10-formyl-5-aminoimidazole-4-carboxamide ribonucleotide; *FH2*, dihydrofolate; *FH4*, tetrahydrofolate; *FPGS*, polyglutamate synthetase; *GGH*, γ-glutamyl hydrolase; *IMP*, inosine-5-monophosphate; *MTHFR*, methylenetetrahydrofolate; *MTR*, methyl tetrahydrofolate reductase; *MTRR*, methionine synthase reductase; *MTX*, methotrexate; *MTX-PG*, methotrexate polyglutamate; *SLC19A1*, solute carrier family 19 member 1; *TYMS*, thymidylate synthase.

Taylor et al., 2005) that is involved in the de novo pathway of purine synthesis and is upregulated in activated B and T cells (Hartono et al., 2013; Sugiyama et al., 2005; Quéméneur et al., 2003; Taylor et al., 2005; Quemeneur et al., 2002) (Fig. 7). IMPDH-II is a rate-limiting enzyme that converts inosine monophosphate (IMP) to guanosine monophosphate (GMP), and its inhibition decreases the size of the guanosine and GMP pools, preventing DNA synthesis, blocking cell proliferation, and inhibiting cytotoxic T-cell proliferation and differentiation into $CD4^{+}$ and $CD8^{+}$ cells (Quéméneur et al., 2003; Allison, 2000; Allison and Eugui, 1993;

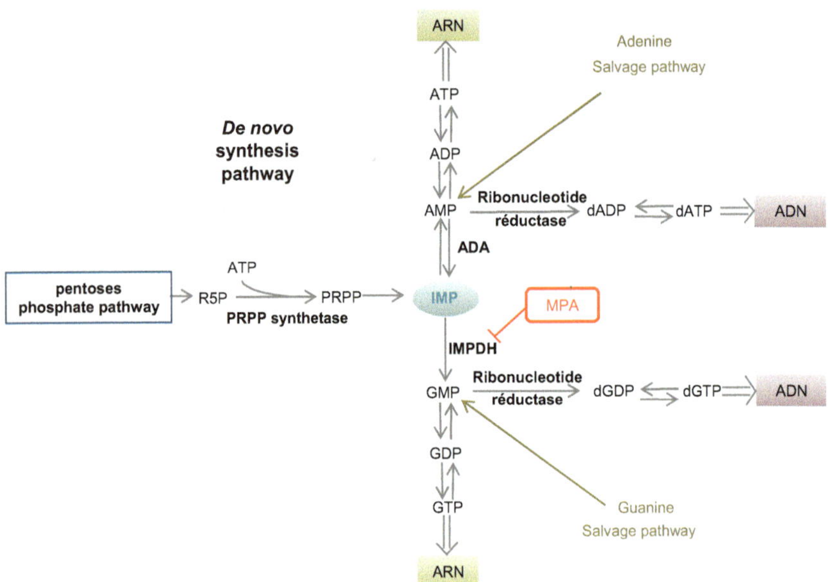

Figure 7 ***Mechanisms of action of MPA.*** The pentose phosphate pathway can give R5P, which because of the action of PRPP synthetase is transformed into PRPP and then into IMP. IMP is transformed into GMP by IMPDH (MPA pharmacological target) or AMP by ADA. By different reactions, GDP, GTP, ADP, and ATP are produced to form RNA, and dGDP, dGTP, dADP, and dATP are produced to form DNA. In addition, the salvage pathway can give adenine or guanine to generate AMP or GMP. *ADA*, adenosine deaminase; *ADP*, adenosine diphosphate; *AMP*, adenosine monophosphate; *ATP*, adenosine triphosphate; *dADP*, deoxyadenosine diphosphate; *dATP*, deoxyadenosine triphosphate; *dGDP*, deoxyguanine diphosphate; *dGTP*, deoxyguanine triphosphate; *GDP*, guanosine diphosphate; *GMP*, guanosine monophosphate; *GTP*, guanosine triphosphate; *IMP*, inosine monophosphate; *IMPDH*, IMP dehydrogenase; *MPA*, mycophenolic acid; *PP pathway*, pentose phosphate pathway; *PRPP synthetase*, 5-phosphoribosyl-1-pyrophosphate synthetase; *PRPP*, 5-phosphoribosyl-1-pyrophosphate; *R5P*, ribose-5-phosphate.

Quemeneur et al., 2002; Wiseman, 2015; Sugiyama et al., 2005; Dayton et al., 1992; Maripuri and Kasiske, 2014). Specifically MPA blocks cell cycle progression in S phase and promotes apoptosis (Allison, 2000). MPA suppresses glycosylation by inhibiting the transfer of fucose and mannose to glycoproteins because nucleotide intermediates are required in these reactions. Inhibition of glycosylation, particularly adhesion molecules, leads to a decrease in lymphocyte recruitment in sites of inflammation (Allison and Eugui, 2000). MMF and MPS are indicated in the prevention of rejections after solid organ transplantation (kidney, heart, and liver). MMF is also prescribed in some autoimmune diseases such as SLE and in neuromuscular diseases (Appel and Appel, 2009; Chaundry et al., 2001).

3.5.2 Effects on Cellular Metabolism

MPA induces T-cell anergy and metabolic reprogramming by suppressing the Akt/mTOR pathway (Dun et al., 2014; He et al., 2011), but how it affects metabolic checkpoints and cell energetic metabolism has not been systematically addressed. A genome-wide association study indicates that treatment with MPA affects Myc- and HIF1α-regulated pathways in endothelial cells, which may explain the antiangiogenic and antitumor effects of this drug (Domhan et al., 2008). The expression of Myc and HIF1α is downregulated in human dermal microvascular endothelial cells after 12 h of MPA but appears to be upregulated in human primary glioblastoma cells, suggesting that the effects of MPA on metabolic checkpoints may vary depending on the cell type.

In CD4+ T cells, MPA decreases activation of the Akt-mTOR pathway, glycolysis, and oxygen uptake (Morel, 2017), and MPA modifies the membrane potential of mitochondria, but whether MPA affects mitochondrial metabolism has not been addressed (Choi et al., 2011; Takebe et al., 2006). MPA profoundly impacts the cellular metabolism of Jurkat T cells by generating an energetic distress, decreasing the glycolytic and glutaminolytic fluxes, and targeting HIF1α and Myc. MPA produces an early and transient drop in the intracellular ATP content related to the inhibition of de novo synthesis of purines, leading to the activation of the energy sensor AMPK. The expression of HIF1α and Myc, promoting the activation of glycolysis and glutaminolysis, is inhibited by MPA (Fernandez-Ramos et al., 2017b). Together, these results indicate that MPA affects T cell's energetic metabolism, which, in turn, could participate in T-cell anergy. However, the relevance of these findings remains to be established in activated naïve T cells stimulated with antigen-presenting cells.

3.6 Overview of the Impact of Immunosuppressive Drugs on T-Cell Metabolism

Taken together, these examples indicate that the biological effects of immunosuppressive drugs go beyond the only field of activating signals inhibition, and emerging evidence supports a critical impact on cell energetic metabolism. Overall, immunosuppressive drugs modify the metabolic reprogramming in T cells by interfering with the metabolic checkpoints HIF1α, AMPK, mTOR, and Myc, as well as with glycolytic and glutaminolytic fluxes (Fig. 9). This inhibition may affect T-cell differentiation in promoting regulatory T cell's generation and decreasing effector T cells. However, further studies are needed to clarify the metabolic modifications induced by these drugs in T cells and to provide a comprehensive understanding of the cellular effects. More relevant models of activated T cells are required to determine the real impact of these immunosuppressive drugs on T-cell metabolism.

4. PHARMACOLOGICAL MODULATION OF ENERGY METABOLISM TO PROMOTE IMMUNOSUPPRESSION

Given the importance of metabolism in the fate of cancer cells and T lymphocytes, metabolic checkpoints and their associated metabolic pathways are crucial in the search for new therapeutic targets in immune system diseases. The most studied case is that of rapamycin, a direct inhibitor of the metabolic checkpoint mTOR. The history of rapamycin exemplifies how a molecule initially developed as an immunosuppressant to block lymphocyte activation (signal 3) presumably controls lymphocyte proliferation through multiple effects on T-cell metabolism.

Other primarily nonimmunosuppressive molecules with effects on metabolic reprogramming may be used to promote immunosuppression (Fig. 8). 2-Deoxy-glucose (2-DG) is a glucose analog, which is phosphorylated by hexokinase (HK) but not metabolized further. Therefore it blocks glycolysis and is used as an HK inhibitor (Allison et al., 2017). Another example is metformin, a mitochondrial complex I inhibitor indicated in type II diabetes. Inhibition of the respiratory chain results in a reduction of ATP concentrations, resulting in an increase in the AMP:ATP ratio and activation of AMPK (Currie et al., 2012; Decensi et al., 2010). Other molecules are currently being evaluated to specifically interfere with metabolic pathways, including inhibitors of glucose transport,

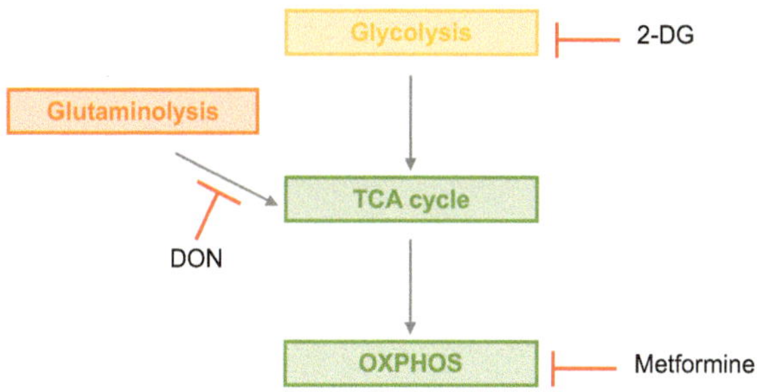

Figure 8 ***Inhibitory effects of chemical agents on cell metabolism undergoing clinical evaluation.*** 2-DG inhibits glycolysis, DON inhibits glutaminolysis, and metformin acts as a complex I inhibitor and prevents oxidative mitochondrial metabolism (OXPHOS). *2-DG*, 2-deoxyglucose; *DON*, 6-diazo-5-oxo-L-norleucine; *OXPHOS*, oxidative phosphorylation.

phosphofructokinase, pyruvate kinase, lactate dehydrogenase, and glutaminase inhibitors, which inhibit lipid synthesis, nucleotides, TCA cycle, and OXPHOS (Heiden, 2011). While metabolic reprogramming is closely related to activation, differentiation, and lymphocyte function, there is a great deal on how targeting cellular metabolism can modulate the immune response by promoting or blocking the emergence of lymphocyte subtypes (Patel and Powell, 2017). This area of research has led to the concept of cellular-specific pharmacological targeting to treat immune-related pathologies.

Cells responsible for graft rejection mainly use aerobic glycolysis for their energy production. In mouse models of skin and heart transplantation, the combination of 2-DG with metformin inhibits activated T cells glycolysis and decreases cytokine production and cell proliferation. The addition of 6-diazo-5-oxo-L-norleucine (DON), a glutamine analog that inhibits its uptake and metabolism, induces a greater decrease in the immune response. This triple combination inhibits the proliferation of CD4+ and CD8+ effector T cells and reduces the occurrence of rejection. In addition, this association promotes the generation of regulatory T cells preferentially using the oxidation of fatty acids, the differentiation of which is therefore not affected by this combination of molecules (Bettencourt and Powell, 2017; Lee et al., 2015; Yap et al., 2015).

Modulation of metabolism in patients with SLE is a therapeutic alternative as there are metabolic alterations in CD4+ T cells. Indeed, SLE is characterized by a hyperactivation of CD4+ T lymphocytes with increased glycolytic and mitochondrial oxidative metabolism accompanied by

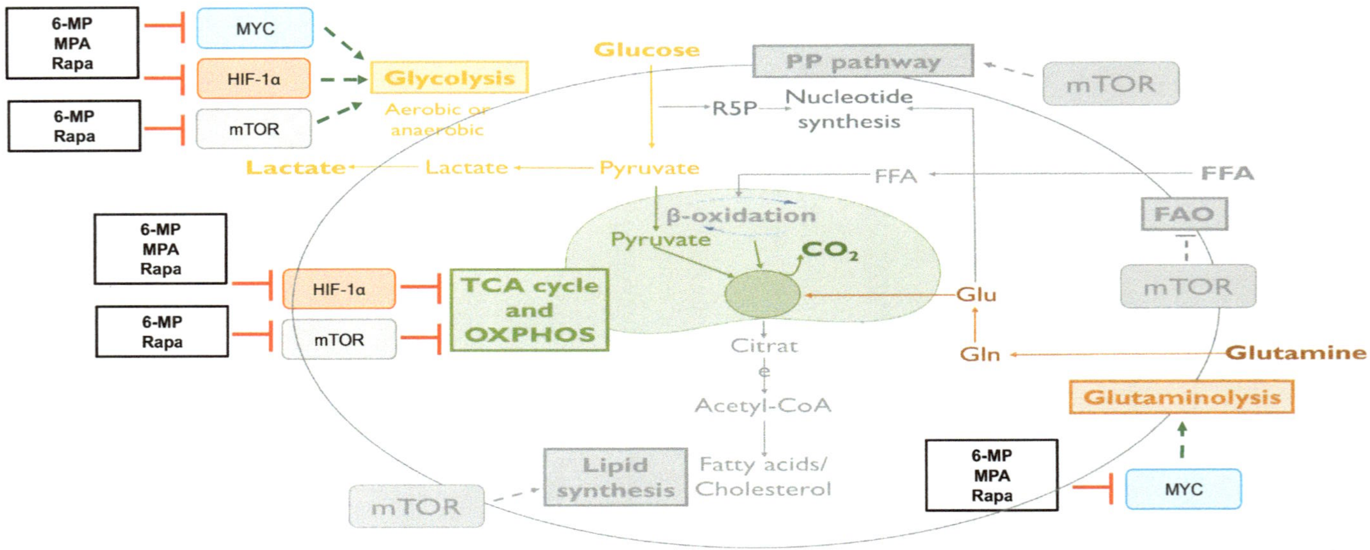

Figure 9 ***Graphical interpretation of the effects of immunosuppressive drugs (rapamycin, 6-MP, and MPA) on metabolic checkpoints and T-cell metabolism.*** In proliferating T cells, 6-MP, MPA, and rapamycin inhibit Myc and HIF-1α, and 6-MP and rapamycin inhibit mTOR. This diagram represents the hypothetical consequences of these three drugs on cellular metabolism. *6-MP*, 6-mercaptopurine; *acetyl-CoA*, acetyl coenzyme A; *FFA*, free fatty acids; *Gln*, glutamine; *Glu*, glutamic acid; *HIF1α*, hypoxia inducible factor 1 α; *MPA*, mycophenolic acid; *mTOR*, mammalian target of rapamycin; *OXPHOS*, oxidative phosphorylation; *R5P*, ribose-5-phosphate; *Rapa*, rapamycin.

elevated oxidation of glutamine and fatty acids. Activation of mTORC1 in CD4+ T cells promotes glycolysis and blocks autophagy. In addition, glucose transporters expression and glucose uptake are increased, together with the activation of the pentose phosphate pathway, leading to cell proliferation and glucose oxidation (Choi et al., 2016; Perl, 2016; Morel, 2017). In a transgenic mouse model of SLE, treatment combining 2-DG and metformin allowed to simultaneously reduce glycolysis and OXPHOS and to normalize metabolism in hyperactivated CD4+ T cells, leading to a reversion of the disease phenotype, including T-cell activation, autoantibody production, and renal impairment. In this model the 2-DG/metformin combination reduced mitochondrial glycolysis and oxygen uptake and inhibited the mTOR pathway. It is interesting to note that administration of 2-DG or metformin at equivalent doses in monotherapy did not achieve the same efficacy (Bettencourt and Powell, 2017; Lee et al., 2015; Mehta and Chandel, 2015; Yin et al., 2015). Although adverse effects or long-term efficacy remain to be evaluated in clinical trials including SLE patients, these findings offer new treatment perspectives in SLE.

The induction of cellular lipid biosynthesis pathways is an integral part of the T-cell activation program (Almeida et al., 2016). The first step of fatty acids synthesis begins with the carboxylation of acetyl-CoA by the enzyme acetyl-CoA carboxylase 1 (ACC1), generating malonyl-CoA. Blocking ACC1 activity strongly restrains Th17 and Th1 differentiation and enhances Treg development (Berod et al., 2014). These findings may have important implications for the treatment of inflammatory diseases as treatment with ACC-specific inhibitors protected mice from experimental autoimmune encephalomyelitis and acute graft-versus-host disease (Raha et al., 2016).

5. CONCLUSION

The comprehensive understanding of the effects of the current immunosuppressive drugs on T-cell metabolic reprogramming not only provides valuable information on how their therapeutic and adverse effects may occur but also offers new opportunities to improve therapeutic strategies both in terms of safety and efficacy. Further investigation of the metabolic checkpoints and their effects on the metabolism and differentiation of T cells could lead to the discovery of new therapeutic targets and a better understanding of the mechanisms underlying the efficacy of the immunosuppressive drugs that are currently available. Indeed, elucidating the effects of immunosuppressive drugs on the metabolic reprogramming of T cells is the next step in optimizing their therapeutic use. In addition, this knowledge will facilitate the development of new drugs and the adaptation of commercialized drugs to target specific metabolic reprogramming checkpoints or metabolic intermediates. Emerging therapeutic strategies will be able to evaluate new combinations of molecules to improve the efficacy and increase the specificity of the treatments, thus reducing the side effects of immunosuppressive drugs.

REFERENCES

Allison, A.C., 2000. Immunosuppressive drugs: the first 50 years and a glance forward. Immunopharmacology 47, 63–83.
Allison, A.C., Eugui, E.M., 1993. The design and development of an immunosuppressive drug, mycophenolate mofetil. Springer Semin. Immunopathol. 14, 353–380.
Allison, A.C., Eugui, E.M., 2000. Mycophenolate mofetil and its mechanisms of action. Immunopharmacology 47, 85–118.
Allison, K.E., Coomber, B.L., Bridle, B.W., 2017. Metabolic reprogramming in the tumour microenvironment: a hallmark shared by cancer cells and T lymphocytes. Immunology 152 (2), 175–184.
Almeida, L., Lochner, M., Berod, L., Sparwasser, T., 2016. Metabolic pathways in T cell activation and lineage differentiation. Semin. Immunol. 28 (5), 514–524.
Appel, A.S., Appel, G.B., 2009. An update on the use of mycophenolate mofetil in lupus nephritis and other primary glomerular diseases. Nat. Rev. Nephrol. 5, 132–142.
Barbarino, J.M., Staatz, C.E., Venkataramanan, R., Klein, T.E., Altman, R.B., 2013. PharmGKB summary: cyclosporine and tacrolimus pathways. Pharmacogenet. Genomics 23 (10), 563–585.
Basu, A., Banerjee, P., Contreras, A.G., Flynn, E., Pal, S., 2011. Calcineurin inhibitor-induced and ras-mediated overexpression of VEGF in renal cancer cells involves mTOR through the regulation of PRAS40. PLoS One 6 (8), e23919.
Basu, A., Banerjee, P., Pal, S., 2012. Critical role of mTOR in calcineurin inhibitor-induced renal cancer progression. Cell Cycle 11 (4), 633–634.
Berod, L., Friedrich, C., Nandan, A., Freitag, J., Hagemann, S., Harmrolfs, K., Sandouk, A., Hesse, C., Castro, C.N., Bahre, H., Tschirner, S.K., Gorinski, N., Gohmert, M., Mayer, C.T., Huehn, J., Ponimaskin, E., Abraham, W.R., Muller, R., Lochner, M., Sparwasser, T., 2014. De novo fatty acid synthesis controls the fate between regulatory T and T helper 17 cells. Nat. Med. 20 (11), 1327–1333.
Bettencourt, I.A., Powell, J.D., 2017. Targeting metabolism as a novel therapeutic approach to autoimmunity, inflammation, and transplantation. J. Immunol. 198, 999–1005.
Bleyer, W.A., 1978. The clinical pharmacology of methotrexate. Cancer 41, 36–51.
Bronk, J., Lister, N., Shaw, M., 1988. Transport and metabolism of 6-thioguanine and 6-mercaptopurine in mouse small intestine. Clin. Sci. (Lond.) 74 (6), 629–638.
Brown, P.M., Pratt, A.G., Isaacs, J.D., 2016. Mechanism of action of methotrexate in rheumatoid arthritis, and the search for biomarkers. Nat. Rev. Rheumatol. 12, 731–742.
Buchholz, M., Schatz, A., Wagner, M., Michl, P., Linhart, T., Adler, G., Gress, T.M., Ellenrieder, V., 2006. Overexpression of c-myc in pancreatic cancer caused by ectopic activation of NFATc1 and the Ca+2/calcineurin signaling pathway. EMBO J. 25, 3714–3724.

Cara, C.J., Pena, A.S., Sans, M., Rodrigo, L., Guerrero-Esteo, M., Hinojosa, J., García-Paredes, J., Guijarro, L.G., 2004. Reviewing the mechanism of action of thiopurine drugs: towards a new paradigm in clinical practice. Med. Sci. Monit. 10 (11), RA247–RA254.

Chaundry, V., Cornblath, D., Griffin, J., O'Brien, R., Drachman, D., 2001. Mycophenolate mofetil: a safe and promising immunosuppressant in neuromuscular diseases. Neurology 56, 94–96.

Chi, H., 2012. Regulation and function of mTOR signalling in T cell fate decisions. Nat. Rev. Immunol. 12 (5), 325–338.

Choi, S.C., Titov, A.A., Sivakumar, R., Li, W., Morel, L., 2016. Immune cell metabolism in systemic lupus erythematosus. Curr. Rheumatol. Rep. 18 (11), 66.

Choi, S.J.N., Lee, H.K., Kim, N.H., Chung, S.Y., 2011. Mycophenolic acid mediated mitochondrial membrane potential transition change lead to T lymphocyte apoptosis. J. Korean Surg. Soc. 81, 235–241.

Chouchana, L., Narjoz, C., Beaune, P., Loriot, M., Roblin, X., 2012. Review article: the benefits of pharmacogenetics for improving thiopurine therapy in inflammatory bowel disease. Aliment. Pharmacol. Ther. 35, 15–36.

Currie, C., Poole, C., Jenkins-Jones, S., Gale, E., Johnson, J., Morgan, C., 2012. Mortality after incident cancer in people with and without type 2 diabetes: impact of metformin on survival. Diabetes Care 35, 299–304.

Dayton, J.S., Turka, L.A., Thompson, C.B., Mitchell, B.S., 1992. Comparison of the effects of mizoribine with those of azathioprine, 6-mercaptopurine, and mycophenolic acid on T lymphocyte proliferation and purine ribonucleotide metabolism. Mol. Pharmacol. 41 (4), 671–676.

Decensi, A., Putoni, M., Goodwin, P., Cazzaniga, M., Gennari, A., Bonanni, B., Gandini, S., 2010. Metformin and cancer risk in diabetic patients: a systematic review and meta-analysis. Cancer Prev. Res. (Phila.) 3 (11), 1451–1461.

Dogra, S., Mahajan, R., 2013. Systemic methotrexate therapy for psoriasis: past, present and future. Clin. Exp. Dermatol. 38, 573–588.

Domhan, S., Muschal, S., Schwager, C., Morath, C., Wirkner, U., Ansorge, W., Maercker, C., Zeier, M., Huber, P.E., Abdollahi, A., 2008. Molecular mechanisms of the antiangiogenic and antitumor effects of mycophenolic acid. Mol. Cancer Ther. 7 (6), 1656–1668.

Dumont, F.J., Su, Q., 1996. Mechanism of action of the immunosuppressant rapamycin. Life Sci. 58 (5), 373–395.

Dun, B., Sharma, A., Xu, H., Liu, H., Bai, S., Zeng, L., She, J.-X., 2014. Transcriptomic changes induced by mycophenolic acid in gastric cancer cells. Am J Transl Res 6 (1), 28–42.

Elion, G.B., 1972. Significance of azathioprine metabolites. Proc. R. Soc. Med. 65 (3), 257–260.

Elion, G.B., 1989. The purine path to chempotherapy. Science 244, 41–47.

Fernandez-Ramos, A.A., Marchetti-Laurent, C., Poindessous, V., Antonio, S., Laurent-Puig, P., Bortoli, S., Loriot, M.A., Pallet, N., 2017a. 6-mercaptopurine promotes energetic failure in proliferating T cells. Oncotarget 8 (26), 43048–43060.

Fernandez-Ramos, A.A., Marchetti-Laurent, C., Poindessous, V., Antonio, S., Petitgas, C., Ceballos-Picot, I., Laurent-Puig, P., Bortoli, S., Loriot, M.A., Pallet, N., 2017b. A comprehensive characterization of the impact of mycophenolic acid on the metabolism of Jurkat T cells. Sci. Rep. 7 (1), 10550.

Fodor, T., Szántó, M., Abdul-Rahman, O., Nagy, L., Dér, Á., Kiss, B., Bai, P., 2016. Combined treatment of MCF-7 cells with AICAR and methotrexate, arrests cell cycle and reverses warburg metabolism through AMP-activated protein kinase (AMPK) and FOXO1. PLoS One 11 (2), e0150232.

Fong, S.C.M., Blaker, P.A., Arenas-Hernandez, M., Marinaki, A.M., Sanderson, J.D., 2015. Getting the best out of thiopurine therapy: thiopurine S-methyltransferase and beyond. Biomark. Med. 9 (1), 51–65.

Fotoohi, A.K., Wrabel, A., Moshfegh, A., Peterson, C., Albertioni, F., 2006. Molecular mechanisms underlying the enhanced sensitivity of thiopurine-resistant T-lymphoblastic cell lines to methyl mercaptopurineriboside. Biochem. Pharmacol. 72 (7), 816–823.

Fuhrmann, A., Lopes, P., Sereno, J., Pedro, J., Espinoza, D., Pereira, M., Reis, F., Eriksson, J., Carvalho, E., 2014. Molecular mechanisms underlying the effects of cyclosporin A and sirolimus on glucose and lipid metabolism in liver, skeletal muscle and adipose tissue in an in vivo rat model. Biochem. Pharmacol. 88, 216–228.

Furst, D., 1997. The rational use of methotrexate in rheumatoid arthritis and other rheumatic diseases. Br. J. Rheumatol. 36, 1196–1204.

Giverhaug, T., Loennechen, T., Aarbakke, J., 1999. The interaction of 6-mercaptopurine (6-MP) and methotrexate (MTX). Gen. Pharmacol. 33, 341–346.

Green, D.R., Galluzzi, L., Kroemer, G., 2014. Metabolic control of cell death. Science 345 (6203), 1250256.

Grim, J., Chládek, J., Martínková, J., 2003. Pharmacokinetics and pharmacodynamics of methotrexate in non-neoplastic diseases. Clin. Pharmacokinet. 42 (2), 139–151.

Gwinn, D.M., Shackelford, D.B., Egan, D.F., Mihaylova, M.M., Mery, A., Vasquez, D.S., Turk, B.E., Shaw, R.J., 2008. AMPK phosphorylation of Raptor mediates metabolic checkpoint. Mol. Cell 30, 214–226.

Halloran, P.F., 2004. Immunosuppressive drugs for kidney transplantation. N. Engl. J. Med. 351 (26), 2715–2729.

Hartman, S., Buchanan, J., 1959. Nucleic acids, purines, pyrimidines (nucleotide synthesis). Ann. Rev. Biochem. 28, 365–410.

Hartono, C., Muthukumar, T., Suthanthiran, M., 2013. Immunosuppressive drug therapy. Cold Spring Harb. Perspect. Med. 3, 1–15.

He, X., Smeets, R., Koenen, H., Vink, P., Wagenaars, J., Boots, A., Joosten, I., 2011. Mycophenolic acid-mediated suppression of human CD4+ T cells: more than mere guanine nucleotide deprivation. Am. J. Transplant. 11, 439–449.

Heiden, M.G.V., 2011. Targeting cancer metabolism: a therapeutic window opens. Nat. Rev. Drug Discov. 10, 671–684.

Herfarth, H.H., 2016. Methotrexate for inflammatory bowel diseases - new developments. Dig. Dis. 34 (1–2), 140–146.

Huang, H.-Y., Chang, H.-F., Tsai, M.-J., Chen, J.-S., Wang, M.-J., 2016. 6-mercaptopurine attenuates tumor necrosis factor-α production in microglia through Nur77-mediated transrepression and PI3K/Akt/mTOR signaling-mediated translational regulation. J. Neuroinflamm. 13 (78), 1–20.

Illsinger, S., Göken, C., Brockmann, M., Thiemann, I., Bednarczyk, J., Schmidt, K.-H., Lücke, T., Hoy, L., Janzen, N., Das, A.M., 2011a. Effect of tacrolimus on energy metabolism in human umbilical endothelial cells. Ann. Transplant. 16 (2), 68–75.

Illsinger, S., Janzen, N., Lücke, T., Bednarczyk, J., Schmidt, K., Hoy, L., Sander, J., Das, A., 2011b. Cyclosporine A: impact on mitochondrial function in endothelial cells. Clin. Transplant. 25, 584–593.

Inoki, K., Kim, J., Guan, K.L., 2012. AMPK and mTOR in cellular energy homeostasis and drug targets. Annu. Rev. Pharmacol. Toxicol. 52, 381–400.

Inoue, K., Yuasa, H., 2014. Molecular basis for pharmacokinetics and pharmacodynamics of methotrexate in rheumatoid arthritis therapy. Drug Metab. Pharmacokinet. 29 (1), 12–19.

Iurlaro, R., León-Annicchiarico, C.L., Muñoz-Pinedo, C., 2014. Regulation of cancer metabolism by oncogenes and tumor suppressors. Methods Enzymol. 542, 59–80.

Karmaus, P.W., Chi, H., 2014. c-Myc and AP4: a relay team for metabolic reprogramming of $CD8^+$ T cells. Nat. Immunol. 15 (9), 828–829.
Khan, A., Nasr, P., El-Charataby, E., El-Sayegh, S., 2016. An insight into the immunologic events and risk assessment in renal transplantation. J. Clin. Med. Res. 8 (5), 367–372.
Khan, S., Sewell, W.C., 2006. Oral immunosuppressive drugs. Clin. Med. 6, 352–355.
Kho, M., Cransberg, K., Weimar, W., Gelder, T.V., 2011. Current immunosuppressive treatment after kidney transplantation. Expert Opin. Pharmacother. 12 (8), 1217–1231.
Kirchner, G.I., Meier-Wiedenbach, I., Manns, M.P., 2004. Clinical pharmacokinetics of everolimus. Clin. Pharmacokinet. 43 (2), 83–95.
Lee, C.-F., Lo, Y.-C., Cheng, C.-H., Furtmüller, G.J., Oh, B., Andrade-Oliveira, V., Thomas, A.G., Bowman, C.E., Slusher, B.S., Wolfgang, M.J., Brandacher, G., Powell, J.D., 2015. Preventing allograft rejection by targeting immune metabolism. Cell Rep. 13 (4), 760–770.
Li, J., Kim, S.G., Blenis, J., 2014. Rapamycin: one drug, many effects. Cell Metab. 19, 373–379.
Lim, M.A., Kohli, J., Bloom, R.D., 2017. Immunosuppression for kidney transplantation: where are we now and where are we going? Transplant. Rev. 31, 10–17.
Liu, Y.V., Hubbi, M.E., Pan, F., McDonald, K.R., Mansharamani, M., Cole, R.N., Liu, J.O., Semenza, G.L., 2007. Calcineurin promotes hypoxia-inducible factor 1a expression by dephosphorylating RACK1 and blocking RACK1 dimerization. J. Biol. Chem. 282 (51), 37064–37073.
Loftus, R.M., Finlay, D.K., 2016. Immunometabolism: Cellular metabolism turns immune regulator. J. Biol. Chem. 291 (1), 1–10.
Macintyre, A., Gerriets, V., Nichols, A., Michalek, R., Rudolph, M., Deoliveira, D., Anderson, S., Abel, E., Chen, B., Hale, L., Rathmell, J., 2014. The glucose transporter Glut1 is selectively essential for CD4 T cell activation and effector function. Cell Metab. 20 (1), 61–72.
Macintyre, A.N., Rathmell, J.C., 2013. Activated lymphocytes as a metabolic model for carcinogenesis. Cancer Metab. 1 (5), 1–12.
Maciolek, J.A., Pasternak, J.A., Wilson, H.L., 2014. Metabolism of activated T lymphocytes. Curr. Opin. Immunol. 27, 60–74.
MacIver, N.J., Michalek, R.D., Rathmell, J.C., 2013. Metabolic regulation of T lymphocytes. Annu. Rev. Immunol. 31, 259–283.
Mahalati, K., Kahan, B.D., 2001. Clinical pharmacokinetics of sirolimus. Clin. Pharmacokinet. 40 (8), 573–585.
Maripuri, S., Kasiske, B.L., 2014. The role of mycophenolate mofetil in kidney transplantation revisited. Transplant. Rev. 28 (1), 26–31.
Mehta, M.M., Chandel, N.S., 2015. Targeting metabolism for lupus therapy. Sci. Transl. Med. 7 (274fs5), 1–2.
Moes, D.J.A., Guchelaar, H.-J., Fijter, J.W.D., 2015. Sirolimus and everolimus in kidney transplantation. Drug Discov. Today 20 (10), 1243–1249.
Mognol, G.P., Araujo-Souza, P.S. d, Robbs, B.K., Teixera, L.K., Viola, J.P., 2012. Transcriptional regulation of the c-myc promoter NFAT1 involvers negative and positive NFAT-responsive elements. Cell Cycle 11 (5), 1014–1028.
Morel, L., 2017. Immunometabolism in systemic lupus erythematosus. Nat. Rev. Rheumatol. 13 (5), 280–290.
Morita, M., Gravel, S.-P., Chénard, V., Sikström, K., Zheng, L., Alain, T., Gandin, V., Avizonis, D., Arguello, M., Zakaria, C., McLaughlan, S., Nouet, Y., Pause, A., Pollak, M., Gottlieb, E., Larsson, O., St-Pierre, J., Topisirovic, I., Sonenberg, N., 2013. mTORC1 controls mitochondrial activity and biogenesis through 4E-BP-dependent translational regulation. Cell Metab. 18 (5), 698–711.

Naesens, M., Sarwal, M.M., 2010a. Harnessing the diversity of the human T-cell repertoire: a monitoring tool for transplantation tolerance? Eur. J. Immunol. 40, 2986–2989.
Naesens, M., Sarwal, M.M., 2010b. Monitoring calcineurin inhibitor therapy: localizing the moving target. Transplantation 89 (11), 1308–1309.
Nizet, V., Johnson, R.S., 2009. Interdependence of hypoxic and innate immune responses. Nat. Rev. Immunol. 9, 609–617.
Palsson-McDermott, E.M., O'Neill, L.A., 2013. The Warburg effect then and now: from cancer to inflammatory diseases. Bioessays 35 (11), 965–973.
Park, H.G., Yi, H., Kim, S.H., Yu, H.S., Ahn, Y.M., Lee, Y.H., Roh, M.-S., Kim, Y.S., 2011. The effect of cyclosporine A on the phosphorylation of the AMPK pathway in the rat hippocampus. Prog. Neuro-Psychopharmacol. Biol. Psychiatry 35, 1933–1937.
Patel, C.H., Powell, J.D., 2017. Targeting T cell metabolism to regulate T cells activation, differentiation and function in disease. Curr. Opin. Immunol. 46, 82–88.
Patel, J.K., Kobashigawa, J.A., 2006. Everolimus: an immunosuppressive agent in transplantation. Expert Opin. Pharmacother. 7 (10), 1347–1355.
Patsoukis, N., Bardhan, K., Chatterjee, P., Sari, D., Liu, B., Bell, L.N., Karoly, E.D., Freeman, G.J., Petkova, V., Seth, P., Li, L., Boussiotis, V.A., 2015. PD-1 alters T-cell metabolic reprogramming by inhibiting glycolysis and promoting lipolysis and fatty acid oxidation. Nat. Commun. 6, 6692.
Pearce, E.L., 2010. Metabolism in T cell activation and differentiation. Curr. Opin. Immunol. 22 (3), 314–320.
Pearce, E.L., Pearce, E.J., 2013. Metabolic pathways in immune cell activation and quiescence. Immunity 38 (4), 633–643.
Pearce, E.L., Poffenberger, M.C., Chang, C.H., Jones, R.G., 2013. Fueling immunity: insights into metabolism and lymphocyte function. Science 342 (6155), 1242454.
Pereira, M.J., Palming, J., Rizell, M., Aureliano, M., Carvalho, E., Svensson, M.K., Eriksson, J.W., 2014. Cyclosporine A and tacrolimus reduce the amount of GLUT4 at the cell surface in human adipocytes: increased endocytosis as a potential mechanism for the diabetogenic effects of immunosuppressive agents. J. Clin. Endocrinol. Metab. 99 (10), E1885–E1894.
Perl, A., 2016. Activation of mTOR (mechanistic target of rapamycin) in rheumatic diseases. Nat. Rev. Rheumatol. 12 (3), 169–182.
Pirkmajer, S., Kulkarni, S.S., Tom, R.Z., Ross, F.A., Hawley, S.A., Hardie, D.G., Zierath, J.R., Chibalin, A.V., 2015. Methotrexate promotes glucose uptake and lipid oxidation in skeletal muscle via AMPK activation. Diabetes 64, 360–369.
Pollizzi, K.N., 2015. mTORC1 and mTORC2 selectively regulate $CD8^+$ T cell differentiation. J. Clin. Invest. 125 (5), 2090–2108.
Pollizzi, K.N., Powell, J.D., 2014. Integrating canonical and metabolic signalling programmes in the regulation of T cell responses. Nat. Rev. Immunol. 14 (7), 435–446.
Pollizzi, K.N., Powell, J.D., 2015. Regulation of T cells by mTOR: the known knowns and the known unknowns. Trends Immunol. 36 (1), 13–20.
Porceddu, M., Buron, N., Roussel, C., Labbe, G., Fromenty, B., Borgne-Sanchez, A., 2012. Prediction of liver injury induced by chemicals in human with a multiparametric assay on isolated mouse liver mitochondria. Toxicol. Sci. 129 (2), 332–345.
Powell, J.D., Delgoffe, G.M., 2010. The mammalian target of rapamycin: linking T cell differentiation, function, and metabolism. Immunity Rev. 33, 301–311.
Quemeneur, L., Flacher, M., Gerland, L.M., Ffrench, M., Revillard, J.P., Bonnefoy-Berard, N., 2002. Mycophenolic acid inhibits IL-2-dependent T cell proliferation, but not IL-2-dependent survival and sensitization to apoptosis. J. Immunol. 169 (5), 2747–2755.

Quéméneur, L., Gerland, L., Flacher, M., Ffrench, M., Revillard, J., Genestier, L., 2003. Differential control of cell cycle, proliferation, and survival of primary T lymphocytes by purine and pyrimidine nucleotides. J. Immunol. 170 (10), 4986–4995.
Radu, R.G., Fujimoto, S., Mukai, E., Takehiro, M., Shimono, D., Nabe, K., Shimodahira, M., Kominato, R., Aramaki, Y., Nishi, Y., Funakoshi, S., Yamada, Y., Seino, Y., 2004. Tacrolimus suppresses glucose-induced release from pancreatic islets by reducing glucokinase activity. Am. J. Physiol. Endocrinol. Metab. 288, E365–E371.
Raha, S., Raud, B., Oberdorfer, L., Castro, C.N., Schreder, A., Freitag, J., Longerich, T., Lochner, M., Sparwasser, T., Berod, L., Koenecke, C., Prinz, I., 2016. Disruption of de novo fatty acid synthesis via acetyl-CoA carboxylase 1 inhibition prevents acute graft-versus-host disease. Eur. J. Immunol. 46 (9), 2233–2238.
Ramanathan, A., Schreiber, S.L., 2009. Direct control of mitochondrial function by mTOR. Proc. Natl. Acad. Sci. U.S.A. 106 (52), 22229–22232.
Rodríguez-Enríquez, S., Marín-Hernández, A., Gallardo-Pérez, J.C., Carreño-Fuentes, L., Moreno-Sánchez, R., 2009. Targeting of cancer energy metabolism. Mol. Nutr. Food Res. 53, 29–48.
Schieke, S.M., Phillips, D., McCoy, J.P., Aponte, A.M., Shen, R.-F., Balaban, R.S., Finkel, T., 2006. The mammalian target of rapamycin (mTOR) pathway regulates mitochondrial oxygen consumption and oxidative capacity. J. Biol. Chem. 281 (37), 27643–27652.
Sehgal, S., 2003. Sirolimus: its discovery, biological properties, and mechanism of action. Transpl. Proc. 35 (Suppl. 3A), 7S–14S.
Serra, J.M., Gutiérrez, A., Alemany, R., Navarro, M., Ros, T., Saus, C., Ginés, J., Sampol, A., Amat, J.C., Serra-Moisés, L., Martín, J., Galmés, A., Vögler, O., Besalduch, J., 2008. Inhibition of c-myc down-regulation by sustained extracellular signal-regulated kinase activation prevents the antimetabolite methtrexate- and gemcitabine-induced differentiation in non-small-cell lung cancer cells. Mol. Pharmacol. 73, 1679–1687.
Simon, N., Morin, C., Urien, S., Tillement, J.-P., Bruguerolle, B., 2003. Tacrolimus and sirolimus decrease oxidative phosphorylation of isolated rat kidney mitochondria. Br. J. Pharmacol. 138, 369–376.
Sinclair, L.V., Rolf, J., Emslie, E., Shi, Y.B., Taylor, P.M., Cantrell, D.A., 2013. Control of amino-acid transport by antigen receptors coordinates the metabolic reprogramming essential for T cell differentiation. Nat. Immunol. 14 (5), 500–508.
Singh, G., Singh, S.K., König, A., Reutlinger, K., Nye, M.D., Adhikary, T., Eilers, M., Gress, T.M., Fernandez-Zapico, M.E., Ellenrieder, V., 2010. Sequential activation of NFAT and c-myc transcription factors mediates the TGF-B switch from a suppressor to a promoter of cancer cell proliferation. J. Biol. Chem. 285 (35), 27241–27250.
Sugiyama, K., Satoh, H., Saito, K., Takahashi, K., Saito, N., Hirano, T., 2005. Immunosuppressive efficacy of mycophenolate mofetil when compared with azathioprine and mizoribine against peripheral lymphocytes from renal transplant recipients. Transpl. Int. 18 (5), 590–595.
Takebe, N., Cheng, X., Fandy, T.E., Srivastava, R.K., Wu, S., Shankar, S., Bauer, K., Shaughnessy, J., Tricot, G., 2006. IMP dehydrogenase inhibitor mycophenolate mofetil induces caspase-dependent apoptosis and cell cycle inhibition in multiple myeloma cells. Mol. Cancer Ther. 5 (2), 457–466.
Tapner, M.J., Jones, B.E., Wu, W.M., Farrell, G.C., 2004. Toxicity of low dose azathioprine and 6-mercaptopurine in rat hepatocytes. Roles of xanthine oxidase and mitochondrial injury. J. Hepatol. 40, 454–463.

Taylor, A.L., Watson, C.J., Bradley, J.A., 2005. Immunosuppressive agents in solid organ transplantation: mechanisms of action and therapeutic efficacy. Crit. Rev. Oncol. Hematol. 56 (1), 23–46.

Tedesco, D., Haragsim, L., 2012. Cyclosporine: a review. J. Transplant. 2012, 1–7.

Thornton, C.C., Al-Rashed, F., Calay, D., Birdsey, G.M., Bauer, A., Mylroie, H., Morley, B.J., Randi, A.M., Haskard, D.O., Boyle, J.J., Mason, J.C., 2016. Methotrexate-mediated activation of an AMPK-CREB-dependent pathway: a novel mechanism for vascular protection in chronic systemic inflammation. Ann. Rheum. Dis. 75 (2), 439–448.

Vaeth, M., Maus, M., Klein-Hessling, S., Freinkman, E., Yang, J., Eckstein, M., Cameron, S., Turvey, S.E., Serfling, E., Berberich-Siebelt, F., Possemato, R., Feske, S., 2017. 'Store-Operated Ca(2+) entry controls clonal expansion of T cells through metabolic reprogramming. Immunity 47 (4), 664–679 e6.

Waickman, A.T., Powell, J.D., 2012. mTOR, metabolism, and the regulation of T-cell differentiation and function. Immunol. Rev. 249, 43–58.

Walczak-Drzewiecka, A., Ratajewski, M., Wagner, W., Dastych, J., 2008. HIF-1A is upregulated in activated mast cells by a process that involves calcineurin and NFAT. J. Immunol. 181, 1665–1672.

Wang, R., Dillon, C.P., Shi, L.Z., Milasta, S., Carter, R., Finkelstein, D., McCormick, L.L., Fitzgerald, P., Chi, H., Munger, J., Green, D.R., 2011. The transcription factor Myc controls metabolic reprogramming upon T lymphocyte activation. Immunity 35 (6), 871–882.

Wang, R., Green, D.R., 2012. Metabolic reprogramming and metabolic dependency in T cells. Immunol. Rev. 249, 14–26.

Wang, S., Kang, X., Cao, S., Cheng, H., Wang, D., Geng, J., 2012. Calcineurin/NFATc1 pathway contributes to cell proliferation in hepatocellular carcinoma. Dig. Dis. Sci. 57, 3184–3188.

Wiseman, A.C., 2016. Immunosuppressive medications. CJASN 11 (2), 332–343.

Wouters, B.G., Koritzinsky, M., 2008. Hypoxia signaling through mTOR and the unfolded protein response in cancer. Nat. Rev. Cancer 8, 851–864.

Yang, K., Chi, H., 2015. AMPK helps T cells survive nutrient starvation. Immunity 42, 4–6.

Yap, M., Brouard, S., Pecqueur, C., Degauque, N., 2015. Targeting CD8 T-cell metabolism in transplantation. Front. Immunol. 6 (547), 1–8.

Yin, Y., Choi, S., Xu, Z., Perry, D., Seay, H., Croker, B., Sobel, E., Brusko, T., Morel, L., 2015. Normalisation of CD4+ T cell metabolism reverses lupus. Sci. Transl. Med. 7, 274ra18.

Yoo, Y., Na, T., Yang, W., Kim, H., Lee, I., Kong, G., Chung, J., Lee, M., 2007. 6-mercaptopurine, an activator of Nur77, enhances transcriptional activity of HIF-1a resulting in new vessel formation. Oncogene 26, 3823–3834.

Zaza, G., Cheok, M., Yang, W., Panetta, J., Pui, C., Relling, M., Evans, W., 2005. Gene expression and thioguanine nucleotide disposition in acute lymphoblastic leukemia after in vivo mercaptopurine treatment. Blood 106 (5), 1778–1785.

Zeiser, R., Negrin, R.S., 2008. Interleukin-2 receptor downstream events in regulatory T cells: implications for the choice of immunosuppressive drug therapy. Cell Cycle 7 (4), 458–462.

Zhang, F., Fu, L., Wang, Y., 2013. 6-thioguanine induces mitochondrial dysfunction and oxidative DNA damage in acute lymphoblastic leukemia cells. Mol. Cell Proteom. 12, 3803–3811.

CHAPTER FIVE

T Lymphocyte—Based Cancer Immunotherapeutics

Adrienne E. Borrie[*,§] **and Saman Maleki Vareki**[¶,||,1]
[*]Department of Physiology and Pharmacology, Western University, London, ON, Canada
[§]Division of Clinical Pharmacology, Western University, London, ON, Canada
[¶]Department of Oncology, Western University, London, ON, Canada
[||]Cancer Research Laboratory Program, Lawson Health Research Institute, London, ON, Canada
[1]Corresponding author: Email: saman.malekivareki@lhsc.on.ca

Contents

International Review of Cell and Molecular Biology, Volume 342
ISSN 1937-6448
https://doi.org/10.1016/bs.ircmb.2018.05.010

Abstract

Immunotherapy represents a breakthrough in cancer treatment in the past decade. It is a powerful treatment modality that can help transform cancer into a manageable disease by enabling the immune system to eliminate tumor cells. Advances in the use of immune checkpoint inhibitors, adoptive cell therapy, and oncolytic viruses have led to exceptional and durable clinical responses in patients with late-stage disease. Here we discuss some approved or under investigation immunotherapeutics that either directly involve T-cells such as chimeric antigen receptor T-cells and checkpoint inhibitors or lead to the activation of T-cells such as oncolytic viruses.

1. INTRODUCTION

Tumor cells are continually evolving in the most hostile environment that is programmed to eliminate neoplastic events in a multicellular organism. Cancer cells struggle to escape internal and external tumor suppressor mechanisms, such as P53 and antitumor immunity, respectively, and this makes them evolutionary fit and hard to eradicate with traditional treatment modalities that cannot adapt as cancer cells can (Dagogo-Jack and Shaw, 2018). For years, tumor immunologists have been studying the interactions between the immune system and tumor cells and have identified the ability of the immune system to recognize and destroy cancer cells under optimum conditions. However, the conditions of tumor microenvironment in clinically manifested cancers are far from optimum for T-cell function (Schaer et al., 2011; Sukari et al., 2016). Cancer cells actively suppress T-cells by various mechanisms, including the production of immunosuppressive factors, such as indoleamine 2,3-dioxygenase (IDO) and transforming growth factor β (TGFβ), and recruiting immunosuppressive cells, such as T regulatory cells (Tregs) and myeloid-derived suppressor cells (MDSCs), to the tumor microenvironment (Gajewski et al., 2006).

Cancer immunotherapy is translating the knowledge we have learned from cancer immunology to improve antitumor immunity in cancer patients. Perhaps the most successful attempt in using patient's own immune power against neoplastic cells comes from the pioneering work of Dr. James Allison who demonstrated that blocking checkpoint molecules on the surface of T-cells leads to a robust antitumor immune response that is capable of destroying established tumors (Sharma and Allison, 2015b). These efforts led to the development of a new class of anticancer drugs called immune checkpoint inhibitors (ICIs) that primarily target the immune system than cancer (Sharma and Allison, 2015b). Over the last few years, ICIs and other forms of cancer immunotherapy have shown promising advances in the management of patients with difficult-to-treat cancers, including advanced melanoma, non–small cell lung cancer (NSCLC), and relapsed/refractory acute lymphoblastic leukemia (ALL; Farkona et al., 2016). The main advantages of immunotherapy over other forms of cancer treatment are (1) the ability of the immune system to evolve and adapt to the tumor's conditions and (2) the potential of developing long-lasting immunity in treated patients against cancer (Farkona et al., 2016).

Here we discuss various forms of immunotherapy agents that are already approved or are under development and either directly target T-cells or activate T-cells as a secondary mechanism of action, such as oncolytic viruses. We will provide an overview of the mechanism of action of these drugs, their status of clinical development, efficacy, and toxicity. Our goal is to provide information that can be used by clinicians and scientists who are already working in this area or are about to enter to the world of cancer immunotherapy.

2. CHECKPOINT INHIBITORS

The principal role of T-cells is to distinguish healthy cells from pathogens or tumor cells through the activation or deactivation of various receptors on their surface. Tumor cells express cell surface molecules that interact with the receptors on T-cells mimicking the signals expressed by healthy cells. This helps malignant cells remain hidden from the activated immune system, allowing for their uncontrolled growth. These molecules and their associated receptors on T-cells keep the immune system "in check," by impeding excessive immune functions avoiding autoimmunity. For this reason, they are called "checkpoint" proteins. ICIs disrupt the effects of these checkpoint proteins, and this effectively "releases the brakes" on the immune system. The inhibitory checkpoint targets are illustrated in Fig. 1.

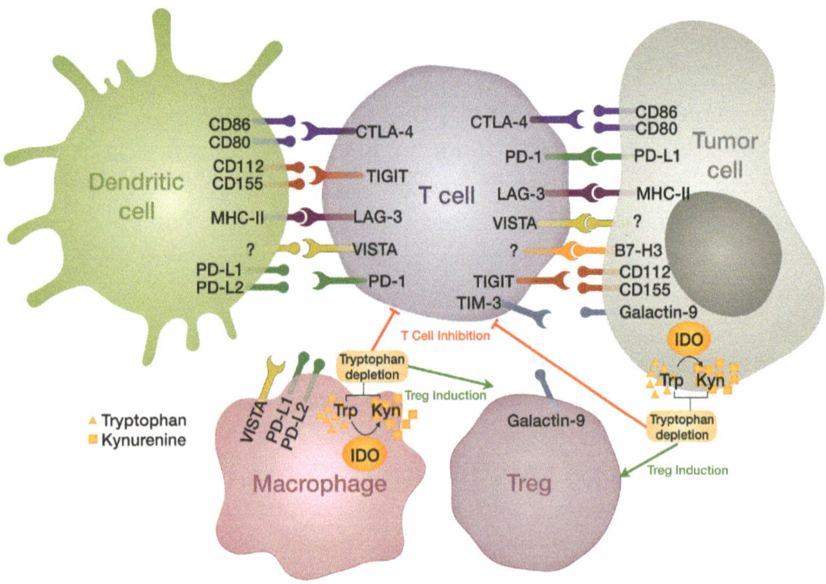

Figure 1 ***Inhibitory receptors and their ligands.*** CTLA-4, PD-1, LAG-3, VISTA, TIM-3, and TIGIT all bind to their respective ligands on the surface of tumor cells and dendritic cells. VISTA and PD-L1/PD-L2 are also expressed on the surface of macrophages. B7-H3 is an inhibitory ligand, which binds to an unknown receptor on the surface of T-cells. IDO expressed on macrophages and tumor cells, converts tryptophan (Trp) to kynurenine (Kyn), which leads to tryptophan depletion, T-cell inhibition, and Treg induction. *CTLA-4*, Cytotoxic T-lymphocyte protein 4; *IDO*, indoleamine 2,3-dioxygenase; *LAG-3*, Lymphocyte-activation gene 3; *PD-1*, Programmed cell death 1; *TIGIT*, T-cell immunoreceptor with Ig and ITIM domains; *TIM-3*, T-cell immunoglobulin and mucin domain-3; *VISTA*, V-domain Ig suppressor of T-cell activation.

2.1 Cytotoxic T-Lymphocyte Protein-4–Targeting Antibodies

Cytotoxic T-lymphocyte protein 4 (CTLA-4), also known as CD152, is an immune checkpoint that negatively regulates activation of conventional T-cells (Walunas et al., 1994). T-cell receptor (TCR) recognition of antigenic peptides in the context of self-major histocompatibility complex (MHC) molecules on the surface of professional antigen-presenting cells (APCs), such as dendritic cells (DCs), provides the first signal for naïve T-cell activation. However, this interaction of TCR with a peptide/MHC complex, known as signal one, is not sufficient for proper activation of naïve T-cells (Bonifaz et al., 2002). Efficient naïve T-cell activation requires a second signal—also known as the costimulatory signal—via CD28. CD28, a protein receptor expressed on the surface of T-cells, binds to either CD80 (B7-1) or CD86 (B7-2) expressed on professional APCs, resulting in T-cell activation, proliferation, increased survival, and cytokine production. CTLA-4 competes with CD28 for binding to CD80 and CD86 with higher affinity, thus limiting T-cell activation (Walunas et al., 1994).

CTLA-4 is not expressed in resting naïve $CD4^+$ or $CD8^+$ T-cells but stored in submembrane vesicles (Freeman et al., 1992). A strong TCR stimulus from an antigenic peptide induces rapid trafficking of inhibitory CTLA-4 molecules to the cell surface, which then negatively regulate T-cell activation via limiting interleukin-2 (IL-2) production and T-cell proliferation. The importance of CTLA-4 in controlling T-cell activation was demonstrated in a study that showed *Ctla4*-knockout mice die at 3–4 weeks because of uncontrolled T-cell proliferation (Waterhouse et al., 1995).

Interestingly CTLA-4 plays an entirely different role in the function of Tregs. Unlike conventional T-cells, CTLA-4 is constitutively expressed on naturally occurring Tregs ($CD4^+$ $FoxP3^+$ T-cells), which is crucial to their suppressive function (Tang et al., 2008). Indeed, $CD4^+CD25^+$ regulatory T-cells constitutively expressing CTLA-4 maintain the immunologic self-tolerance in the body (Takahashi et al., 2000). The CTLA-4 expression on these cells depends on FoxP3 (Hori et al., 2003), and loss of CTLA-4 impairs Tregs' suppressive function (Takahashi et al., 2000). CTLA-4 molecules on the surface of Tregs can induce downregulation of CD80 and CD86 costimulatory molecules on APCs thereby suppressing efficient T-cell activation (Wing et al., 2008; Qureshi et al., 2011).

Targeting CTLA-4 represents a unique and novel approach to cancer therapy because it does not target tumor cells as traditional cancer treatment modalities do. Instead, it removes the brakes from tumor-specific T lymphocytes unleashing anti-tumor immunity. Currently, there are two anti-CTLA-4 monoclonal antibodies—ipilimumab and tremelimumab—which are approved and under investigation, respectively, in the context of monotherapy and combination regimens. Table 1 summarizes the Food and Drug Administration (FDA)-approved checkpoint inhibitors.

2.1.1 Ipilimumab

Ipilimumab, also known as MDX-010 (trade name: Yervoy), is a human immunoglobulin G1 (IgG1) monoclonal antibody targeting CTLA-4 developed by Bristol-Myers Squibb.

The first promising results came from a phase III clinical trial (CA184-002, NCT00094653) in which previously treated unresectable stages III and IV melanoma patients were randomly assigned to receive ipilimumab, gp-100 peptide vaccine, or both agents (Hodi et al., 2010). This trial showed significantly improved overall survival (OS) for patients in the ipilimumab arm when compared to the gp-100 peptide vaccine (10.1 vs. 6.4 months mean OS; hazard ratio [HR] 6.6 [$P = .0026$]). Furthermore, it showed that the addition of the vaccine to ipilimumab (10.0 months mean OS) did not improve OS (HR 0.68 for ipilimumab plus gp100 vs. gp100 alone [$P = .0004$]). These results represented the first evidence that immune checkpoint blockade alone can improve survival for patients with advanced cancer (stage III/IV). This study resulted in the FDA approval of ipilimumab in 2011 for treatment of patients with unresectable or metastatic melanoma. Long-term follow-up studies of patients treated with ipilimumab demonstrated the durability of long-term survival in treated patients (McDermott et al., 2013; Schadendorf et al., 2015). Moreover a subsequent phase III trial (CA184-024, NCT00324155) showed that ipilimumab plus dacarbazine improved OS when compared with treatment with dacarbazine alone in treatment-naïve patients with advanced melanoma (Robert et al., 2011). In a milestone analysis of patients in this study, authors found that the 5-year survival rate in patients treated with ipilimumab plus dacarbazine was more than twice that (18.2% vs. 8.8%) of patients treated with dacarbazine alone (Maio et al., 2015).

Ipilimumab is given intravenously at a dose of 3 mg/kg every 3 weeks (q3wk) for a maximum of four doses. It has a relatively low clearance of 14.9 mL/h and a mean terminal elimination half-life of 12.5 h

Table 1 FDA-Approved Checkpoint Inhibitors

Drug	Brand Name	Indication	Recommended Dosage
Ipilimumab	Yervoy	Unresectable or metastatic melanoma	3 mg/kg IV, once every 3 weeks, for four doses
Ipilimumab	Yervoy	BRAF V600 wild-type, unresectable, or metastatic melanoma (in combination with nivolumab)	3 mg/kg IV, once every 3 weeks, for four doses
Ipilimumab	Yervoy	Adjuvant treatment of high-risk melanoma following surgery	10 mg/kg IV, once every 3 weeks, for four doses, followed by 10 mg/kg IV every 3 months for up to 3 years
Ipilimumab	Yervoy	Unresectable or metastatic melanoma in pediatric patients 12 and older	3 mg/kg IV, once every 3 weeks, for four doses
Nivolumab	Opvido	Unresectable or metastatic melanoma	3 mg/kg IV, once every 2 weeks until disease progression or toxicity
Nivolumab	Opvido	Metastatic non–small cell lung cancer (NSCLC)	3 mg/kg IV, once every 2 weeks until disease progression or toxicity
Nivolumab	Opvido	BRAF V600 wild-type, unresectable, or metastatic melanoma (in combination with ipilimumab)	When in combination with ipilimumab, 1 mg/kg IV, once every 3 weeks for four doses. When used alone, 3 mg/kg IV, once every 2 weeks until disease progression or toxicity[1]
Nivolumab	Opvido	Previously untreated advanced renal cell carcinoma (in combination with ipilimumab)	Nivolumab (3 mg/kg) plus ipilimumab (1 mg/kg) every 3 weeks for four doses followed by nivolumab monotherapy (3 mg/kg) every 2 weeks
Nivolumab	Opvido	Metastatic non–small cell lung cancer (NSCLC)	240 mg IV, once every 2 weeks until disease progression or toxicity[1]
Nivolumab	Opvido	Advanced renal cell carcinoma	240 mg IV, once every 2 weeks until disease progression or toxicity[1]

(Continued)

Table 1 FDA-Approved Checkpoint Inhibitors—cont'd

Drug	Brand Name	Indication	Recommended Dosage
Nivolumab	Opvido	Classical Hodgkin lymphoma	3 mg/kg IV, once every 2 weeks until disease progression or toxicity[1]
Nivolumab	Opvido	Recurrent or metastatic squamous cell carcinoma of the head and neck (SCCHN)	240 mg IV, once every 2 weeks until disease progression or toxicity[1]
Nivolumab	Opvido	Locally advanced or metastatic urothelial carcinoma	240 mg IV, once every 2 weeks until disease progression or toxicity[1]
Nivolumab	Opvido	Mismatch repair deficient (dMMR) and microsatellite instability high (MSI-H) metastatic colorectal cancer	240 mg IV, once every 2 weeks until disease progression or toxicity
Nivolumab	Opvido	Hepatocellular carcinoma	240 mg IV, once every 2 weeks until disease progression or toxicity[1]
Nivolumab	Opvido	Adjuvant treatment of patients with melanoma with lymph node involvement or metastatic disease	240 mg IV, once every 2 weeks until disease progression or toxicity for 1 year[1]
Pembrolizumab	Keytruda	Unresectable or metastatic melanoma	2 mg/kg IV, once every 3 weeks until disease progression or toxicity
Pembrolizumab	Keytruda	Recurrent or metastatic SCCHN	200 mg IV, once every 3 weeks until disease progression, toxicity, or up to 2 years
Pembrolizumab	Keytruda	Metastatic NSCLC	200 mg IV, once every 3 weeks until disease progression, toxicity, or up to 2 years
Pembrolizumab	Keytruda	Refractory classical Hodgkin lymphoma	200 mg IV, once every 3 weeks until disease progression, toxicity, or up to 2 years
Pembrolizumab	Keytruda	Locally advanced or metastatic urothelial carcinoma	200 mg IV, once every 3 weeks until disease progression, toxicity, or up to 2 years

Pembrolizumab	Keytruda	Unresectable or metastatic, MSI-H or dMMR solid tumors	200 mg IV, once every 3 weeks until disease progression, toxicity, or up to 2 years
Pembrolizumab	Keytruda	Locally advanced or metastatic, gastric or gastroesophageal junction adenocarcinoma	200 mg IV, once every 3 weeks until disease progression, toxicity, or up to 2 years
Atezolizumab	Tecentriq	Metastatic urothelial carcinoma	1200 mg IV, once every 3 weeks until disease progression or toxicity
Atezolizumab	Tecentriq	Metastatic NSCLC	1200 mg IV, once every 3 weeks until disease progression or toxicity
Avelumab	Bavencio	Merkel cell carcinoma	10 mg/kg IV, once every 2 weeks until disease progression or toxicity
Avelumab	Bavencio	Metastatic urothelial carcinoma	10 mg/kg IV, once every 2 weeks until disease progression or toxicity
Durvalumab	Imfinzi	Metastatic urothelial carcinoma	10 mg/kg IV, once every 2 weeks until disease progression or toxicity
Durvalumab	Imfinzi	Unresectable stage III NSCLC	10 mg/kg IV, once every 2 weeks until disease progression or toxicity

[1]Non–small cell lung cancer (NSCLC), head and neck squamous cell carcinoma (HNSCC), mismatch repair deficient (dMMR), microsatellite instability high (MSI-H). On March 6, 2018, the dosing schedule for nivolumab was updated to include 480 mg infused every 4 weeks (Q4W) for a majority of approved indications.

(Cameron et al., 2011). Though the introduction of ipilimumab was promising as well as one of the first new hopes for patients with metastatic melanoma, it was not without challenges. Treatment with ipilimumab generated severe (grades 3 and 4) immune-related adverse events (irAEs) and even death (grade 5). These include rash, diarrhea, colitis, hepatitis, and endocrinopathy and have been observed in 10%–35% of patients treated with ipilimumab (Weber et al., 2012). The endocrinopathies, such as hypophysitis and thyroiditis, are nonreversible, and management includes long-term use of hormone replacement (Weber et al., 2012). The EORTC 18,071 trial was a large phase III study, which compared ipilimumab to placebo in 951 stage III patients who had undergone complete resection of their tumors. Ipilimumab, dosed at 10 mg/kg, was given q3wk for four doses and then once every 3 months for up to 3 years (Eggermont et al., 2015). The 3-year recurrence-free survival was significantly improved for patients treated with ipilimumab (46.5% [95% confidence interval (CI) 41.5–51.3]) versus placebo (34% [95% CI 30.1–39.5]). This finding led to the FDA approval of ipilimumab as the first adjuvant immunotherapy treatment for advanced melanoma in 2015 (Eggermont and Dummer, 2017).

On a mechanistic level, ipilimumab increases the frequency of inducible T-cell costimulator $(ICOS)^+$ $CD4^+$ T-cells in multiple types of tumors, including melanoma, prostate, and bladder cancer (Chaput et al., 2017; Chen et al., 2009; Liakou et al., 2008). $ICOS^+$ $CD4^+$ T-cells from bladder cancer patients treated with ipilimumab produced high levels of interferon gamma (IFNγ) upon stimulation in vitro and were able to recognize the NY-ESO-1 antigen. Increase in this cell population also resulted in the increased ratio of $ICOS^+$ $CD4^+$ to $FOXP3^+$ $CD4^+$ Tregs in patients after treatment with ipilimumab (Liakou et al., 2008). Expansion of $ICOS^+$ $CD4^+$ T-cells following treatment with ipilimumab has also been associated with better clinical outcome, including higher OS in melanoma patients (Carthon et al., 2010). Recent studies have suggested an alternate hypothesis for the mechanism of anti-CTLA-4 mAbs (Du et al., 2018a, 2018b). Studies in mice have suggested that the therapeutic benefit of ipilimumab may be the result of local Treg cell depletion in the tumor microenvironment rather than the blockade of the B7-CTLA-4 interaction (Du et al., 2018b).

Melanoma patients responding to ipilimumab have shown to have high baseline frequencies of circulating nonclassical $CD14^+CD16^+$ monocytes expressing FcγRIIIA (CD16) and were able to deplete Tregs *ex* vivo in the presence of ipilimumab via antibody-dependent cell-mediated cytotoxicity (ADCC) (Romano et al., 2015).

2.1.2 Tremelimumab

Tremelimumab, also known as CP-675,206, is a fully human IgG2 monoclonal antibody targeting CTLA-4 developed by Pfizer. Initial phase I and II studies of tremelimumab were promising; however, the phase III trial (NCT00257205) in patients with metastatic melanoma was stopped after the review of early interim data revealed that treatment with tremelimumab would not demonstrate superiority to standard chemotherapy (Ribas et al., 2013). The objective response rates were 10.7% in the tremelimumab arm, and 9.8% in the chemotherapy arm. Despite no significant differences between study arms in complete or partial response rates, interestingly, responses to tremelimumab were significantly more durable with a median response duration of 35.8 months for patients in the tremelimumab arm compared with13.7 months for patients in the chemotherapy arm (Ribas et al., 2013).

Tremelimumab has a similar toxicity profile as ipilimumab, with most common toxicities being skin rash, fatigue, nausea/vomiting, and diarrhea/colitis. A small percentage of patients experience hepatitis, pancreatitis, and endocrine abnormalities such as thyroid, hypothalamus, and pituitary disorders. In general, anti-CTLA-4 therapy can induce latent effects on the endocrine system. Therefore, it is essential to monitor the thyroid function in patients receiving these drugs (Gan et al., 2017). Tremelimumab has a dosing regimen of 15 mg/kg once every 3 months. Pharmacokinetic studies revealed that tremelimumab has a long mean terminal elimination half-life of 22 days (Ribas et al., 2005).

While tremelimumab was initially perceived to be a failure, it has been shown to bind to the same region on CTLA-4 with similar binding affinity as ipilimumab (He et al., 2017a). Tremelimumab is currently under investigation by MedImmune (a subsidiary of AstraZeneca) for multiple indications as a combination therapy with a PD-L1 antibody, durvalumab. This combination is being tested in various indications, including in patients with head and neck squamous cell carcinoma (HNSCC; NCT02369874) and NSCLC (NCT02453282) (see also Section 2.3.2).

2.2 Programmed Cell Death-1–Targeting Antibodies

Programmed cell death 1 (PD-1), also known as CD279, is a member of the CD28 transmembrane protein family expressed primarily on the surface of activated T-cells, myeloid cells, and B-cells (Parry et al., 2005). PD-1 has two known ligands, programmed cell death ligand 1 (PD-L1) also known as B7-H1 and programmed cell death ligand 2 (PD-L2) also known as B7-DC (Keir et al., 2008). PD-L1 is broadly expressed on a wide variety of hematopoietic (professional and nonprofessional APCs) and nonhematopoietic organs, such as vascular endothelium, epithelium, heart, muscle, liver, pancreas, placenta, and eye (Keir et al., 2006). On the other hand, the PD-L2 expression is much more restricted to few subsets of hematopoietic cells, such as DCs, macrophages, B-cells, Th2 cells, and bone marrow—derived mast cells (Francisco et al., 2010). Outside the hematopoietic system, PD-L2 is expressed on the airway epithelia, and its expression on bronchial biopsies was associated with the severity of asthma in patients and mice (Lewkowich et al., 2013).

Upon binding to its ligands (PD-L1 or PD-L2), PD-1 delivers a coinhibitory signal in the proximity of the TCR signaling complex. PD-1 ligand engagement phosphorylates its cytoplasmic immunoreceptor tyrosine—based inhibition motif (ITIM) and immunoreceptor tyrosine—based switch motif recruiting SHP2 phosphatase (Okazaki et al., 2001). SHP2 then dephosphorylates TCR proximal signaling molecules, including CD3ζ, ZAP70, and PKCθ. This dephosphorylation will attenuate the TCR/CD28 signaling, reducing T-cell functions, including effector cytokine production and target cell killing (Freeman, 2008).

PD-1 is an immune checkpoint molecule, and similar to CLTA-4, inhibits cytotoxic T-cell activity; however, the two pathways are not functionally redundant (Parry et al., 2005). Tumors can upregulate PD-L1, and this inhibits anti-tumor responses from T-cells. PD-L1 can also directly bind to CD80 (B7-1) on APCs and inhibit T-cell activation and cytokine production (Butte et al., 2007).

Antibodies that target and block the interaction between PD-1 and PD-L1 reinvigorate tumor-specific T-cells. This class of agents has shown promising clinical results in various solid and hematological cancers as monotherapy or in combination with other checkpoint inhibitors (Chowdhury et al., 2017). Furthermore, the PD-1 expression on a subset of human melanoma cells has been shown to promote tumorigenesis through a melanoma cell-intrinsic mechanism (Kleffel et al., 2015).

Blocking PD-1 with therapeutic antibodies can result in various irAEs, including colitis, hepatitis, pneumonitis, rash, and endocrinopathies (Hahn et al., 2017). A recent meta-analysis of phase II and III clinical trials for ipilimumab, nivolumab, pembrolizumab, and atezolizumab revealed that the incidence of irAEs, specifically of immune-related rash and colitis, were significantly higher with ipilimumab than with PD-1/PD-L1 inhibitors (De Velasco et al., 2017).

2.2.1 Nivolumab

Nivolumab (trade name: Opvido) is a fully human IgG4 kappa monoclonal antibody that selectively targets the PD-1 receptor on $CD8^+$ T-cells. Nivolumab blocks the PD-1 inhibitory signaling pathway in $CD8^+$ T-cells, allowing them to produce effector cytokines (IFN-γ, tumor necrosis factor [TNF]-α) and induce apoptosis in tumor cells (Wang et al., 2014). Because of its IgG4 structure, Nivolumab is a nondepleting antibody as opposed to ipilimumab (IgG1; Wang et al., 2014).

Nivolumab showed promising results in two phase III trials, which looked at both advanced squamous (CheckMate 017; NCT01642004) and nonsquamous (CheckMate 057; NCT01673867) NSCLC. In Checkmate 017, the 2-year OS rates were significantly better with nivolumab (23% [95% CI, 16%–30%]) than docetaxel (8% [95% CI, 4%–13%]). Similarly, the CheckMate 057 found 2-year survival was improved in patients on nivolumab (29% [95% CI, 24%–34%]) versus docetaxel (16% [95% CI, 12%–20%]; Horn et al., 2017).

Currently, nivolumab is approved for the treatment of various malignancies, including unresectable or metastatic melanoma, metastatic NSCLC, advanced renal cell carcinoma, classical Hodgkin lymphoma, recurrent or metastatic HNSCC, locally advanced or metastatic urothelial carcinoma, mismatch repair deficient (dMMR) and microsatellite instability-high (MSI-H) metastatic colorectal cancer, and hepatocellular carcinoma (Nivolumab, 2014). Nivolumab is also approved for the use in combination with ipilimumab in patients with BRAF V600 wild-type, unresectable, or metastatic melanoma (Nivolumab, 2014).

Pharmacokinetic studies of nivolumab demonstrated that steady-state concentrations of nivolumab were reached by 12 weeks when administered at a dose of 3 mg/kg q2wk (Agrawal et al., 2016). Nivolumab has a low clearance time (9.5 mL/h) and a mean elimination half-life of 26.7 days (Nivolumab, 2014). When administered in combination with ipilimumab, the clearance time of nivolumab was increased by 24%, whereas there was no effect on the clearance of ipilimumab (Nivolumab, 2014).

When nivolumab is given as a single agent, common adverse drug reactions include fatigue, rash, and loss of appetite and nausea. Severe adverse drug reactions (grades 3 and 4) include pneumonitis (0.7%), colitis (0.4%), hepatitis (1.7%), severe rash (0.4%) hypophysitis (0.2%), and nephritis (0.2%; Hahn et al., 2017). Autoimmune fulminant myocarditis is a rare (0.09%) but severe and potentially fatal side effect, which has been recently reported following treatment with nivolumab (Frigeri et al., 2018; Chen et al., 2017; Tajiri et al., 2018). When given in combination with ipilimumab, the most common adverse drug reactions include rash, pruritus, headache, vomiting, and colitis (Larkin et al., 2015). Combination therapy of ipilimumab and nivolumab results in a significantly increased incidence of adverse drug reactions than when patients are treated with either agent alone (Spain et al., 2016). Management of grade 1 or 2 colitis includes the use of antimotility agents, such as loperamide or diphenoxylate/atropine and dietary modifications. Severe colitis may be treated with prednisone 1–2 mg/kg by mouth daily or methylprednisolone 2 mg/kg intravenously twice daily (Postow, 2015). Topical corticosteroid creams can be used to treat mild rash, while severe rashes (grade 3/4) should be managed with oral corticosteroids (Postow, 2015).

PD-L1 testing is not required for the use of nivolumab; however, nivolumab does have a complementary diagnostic (Dako PD-L1 IHC 28-8 pharmDx). A retrospective analysis of the tumor specimens from the Checkmate 057 demonstrated a higher efficacy for nivolumab in patients whose tumors expressed PD-L1 in more than 10% of cells than the groups with less than 1% or 5% PD-L1 expression (Borghaei et al., 2015). While the Dako PD-L1 IHC 28-8 assay has not obtained regulatory status as a companion diagnostic, it may help to determine to what degree a patient's tumor expresses the PD-L1 protein.

In a randomized phase III study of 906 melanoma patients who underwent total resection of their tumors, nivolumab treatment (3 mg/kg q2wk) was compared to ipilimumab (10 mg/kg q3wk for four doses and then q12wk) in an adjuvant setting for a period of up to 1 year or until disease progression or emergence of unacceptable toxicity or withdrawal of consent (Weber et al., 2017). Patients in the nivolumab group experienced 70.5% (95% CI, 66.1–74.5) recurrence-free survival compared to 60.8% (95% CI, 56.0–65.2) in the ipilimumab group. Nivolumab treatment was also associated with less toxicity and discontinuation of treatment because of any adverse event (Weber et al., 2017). Findings from this study resulted in the FDA approval of nivolumab as adjuvant therapy for patients with completely resected melanoma with lymph node involvement or metastatic disease (Nivolumab, 2014).

2.2.2 Pembrolizumab

Pembrolizumab (trade name: Keytruda), also known as MK-3475 and lambrolizumab, is a humanized IgG4 kappa monoclonal antibody developed by Merck & Co. Similar to nivolumab, pembrolizumab also reinvigorates $CD8^+$ T-cells by blocking PD-1 signaling pathway, allowing them to induce apoptosis in tumor cells (Hamid et al., 2013). Pembrolizumab is also a nondepleting antibody similar to nivolumab (Homet Moreno and Ribas, 2015).

Pembrolizumab first showed promising results in the KEYNOTE-001 study (NCT01295827) in which it was administered to patients with advanced melanoma who had previously received ipilimumab (Garon et al., 2015). It is currently approved for the treatment of patients with metastatic NSCLC, recurrent or metastatic HNSCC, and refractory classical Hodgkin lymphoma, and locally advanced or metastatic urothelial carcinoma (Pai-Scherf et al., 2017). Pembrolizumab is also approved for use in unresectable or metastatic, MSI-H or dMMR solid tumors, which make pembrolizumab the first oncology drug, in the era of precision medicine, that has received a tissue/site-agnostic approval from FDA. Most recently, pembrolizumab was approved for the treatment of patients with locally advanced or metastatic, gastric, or gastroesophageal junction adenocarcinoma (Pai-Scherf et al., 2017). Multiple phase III trials are underway for patients with triple negative breast cancer (NCT02768701), renal cell carcinoma (NCT02853344, NCT03142334), and several other hematological malignancies and solid tumors.

Pembrolizumab is given intravenously at a dose of 200 mg/kg, q3wk until disease progression or reports of unacceptable toxicity. It has a low pharmacokinetic clearance of 9.2 mL/h, a limited volume of distribution, and an elimination half-life of 27.3 days (Ahamadi et al., 2017). The toxicity profile of pembrolizumab is far less severe than ipilimumab, making it a more attractive option for patients (Spain et al., 2016). Adverse drug reactions include body pain, chills, constipation, cough, fever, headache, loss of voice, rapid weight gain, and bleeding. Serious adverse drug reactions (grades 3 and 4) include pneumonitis (1.3%), colitis (1.1%), hepatitis (0.5%), hypothyroidism (0.1%), hyperthyroidism (0.1%), hypophysitis (0.3%), and nephritis (0.1%) (Hahn et al., 2017). Clinical studies have shown that many of pembrolizumab's adverse drug reactions can be managed with supportive care or immune suppression from steroid therapy (Spain et al., 2016). For grades 3 and 4 hepatitis and pneumonitis, therapy should be discontinued, and systemic steroids (prednisone 1–2 mg/kg by mouth daily or methylprednisolone 2 mg/kg intravenously twice daily) should be started (Hahn et al., 2017). Some immune-mediated endocrinopathies such as hypophysitis, thyroiditis, hypothyroidism, and hyperthyroidism can become permanent and won't resolve with steroid therapy (Byun et al., 2017). In the rare event of an endocrinopathy, the patient should be treated specifically for the type of endocrinopathy they are experiencing, such as starting levothyroxine for hypothyroidism or treatment with a β-blocker for hyperthyroidism (Byun et al., 2017).

Pembrolizumab is the first anti–PD-1 therapy associated with an FDA-approved companion diagnostic assay, the anti–PD-L1 immunohistochemical (IHC) 22C3 PharmDx (Agilent Technologies, Santa Clara, CA) using the Dako Autostainer (Dako, Carpinteria, CA). This diagnostic was first developed during the Keynote-001 trial when study results demonstrated that PD-L1 expression in at least 50% of tumor cells was associated with improved efficacy of pembrolizumab and was, therefore, a biomarker for predicting the treatment effectiveness (Garon et al., 2015). The results of the Keynote-010 trial revealed that patients with tumors that are positive for PD-L1 expression had significantly higher progression-free survival (PFS) and OS on pembrolizumab than patients on docetaxel. Furthermore, patients with at least 50% of tumor cells expressing PD-L1 had the most significant benefit from pembrolizumab, with a HR of 0.59 (95% CI, 0.46–0.74; Herbst et al., 2016). These findings validated 22C3 PharmDx as a biomarker assay and represent a significant step forward in identifying which patients are most likely to respond to pembrolizumab therapy.

In the adjuvant setting, pembrolizumab in phase III EORTC1325/KEYNOTE-054 interim analysis showed significantly improved recurrence-free survival over placebo (HR = 0.57; [98.4% CI, 0.43–0.74]; $P < .0001$; Merck, 2018b). Notably the toxicity profile of adjuvant pembrolizumab was consistent with that observed in previous studies in patients with advanced melanoma (Merck, 2018b). In January 2018, Merck announced that in a phase III study (KEYNOTE-189), pembrolizumab in combination with pemetrexed and cisplatin or carboplatin, in first-line treatment, has significantly increased PFS and OS in nonsquamous NSCLC patients compared with pemetrexed plus platinum chemotherapy alone (Merck, 2018a). The results of this trial will potentially be practice changing in the management of patients with NSCLC.

2.3 PD-L1–Targeting Antibodies

2.3.1 Atezolizumab

Atezolizumab (trade name: Tecentriq) is a humanized monoclonal antibody of IgG1 isotype produced in Chinese hamster ovary cells and developed by Genentech. It binds to the PD-1 ligand, PD-L1, and prevents its interaction with PD-1 without affecting the binding of PD-L2 (B7-DC or CD273) to PD-1 (Herbst et al., 2014). Atezolizumab also blocks the interaction between PD-L1 and the receptor B7-1 (or CD80). PD-L1 is known to be also present on activated T-cells; thus atezolizumab was engineered to have a modification in the Fc domain that eliminates ADCC at clinically relevant doses. This alteration prevents the depletion of T-cells expressing PD-L1 (Deng et al., 2016). Clinical activity of atezolizumab was first assessed in a large phase I study (NCT01375842) in 277 patients across multiple cancer types. In this trial, clinical response as assessed by Response Evaluation Criteria in Solid Tumors, version 1.1 (RECIST v1.1), was significantly associated ($P = .007$) with the PD-L1 expression of tumor-infiltrating immune cells (Herbst et al., 2014). The IMVigor 210 trial (NCT02108652) was the landmark phase II trial, which demonstrated the efficacy and safety of this drug in patients with urothelial cancer (Rosenberg et al., 2016). Recently published findings from the IMVigor 210 trial demonstrated objective response rate (ORR) of 23% and complete response (CR) of 9% at a median follow-up time of 17.2 months (Balar et al., 2017). FDA has approved atezolizumab for the treatment of metastatic urothelial carcinoma and metastatic NSCLC as second-line therapy in patients whose disease progressed during or following platinum-containing chemotherapy (Hamilou et al., 2017). However, atezolizumab failed to improve OS in the second line over the investigator's choice of chemotherapy (Powles et al., 2017). Nevertheless, FDA has not removed the approval of atezolizumab for the second line in bladder cancer, and some experts in the field still believe patients could benefit from treatment with atezolizumab. While the phase III trial failure was a biomarker failure, it was not a drug failure (personal communication with Dr. Bishoy M. Faltas, Department of Medical Oncology, Weill-Cornell Medical College, NY). In a phase III study (IMvigor130), atezolizumab in combination with chemotherapy is being compared with chemotherapy alone (NCT02807636). A recent meta-analysis, which indirectly compared nivolumab, pembrolizumab, and atezolizumab, revealed that nivolumab and pembrolizumab were associated with a significant increase of ORR (1.66 [95% CI, 1.07–2.58] and 1.94 [95%, CI 1.30–2.90], respectively) when compared with atezolizumab (Passiglia et al., 2018). These findings suggest that while atezolizumab is approved for use in NSCLC, other ICIs may offer better treatment outcomes for that specific tumor type.

Atezolizumab is also under phase II and phase III investigation for use in metastatic renal cell carcinoma (NCT03024996), relapsed or refractory Hodgkin lymphoma (NCT03120676), triple negative breast cancer (NCT03164993, NCT02530489), and colorectal cancer (NCT02982694).

Atezolizumab is given at a dose of 1200 mg once q3wk in patients with urothelial carcinoma. Other trials have shown that administering 1–20 mg/kg produced a dose-proportional increase in exposure to the drug (Herbst et al., 2014). Atezolizumab has a long median terminal half-life of 27 days and a clearance of 8.3 mL/h (Atezolizumab, 2016).

Safety data from 315 patients registered in the IMVigor 210 trial (NCT02108652) revealed that 98% experienced an adverse drug reaction on atezolizumab, with the most common being fatigue, nausea and decreased appetite, constipation, diarrhea, rash, and musculoskeletal effects. Severe adverse drug reactions include immune-related adverse reactions such as hepatitis (1%), colitis (0.5%), hypophysitis, and dysfunction in the thyroid (Perez-Gracia et al., 2017). Atezolizumab was temporarily interrupted in 30% of the patients due to adverse drug reactions, while 4% of the patients permanently discontinued the treatment. Twenty-two percent of patients were treated with steroid therapy to help mitigate treatment-related adverse drug reactions (Perez-Gracia et al., 2017). Management of the irAEs was achieved with the use of topical corticosteroids for rash and systemic steroid treatment (prednisone 1–2 mg/kg by mouth daily or methylprednisolone 2 mg/kg intravenously twice daily) for other severe immune-related adverse reactions (Hahn et al., 2017).

An ongoing question in the Immuno-Oncology (IO) field is whether the PD-L1 expression is a relevant biomarker for predicting atezolizumab treatment effectiveness. Initial studies implied that PD-L1 expression on tumor cells and tumor-infiltrating immune cells correlated with better clinical response (Herbst et al., 2014). Genentech has developed a companion diagnostic immunohistochemistry assay for PD-L1 (Vennapusa et al., 2016). However, subsequent studies have shown that up to 15% of patients with less than 1% PD-L1 expression still benefit from atezolizumab treatment (Rosenberg et al., 2016). While PD-L1 expression has the potential to be a predictive factor for treatment response to atezolizumab, more research needs to be done to establish a cut-off reference for PD-L1—positive status (Festino et al., 2016).

2.3.2 Durvalumab

Durvalumab (trade name: Imfinzi), first known as MEDI4736, is a human IgG1 monoclonal antibody and potent antagonist of PD-L1 being developed by MedImmune/AstraZeneca. It blocks the binding of PD-L1 to both PD-1 and CD80 (B7-1) found on T-cells and APCs (Stewart et al., 2015). It is also highly specific and does not bind to PD-L2, which may help avoid potential immune-related toxicity due to the PD-L2 blockade. The IgG1 heavy chain in the Fc region antibody domain was engineered to eliminate ADCC and complement-dependent cytotoxicity to prevent depletion of effector T-cells that express PD-L1 (Stewart et al., 2015). In May 2017, Durvalumab received approval for the treatment of metastatic urothelial cancer. Durvalumab is also currently being investigated in phase III trials for the use in the treatment of locally advanced, unresectable NSCLC (PACIFIC trial—NCT02125461). Early results from this study demonstrated that PFS was significantly longer with durvalumab (16.8 months [95% CI, 13.0—18.1]) than with placebo (5.6 months [95% CI, 4.6—7.8]; Antonia et al., 2017). Further studies are needed to determine the impact of durvalumab on OS and long-term toxicity. Durvalumab is also being studied in combination with tremelimumab in phase II and III trials, discussed in detail in Section 6.1.2.

Though it is not required for the use of durvalumab, Roche developed a complementary diagnostic assay, the VENTANA PD-L1 (SP263) assay, for the qualitative detection of PD-L1 expression. The FDA has approved the assay for the use in patients with locally advanced or metastatic urothelial carcinoma who are being considered for treatment with durvalumab in May 2017 (Syed, 2017).

The recommended dose of durvalumab is 10 mg/kg for intravenous infusion over 60 min q2wk. Dose-proportional pharmacokinetics is achieved at 3 mg/kg or higher, and the mean terminal half-life is 21 days. Durvalumab clearance time is 9.6 mL/h (Baverel et al., 2017). A phase I/II trial (Study 1108; NCT01693562) of 191 patients with urothelial carcinoma found that treatment with durvalumab needed to be delayed or interrupted due to adverse drug reactions in more than 30% of patients with immune-related side effects occurring in 46% of patients. Serious (grade 3/4) adverse drug reactions occurred in 6.8% of patients, and grade 3/4 irAEs occurred in 2.1% of patients. More extensive studies of durvalumab have found that it is associated with adverse drug reactions such as fatigue, hypothyroidism, colitis, pneumonitis, nephritis, and rash at a higher rate when compared with placebo (Antonia et al., 2017). irAEs are treated with topical corticosteroids, systemic glucocorticoids, and endocrine therapy (Antonia et al., 2017).

2.3.3 Avelumab

Avelumab (trade name: Bavencio) is a human IgG1 lambda monoclonal antibody that binds PD-L1 and blocks the interaction between PD-L1 and PD-1 and CD80 (B7-1) developed by Merck and Pfizer. Unlike durvalumab, in which the Fc region of the antibody was engineered to eliminate ADCC, avelumab retains its native Fc component (Boyerinas et al., 2015). ADCC has been previously associated with higher toxicity; however, preclinical studies have demonstrated that it is an essential mechanism of action for other immunotherapy drugs, such as trastuzumab and inebilizumab (Petricevic et al., 2013; Ward et al., 2011).

Avelumab has received accelerated approval by the FDA for the treatment of metastatic Merkel cell carcinoma (MCC) as first-line therapy and as the first FDA-approved product to treat this type of cancer (Kim, 2017). Avelumab offers an exciting new therapeutic option for patients with MCC, an aggressive skin cancer that is associated with old age, poor prognosis, and lower survival compared with other skin malignancies (Poulsen, 2004). Previously there were no targeted therapies for MCC and treatment with traditional chemotherapy rarely resulted in a durable response (Kotteas and Pavlidis, 2015). MCC tumor cells have been shown to express PD-L1, and there is the expression of PD-L1 on tumor-infiltrating immune cells in the tumor microenvironment of MCC patients, which provides a rationale for targeting the PD-1/PD-L1 pathway (Lipson et al., 2013). Approval for avelumab was based on results from the JAVELIN Merkel 200 trial (NCT02155647) a single-arm, open-label study in 88 patients with chemotherapy-refractory MCC. Preliminary results from this trial demonstrated an objective response rate of 31.8% and disease control rate of 42% (Kaufman et al., 2016).

Avelumab has also received accelerated approval for the second-line therapy in patients with locally advanced or metastatic urothelial carcinoma whose disease progressed during or following platinum-containing chemotherapy (Lalani et al., 2017). Results from the NCT01772004 trial of 161 advanced metastatic urothelial carcinoma patients previously treated with platinum agents demonstrated objective response rate for 27 patients

(17% [95% CI, 11%–24%]), including nine (6%) complete responses and 18 (11%) partial responses (Patel et al., 2017).

Javelin is an international clinical development program for avelumab. It is currently investigating the use of avelumab in more than 30 settings including in treatment of patients with renal cell cancer (NCT02493751), refractory Hodgkin lymphoma (NCT02603419), NSCLC (NCT02584634, NCT02395172), ovarian cancer (NCT02580058), and HNSCC (NCT02952586). Merck and Pfizer are currently working with Agilent Technologies to develop a companion diagnostic test for avelumab (Kim, 2017).

Avelumab is dosed intravenously at 10 mg/kg q2wk (Kaufman et al., 2016). It has a mean terminal half-life of 6.7 days, which is relatively shorter than other PD-L1 monoclonal antibodies. Avelumab has a clearance of 24.6 mL/h (Avelumab, 2017). Low-grade (grade 1/2) adverse drug reactions include fatigue (50%), musculoskeletal pain (32%), diarrhea (23%), infusion-related reaction (22%), rash (22%), and nausea (22%; Avelumab, 2017). Severe adverse drug reactions occur very rarely (less than 5%) in patients and include severe fatigue, enterocolitis, increased aminotransferases, and intestinal nephritis (Kaufman et al., 2016). Adverse drug reactions to avelumab are managed with the use of topical corticosteroids for rash, acetaminophen, and antihistamine for infusion-related reactions, antiemetic medication for nausea, and supportive care for fatigue and diarrhea (Bommareddy and Kaufman, 2017).

2.4 Additional Inhibitory Targets

The success of CLTA-4 and PD-1/PD-L1 blocking therapy has led to multiple FDA drug approvals over the past 5 years and paved the way for the development of additional checkpoint inhibitor targets. These checkpoint molecules represent the lesser-known pathways of immune checkpoint molecules with therapeutic potential. The clinical trials that are associated with drugs and antibodies that target these molecules are summarized in Table 2.

2.4.1 T-Cell Immunoglobulin and Mucin Domain-3

T-cell immunoglobulin and mucin domain-3 (TIM-3) is an immune checkpoint receptor, which plays a role in tumor immunological tolerance and T-cell exhaustion (Monney et al., 2002). TIM-3 is expressed on the surface of T helper 1-type (Th1) $CD4^+$ T effector cells, DCs, microglia, macrophages, mast cells, natural killer (NK) cells, and activated $CD8^+$

Table 2 Selected Clinical Trials for Additional Immune Checkpoint Inhibitors

Checkpoint Target	Drug	Single Agent or in Combination	Indication	Phase of Trial	Clinical Trial Identifier
TIM-3	TSR-022	Anti–PD-1 monoclonal antibody	Advanced solid tumors	I	NCT02817633
TIM-3	LY3321367	Anti–PD-L1 monoclonal antibody	Advanced solid tumors	I	NCT03099109
TIM-3	MBG453	Anti–PD-1 monoclonal antibody	Advanced and metastatic solid tumors	I	NCT02608268
TIM-3	MBG453	Anti–PD-1 monoclonal antibody, decitabine	Relapsed/refractory AML	Ib/II	NCT03066648
TIGIT	OMP-313M32	Single agent	Advanced and metastatic solid tumors	I	NCT03119428
TIGIT	MTIG7192A	Atezolizumab	Advanced and metastatic solid tumors	I	NCT02794571
VISTA	CA-170	Single agent	Advanced tumors and lymphomas	I	NCT02812875
LAG-3	Relatlimab	Ipilimumab and nivolumab	Advanced renal cell carcinoma	II	NCT02996110
LAG-3	Relatlimab	Ipilimumab and nivolumab	Advanced gastric cancer	II	NCT02935634
LAG-3	Relatlimab	Nivolumab	Advanced solid tumors	I	NCT02966548
LAG-3	Relatlimab	Nivolumab	Advanced solid tumors	I/II	NCT01968109
IDO	Indoximod	Ipilimumab, nivolumab, pembrolizumab	Advanced melanoma	I/II	NCT02073123
IDO	Indoximod	Docetaxel and paclitaxel	Metastatic breast cancer	II	NCT01792050
IDO	Epacadostat	Pembrolizumab	Advanced thymic carcinoma	II	NCT02364076
IDO	Epacadostat	Pembrolizumab	Unresectable or metastatic melanoma	III	NCT02752074

IDO	Epacadostat	Pembrolizumab	Unresectable or metastatic gastric cancer	II	NCT03196232
IDO	Epacadostat	Pembrolizumab	Recurrent or progressive metastatic urothelial carcinoma	III	NCT03374488
IDO	Epacadostat	Pembrolizumab and azacitidine	Advanced solid tumors	I/II	NCT02959437
IDO	Epacadostat	Pembrolizumab and platinum-based chemotherapy	Advanced solid tumors	I	NCT02862457
IDO	Epacadostat	Durvalumab	Advanced solid tumors	I/II	NCT02318277
IDO	Epacadostat	Nivolumab	Advanced solid tumors	I/II	NCT02327078
B7–H3 (CD276)	Enoblituzumab	Single agent	Children with B7-H3–expressing solid tumors	I	NCT02982941
B7–H3 (CD276)	Enoblituzumab	Single agent	Localized intermediate and high-risk prostate cancer	I	NCT02923180
B7–H3 (CD276)	Enoblituzumab	Pembrolizumab	Refractory solid tumors which express B7-H3	I	NCT02475213
B7–H3 (CD276)	Enoblituzumab	Ipilimumab	Refractory solid tumors which express B7-H3	I	NCT02381314

T-cells (Gorman et al., 2014). Galectin-9, the primary ligand for TIM-3, is produced by some immune cells and also secreted by tumor cells in the tumor microenvironment (Zhu et al., 2005; Oomizu et al., 2012). Galectin-9 induces apoptosis in TIM-3^+ Th1 cells, and together with TIM-3, they play a critical role in inhibiting Th1 responses (Zhu et al., 2005; Oomizu et al., 2012). TIM-3 signaling has also been shown to lead to the exhaustion of interferon-producing T-cells, which ultimately results in Th1 suppression and the development of immune tolerance (Sakuishi et al., 2010). TIM-3 attenuates activation signals mediated by CD3/CD28 costimulation and affects several TCR-mediated phenotypes: (1) it inhibits CD69 expression, which is an early marker of T-cell activation; (2) it suppresses nuclear factor kappa-light-chain-enhancer of activated B cells/nuclear factor of activated T-cells (NF-κB/NFAT) activation; and (3) it also suppresses IL-2 secretion (Tomkowicz et al., 2015).

Upregulation of TIM-3 expression on T-cells is associated with worse prognosis in patients with gastric cancer, which is thought to be due to the immunosuppressive capacity of TIM-3 (Shen et al., 2016). More recently, serum soluble TIM-3 has been suggested as a novel potential diagnostic and prognostic biomarker for osteosarcoma patients (Ge et al., 2017). Patients with higher disease burden and distant metastasis had high levels of soluble TIM-3 in their serum (Ge et al., 2017). One study showed that coexpression of PD-1 and TIM-3 caused a more dysfunctional phenotype in tumor antigen-specific $CD8^+$ T-cells, which was reversible via TIM-3 pathway blockade alone or in combination with PD-1/PD-L1 pathway blockade (Fourcade et al., 2010). The high number of TIM-3^+ tumor-associated macrophages was linked to poor prognosis in renal cell carcinoma (Komohara et al., 2015).

Preclinical studies of TIM-3 inhibitory antibodies showed limited antitumor effectiveness in mice when administered alone (Ngiow et al., 2011). However, combining the use of PD-1 and TIM-3 monoclonal antibodies significantly suppressed established tumor growth and even resulted in cures in a small proportion of treated animals (Ngiow et al., 2011). Currently, Tesero has an anti–TIM-3 monoclonal antibody, TSR-022, in a phase I trial in combination with an anti–PD-1 monoclonal antibody in patients with advanced solid tumors (NCT02817633). Eli Lilly is also developing an anti–TIM-3 monoclonal antibody, LY3321367, in a phase Ia/b trial in combination with an anti–PD-L1 monoclonal antibody in patients with advanced solid tumors (NCT03099109). Novartis is also conducting clinical trials for their anti–TIM-3 antibody, MBG-453, as a single agent or in combination with a PD-1 monoclonal antibody for patients with advanced solid tumors (NCT02608268). A second phase I trial is investigating the safety and tolerability of MBG453 and a PD-1 monoclonal antibody in combination with decitabine in relapsed/refractory AML patients (NCT03066648).

2.4.2 T-Cell Immunoreceptor With Ig and ITIM Domains

T-cell immunoreceptor with Ig and ITIM domains (TIGIT) is a protein receptor, which belongs to the CD28 family, and expressed on the surface of $CD4^+$ T-cells, dysfunctional $CD8^+$ T-cells, NK T-cells (NKT cells), Tregs, and NK cells (Dougall et al., 2017). TIGIT has two primary ligands, CD155 and CD112, which are expressed on DCs, T-cells, and multiple other cell types, including tumor cells, which TIGIT binds to with high and low affinity, respectively (Chauvin et al., 2015). CD155 binds to CD226, a receptor expressed on T-cells and NK cells, leading to activation of the effector functions in these cells (Yu et al., 2009). However, TIGIT has a higher affinity for CD155 than CD226 and can, therefore, outcompete CD226 for binding to CD155, resulting in immunosuppressive actions (Yu et al., 2009). TIGIT is upregulated on tumor antigen-specific T-cells and tumor-infiltrating lymphocytes (TILs) in melanoma patients (Chauvin et al., 2015). In gastric cancer patients, TIGIT-expressing $CD8^+$ T-cells showed metabolic dysfunction, impaired glucose uptake, and decreased lactate production (He et al., 2017b). Interestingly, exogenous glucose was able to reverse TIGIT-mediated dysfunction in $CD8^+$ T-cells isolated from cancer patients' PBMCs. This phenotype reversal was accompanied by increased glut1 expression (essential for glucose uptake) and enhanced T-cell effector function (IFN-γ production) in these cells (He et al., 2017b).

PD-1 and TIM-3 checkpoint molecules have higher expressions levels on TIGIT positive $CD8^+$ T-cells isolated from cancer patients' PBMCs (He et al., 2017b). In preclinical models, anti-TIGIT and anti–PD-1 antibodies administered together were shown to enhance $CD8^+$ T-cell effector function, resulting in significant tumor reduction in mice (Johnston et al., 2014; He et al., 2017b). Interestingly, expression of TIGIT on Tregs was shown to upregulate TIM-3 expression on tumor cells in mice. In the same study, TIM-3 and TIGIT were shown to suppress antitumor immune responses synergistically (Kurtulus et al., 2015).

Currently, there are two anti-TIGIT molecules in phase I clinical trials: OMP-313M32 is an anti-TIGIT antibody, being developed by OncoMed Pharmaceuticals, as a single agent administered to patients with locally advanced or metastatic solid tumors (NCT03119428). GeneTech is also developing an anti-TIGIT monoclonal antibody, MTIG7192A, which is currently being investigated in a phase I trial for the use as a single agent and in combination with atezolizumab in patients with locally advanced or metastatic tumors (NCT02794571).

2.4.3 V-Domain Ig Suppressor of T-Cell Activation

V-domain Ig suppressor of T-cell activation (VISTA), also known as B7-H5, C10orf54, differentiation of ESC-1 (Dies1), platelet receptor Gi24 precursor, or PD1 homolog (PD1H), is expressed on neutrophils, monocytes, APCs, T-cells, and activated Tregs, but not on B-cells (Wang et al., 2011). Interestingly, VISTA has been shown to act as both a ligand and a receptor expressed on T-cells to inhibit T-cell effector functions (Lines et al., 2014a; Flies et al., 2014). VISTA signaling has been shown to block T-cell proliferation, decrease IFN-γ and TNF-α, and increase the conversion of naïve T-cells into Tregs (Lines et al., 2014b).

In preclinical studies, monoclonal antibody blockade of VISTA has been shown to decrease MDSCs and increase immune infiltration, proliferation, and effector function of tumor-reactive T-cells in the tumor microenvironment in mice (Le Mercier et al., 2014). VISTA has been shown to have a distinct and nonredundant role in immune regulation from other checkpoint molecules such as the PD-1/PD-L1 (Liu et al., 2015). Combination of anti-VISTA and anti–PD-L1 monoclonal antibodies induced tumor regression and conferred long-term survival in animals harboring the ICI-refractory CT26 colorectal tumors (Liu et al., 2015).

VISTA was recently identified as a potential compensatory inhibitory pathway in prostate tumors after ipilimumab therapy. Tumors from untreated and ipilimumab-treated patients in a presurgical clinical trial were analyzed, and researchers found that VISTA levels increased on independent subsets of macrophages in treated tumors, resulting in adaptive resistance of prostate tumors to ipilimumab treatment (Gao et al., 2017). Interestingly, this phenomenon was not observed in melanoma patients, often responsive to ipilimumab (Gao et al., 2017). Anti-VISTA molecules are being tested in phase I clinical trials; however, the safety and efficacy of these agents have yet to be determined. JNJ-61610588 (Janssen Research & Development) was in development as a human IgG1 Kappa anti-VISTA monoclonal antibody with an intact effector function; however, as of November 2017, the trial has been terminated (NCT02671955). CA-170, an orally administrated small molecule which targets PD-L1, PD-L2, and VISTA, is being developed by Curis Inc. and is now in phase I clinical trials in patients with advanced tumors and lymphomas (NCT02812875).

2.4.4 Lymphocyte-Activation Gene 3

Lymphocyte-activation gene 3 (LAG-3), also known as CD223, is expressed on activated T-cells, Tregs, B-cells, and DCs (Huard et al., 1994). LAG-3 regulates $CD8^+$ T-cell expansion in already initiated immune reactions and increases Treg activity. It can also induce immunosuppression by alternative activation (IFNα-deficient activation) of plasmacytoid DCs (pDCs) in the tumor microenvironment (Scurr et al., 2014; Williams et al., 2017; Castelli et al., 2014). LAG-3 has been found to be a central inhibitory receptor, along with PD-1, which contributes to T-cell exhaustion (Wherry, 2011). Preclinical studies have demonstrated a strong synergy between the PD-1 and LAG-3 inhibitory pathways. The dual blockade of these two pathways restores the effector function of $CD8^+$ T-cells and diminishes Treg populations, resulting in reduced tumor growth in the majority of treated animals that were previously refractory to antibody monotherapy (Woo et al., 2012). The LAG-3 blockade was also shown to upregulate CTLA-4, PD-1, and TIM-3 in a preclinical study, and triple blockade of CTLA-4, PD-1, and LAG-3 led to increased functional cytotoxic T-cell levels and reduced Tregs and MDSCs, which significantly increased the survival of treated animals (Huang et al., 2017).

There are currently several early phase clinical trials examining the use of anti—LAG-3 monoclonal antibodies in combination with anti—CTLA-4 and anti—PD-1 agents. Relatlimab, also known as BMS-986016, is an anti—LAG-3 monoclonal antibody currently in development by Bristol Myers Squibb. Currently, several clinical trials are investigating relatlimab in combination with ipilimumab and nivolumab in patients with advanced renal cell carcinoma (NCT02996110), as well as patients with advanced gastric cancer (NCT02935634). Relatlimab is also being investigated in phase I/II trials, in combination with or without nivolumab in patients with advanced solid tumor (NCT02966548, NCT01968109). Preliminary results from the NCT01968109 trial were presented at ESMO 2017 and showed that relatlimab and nivolumab combination is well tolerated with a safety profile similar to nivolumab monotherapy (Ascierto et al., 2017). This treatment combination also provides encouraging efficacy in melanoma patients previously treated with anti—PD-1/PD-L1 therapy (Ascierto et al., 2017).

2.4.5 Indoleamine 2,3-dioxygenase

Indoleamine-2,3-dioxygenase, also known as IDO, has two isoforms: IDO-1 and IDO-2. IDO-1 (hereafter IDO) is an immune regulatory metabolic checkpoint molecule, which is responsible for the first and the rate-limiting step of tryptophan catabolism within the kynurenine pathway (Munn et al., 1999). Decreased tryptophan is known to suppress T-cell proliferation (Maleki Vareki et al., 2015). In mice, IDO is shown to induce resistance to anti–CTLA-4 therapy (Holmgaard et al., 2013). Other preclinical studies have shown that IDO inhibitors may stimulate the antitumor immune response and overcome resistance to chemotherapy agents such as olaparib and cisplatin, as well as gamma radiation (Maleki Vareki et al., 2014; Zheng et al., 2006). IDO expression by tumor cells is associated with poor clinical prognosis in many cancers, including colon carcinoma, ovarian carcinoma, and endometrial carcinoma (Okamoto et al., 2005; Ino et al., 2006; Brandacher et al., 2006).

Several IDO inhibitors are currently under development. Indoximod (1-methyl-D-tryptophan), which competitively inhibits IDO, has been shown to increase or maintain tryptophan levels and support T-cell function (Soliman et al., 2016). The most common adverse drug reactions found in indoximod safety trials were fatigue (58.6%), anemia (51.7%), hyperglycemia (48.3%), infections (44.8%), and nausea (41.4%; Soliman et al., 2014). Phase II studies of indoximod are currently ongoing, and it is being evaluated in patients with stage III and IV melanoma in combination with ipilimumab, nivolumab, or pembrolizumab (NCT02073123). Preliminary results from this trial from 60 melanoma patients who received pembrolizumab plus indoximod demonstrate an objective response rate of 52% (31 patients), which included five complete responses (8%) and 26 partial responses (43%; Zakharia et al., 2017). Indoximod is also being evaluated in patients with metastatic breast cancer in combination with taxane therapy (NCT01792050).

Epacadostat (also known as INCB024360) is an orally available small molecule inhibitor of IDO, which decreases kynurenine levels in tumor cells, and restores the proliferation and activation of DCs, NK cells, and T-lymphocytes (Yue et al., 2017). Epacadostat is currently under investigation in several phase II and III trials in combination with pembrolizumab in patients with thymic carcinoma (NCT02364076), unresectable or metastatic melanoma (NCT02752074), unresectable or metastatic gastric cancer (NCT03196232), recurrent or progressive metastatic urothelial carcinoma (NCT03374488), and advanced solid tumors (NCT02959437, NCT02862457). Epacadostat is also being studied in advanced cancers in combination with durvalumab (NCT02318277) and nivolumab (NCT02327078).

2.4.6 B7-H3 (CD276)

B7-H3 (CD276) is a transmembrane protein and a member of the B7 immunoregulatory family and shares homology with other B7 ligands such as PD-L1 (B7-H1; Jung and Choi, 2013). B7-H3 protein expression can be induced in T-cells, NK cells, and APCs; however, it is not constitutively expressed on these types of cells (Loos et al., 2010). The B7-H3 expression is not limited to immune cells but is more broadly expressed in other cell types such as osteoblasts, fibroblasts, epithelial cells, as well as in organs such as the liver, lung, bladder, testis, prostate, breast, and placenta (Loos et al., 2010). B7-H3 is known to inhibit APCs while stimulating Treg cells, resulting in IL-2 suppression, though its receptor is unknown (Li et al., 2017). Exogenous IL-2 can revert B7-H3 suppression of IL-2 (Leitner et al., 2009).

B7-H3 overexpression in tumors is associated with poor prognosis in several cancers, including NSCLC (Sun et al., 2006), clear cell renal cell carcinoma (Crispen et al., 2008), prostate cancer (Roth et al., 2007; Zang et al., 2007), and colorectal carcinoma (Sun et al., 2010). However, high tumor B7-H3 expression has also been associated with significantly better postoperative prognosis in patients with pancreatic cancer and gastric cancer (Wu et al., 2006; Loos et al., 2009). While the role of B7-H3 remains unclear and the clinical data appear to conflict, these inconsistencies may be explained by the existence of multiple receptors for B7-H3, but at this time, this is unknown.

Enoblituzumab (also known as MGA271) is an anti-CD276 humanized IgG1κ monoclonal antibody currently in phase I clinical trials (Loo et al., 2012) as monotherapy for children with B7-H3–expressing tumors (NCT02982941) and in men with localized intermediate and high-risk prostate cancer (NCT02923180). It is also being evaluated in combination with pembrolizumab in patients with refractory solid tumors that express B7-H3 (NCT02475213). Enoblituzumab is also being studied in combination with ipilimumab in patients with refractory solid tumors that express B7-H3 (NCT02381314).

3. T-CELL ACTIVATING MOLECULES

Costimulatory TCRs provide the second signal for proper T-cell activation following TCR engagement with an antigenic peptide/MHC complex (Sharpe and Abbas, 2006). Most of these molecules are upregulated following antigenic peptide recognition and TCR signaling. These molecules are under active investigation as potential targets in developing novel IO drugs for cancer treatment (Khalil et al., 2015). Among others, 4-1BB, OX40, GITR, and CD40 are the most studied members of these novel IO targets and are all members of the TNF receptor family (Croft, 2009). The costimulatory TCRs are depicted in Fig. 2.

3.1 T-Cell Antigen 4-1BB Homolog

4-1BB is a surface glycoprotein receptor and a member of the TNF receptor family. It is also known as CD137 and TNF receptor superfamily member 9 (Bartkowiak and Curran, 2015). 4-1BB is expressed on the surface of activated T-cells—to a larger extent $CD8^+$ T-cells—as well as B-cells,

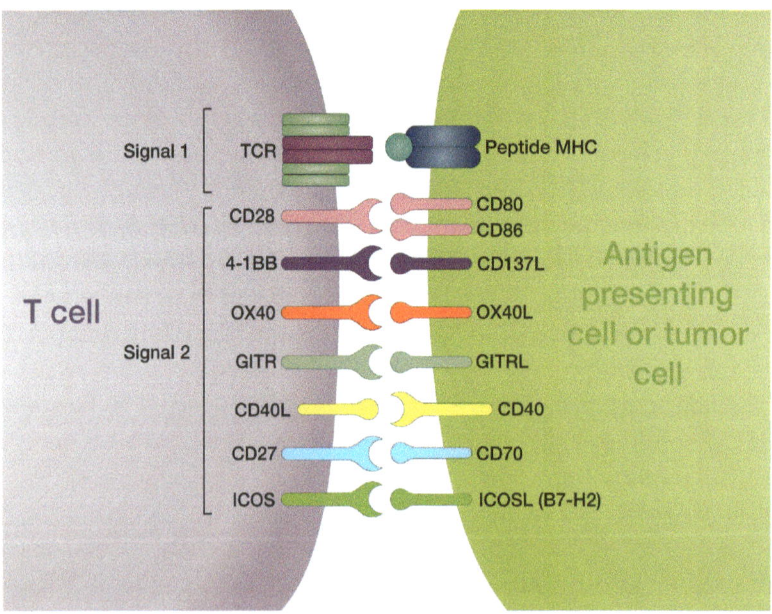

Figure 2 ***Costimulatory receptors and their ligands.*** T-cell activation is mediated by two signals; Signal one comes from the binding of the T-cell receptor (TCR) to MHC. Signal 2 comes from costimulatory molecules, most often through signaling between CD80/CD86 and CD28. *MHC*, major histocompatibility complex.

DCs, NKs, monocytes, and granulocytes (Melero et al., 2008; Bartkowiak and Curran, 2015). Upon binding to its ligand (CD137L), 4-1BB signaling inhibits T-cell apoptosis via polyubiquitination-mediated signals through TNF receptor–associated factor 2 and also enhances T-cell proliferation and effector functions (Sanchez-Paulete et al., 2016).

Urelumab is a human IgG4 monoclonal antibody, which targets 4-1BB, and is being developed by Bristol Myers Squibb. Phase I safety data in patients with either advanced solid tumors or relapsed/refractory B-cell non-Hodgkin lymphoma (NCT01471210) established a maximum tolerated dose of 0.1 mg/kg given q2wk (Segal et al., 2017). The most common side effects from this trial included fatigue (16%), nausea (13%), decreased appetite (8%), and rash (5%), which were managed with topical corticosteroids and supportive care (Segal et al., 2017). Urelumab is currently in phase I/II clinical trials with nivolumab in patients with muscle-invasive urothelial carcinoma of the bladder (NCT02845323), solid tumors, B-cell non-Hodgkin lymphoma (NCT02253992), and advanced or metastatic malignant tumors (NCT02534506). It is also being investigated in early phase I trials in combination with nivolumab and adoptive cell therapy (ACT) (discussed in Section 4) in patients with metastatic melanoma (NCT02652455).

Utomilumab (also known as PF-05082566) is a humanized monoclonal antibody that binds to 4-1BB and stimulates signaling (Fisher et al., 2012). It is currently in phase I clinical trials in combination with rituximab for patients with solid tumors or lymphoma (NCT01307267). Utomilumab is also being investigated in a phase I trial in combination with an anti-OX40 antibody, PF-04518600, in patients with select advanced or metastatic carcinoma (NCT02315066). Other Ib/II trials are underway studying avelumab in combination with utomilumab and PF-04518600 for the treatment of patients with advanced malignancies (NCT02554812) and acute myeloid leukemia (NCT03390296). Utomilumab is also being evaluated as an adjuvant add-on therapy in combination with trastuzumab in patients with HER-2–positive breast cancer (NCT03364348). These molecules and the other costimulatory molecule therapies are detailed in Table 3.

3.2 OX40

OX40 (also known as CD134 and TNFRSF4) is an activating receptor that is absent on naïve T-cells but is expressed on activated T-cells 24–72 h after antigen stimulation (Lei et al., 2013). OX40 has one known ligand called OX40L (also known as CD252), which is a TNF family member and

Table 3 Selected Clinical Trials for Therapies Targeting Costimulatory Molecules

Costimulatory Molecule	Drug	Single Agent or in Combination	Indication	Phase of Trial	Clinical Trial Identifier
4-1BB	Urelumab	Nivolumab	Muscle-invasive urothelial carcinoma of the bladder	II	NCT02845323
4-1BB	Urelumab	Nivolumab	Advanced solid tumors and B-cell non-Hodgkin lymphoma	I/II	NCT02253992
4-1BB	Urelumab	Nivolumab	Advanced or metastatic malignant tumors	I	NCT02534506
4-1BB	Urelumab	Nivolumab and adoptive cell therapy	Metastatic melanoma	I	NCT02652455
4-1BB	Utomilumab	Rituximab	Advanced solid tumors or lymphoma	I	NCT01307267
4-1BB	Utomilumab	PF-04518600	Advanced or metastatic carcinoma	I	NCT02315066
4-1BB	Utomilumab	Avelumab, PF-04518600	Advanced malignancies	Ib/II	NCT02554812
4-1BB	Utomilumab	Avelumab, PF-04518600	Advanced malignancies	I/II	NCT03217747
4-1BB	Utomilumab	Avelumab, PF-04518600, cisplatin	Acute myeloid leukemia	II	NCT03390296
4-1BB	Utomilumab	Trastuzumab	HER-2 positive breast cancer	I	NCT03364348
OX40	GSK3174998	Pembrolizumab	Advanced solid tumors	I	NCT02528357
OX40	MOXR0916	Single agent	Locally advanced or metastatic solid tumors	I	NCT02219724
OX40	MOXR0916	Atezolizumab	Locally advanced or metastatic solid tumors	I	NCT02410512
OX40	MOXR0916	Atezolizumab	Metastatic urothelial carcinoma	II	NCT03029832

OX40	PF-04518600	Utomilumab	Advanced or metastatic carcinoma	I	NCT02315066
OX40	PF-04518600	Avelumab, utomilumab	Advanced malignancies	Ib/II	NCT02554812
OX40	PF-04518600	Avelumab, utomilumab	Advanced malignancies	I/II	NCT03217747
OX40	PF-04518600	Avelumab, utomilumab, cisplatin	Acute myeloid leukemia	II	NCT03390296
OX40	MEDI0562	Single agent	Advanced solid tumors	I	NCT02318394
OX40	MEDI0562	Tremelimumab, durvalumab	Advanced solid tumors	I	NCT02705482
OX40	MEDI6469	Single agent	Breast cancer patients with metastatic lesions	I/II	NCT01862900
OX40	MEDI6469	Single agent	Advanced or metastatic HNSCC	I	NCT02274155
OX40	MEDI6469	Cyclophosphamide	Metastatic prostate cancer	I/II	NCT01303705
OX40	MEDI6383	Durvalumab	Advanced solid tumors	I	NCT02221960
GITR	TRX518	Single agent	Advanced melanoma or other solid tumors	I	NCT01239134
GITR	TRX518	Gemcitabine, pembrolizumab, nivolumab	Advanced solid tumors	I	NCT02628574
GITR	MEDI1873	Single agent	Advanced solid tumors	I	NCT02583165
GITR	INCAGN01876	Single agent	Advanced or metastatic solid tumors	I/II	NCT02697591
GITR	INCAGN01876	Ipilimumab and nivolumab	Advanced or metastatic malignancies	I/II	NCT03126110
GITR	INCAGN01876	Pembrolizumab and epacadostat	Advanced or metastatic malignancies	I/II	NCT03277352

(Continued)

Table 3 Selected Clinical Trials for Therapies Targeting Costimulatory Molecules—cont'd

Costimulatory Molecule	Drug	Single Agent or in Combination	Indication	Phase of Trial	Clinical Trial Identifier
GITR	MK-1248	Pembrolizumab	Advanced solid tumors	I	NCT02553499
GITR	MK-4166	Pembrolizumab	Advanced solid tumors	I	NCT02132754
GITR	BMS-986156	Nivolumab	Advanced solid cancers or cancers that have spread	I/II	NCT02598960
CD40	CP-870,893	Tremelimumab	Metastatic melanoma	I	NCT01103635
CD27	Varlilumab	Single agent	Advanced solid tumors or hematologic cancers		NCT01460134
CD27	Varlilumab	Nivolumab	Advanced refractory solid tumors		NCT02335918
CD27	Varlilumab	Nivolumab	Relapsed or refractory aggressive B-cell lymphomas		NCT03038672

is expressed on activated APCs (Croft, 2010). OX40L is known to be expressed throughout the body at low levels in healthy individuals and expressed to a higher degree in patients with autoimmune disorders (Sugamura et al., 2004). OX40L-deficient DCs are not able to provide enough costimulatory signal to T-cells for cytokine production (Chen et al., 1999). OX40 signaling results in NF-κB1 activation in $CD4^+$ T-cells (Song et al., 2008). It also controls the expression of survivin, which is an inhibitor of apoptosis in T-cells (Song et al., 2005).

In preclinical studies, OX40 engagement by an anti-OX40 antibody was shown to produce robust tumor regression associated with costimulation of effector T-cells and reduction of Tregs (Weinberg et al., 2000).

Several molecules targeting OX40 are under development. GlaxoSmithKline initiated the first in human clinical trial for an anti-OX40 antibody in 2015 with their molecule, GSK3174998. GSK3174998 is a humanized IgG1 monoclonal antibody that is currently in phase I trials alone and in combination with pembrolizumab in patients with selected advanced or recurrent solid tumors (NCT02528357). MOXR0916, an OX40 agonist, has shown early signs of antitumor activity when combined with atezolizumab in patients with locally advanced or metastatic solid tumors (NCT02410512) and in patients with metastatic urothelial carcinoma (NCT03029832). MOXR0916 is also being investigated as a monotherapy in a phase I trial in patients with locally advanced or metastatic solid tumors (NCT02219724). Pfizer is currently developing an anti-OX40 antibody, PF-04518600, and is currently testing it in a phase II clinical trial in combination with the small molecule tyrosine kinase, axitinib, in patients with RCC (NCT03092856). PF-04518600 is also being investigated in a phase I trial in combination with utomilumab in patients with select advanced or metastatic carcinoma (NCT02315066). Studies of triple therapy with avelumab, utomilumab, and PF-04518600 are also currently underway in patients with AML (NCT03390296) and other advanced malignancies (NCT02554812, NCT03217747). AstraZeneca is developing three anti-OX40 antibodies, known as MEDI0562, MEDI6469, MEDI6383, and they are each in phase I/II trials in patients with various advanced or metastatic tumors (NCT02318394, NCT02705482, NCT01862900, NCT02274155, NCT01303705, NCT02221960).

3.3 Glucocorticoid-Induced TNFR Protein

Glucocorticoid-induced TNFR-related protein (GITR), also known as CD357 and TNFRSF18, is a costimulatory molecule that is a member of the TNF receptor family. GITR is broadly expressed on activated effector T-cells, NK cells, and B-cells (Shimizu et al., 2002). GITR is constitutively expressed on Tregs which express the transcription factor forkhead box P3 ($Foxp3^+$; Ko et al., 2005). Its ligand, GITRL, is expressed on APCs, B-cells, monocytes, macrophages, endothelial cells, and activated T-cells (Kwon et al., 1999; Gurney et al., 1999). Recent studies suggest that GITR costimulation mediates antitumor immunity by enhancing the differentiation of IL-9 producing $CD4^+$ T helper cells and increasing cytotoxic T lymphocyte responses (Kim et al., 2015). GITR signaling similar to other TNFR family members activates NF-κB and mitogen-activated protein kinases (MAPKs), which are essential to T-cell survival (Song et al., 2005).

Several drug candidates that target GITR are under development. TRX518 is a monoclonal antibody that targets GITR, and it is currently under investigation in a phase I trial as monotherapy for patients with unresectable stage III or stage IV melanoma or other solid tumor malignancies (NCT01239134). TRX518 is also being investigated in phase I clinical trials in combination with gemcitabine, pembrolizumab, nivolumab in patients with advanced solid tumors (NCT02628574). MEDI1873 is a GITR agonist developed by MedImmune LLC, which is currently in phase I trials for use as a single agent in adult patients with selected advanced solid tumors (NCT02583165). INCAGN01876 is an anti-GITR humanized monoclonal antibody being developed by Incyte Corporation and is currently being tested in phase I/II trials as a monotherapy in patients with advanced or metastatic solid tumors (NCT02697591). It is also being evaluated in combination with ipilimumab and nivolumab (NCT03126110) and with pembrolizumab and epacadostat (NCT03277352) in patients with advanced or metastatic malignancies. Merck is currently developing two GITR agonists, MK-1248 (NCT02553499) and MK-4166 (NCT02132754), and evaluating them in phase I trials in combination with pembrolizumab in patients with advanced solid tumors. Finally, Bristol-Myers Squibb is also developing an agonistic monoclonal antibody, which targets GITR called BMS-986,156. It is currently being evaluated in phase I/II trials in combination with nivolumab in patients with solid cancers that are advanced or cancers that have spread (NCT02598960).

3.4 CD40

CD40, also known as TNFRSF5, is a TNF receptor family member, which is known to be expressed on APCs, such as DCs, B-cells, and monocytes, and some nonimmune cells and a range of tumors (Vonderheide, 2007; Eliopoulos and Young, 2004). CD40 has a known ligand, CD40L or CD154, which is expressed predominately on activated T-cells (van Kooten and Banchereau, 2000). In preclinical studies, anti-CD40 monoclonal antibodies have been shown to activate APCs and generate antitumor T-cell responses (French et al., 1999). Anti-CD40 antibody treatment cured the majority of lymphoma-bearing mice and induced $CD8^+$ T-cell—dependent long-term immunity against lymphoma cells in cured animals (Tutt et al., 2002). Upon ligation with CD40L, CD40 recruits a group of adapter proteins known as TNFR-associated factors (TRAFs) to its cytoplasmic domain. The TRAF protein then activates several signaling pathways, including phosphoinositide 3-kinase (PI3K), the phospholipase Cγ (PLCγ) pathway, NF-κB, as well as MAPKs (Elgueta et al., 2009).

Pfizer is developing a fully human IgG2 isotype monoclonal antibody CP-870,893 or RO7009789, currently in clinical trials in combination with other immune checkpoint blockade inhibitors. Tremelimumab and CP-870,893 are currently being studied in patients with metastatic melanoma (NCT01103635). Early phase Ib results showed that CP-870,893 was safe and well-tolerated when given in combination with cisplatin and pemetrexed in patients with malignant pleural mesothelioma (Nowak et al., 2015). Dacetuzumab (also known as SGN-40 or huS2C6) is a humanized IgG1 monoclonal antibody being developed by Seattle Genetics and is known to be a weaker agonist than CP-870,893 (Vonderheide, 2007). In clinical trials, early results of dacetuzumab monotherapy showed activity in 25% of patients with relapsed/refractory diffuse large B-cell lymphoma. However, larger phase IIb studies failed to show a difference in complete response (trial endpoint) between dacetuzumab and placebo. Therefore further patient enrolment was stopped (Fayad et al., 2015).

3.5 CD27

CD27, also known as TNFRSF7, is a member of the TNF receptor superfamily and is expressed on the surface of naïve $CD4^+$ and $CD8^+$ T-cells (Nolte et al., 2009). CD27 acts as a costimulatory molecule for B-cells and T-cells. CD27 signaling activates NF-κB, which promotes B-cell and T-cell survival, proliferation, effector functions (Riether et al., 2012). Interestingly, there is evidence that CD27 is shed from the cell surface following T-cell activation, which is different from other TNFR family members (Huang et al., 2013). It has a single ligand, CD70, which is expressed on APCs following immune activation (Denoeud and Moser, 2011).

Varlilumab, previously known as CDX-1127, is a fully human monoclonal antibody, which targets CD27. It is currently being investigated as a monotherapy (NCT01460134) in phase I trials and in combination with nivolumab (NCT02335918, NCT03038672). Preliminary safety data have established that 10 mg/kg given intravenously once q4wk is the maximum tolerated dose (Burris et al., 2017). Adverse drug reactions were grade 1 and 2 in severity and included fatigue (54%), nausea (30%), dyspnea (25%), rash (20%), and diarrhea (18%) as the most common side effects (Burris et al., 2017).

3.6 Inducible T-Cell Costimulator

ICOS, also known as CD278, is expressed on the surface of activated T-cells. It is not a member of the TNF receptor family. However, it does act as a costimulatory molecule and expressed on T-cells (Yoshinaga et al., 1999). ICOS has a known ligand; B7-H2 (also known as ICOSL and CD275) that is expressed on activated monocytes and DCs (Aicher et al., 2000). ICOS is expressed on T-cells after TCR activation, and ICOS expression can also be stimulated by CD28 signaling (Jung and Choi, 2013). ICOS expression is upregulated in the presence of CTLA-4—blocking antibodies, upon T-cell activation and animals lacking ICOS or its ligand show deficient antitumor activity in the presence of CTLA-4—blocking antibodies (Fan et al., 2014).

GSK3359609 is a humanized IgG4 anti-ICOS receptor agonist, and it is currently in phase I clinical trials and being evaluated as a monotherapy and in combination with pembrolizumab in patients with selected advanced solid tumors (NCT02723955). Early safety data results show that this antibody is well-tolerated without dose-limiting toxicities (Angevin et al., 2017).

4. ADOPTIVE CELL THERAPY

Adoptive cell therapy is unique when compared with other forms of immunotherapy, as it involves isolating T-cells from patient's body, modifying and expanding them ex vivo, and transferring them back into the patient after lymphodepleting conditioning. It is also exceptional in the way that it is highly personalized to the patient, reprogramming their cells to combat their cancer. Furthermore, this type of treatment is remarkable as it has the potential to eliminate disease with a durable response after only one treatment (Rosenberg and Restifo, 2015). T-cells can be genetically manipulated to target cancer antigens through the insertion of a tumor-specific chimeric antigen receptor (CAR) or an antitumor TCR (Rosenberg and Restifo, 2015). Adoptive cell therapy can also utilize autologous TILs to target cancer cells (Rosenberg, 2011). In this section, we review the latest advances in adoptive cell therapy.

4.1 CAR T-Cell Therapy

CAR T-cell therapy represents one of the most promising advances in cancer treatments today. The American Society of Clinical Oncology named CAR T-cell therapy the advance of the year in oncology in its annual report in January 2018 (ASCO, 2018). CARs are genetically engineered receptors, which can selectively identify and bind to targets on the surface of tumor cells (Kochenderfer and Rosenberg, 2013). Early preclinical studies focused on targeting CD19 as an antigen because it is uniformly expressed on B-cell lymphomas, and its expression in healthy cells is restricted to B-cell precursors, mature B-cells, and plasma cells (Nadler et al., 1983; Scheuermann and Racila, 1995). Researchers demonstrated that CD19 CARs were capable of activating T-cells in a CD19-specific manner and subsequently eliminating $CD19^+$ primary B-ALL blast cells in vitro (Cooper et al., 2003). Mouse studies demonstrated that CD19 CARs could destroy CD19-expressing target cells in xenograft models (Brentjens et al., 2003).

The first use of an anti-CD19 CAR was in 2010 (Kochenderfer et al., 2010).

4.1.1 Structure and Mechanism of CAR T-Cells

CAR T-cells are composed of four main parts. First, there is an extracellular component called the single-chain variable fragment (scFv). The scFv is an antigen-recognition molecule incorporating the variable domains of the light and heavy chains from the structure of a monoclonal antibody (Eshhar et al., 1993). Next, there is the hinge for flexibility and distance and a transmembrane spacer, which anchors the CAR to the membrane. Third, there is a costimulatory molecule, located intracellularly, which has been comprised of 4-1BB and/or 28CD signaling domains (Kochenderfer et al., 2009; Carpenito et al., 2009). Finally, there is the TCR T3 zeta chain (CD3ζ) subunit, which is an activation-signaling domain, also located intracellularly (Irving and Weiss, 1991; Sadelain et al., 2013).

Fig. 3 illustrates the differences between the generations of CARs. The first generation of CAR T-cells only had a CD3ζ signaling domain and no costimulatory molecule. However, without a costimulatory molecule, the

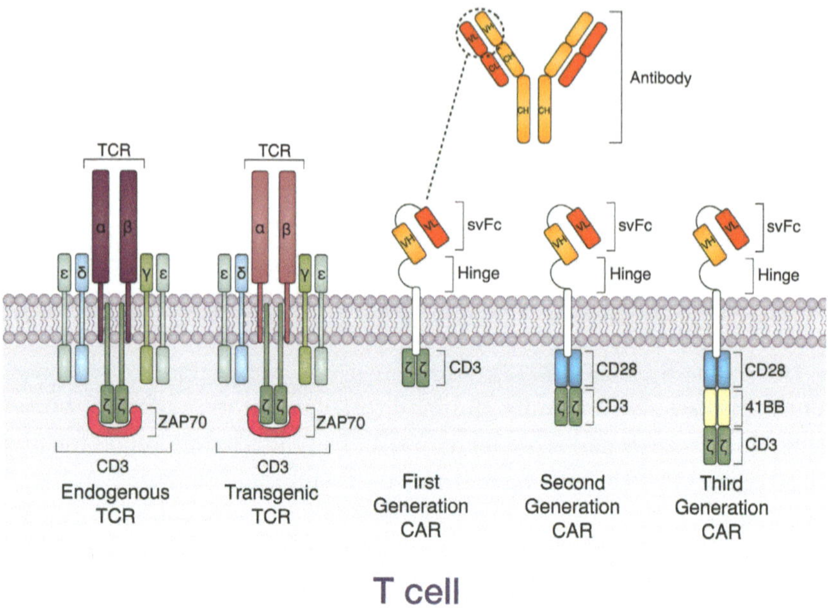

Figure 3 ***The structure of endogenous T-cell receptor (TCR), transgenic TCR, and the three generations of chimeric antigen receptors (CARs).*** In the transgenic TCR, the alpha and beta components are engineered to have retargeted specificity for tumors. The scFv is derived from the variable heavy (VH) and light (VL) antigen-binding domain of antibodies. A hinge connects the scFv to the CD3ζ transmembrane and intracellular signaling domains from the T-cell receptor complex in first-generation CARs. Second and third generations of CARs are produced by the addition of costimulatory signaling domains, including CD28 and/or 4-1BB signaling domains.

CAR T-cells were unable to achieve objective clinical antilymphoma responses (Jensen et al., 2010). The second generation of CAR T-cells combined the CD3ζ signaling domain with either the 4-1BB or CD28 signaling domains (Kochenderfer et al., 2009; Carpenito et al., 2009). Finally, the third generation of CAR T-cells included CD3ζ plus both CD28 and either 4-1BB or OX40 signalling domains (Milone et al., 2009).

One hallmark of the CAR T-cells is their MHC independence. CARs do not require the engagement of HLA-peptide complexes and can recognize target molecules without further peptide processing or HLA expression (Sadelain et al., 2009). CARs have the ability to recognize a tumor-associated antigen on any HLA background. This is in contrast to TCRs, which need to be matched to the patient's haplotype (Sadelain et al., 2013).

The gene encoding the CAR construct is inserted into the genomes of T-cells using a gene therapy vector. The two most common vectors are lentiviruses and gammaretroviruses (Suerth et al., 2012). Normally, T-cells are activated and incubated with the vectors encoding the CAR molecules for several days. Extra vectors are then removed from the media by dilution and/or media exchange. Upon cell entry, vectors express the CAR genetic information as RNA, which is then reverse transcribed into DNA and followed by permanent integration into the T-cells' genome (Levine et al., 2017). Lentivirus vectors are commonly used for CAR T-cell clinical trials because of their safer integration profile sites. In the CAR T-cell manufacturing process, the final T-cells are considered the final product, but the vectors are considered raw material that can be mass produced and stored in −80°C for years before the actual usage (Levine et al., 2017). Manufacturing the vectors takes place under Good Manufacturing Practice (GMP) conditions, and vectors need to be kept sterile since the final CAR T-cell product cannot be filtered for sterilization (Levine et al., 2017).

Fig. 4 illustrates production and clinical treatment using CAR T-cells. The first step in CAR T-cell therapy is leukapheresis, which is a procedure to separate and collect white blood cells. Patient blood is acquired through a needle that is placed into one arm and goes through a machine, and white blood cells are removed. The plasma and red cells are returned to the patient through a second needle in the other arm. The procedure takes approximately 4–6 h to complete. Next, the T-cells are cryofrozen and sent to a specialized GMP manufacturing facility. They are thawed, activated, and genetically engineered with a retrovirus or lentivirus encoding the CAR. CAR T-cells are further expanded and harvested. The patient undergoes a lymphodepletion regimen in preparation for the reinfusion. Finally, the patient is infused with the CAR T-cell treatment. The patient is carefully monitored in hospital for at least a week to watch for signs of cytokine release syndrome.

4.1.2 Tisagenlecleucel

Tisagenlecleucel (trade name: Kymriah, also known as CTL019) is an anti-CD19 CAR T-cell therapy developed by Novartis and the University of Pennsylvania. It was first FDA approved on August 30, 2017, for children and young adults up to age 25 years with B-cell precursor ALL that is refractory or in second or later relapses. Tisagenlecleucel is the first CAR T-cell

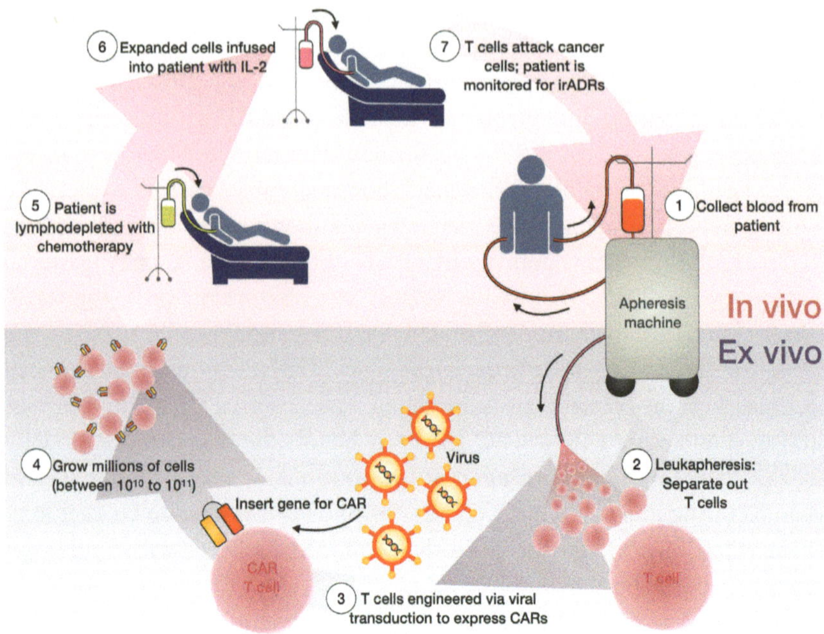

Figure 4 ***The production and clinical use of CAR T-cells.*** (1) Patient blood is collected, the leukocytes are removed while red blood cells, and plasma are returned to the patient through the process of leukapheresis. (2) T-cells are isolated. (3) Genetic information encoding the CARs is inserted into the T-cell via viral transduction. (4) Cells are grown in culture with IL-2 to yield millions of T-cells. (5) Patient has undergone lymphodepletion with chemotherapy, such as cyclophosphamide. (6) The patient is infused with the CAR T-cells. and (7) The patient is carefully monitored in hospital for the development of immune-related adverse drug reactions (irADRs). *CARs*, chimeric antigen receptors; *IL*, interleukin.

immunotherapy approved by the FDA and generated through a 22-day process. It targets CD19 using a scFv domain derived from a FMC63 hybridoma and utilizes lentiviral transduction (Kalos et al., 2011). The hinge and transmembrane domains are derived from human CD8-alpha. The costimulatory domain of tisagenlecleucel is derived from human 4-1BB (Milone et al., 2009).

The initial results of a global multicentric registration trial of tisagenlecleucel demonstrated 83% complete response rate in 63 pediatric and young adult patients with relapsed and refractory B-ALL (Buechner et al., 2017).

A small case study cohort of adult patients with lymphoma received tisagenlecleucel, and 18 of 28 had a response (64% [95% CI, 44%–81%])

(Schuster et al., 2017). Tisagenlecleucel continues to be investigated in a phase II clinical trial of pediatric ALL patients (NCT02228096).

4.1.3 *Axicabtagene Ciloleucel*

Axicabtagene ciloleucel (trade name: Yescarta, also known as KTE-C19 and axi-cel) is an anti-CD19 CAR T-cell therapy developed by Kite Pharma. It is the first CAR T-cell approved by FDA for adult patients with relapsed or refractory large B-cell lymphoma (October 18, 2017). Axicabtagene ciloleucel is generated over 17 days (Neelapu et al., 2017). The extracellular anti-CD19 antibody domain is derived from FMC63 (Kochenderfer et al., 2009). The hinge, transmembrane domain, and costimulatory domain of axicabtagene ciloleucel are derived from human CD28 (Kochenderfer et al., 2009).

Axicabtagene ciloleucel is currently in phase II clinical trials for adult patients with refractory aggressive non-Hodgkin Lymphoma (ZUMA-1, NCT02348216). Results from the multicenter, ZUMA-1 trial of 111 patients demonstrated clinical significance of CD19-directed CAR in patients with refractory lymphomas (Neelapu et al., 2017). Axicabtagene ciloleucel is also under investigation in adult patients with relapsed/refractory indolent non-Hodgkin lymphoma (NCT03105336), relapsed/refractory ALL (ZUMA-3, NCT02614066), and relapsed/refractory diffuse large B-cell lymphoma (DLBCL; ZUMA-7, NCT03391466). It is also being investigated for use in pediatric and adolescent patients with relapsed/refractory ALL (ZUMA-4, NCT02625480). There is a study, which is investigating the use of atezolizumab in combination with axicabtagene ciloleucel for patients with DLBCL (ZUMA-6, NCT02926833).

4.1.4 *Additional CAR T-Cell Targets*

CD22 is an additional B-cell marker, which may have therapeutic potential in CAR T-cells. Preclinical studies have demonstrated that lymphoblasts from patients with precursor B-cell ALL express CD22 and that anti-CD22 CAR T-cells demonstrated antileukemic activity (Haso et al., 2013). Anti-CD22 CAR T-cells are currently in a phase I clinical trial in pediatric and young adults with recurrent or refractory CD22-expressing B-cell malignancies (NCT02315612).

(Additional trials for CD22 - NCT03398967, NCT03330691, NCT02935153, NCT03262298)

CD30 is a transmembrane receptor, which is highly expressed on malignant cells in Hodgkin Lymphoma (Durkop et al., 1992). Anti-CD30 CAR T-cells are currently under investigation in phase I clinical trials in patients with $CD30^+$ lymphoma (NCT03049449, NCT02274584, and NCT02690545).

(Additional trials for CD30 - NCT02917083, NCT02958410, NCT02259556, NCT03383965)

4.1.5 CAR T-Cells in Solid Tumors

While CAR T-cells have had recent success in blood-based cancers, there has been less progress in solid tumors due to various barriers that exist in solid tumors that are absent in hematological malignancies. It has been more challenging to identify specific, highly expressed antigens in solid tumors that are not expressed on the surface of normal tissues. Furthermore, there are more barriers for CAR T-cells trafficking to a tumor and infiltrating tumor stroma. Additionally, the hostile tumor microenvironment and presence of immunosuppressive molecules and immunoregulatory cells such as Tregs and MDSCs will further diminish the ability of CAR T-cells to successfully exert cytotoxicity against tumor cells (Newick et al., 2017). Antigen specificity is crucial to the safety of these powerful live drugs. Most solid tumor neoantigens are intracellular and not accessible to CARs that recognize cell surface antigens (Newick et al., 2017). Therefore, finding a proper antigen for CAR targeting is limited to cell surface antigens such as oncofetal antigens and aberrantly expressed tumor-selective antigens that are overexpressed on malignant cells but have lower expression levels on normal cells. For example, epidermal growth factor receptor (EGFR) variant 3 is an oncofetal protein that is expressed on glioblastoma cells and not normal cells and EGFRvIII CAR T-cells are currently being tested in patients with glioblastoma (NCT02209376 and NCT01454596; Newick et al., 2017).

Mesothelin is an example of a glycoprotein that is overexpressed in several cancers, including mesothelioma, pancreatic, and ovarian carcinomas and mesothelin-specific CARs are in clinical studies (NCT02159716 and NCT02414269). A significant limitation to targeting tumor-selective antigens, however, is the degree of antigen heterogeneity in the tumor population and the potential escape clones with low or lack of antigen expression. Nevertheless, there are many ongoing studies to try address all the potential issues with bringing CAR T-cell into the realm of solid tumors (Newick et al., 2017).

4.2 Transducing Cells with T-cell Rreceptor Therapy

Another promising avenue of adoptive cell therapy is genetically engineered TCR therapy, also known as transgenic TCR therapy. These are synthetic receptors that maintain the structure of the native TCR but are designed for antigen selectivity and high affinity. In TCR therapy, T-cells are genetically modified by introducing genes which encode natural αβ TCRs in order to generate antigen-specific T-cells.

The endogenous TCR is expressed on the surface of T lymphocytes. TCR is responsible for recognizing antigenic peptides presented by MHC molecules. In most (95%) human T-cells, the TCR is comprised of an α chain and a β chain, while in the other 5%, the TCR is comprised of a γ chain and a δ chain. TCR therapy utilizes the native α and β chains of a TCR specific for the tumor antigen. The TCR is comprised of an α/β heterodimer and has the ability to redirect the tumor antigens in an MHC-dependent manner (Rudolph et al., 2006).

Intracellularly processed antigens can be recognized by a transgene-encoded TCR expressed by T-cells, which is an advantage for TCR therapy over CAR T-cells in the treatment of solid tumors. On the other hand, the MHC dependence of TCR therapy could be a disadvantage when facing tumor cells that downregulate their MHC I molecules. Another advantage of TCR T-cells is their high sensitivity to low levels of surface antigens as oppose to CAR T-cells need of high levels of surface antigen (Ping et al., 2017; Fry et al., 2018).

Tumor antigen specificity is crucial for the safety of TCR therapies, as "*on-target off-tumor*" toxicities can be fatal. Tumor-specific antigens and tumor neoantigens can be used as targets for TCR therapy. Tumor antigens, such as melanoma-associated antigen (MAGE)-A3, MAGE-A4, melanoma-associated antigen recognized by T-cells, and New York esophageal squamous cell carcinoma (NY-ESO)-1, have been tested in preclinical and clinical studies of TCR therapy (Ping et al., 2017). Often to identify naturally occurring neoantigen-specific TCRs patient tumors are subjected to whole exome sequencing and RNA sequencing to determine neoantigen-forming mutations. Often mutated gene products (12–24 mutated amino acids flanked by several amino acids of the wild-type protein) are synthesized using tandem minigene constructs. Those tandem minigenes are transfected into patient's autologous DCs, which are then cultured—in vitro—with patient's TILs. TILs producing IFN-γ and expressing CD137 (a marker of recent activation) are sorted and expanded in vitro.

The α and β TCRs of the enriched T-cell population are isolated and identified using various genomics and sequencing methods. The patient's PBMCs are then transduced with those TCRs using retroviral or lentiviral vectors then evaluated for the recognition of the relevant neoantigens. These cells are then expanded and infused back to the patient (Parkhurst et al., 2017).

4.3 Tumour-infiltrating Llymphocytes Therapy

TILs are immune cells, which have migrated into the microenvironment of a tumor. Early studies demonstrated that the presence TILs in tumors was often associated with better clinical outcomes (Vanky et al., 1986; Zhang et al., 2003). TILs were identified and isolated within surgically resected melanoma samples, and this led to the use of antigen-targeted T-cells for immunotherapy (van der Bruggen et al., 1991; Rosenberg, 1999).

The process for generating TIL therapy begins by resecting a tumor sample and growing small fragments of it in an IL-2 solution (Rosenberg and Restifo, 2015). IL-2 causes the lymphocytes to grow and overcome the tumor cells until only lymphocytes remain in culture. These cells are then tested to see if they are reactive to specific tumor cells and then selected and grown further to between 10^{10} and 10^{11} cells. The whole process takes approximately 5–6 weeks to generate the cells (Rosenberg and Restifo, 2015). Patients are then given lympho-depleting chemotherapy and then reinfused with the TIL therapy and IL-2 (Dudley et al., 2002).

The first use of TIL therapy and IL-2 was in a clinical trial of 20 patients with metastatic melanoma and produced higher rates of objective response than treatment with IL-2 administered alone or with lymphokine-activated killer cells (Rosenberg et al., 1988). Other early clinical trials demonstrated that the treatment of metastatic melanoma with autologous TIL transfer and IL-2 resulted in objective response rates of 50%–70% in patients whose disease was refractory to standard therapies (Dudley et al., 2008). While TIL therapy has resulted in durable and complete responses in patients with metastatic melanoma, and it presents exciting treatment options for patients with unresectable and refractory tumors, TIL therapy antitumor effects may be limited due to T-cell exhaustion (Wherry and Kurachi, 2015). Recently researchers have developed innovative approaches to select tumor-specific TILs (Parkhurst et al., 2017).

While TIL therapy has had the most significant success in patients with metastatic melanoma, TIL therapy clinical trials are currently underway in multiple different tumor types. TIL therapy is being evaluated in phase I/II trials in patients with metastatic RCC (NCT02926053) and pleural mesothelioma (NCT02414945). LN-145 is an autologous TIL therapy developed by Iovance Biotherapeutics currently in phase II trials in patients with recurrent or metastatic HNSCC (NCT03083873) and recurrent, metastatic, or persistent cervical carcinoma (NCT03108495). Several immunotherapies are being evaluated in combination with TIL therapy (summarized in Section 6.4).

4.4 Adoptive T-Cell Therapy Toxicity

Manipulating the immune system through CAR T-cells has given rise to novel and potentially life-threatening toxicities.

4.4.1 Cytokine-Release Syndrome

Cytokine-release syndrome (CRS) is characterized by high fever, hypotension, hypoxia, and/or multiorgan toxicity (Davila et al., 2014). CRS is the most common acute toxicity of CAR T-cells, and it is largely reversible within 2 weeks. Increased CRS grade has been associated with increased serum levels of many cytokines including IL-2R, IL-6, and with and higher levels of blood CAR T-cells (Maude et al., 2014). Patients who develop CRS symptoms within 3 days of infusion with CAR T-cells, those with comorbidities, and patients a high disease burden at the time of treatment show an association with severity of CRS (Maude et al., 2014; Lee et al., 2015).

4.4.2 CAR T-Cell–Related Encephalopathy Syndrome

CAR T-cell–related encephalopathy syndrome (CRES) is characterized by a toxic encephalopathic state with symptoms of confusion and delirium, occasionally seizures, and cerebral edema (Davila et al., 2014). The early symptoms of CRES are language disturbance, being decreased attention, and difficulties with handwriting. CRES symptoms can be biphasic, with the first phase coinciding with the onset of CRS and the second phase after the CRS symptoms have subsided.

4.4.3 Fulminant Hemophagocytic Lymphohistiocytosis

Hemophagocytic lymphohistiocytosis (HLH) also known as the macrophage-activation syndrome (MAS) is a rare side effect of CAR T-cell therapy (happens to ~1% of treated patients) that is characterized by severe immune activation, lymphohistiocytic tissue infiltration, and immune-mediated multiorgan failure. HLH is caused by hyperactivation of immune cells, mainly macrophages and lymphocytes, immune tissue infiltration, and proinflammatory cytokine production. Patients with HLH often have similar symptoms as CRS (Neelapu et al., 2018).

4.4.4 Management of ACT Toxicity

The same day as the FDA approved kymriah, tocilizumab (ACTEMRA, Genentech Inc.) was also approved for the treatment of patients 2 years of age or older with CRS that occurs with CAR T-cell therapy. In an analysis of data from clinical trials of CAR T-cells, 69% of patients with severe or life-threatening CRS had resolution of CRS within 2 weeks following one or two doses of tocilizumab. Tocilizumab is an anti—IL-6 receptor antagonist (Neelapu et al., 2018). Prophylactic administration of tocilizumab has been suggested before (Maude et al., 2015). Currently, researchers at the University of Pennsylvania studying the effect of early treatment with tocilizumab on CAR T-cell associated CRS and efficacy (NCT02906371). Corticosteroids are also effective in managing CRS. However, they can be toxic to the infused cells and may limit the efficacy of the therapy. CRES and MAS are also generally managed similarly to CRS.

4.5 Financial Toxicity of Adoptive Cell Therapy

The financial burden of cancer treatment is a reality in our world and one that needs to get more attention. A cancer patient may survive cancer due to the advances in technology and novel drugs but then may experience long-lasting adverse effects of financial responsibility/debt for the cost of the treatment.

ACT can be indeed a useful tool in treating some malignancies; however, these treatments are very costly, which makes them challenging to become mainstream for oncology patient management in most countries or even in the United States where many patients do not have insurance. Moreover, patients who do have insurance coverage still might need to pay a copayment cost that could be high for many low-income families. A single injection of tisagenlecleucel costs $475,000 USD before adding the hospital costs. In response to criticism of the high cost, Novartis has stated that this is cheaper than some bone marrow transplants. Novartis has also stated that it will not charge patients who do not respond to the therapy within a month. However, this will not include the patients whose cancer responds to the treatment within the first month and then fails therapy. A single injection of axicabtagene ciloleucel costs $373,000 USD. Since TCR therapy and TIL therapy are still experimental, there is no actual price for these treatments. There is a substantial need to reduce the price of these promising cancer treatments to make them accessible to patients. We need to keep in mind that an unaffordable effective drug is similar to not having a drug.

5. ONCOLYTIC VIRUS CELL THERAPY

Oncolytic immunotherapy employs the use of modified viruses to target and lyse cancer cells (Liu et al., 2007). Oncolytic viruses can kill cancer cells through direct oncolysis (Mullen and Tanabe, 2002). They also trigger apoptosis in neighboring, uninfected cancer tissue (Stanziale et al., 2004). Oncolytic viruses have also been shown to target cancer cells through the activation of immune response, particularly T-cells.

5.1 Talimogene Laherparepvec

Talimogene laherparepvec (T-VEC; trade name: Imlygic) is an attenuated herpes simplex virus, type 1 (HSV-1), which is engineered to express granulocyte-macrophage colony-stimulating factor (GM-CSF; Liu et al., 2003). T-VEC is modified in a number of ways, including deletion of the herpes virus neurovirulence factor gene (ICP34.5), which eliminates the neuropathogenicity typically associated with HSV-1 infection, while also increasing the oncolytic potential of the virus (Liu et al., 2003). The infected cell protein 47 gene was also removed as it is known to block $CD8^+$ T-cell recognition of infected cells by inhibiting antigen presentation (Goldsmith et al., 1998). By adding the gene which encodes GM-CSF, local GM-CSF production is increased within the tumor microenvironment, and this recruits and activates APCs, which leads to tumor-specific T-cell responses (Kaufman et al., 2014). T-VEC is injected directly into the lesion and replicates within tumor cells. As the tumor cells are infected, they lyse, causing necrosis and cell death. This causes tumor-associated antigens to be released, leading to priming of endogenous anti-tumor T-cell responses (Kaufman et al., 2014).

The first phase I trial on T-VEC was conducted in 2006 on 30 patients with different cancers including breast, head and neck, colorectal, and melanoma who had tumors that were amenable to direct injection of the virus. Mild pyrexia was the most common adverse event followed by local inflammation at the site of injection. In a phase II study of T-VEC in 50 melanoma patients, five patients achieved a partial response, and eight patients showed a complete response to treatment (26% overall response based on RECIST; Rehman et al., 2016). In October 2015 based on the positive results (26% ORR) of phase III study of T-VEC in unresectable melanoma patients (OPTiM trial), US FDA approved T-VEC as the first oncolytic virus therapy for melanoma (Greig, 2016). This was a historic approval for the field of oncolytic virus therapy in cancer. T-VEC therapy had a favorable safety profile with only 2.1% of patients treated with T-VEC experiencing cellulitis as grade 3 and 4 adverse events (Rehman et al., 2016).

Combination of T-VEC with pembrolizumab has shown to be safe and associated with high response rates (62% overall response) in a phase Ib study (Ribas et al., 2017). Importantly, response to combination therapy was not limited to tumors with high $CD8^+$ infiltration and IFN-γ signature (Ribas et al., 2017). This indicates that T-VEC therapy can be used to treat noninflamed tumors (normally not responsive to anti–PD-1 therapy) in combination with an anti–PD-1 agent such as pembrolizumab.

5.2 Additional Oncolytic Viruses

Currently, there are also several promising nonherpes virus oncolytics under investigation in clinical trials. Pexastimogene devacirepvec, also known as Pexa-Vec and JX-594, is an oncolytic virus derived from Wyeth strain vaccinia virus. It is engineered to express transgenes encoding GM-CSF (Kim et al., 2006). Early phase I trial data demonstrated that Pexa-Vec has a tolerable safety profile (Breitbach et al., 2011), and it is now being investigated in phase I and II trials in combination with various ICIs. Reolysin, also known as pelareorep, is a live, replication competent, unmodified human reovirus which has selectively cytotoxic effects in tumor cells with an RAS signaling pathway (Chakrabarty et al., 2015). As RAS mutations are observed in up to 90% of pancreatic cancer, Reolysin may offer a new treatment modality for this difficult to treat tumor type (Jones et al., 2008). Phase I and II trials of Reolysin have shown safety and tolerability, but additional studies are underway to establish effectiveness when used in combination with other treatments (Noonan et al., 2016). Finally, the MG1 Maraba/MAGE-A3 (MG1MA3) is a new therapeutic virus, which has shown promising results in preclinical models of sarcoma (Le Boeuf et al., 2017). Phase I clinical trials examining the Maraba virus are currently recruiting patients with NSCLC (NCT02879760) and other advanced and metastatic tumors (NCT02285816). Table 4 summarizes current trials of oncolytic viruses.

Table 4 Selected Clinical Trials of Oncolytic Viruses

Virus Type	Oncolytic Virus	Single Agent or in Combination	Indication	Phase of Trial	Clinical Trial Identifier
HSV	T-VEC	Single agent	Pancreatic cancer	I	NCT03086642
HSV	T-VEC	Single agent	Children with non-CNS tumors	I	NCT02756845
HSV	T-VEC	Single agent	Breast cancer local recurrence	II	NCT02658812
HSV	T-VEC	Paclitaxel	Triple negative breast cancer	I/II	NCT02779855
HSV	T-VEC	Ipilimumab	Unresected melanoma	Ib/II	NCT01740297
HSV	T-VEC	Pembrolizumab	Metastatic or locally advanced sarcoma	II	NCT03069378
HSV	T-VEC	Pembrolizumab	Recurrent or metastatic HNSCC	Ib	NCT02626000
Vaccinia	Pexa-Vec	Nivolumab	Hepatocellular carcinoma	I/II	NCT03071094
Vaccinia	Pexa-Vec	Tremelimumab and durvalumab	Refractory colorectal cancer	I/II	NCT03206073
Vaccinia	Pexa-Vec	Ipilimumab	Metastatic and advanced solid tumors	I	NCT02977156
Reovirus	Reolysin	Chemotherapy and pembrolizumab	Advanced pancreatic adenocarcinoma	I	NCT02620423
Reovirus	Reolysin	Lenalidomide or pomalidomide	Relapsed and refractory multiple myeloma	I	NCT03015922
Reovirus	Reolysin	Bevacizumab	KRAS mutant metastatic colorectal cancer	I	NCT01274624
Reovirus	Reolysin	Sargramostim	High-grade relapsed or refractory brain tumors	I	NCT02444546
Maraba	MG1-MAGEA3	Single agent	Advanced and metastatic solid tumors	I	NCT02879760
Maraba	MG1-MAGEA3	Pembrolizumab	NSCLC	I	NCT02285816

6. COMBINATION THERAPY

Combination therapy may result in the superior patient outcome, as it can target multiple aspects of the biology of the tumor. However, safety is the primary concern when considering combination treatment of two or more drugs in oncology. Combination treatments should be based on scientific rationale and guided by carefully designed preclinical studies. However, due to the limitations of animal models to entirely recapitulate human cancers and antitumor immunity, such combinations need to be tested in human trials for safety and efficacy before widespread use in community settings. There are several factors to consider when designing combination studies, including the type of drugs, their doses, and scheduling. For example, if a chemotherapy agent is used to kill cancer cells and release tumor antigens to be recognized by the immune system, then that chemotherapy drug should be administered before the immunotherapy agent.

6.1 Combination of Checkpoint Inhibitors

Therapy that utilizes a combination of checkpoint inhibitors can be promising if designed based on scientific evidence and not merely due to the availability of the drugs and the desire of the drug companies to test the combination in the clinic. There is a great deal of complexity in the regulation and expression of multiple immune checkpoints and their ligands, and it is possible that antitumor immune response will involve more than one checkpoint pathway. There is also evidence that combination therapies may activate previously "noninflamed" or nonimmunogenic tumors.

6.1.1 CTLA-4 and PD-1 Combination

CTLA-4 antibodies are known to increase the number of T-cells within tumors and induce an IFN-γ signature in a tumor. This can stimulate expression of PD-L1 in the tumor microenvironment and increase the chance of benefit from PD-1/PD-L1 antibody therapies (Sharma and Allison, 2015a). A recent meta-analysis found that the combination of ipilimumab and nivolumab in melanoma and NSCLC patients showed a greater benefit in PFS and ORR but was also associated with higher toxicity (Wu et al., 2017).

There are currently more than 100 clinical trials, which are examining the combination of ipilimumab and nivolumab. CheckMate 067 study showed that combination of ipilimumab with nivolumab had a significantly better OS than nivolumab alone in patients with metastatic melanoma. While the combination arm had higher response rate, it was also associated with higher levels of toxicity (Wolchok et al., 2017). The CheckMate 227 trial (NCT02477826) that compared the combination effect of ipilimumab and nivolumab in stage 4 NSCLC patients with high tumor mutation burden (TMB) showed better 1-year PFS among patients in the combination immunotherapy arm (42.6%) than patients who received chemotherapy (13.2%; Hellmann et al., 2018). Moreover, this trial indicated that high TMB could be used as a clinically relevant biomarker of response to the combination of immunotherapy. Therefore, the predictive of TMB should be investigated in other immunotherapy trials. CheckMate 214 trial (NCT02231749) showed significantly higher OS for the combination of ipilimumab and nivolumab in untreated advanced RCC patients with intermediate- and poor-risk status compared with sunitinib (Motzer et al., 2018).

Several phase III trials of nivolumab and ipilimumab are currently being done in patients with advanced gastric cancer (NCT02872116), advanced or metastatic melanoma (NCT03068455, NCT01844505, NCT02905266), recurrent glioblastoma (NCT02017717), recurrent or metastatic HNSCC (NCT02741570), and inoperable or metastatic urothelial cancer (NCT03036098).

6.1.2 CTLA-4 and PD-L1 Combination

A phase IB study evaluated durvalumab combined with tremelimumab (Study 006, NCT02000947) in 102 treatment naïve patients with locally advanced or metastatic NSCLC. The maximum tolerated dose of the combination therapy was 20 mg/kg of durvalumab with 1 mg/kg of tremelimumab, q4wk (Antonia et al., 2016). Initial results of the trial were promising; 6 of 26 patients (23% [95% CI 9–44]) patients in the combined tremelimumab 1 mg/kg cohort achieved objective responses by investigator assessment (Antonia et al., 2016). The combination of tremelimumab and durvalumab is currently being evaluated as a first-line therapy in the open-label, multicenter, global Phase III NEPTUNE trial (NCT02542293) with the goal of enrolling 800 treatment-naïve patients with advanced or metastatic NSCLC. There is also a multicenter, global phase III trial of tremelimumab and durvalumab in patients with stage 4 urothelial cancer, which is currently enrolling (NCT02516241). Tremelimumab and durvalumab combination therapy is also being evaluated in phase III trials in patients with recurrent or metastatic HNSCC (NCT02369874). The combination is also being evaluated in phase III trials as a first-line treatment in patients with unresectable hepatocellular carcinoma (NCT03298451) and metastatic pancreatic cancer (NCT02879318).

6.2 Combination of Checkpoint Inhibitors With Traditional Chemotherapy

Resistance to traditional chemotherapy is a major issue in treating cancer, especially in advanced and metastatic tumors. T-cells have the potential to assist in addressing this problem. There is also evidence that chemotherapy may improve the effectiveness of immunotherapies by inducing immunogenic cell death in tumor cells as well as decreasing the number of tumor cells, Tregs, and MDSCs thus reducing the competition of these cells with T-cells for shared nutrients (Bracci et al., 2014). In October 2017, US FDA has approved the combination of pembrolizumab (anti–PD-1) and platinum-doublet chemotherapy (carboplatin and pemetrexed) as first-line treatment for patients with metastatic nonsquamous NSCLC. This approval was based on the phase II KEYNOTE-021 study that showed combination was safe and more effective than chemotherapy alone. Indeed, 55% of patients in the combination arm have achieved an objective response compared with 29% response from patients in the chemotherapy arm (Langer et al., 2016). A phase III study (KEYNOTE-189) in 616 patients with metastatic nonsquamous NSCLC with wild-type *EGFR* and *ALK* who received combination of chemotherapy (pemetrexed and a platinum-based drug) and 200 mg of pembrolizumab or placebo demonstrated superior survival benefit in patients in pembrolizumab combination arm (12 months OS of 69.2%; 95% CI, 64.1–73.8) versus placebo combination (12 months OS of 49.4%; 95% CI, 42.1–56.2). Notably, OS benefit was observed across all PD-L1 categories (Gandhi et al., 2018).

6.3 Combination of Checkpoint Inhibitors With NK-Activating Molecules

6.3.1 Killer-Cell Immunoglobulin-Like Receptor and PD-1

The killer-cell immunoglobulin-like receptor (KIR) family proteins are expressed on the membrane of NK cells and T-cells (Bashirova et al., 2006). KIR proteins interact with MHC class I molecules and help to regulate the function of NK cells (Handgretinger et al., 2016). Some KIRs have been found to have both inhibitory (MHC molecules suppress cytotoxic

NK cells) and while others have shown activating (MHC molecules activate the cytotoxic activity of NK cells) functions (Long et al., 2013). NK cells use inhibitory KIRs for "self"-MHC class I molecules to maintain a state of responsiveness and to kill cells which have lost MHC class I molecules. Healthy cells are spared from destruction as they expressed the self-MHC class I molecules (Kim et al., 2005). Control of NK cell activity is regulated by the balance of inhibitory, activating, and costimulatory receptors (Long et al., 2013).

Lirilumab is an anti-KIR human monoclonal antibody developed by Bristol Myers Squibb, which targets the KIR2DL-1, -2, -3 proteins, all inhibitory receptors, specifically and blocks interaction with their HLA ligands. The strategy of lirilumab is to disrupt the suppressive activity of KIR-HLA binding using antibody blockade and increase the number of activated effector cells (Romagne et al., 2009). It is currently being investigated in a phase II study in combination with azacytidine in patients with relapsed AML (NCT02399917), and early results have shown that the combination is well-tolerated (Daver et al., 2017). Lirilumab is also being evaluated in combination with nivolumab as an adjuvant treatment for patients with relapsed, resectable squamous cell carcinoma of the head and neck (NCT03341936). Lirilumab was evaluated in a double-blind placebo-controlled phase II trial as maintenance therapy in patients who were in first complete remission. The results of this study were presented at the 2017 American Society of Hematology meeting, and the authors found that while lirilumab was well tolerated, lirilumab treatment did not result in significantly increased leukemia-free survival (Vey et al., 2017).

6.4 Combination of Checkpoint Inhibitors With Adoptive Cell Therapy

Currently, there are several trials, which pair checkpoint inhibitors with transgenic TCR therapy and TIL therapy.

One study is investigating the use of NY-ESO-1 reactive TCR therapy with nivolumab in patients with stage IV or locally advanced solid tumors expressing NY-ESO-1 (NYM; NCT02775292).

A phase I/II trial is investigating TIL therapy in combination with nivolumab and ipilimumab in patients with metastatic ovarian cancer (NCT03287674). A phase I trial is underway for TIL therapy in combination with nivolumab in patients with advanced NSCLC (NCT03215810).

7. CONCLUSION

Cancer immunotherapy is now considered the fourth pillar of oncology and is helping patients across the cancer spectrum. While many patients benefit from treatment with ICIs, the majority of patients still do not respond to these drugs. The challenge facing clinicians and scientists in academia and industry is the identification of novel immunotherapy drugs or combinations of currently available drugs with immunotherapeutics to improve patient's outcome in a larger group of patients.

ICIs are currently the most successful class of immunotherapy drugs that have already received regulatory approvals from the US FDA, Health Canada, European Medicine Agency (EMA), and other regulatory agencies around the world. These drugs mostly enhance the activity of T-cells by removing the negative effect of checkpoint molecules on these cells. Here, we discussed the already approved ICIs and their mechanism of action. We also discussed novel agents that activate T-cells by through the costimulatory signaling pathways, such as 4-1BB and OX40 agonists. We examined novel agents that target other coinhibitory molecules, such as TIM-3 and TIGIT. We then changed our focus to adoptive cell therapy and its poster child CAR T-cell therapy.

Perhaps the highest potential for immunotherapy in the future will materialize from combining CAR T-cell therapy with various forms of immunotherapy and/or other conventional treatment modalities. However, this needs to be tailored to each patient's immune system and their cancer. Most immunotherapies today are focused on T-cells and more specifically $CD8^+$ T-cells. When designing a treatment for a patient that involves immunotherapy, we need to remember that patient's T-cells are the main agent of therapy and treatments need to be designed around them. More importantly, when dealing with a biological agent, a patient's own physiology needs to be in the forefront of our mind if we aim not only to eradicate cancer but also to induce long-term immunity against that cancer in patient's body.

REFERENCES

Agrawal, S., Feng, Y., Roy, A., Kollia, G., Lestini, B., 2016. Nivolumab dose selection: challenges, opportunities, and lessons learned for cancer immunotherapy. J. Immunother Cancer 4, 72.

Ahamadi, M., Freshwater, T., Prohn, M., Li, C.H., de Alwis, D.P., de Greef, R., Elassaiss-Schaap, J., Kondic, A., Stone, J.A., 2017. Model-based characterization of the pharmacokinetics of pembrolizumab: a humanized anti-PD-1 monoclonal antibody in advanced solid tumors. CPT Pharmacometrics Syst. Pharmacol. 6, 49–57.

Aicher, A., Hayden-Ledbetter, M., Brady, W.A., Pezzutto, A., Richter, G., Magaletti, D., Buckwalter, S., Ledbetter, J.A., Clark, E.A., 2000. Characterization of human inducible costimulator ligand expression and function. J. Immunol. 164, 4689–4696.

Angevin, E., Barnette, M.S., Bauer, T.M., Cho, D.C., Ellis, C.E., Gan, H.K., Hansen, A.R., Hoos, A., Jewell, R.C., Katz, J., Martin-Liberal, J., Maio, M., Mayes, P.A., Millward, M.J., Moreno, V., Olszanski, A.J., Rischin, D., Schellens, J.H.M., Yadavilli, S., Zhou, H., 2017. INDUCE-1: a phase I open-label study of GSK3359609, an ICOS agonist antibody, administered alone and in combination with pembrolizumab in patients with advanced solid tumors. J. Clin. Oncol. 35, TPS3113.

Antonia, S., Goldberg, S.B., Balmanoukian, A., Chaft, J.E., Sanborn, R.E., Gupta, A., Narwal, R., Steele, K., Gu, Y., Karakunnel, J.J., Rizvi, N.A., 2016. Safety and antitumour activity of durvalumab plus tremelimumab in non-small cell lung cancer: a multicentre, phase 1b study. Lancet Oncol. 17, 299–308.

Antonia, S.J., Villegas, A., Daniel, D., Vicente, D., Murakami, S., Hui, R., Yokoi, T., Chiappori, A., Lee, K.H., de Wit, M., Cho, B.C., Bourhaba, M., Quantin, X., Tokito, T., Mekhail, T., Planchard, D., Kim, Y.C., Karapetis, C.S., Hiret, S., Ostoros, G., Kubota, K., Gray, J.E., Paz-Ares, L., de Castro Carpeno, J., Wadsworth, C., Melillo, G., Jiang, H., Huang, Y., Dennis, P.A., Ozguroglu, M., Investigators, P, 2017. Durvalumab after chemoradiotherapy in stage III non-small-cell lung cancer. N. Engl. J. Med. 377, 1919–1929.

Ascierto, P.A., Bono, P., Bhatia, S., Melero, I., Nyakas, M.S., Svane, I., Gomez-Roca, C., Schadendorf, D., Dummer, R., Marabelle, A., Hoeller, C., Maurer, M., Harbison, C.T., Mitra, P., Suryawanshi, S., Thudium, K.B., Munoz Couselo, E., 2017. Efficacy of BMS-986016, a monoclonal antibody that targets lymphocyte activation gene-3 (LAG-3), in combination with nivolumab in pts with melanoma. In: ESMO 2017 Congress, 10 September 2017, Madrid, Spain. Annals of Oncology, v605-v649.

ASCO, 2018. CAR T-cell Immunotherapy Named Advance of the Year in Annual ASCO Report. The American Society of Clinical Oncology, Alexandria, VA.

Atezolizumab, F.L., 2016. TECENTRIQ (atezolizumab) injection, for intravenous use. In: Administration, F.D. (Ed.).

Avelumab, F.L., 2017. BAVENCIO (avelumab) injection, for intravenous use. In: Administration, F.D. (Ed.).

Balar, A.V., Galsky, M.D., Rosenberg, J.E., Powles, T., Petrylak, D.P., Bellmunt, J., Loriot, Y., Necchi, A., Hoffman-Censits, J., Perez-Gracia, J.L., Dawson, N.A., Van Der Heijden, M.S., Dreicer, R., Srinivas, S., Retz, M.M., Joseph, R.W., Drakaki, A., Vaishampayan, U.N., Sridhar, S.S., Quinn, D.I., Duran, I., Shaffer, D.R., Eigl, B.J., Grivas, P.D., Yu, E.Y., Li, S., Kadel 3rd, E.E., Boyd, Z., Bourgon, R., Hegde, P.S., Mariathasan, S., Thastrom, A., Abidoye, O.O., Fine, G.D., Bajorin, D.F., Group, I.M.S., 2017. Atezolizumab as first-line treatment in cisplatin-ineligible patients with locally advanced and metastatic urothelial carcinoma: a single-arm, multicentre, phase 2 trial. Lancet 389, 67–76.

Bartkowiak, T., Curran, M.A., 2015. 4-1BB agonists: multi-potent potentiators of tumor immunity. Front. Oncol. 5, 117.

Bashirova, A.A., Martin, M.P., Mcvicar, D.W., Carrington, M., 2006. The killer immunoglobulin-like receptor gene cluster: tuning the genome for defense. Annu. Rev. Genomics Hum. Genet. 7, 277–300.

Baverel, P.G., Dubois, V.F.S., Jin, C.Y., Zheng, Y., Song, X., Jin, X., Mukhopadhyay, P., Gupta, A., Dennis, P.A., Ben, Y., Vicini, P., Roskos, L., Narwal, R., 2017. Population pharmacokinetics of durvalumab in cancer patients and association with longitudinal biomarkers of disease status. Clin. Pharmacol. Ther.

Bommareddy, P.K., Kaufman, H.L., 2017. Avelumab and other recent advances in Merkel cell carcinoma. Future Oncol. 13, 2771–2783.

Bonifaz, L., Bonnyay, D., Mahnke, K., Rivera, M., Nussenzweig, M.C., Steinman, R.M., 2002. Efficient targeting of protein antigen to the dendritic cell receptor DEC-205 in the steady state leads to antigen presentation on major histocompatibility complex class I products and peripheral CD8+ T cell tolerance. J. Exp. Med. 196, 1627–1638.

Borghaei, H., Paz-Ares, L., Horn, L., Spigel, D.R., Steins, M., Ready, N.E., Chow, L.Q., Vokes, E.E., Felip, E., Holgado, E., Barlesi, F., Kohlhaufl, M., Arrieta, O., Burgio, M.A., Fayette, J., Lena, H., Poddubskaya, E., Gerber, D.E., Gettinger, S.N., Rudin, C.M., Rizvi, N., Crino, L., Blumenschein Jr., G.R., Antonia, S.J., Dorange, C., Harbison, C.T., Graf Finckenstein, F., Brahmer, J.R., 2015. Nivolumab versus docetaxel in advanced nonsquamous non-small-cell lung cancer. N. Engl. J. Med. 373, 1627–1639.

Boyerinas, B., Jochems, C., Fantini, M., Heery, C.R., Gulley, J.L., Tsang, K.Y., Schlom, J., 2015. Antibody-dependent cellular cytotoxicity activity of a novel anti-PD-L1 antibody avelumab (MSB0010718C) on human tumor cells. Cancer Immunol. Res. 3, 1148–1157.

Bracci, L., Schiavoni, G., Sistigu, A., Belardelli, F., 2014. Immune-based mechanisms of cytotoxic chemotherapy: implications for the design of novel and rationale-based combined treatments against cancer. Cell Death Differ. 21, 15–25.

Brandacher, G., Perathoner, A., Ladurner, R., Schneeberger, S., Obrist, P., Winkler, C., Werner, E.R., Werner-Felmayer, G., Weiss, H.G., Gobel, G., Margreiter, R., Konigsrainer, A., Fuchs, D., Amberger, A., 2006. Prognostic value of indoleamine 2,3-dioxygenase expression in colorectal cancer: effect on tumor-infiltrating T cells. Clin. Cancer Res. 12, 1144–1151.

Breitbach, C.J., Burke, J., Jonker, D., Stephenson, J., Haas, A.R., Chow, L.Q., Nieva, J., Hwang, T.H., Moon, A., Patt, R., Pelusio, A., Le Boeuf, F., Burns, J., Evgin, L., De Silva, N., Cvancic, S., Robertson, T., Je, J.E., Lee, Y.S., Parato, K., Diallo, J.S., Fenster, A., Daneshmand, M., Bell, J.C., Kirn, D.H., 2011. Intravenous delivery of a multi-mechanistic cancer-targeted oncolytic poxvirus in humans. Nature 477, 99–102.

Brentjens, R.J., Latouche, J.B., Santos, E., Marti, F., Gong, M.C., Lyddane, C., King, P.D., Larson, S., Weiss, M., Riviere, I., Sadelain, M., 2003. Eradication of systemic B-cell tumors by genetically targeted human T lymphocytes co-stimulated by CD80 and interleukin-15. Nat. Med. 9, 279–286.

Buechner, J., Grupp, S.A., Maude, S.L., Boyer, M., Bittencourt, H., Laetsch, T.W., Bader, P., Verneris, M.R., Stefanski, H., Myers, G.D., Qayed, M., Pulsipher, M.A., De Moerloose, B., Hiramatsu, H., Schlis, K., Davis, K., Martin, P.L., Nemecek, E., Peters, C., Wood, P., Taran, T., Mueller, K.T., Zhang, Y., Rives, S., 2017. Global registration trial of efficacy and safety of CTL019 in pediatric and young adult patients with relapsed/refractory (R/R) acute lymphoblastic leukemia (all): update to the interim analysis. Clin. Lymphoma, Myeloma & Leukemia 17, S263–S264.

Burris, H.A., Infante, J.R., Ansell, S.M., Nemunaitis, J.J., Weiss, G.R., Villalobos, V.M., Sikic, B.I., Taylor, M.H., Northfelt, D.W., Carson 3rd, W.E., Hawthorne, T.R., Davis, T.A., Yellin, M.J., Keler, T., Bullock, T., 2017. Safety and activity of varlilumab, a novel and first-in-class agonist anti-CD27 antibody, in patients with advanced solid tumors. J. Clin. Oncol. 35, 2028–2036.

Butte, M.J., Keir, M.E., Phamduy, T.B., Sharpe, A.H., Freeman, G.J., 2007. Programmed death-1 ligand 1 interacts specifically with the B7-1 costimulatory molecule to inhibit T cell responses. Immunity 27, 111–122.

Byun, D.J., Wolchok, J.D., Rosenberg, L.M., Girotra, M., 2017. Cancer immunotherapy - immune checkpoint blockade and associated endocrinopathies. Nat. Rev. Endocrinol. 13, 195–207.

Cameron, F., Whiteside, G., Perry, C., 2011. Ipilimumab: first global approval. Drugs 71, 1093–1104.

Carpenito, C., Milone, M.C., Hassan, R., Simonet, J.C., Lakhal, M., Suhoski, M.M., Varela-Rohena, A., Haines, K.M., Heitjan, D.F., Albelda, S.M., Carroll, R.G., Riley, J.L., Pastan, I., June, C.H., 2009. Control of large, established tumor xenografts with genetically retargeted human T cells containing CD28 and CD137 domains. Proc. Natl. Acad. Sci. U.S.A 106, 3360–3365.

Carthon, B.C., Wolchok, J.D., Yuan, J., Kamat, A., Ng Tang, D.S., Sun, J., Ku, G., Troncoso, P., Logothetis, C.J., Allison, J.P., Sharma, P., 2010. Preoperative CTLA-4 blockade: tolerability and immune monitoring in the setting of a presurgical clinical trial. Clin. Cancer Res. 16, 2861–2871.

Castelli, C., Triebel, F., Rivoltini, L., Camisaschi, C., 2014. Lymphocyte activation gene-3 (LAG-3, CD223) in plasmacytoid dendritic cells (pDCs): a molecular target for the restoration of active antitumor immunity. OncoImmunology 3, e967146.

Chakrabarty, R., Tran, H., Selvaggi, G., Hagerman, A., Thompson, B., Coffey, M., 2015. The oncolytic virus, pelareorep, as a novel anticancer agent: a review. Invest. New Drugs 33, 761–774.

Chaput, N., Lepage, P., Coutzac, C., Soularue, E., Le Roux, K., Monot, C., Boselli, L., Routier, E., Cassard, L., Collins, M., Vaysse, T., Marthey, L., Eggermont, A., Asvatourian, V., Lanoy, E., Mateus, C., Robert, C., Carbonnel, F., 2017. Baseline gut microbiota predicts clinical response and colitis in metastatic melanoma patients treated with ipilimumab. Ann. Oncol. 28, 1368–1379.

Chauvin, J.M., Pagliano, O., Fourcade, J., Sun, Z., Wang, H., Sander, C., Kirkwood, J.M., Chen, T.H., Maurer, M., Korman, A.J., Zarour, H.M., 2015. TIGIT and PD-1 impair tumor antigen-specific CD8(+) T cells in melanoma patients. J. Clin. Invest. 125, 2046–2058.

Chen, A.I., Mcadam, A.J., Buhlmann, J.E., Scott, S., Lupher JR., M.L., Greenfield, E.A., Baum, P.R., Fanslow, W.C., Calderhead, D.M., Freeman, G.J., Sharpe, A.H., 1999. Ox40-ligand has a critical costimulatory role in dendritic cell:T cell interactions. Immunity 11, 689–698.

Chen, H., Liakou, C.I., Kamat, A., Pettaway, C., Ward, J.F., Tang, D.N., Sun, J., Jungbluth, A.A., Troncoso, P., Logothetis, C., Sharma, P., 2009. Anti-CTLA-4 therapy results in higher CD4+ICOShi T cell frequency and IFN-gamma levels in both nonmalignant and malignant prostate tissues. Proc. Natl. Acad. Sci. U.S.A 106, 2729–2734.

Chen, Q., Huang, D.S., Zhang, L.W., Li, Y.Q., Wang, H.W., Liu, H.B., 2017. Fatal myocarditis and rhabdomyolysis induced by nivolumab during the treatment of type B3 thymoma. Clin. Toxicol. 1–5.

Chowdhury, P.S., Chamoto, K., Honjo, T., 2017. Combination therapy strategies for improving PD-1 blockade efficacy: a new era in cancer immunotherapy. J. Intern. Med.

Cooper, L.J., Topp, M.S., Serrano, L.M., Gonzalez, S., Chang, W.C., Naranjo, A., Wright, C., Popplewell, L., Raubitschek, A., Forman, S.J., Jensen, M.C., 2003. T-cell clones can be rendered specific for CD19: toward the selective augmentation of the graft-versus-B-lineage leukemia effect. Blood 101, 1637–1644.

Crispen, P.L., Sheinin, Y., Roth, T.J., Lohse, C.M., Kuntz, S.M., Frigola, X., Thompson, R.H., Boorjian, S.A., Dong, H., Leibovich, B.C., Blute, M.L., Kwon, E.D., 2008. Tumor cell and tumor vasculature expression of B7-H3 predict survival in clear cell renal cell carcinoma. Clin. Cancer Res. 14, 5150–5157.

Croft, M., 2009. The role of TNF superfamily members in T-cell function and diseases. Nat. Rev. Immunol. 9, 271–285.

Croft, M., 2010. Control of immunity by the TNFR-related molecule OX40 (CD134). Annu. Rev. Immunol. 28, 57–78.

Dagogo-Jack, I., Shaw, A.T., 2018. Tumour heterogeneity and resistance to cancer therapies. Nat. Rev. Clin. Oncol. 15, 81–94.

Daver, N.G., Basu, S., Garcia-Manero, G., Cortes, J.E., Ravandi, F., Jabbour, E., Hendrickson, S., Brandt, M., Pierce, S., Gordon, T., Pemmaraju, N., Andreeff, M., Ning, J., Kornblau, S., Kadia, T.M., Dinardo, C.D., Konopleva, M., Allison, J.P., Kantarjian, H.M., Sharma, A., 2017. Phase IB/II study of nivolumab with azacytidine (AZA) in patients with relapsed AML. J. Clin. Oncol. 35, 7026.

Davila, M.L., Riviere, I., Wang, X., Bartido, S., Park, J., Curran, K., Chung, S.S., Stefanski, J., Borquez-Ojeda, O., Olszewska, M., Qu, J., Wasielewska, T., He, Q., Fink, M., Shinglot, H., Youssif, M., Satter, M., Wang, Y., Hosey, J., Quintanilla, H., Halton, E., Bernal, Y., Bouhassira, D.C., Arcila, M.E., Gonen, M., Roboz, G.J., Maslak, P., Douer, D., Frattini, M.G., Giralt, S., Sadelain, M., Brentjens, R., 2014. Efficacy and toxicity management of 19-28z CAR T cell therapy in B cell acute lymphoblastic leukemia. Sci. Transl. Med. 6, 224ra25.

De Velasco, G., Je, Y., Bosse, D., Awad, M.M., Ott, P.A., Moreira, R.B., Schutz, F., Bellmunt, J., Sonpavde, G.P., Hodi, F.S., Choueiri, T.K., 2017. Comprehensive meta-analysis of key immune-related adverse events from CTLA-4 and PD-1/PD-L1 inhibitors in cancer patients. Cancer Immunol. Res. 5, 312−318.

Deng, R., Bumbaca, D., Pastuskovas, C.V., Boswell, C.A., West, D., Cowan, K.J., Chiu, H., Mcbride, J., Johnson, C., Xin, Y., Koeppen, H., Leabman, M., Iyer, S., 2016. Preclinical pharmacokinetics, pharmacodynamics, tissue distribution, and tumor penetration of anti-PD-L1 monoclonal antibody, an immune checkpoint inhibitor. MAbs 8, 593−603.

Denoeud, J., Moser, M., 2011. Role of CD27/CD70 pathway of activation in immunity and tolerance. J. Leukoc. Biol. 89, 195−203.

Dougall, W.C., Kurtulus, S., Smyth, M.J., Anderson, A.C., 2017. TIGIT and CD96: new checkpoint receptor targets for cancer immunotherapy. Immunol. Rev. 276, 112−120.

Du, X., Liu, M., Su, J., Zhang, P., Tang, F., Ye, P., Devenport, M., Wang, X., Zhang, Y., Liu, Y., Zheng, P., 2018a. Uncoupling therapeutic from immunotherapy-related adverse effects for safer and effective anti-CTLA-4 antibodies in CTLA4 humanized mice. Cell Res. 28, 433−447.

Du, X., Tang, F., Liu, M., Su, J., Zhang, Y., Wu, W., Devenport, M., Lazarski, C.A., Zhang, P., Wang, X., Ye, P., Wang, C., Hwang, E., Zhu, T., Xu, T., Zheng, P., Liu, Y., 2018b. A reappraisal of CTLA-4 checkpoint blockade in cancer immunotherapy. Cell Res. 28, 416−432.

Dudley, M.E., Wunderlich, J.R., Robbins, P.F., Yang, J.C., Hwu, P., Schwartzentruber, D.J., Topalian, S.L., Sherry, R., Restifo, N.P., Hubicki, A.M., Robinson, M.R., Raffeld, M., Duray, P., Seipp, C.A., Rogers-Freezer, L., Morton, K.E., Mavroukakis, S.A., White, D.E., Rosenberg, S.A., 2002. Cancer regression and autoimmunity in patients after clonal repopulation with antitumor lymphocytes. Science 298, 850−854.

Dudley, M.E., Yang, J.C., Sherry, R., Hughes, M.S., Royal, R., Kammula, U., Robbins, P.F., Huang, J., Citrin, D.E., Leitman, S.F., Wunderlich, J., Restifo, N.P., Thomasian, A., Downey, S.G., Smith, F.O., Klapper, J., Morton, K., Laurencot, C., White, D.E., Rosenberg, S.A., 2008. Adoptive cell therapy for patients with metastatic melanoma: evaluation of intensive myeloablative chemoradiation preparative regimens. J. Clin. Oncol. 26, 5233−5239.

Durkop, H., Latza, U., Hummel, M., Eitelbach, F., Seed, B., Stein, H., 1992. Molecular cloning and expression of a new member of the nerve growth factor receptor family that is characteristic for Hodgkin's disease. Cell 68, 421−427.

Eggermont, A.M., Chiarion-Sileni, V., Grob, J.J., Dummer, R., Wolchok, J.D., Schmidt, H., Hamid, O., Robert, C., Ascierto, P.A., Richards, J.M., Lebbe, C., Ferraresi, V., Smylie, M., Weber, J.S., Maio, M., Konto, C., Hoos, A., De Pril, V., Gurunath, R.K., De Schaetzen, G., Suciu, S., Testori, A., 2015. Adjuvant ipilimumab versus placebo after complete resection of high-risk stage III melanoma (EORTC 18071): a randomised, double-blind, phase 3 trial. Lancet Oncol. 16, 522–530.

Eggermont, A.M.M., Dummer, R., 2017. The 2017 complete overhaul of adjuvant therapies for high-risk melanoma and its consequences for staging and management of melanoma patients. Eur. J. Cancer 86, 101–105.

Elgueta, R., Benson, M.J., De Vries, V.C., Wasiuk, A., Guo, Y., Noelle, R.J., 2009. Molecular mechanism and function of CD40/CD40L engagement in the immune system. Immunol. Rev. 229, 152–172.

Eliopoulos, A.G., Young, L.S., 2004. The role of the CD40 pathway in the pathogenesis and treatment of cancer. Curr. Opin. Pharmacol. 4, 360–367.

Eshhar, Z., Waks, T., Gross, G., Schindler, D.G., 1993. Specific activation and targeting of cytotoxic lymphocytes through chimeric single chains consisting of antibody-binding domains and the gamma or zeta subunits of the immunoglobulin and T-cell receptors. Proc. Natl. Acad. Sci. U.S.A 90, 720–724.

Fan, X., Quezada, S.A., Sepulveda, M.A., Sharma, P., Allison, J.P., 2014. Engagement of the ICOS pathway markedly enhances efficacy of CTLA-4 blockade in cancer immunotherapy. J. Exp. Med. 211, 715–725.

Farkona, S., Diamandis, E.P., Blasutig, I.M., 2016. Cancer immunotherapy: the beginning of the end of cancer? BMC Med. 14, 73.

Fayad, L., Ansell, S.M., Advani, R., Coiffier, B., Stuart, R., Bartlett, N.L., Forero-Torres, A., Kuliczkowski, K., Belada, D., Ng, E., Drachman, J.G., 2015. Dacetuzumab plus rituximab, ifosfamide, carboplatin and etoposide as salvage therapy for patients with diffuse large B-cell lymphoma relapsing after rituximab, cyclophosphamide, doxorubicin, vincristine and prednisolone: a randomized, double-blind, placebo-controlled phase 2b trial. Leuk. Lymphoma 56, 2569–2578.

Festino, L., Botti, G., Lorigan, P., Masucci, G.V., Hipp, J.D., Horak, C.E., Melero, I., Ascierto, P.A., 2016. Cancer treatment with anti-PD-1/PD-L1 agents: is PD-L1 expression a biomarker for patient selection? Drugs 76, 925–945.

Fisher, T.S., Kamperschroer, C., Oliphant, T., Love, V.A., Lira, P.D., Doyonnas, R., Bergqvist, S., Baxi, S.M., Rohner, A., Shen, A.C., Huang, C., Sokolowski, S.A., Sharp, L.L., 2012. Targeting of 4-1BB by monoclonal antibody PF-05082566 enhances T-cell function and promotes anti-tumor activity. Cancer Immunol. Immunother. 61, 1721–1733.

Flies, D.B., Han, X., Higuchi, T., Zheng, L., Sun, J., Ye, J.J., Chen, L., 2014. Coinhibitory receptor PD-1H preferentially suppresses CD4(+) T cell-mediated immunity. J. Clin. Invest. 124, 1966–1975.

Fourcade, J., Sun, Z., Benallaoua, M., Guillaume, P., Luescher, I.F., Sander, C., Kirkwood, J.M., Kuchroo, V., Zarour, H.M., 2010. Upregulation of Tim-3 and PD-1 expression is associated with tumor antigen-specific CD8+ T cell dysfunction in melanoma patients. J. Exp. Med. 207, 2175–2186.

Francisco, L.M., Sage, P.T., Sharpe, A.H., 2010. The PD-1 pathway in tolerance and autoimmunity. Immunol. Rev. 236, 219–242.

Freeman, G.J., 2008. Structures of PD-1 with its ligands: sideways and dancing cheek to cheek. Proc. Natl. Acad. Sci. U.S.A 105, 10275–10276.

Freeman, G.J., Lombard, D.B., Gimmi, C.D., Brod, S.A., Lee, K., Laning, J.C., Hafler, D.A., Dorf, M.E., Gray, G.S., Reiser, H., et al., 1992. CTLA-4 and CD28 mRNA are coexpressed in most T cells after activation. Expression of CTLA-4 and CD28 mRNA does not correlate with the pattern of lymphokine production. J. Immunol. 149, 3795–3801.

French, R.R., Chan, H.T., Tutt, A.L., Glennie, M.J., 1999. CD40 antibody evokes a cytotoxic T-cell response that eradicates lymphoma and bypasses T-cell help. Nat. Med. 5, 548–553.

Frigeri, M., Meyer, P., Banfi, C., Giraud, R., Hachulla, A.L., Spoerl, D., Friedlaender, A., Pugliesi-Rinaldi, A., Dietrich, P.Y., 2018. Immune checkpoint inhibitor-associated myocarditis: a new challenge for cardiologists. Can. J. Cardiol. 34, 92e1–92e3.

Fry, T.J., Shah, N.N., Orentas, R.J., Stetler-Stevenson, M., Yuan, C.M., Ramakrishna, S., Wolters, P., Martin, S., Delbrook, C., Yates, B., Shalabi, H., Fountaine, T.J., Shern, J.F., Majzner, R.G., Stroncek, D.F., Sabatino, M., Feng, Y., Dimitrov, D.S., Zhang, L., Nguyen, S., Qin, H., Dropulic, B., Lee, D.W., Mackall, C.L., 2018. CD22-targeted CAR T cells induce remission in B-ALL that is naive or resistant to CD19-targeted CAR immunotherapy. Nat. Med. 24, 20–28.

Gajewski, T.F., Meng, Y., Harlin, H., 2006. Immune suppression in the tumor microenvironment. J. Immunother. 29, 233–240.

Gan, E.H., Mitchell, A.L., Plummer, R., Pearce, S., Perros, P., 2017. Tremelimumab-induced graves hyperthyroidism. Eur. Thyroid J. 6, 167–170.

Gandhi, L., Rodriguez-Abreu, D., Gadgeel, S., Esteban, E., Felip, E., de Angelis, F., Domine, M., Clingan, P., Hochmair, M.J., Powell, S.F., Cheng, S.Y., Bischoff, H.G., Peled, N., Grossi, F., Jennens, R.R., Reck, M., Hui, R., Garon, E.B., Boyer, M., Rubio-Viqueira, B., Novello, S., Kurata, T., Gray, J.E., Vida, J., Wei, Z., Yang, J., Raftopoulos, H., Pietanza, M.C., Garassino, M.C., Investigators, K., 2018. Pembrolizumab plus chemotherapy in metastatic non-small-cell lung cancer. N. Engl. J. Med.

Gao, J., Ward, J.F., Pettaway, C.A., Shi, L.Z., Subudhi, S.K., Vence, L.M., Zhao, H., Chen, J., Chen, H., Efstathiou, E., Troncoso, P., Allison, J.P., Logothetis, C.J., Wistuba, I., Sepulveda, M.A., Sun, J., Wargo, J., Blando, J., Sharma, P., 2017. VISTA is an inhibitory immune checkpoint that is increased after ipilimumab therapy in patients with prostate cancer. Nat. Med. 23, 551–555.

Garon, E.B., Rizvi, N.A., Hui, R., Leighl, N., Balmanoukian, A.S., Eder, J.P., Patnaik, A., Aggarwal, C., Gubens, M., Horn, L., Carcereny, E., Ahn, M.J., Felip, E., Lee, J.S., Hellmann, M.D., Hamid, O., Goldman, J.W., Soria, J.C., Dolled-Filhart, M., Rutledge, R.Z., Zhang, J., Lunceford, J.K., Rangwala, R., Lubiniecki, G.M., Roach, C., Emancipator, K., Gandhi, L., Investigators, K., 2015. Pembrolizumab for the treatment of non-small-cell lung cancer. N. Engl. J. Med. 372, 2018–2028.

Ge, W., Li, J., Fan, W., Xu, D., Sun, S., 2017. Tim-3 as a diagnostic and prognostic biomarker of osteosarcoma. Tumour. Biol. 39, 1010428317715643.

Goldsmith, K., Chen, W., Johnson, D.C., Hendricks, R.L., 1998. Infected cell protein (ICP) 47 enhances herpes simplex virus neurovirulence by blocking the CD8+ T cell response. J. Exp. Med. 187, 341–348.

Gorman, J.V., Starbeck-Miller, G., Pham, N.L., Traver, G.L., Rothman, P.B., Harty, J.T., Colgan, J.D., 2014. Tim-3 directly enhances CD8 T cell responses to acute Listeria monocytogenes infection. J. Immunol. 192, 3133–3142.

Greig, S.L., 2016. Talimogene laherparepvec: first global approval. Drugs 76, 147–154.

Gurney, A.L., Marsters, S.A., Huang, R.M., Pitti, R.M., Mark, D.T., Baldwin, D.T., Gray, A.M., Dowd, A.D., Brush, A.D., Heldens, A.D., Schow, A.D., Goddard, A.D., Wood, W.I., Baker, K.P., Godowski, P.J., Ashkenazi, A., 1999. Identification of a new member of the tumor necrosis factor family and its receptor, a human ortholog of mouse GITR. Curr. Biol. 9, 215–218.

Hahn, A.W., Gill, D.M., Agarwal, N., Maughan, B.L., 2017. PD-1 checkpoint inhibition: toxicities and management. Urol. Oncol. 35, 701–707.

Hamid, O., Robert, C., Daud, A., Hodi, F.S., Hwu, W.J., Kefford, R., Wolchok, J.D., Hersey, P., Joseph, R.W., Weber, J.S., Dronca, R., Gangadhar, T.C., Patnaik, A., Zarour, H., Joshua, A.M., Gergich, K., Elassaiss-Schaap, J., Algazi, A., Mateus, C., Boasberg, P., Tumeh, P.C., Chmielowski, B., Ebbinghaus, S.W., Li, X.N., Kang, S.P., Ribas, A., 2013. Safety and tumor responses with lambrolizumab (anti-PD-1) in melanoma. N. Engl. J. Med. 369, 134–144.

Hamilou, Z., Lavaud, P., Loriot, Y., 2017. Atezolizumab in urothelial bladder carcinoma. Future Oncol.

Handgretinger, R., Lang, P., Andre, M.C., 2016. Exploitation of natural killer cells for the treatment of acute leukemia. Blood 127, 3341–3349.

Haso, W., Lee, D.W., Shah, N.N., Stetler-Stevenson, M., Yuan, C.M., Pastan, I.H., Dimitrov, D.S., Morgan, R.A., Fitzgerald, D.J., Barrett, D.M., Wayne, A.S., Mackall, C.L., Orentas, R.J., 2013. Anti-CD22-chimeric antigen receptors targeting B-cell precursor acute lymphoblastic leukemia. Blood 121, 1165–1174.

He, M., Chai, Y., Qi, J., Zhang, C.W.H., Tong, Z., Shi, Y., Yan, J., Tan, S., Gao, G.F., 2017a. Remarkably similar CTLA-4 binding properties of therapeutic ipilimumab and tremelimumab antibodies. Oncotarget 8, 67129–67139.

He, W., Zhang, H., Han, F., Chen, X., Lin, R., Wang, W., Qiu, H., Zhuang, Z., Liao, Q., Zhang, W., Cai, Q., Cui, Y., Jiang, W., Wang, H., Ke, Z., 2017b. CD155T/TIGIT signaling regulates CD8(+) t-cell metabolism and promotes tumor progression in human gastric cancer. Cancer Res. 77, 6375–6388.

Hellmann, M.D., Ciuleanu, T.E., Pluzanski, A., Lee, J.S., Otterson, G.A., Audigier-Valette, C., Minenza, E., Linardou, H., Burgers, S., Salman, P., Borghaei, H., Ramalingam, S.S., Brahmer, J., Reck, M., O'byrne, K.J., Geese, W.J., Green, G., Chang, H., Szustakowski, J., Bhagavatheeswaran, P., Healey, D., Fu, Y., Nathan, F., PAZ-Ares, L., 2018. Nivolumab plus ipilimumab in lung cancer with a high tumor mutational burden. N. Engl. J. Med.

Herbst, R.S., Baas, P., Kim, D.W., Felip, E., Perez-Gracia, J.L., Han, J.Y., Molina, J., Kim, J.H., Arvis, C.D., Ahn, M.J., Majem, M., Fidler, M.J., De Castro Jr., G., Garrido, M., Lubiniecki, G.M., Shentu, Y., Im, E., Dolled-Filhart, M., Garon, E.B., 2016. Pembrolizumab versus docetaxel for previously treated, PD-L1-positive, advanced non-small-cell lung cancer (KEYNOTE-010): a randomised controlled trial. Lancet 387, 1540–1550.

Herbst, R.S., Soria, J.C., Kowanetz, M., Fine, G.D., Hamid, O., Gordon, M.S., Sosman, J.A., Mcdermott, D.F., Powderly, J.D., Gettinger, S.N., Kohrt, H.E., Horn, L., Lawrence, D.P., Rost, S., Leabman, M., Xiao, Y., Mokatrin, A., Koeppen, H., Hegde, P.S., Mellman, I., Chen, D.S., Hodi, F.S., 2014. Predictive correlates of response to the anti-PD-L1 antibody MPDL3280A in cancer patients. Nature 515, 563–567.

Hodi, F.S., O'day, S.J., Mcdermott, D.F., Weber, R.W., Sosman, J.A., Haanen, J.B., Gonzalez, R., Robert, C., Schadendorf, D., Hassel, J.C., Akerley, W., Van Den Eertwegh, A.J., Lutzky, J., Lorigan, P., Vaubel, J.M., Linette, G.P., Hogg, D., Ottensmeier, C.H., Lebbe, C., Peschel, C., Quirt, I., Clark, J.I., Wolchok, J.D., Weber, J.S., Tian, J., Yellin, M.J., Nichol, G.M., Hoos, A., Urba, W.J., 2010. Improved survival with ipilimumab in patients with metastatic melanoma. N. Engl. J. Med. 363, 711–723.

Holmgaard, R.B., Zamarin, D., Munn, D.H., Wolchok, J.D., Allison, J.P., 2013. Indoleamine 2,3-dioxygenase is a critical resistance mechanism in antitumor T cell immunotherapy targeting CTLA-4. J. Exp. Med. 210, 1389–1402.

Homet Moreno, B., Ribas, A., 2015. Anti-programmed cell death protein-1/ligand-1 therapy in different cancers. Br. J. Cancer 112, 1421–1427.
Hori, S., Nomura, T., Sakaguchi, S., 2003. Control of regulatory T cell development by the transcription factor Foxp3. Science 299, 1057–1061.
Horn, L., Spigel, D.R., Vokes, E.E., Holgado, E., Ready, N., Steins, M., Poddubskaya, E., Borghaei, H., Felip, E., Paz-Ares, L., Pluzanski, A., Reckamp, K.L., Burgio, M.A., Kohlhaeufl, M., Waterhouse, D., Barlesi, F., Antonia, S., Arrieta, O., Fayette, J., Crino, L., Rizvi, N., Reck, M., Hellmann, M.D., Geese, W.J., Li, A., Blackwood-Chirchir, A., Healey, D., Brahmer, J., Eberhardt, W.E.E., 2017. Nivolumab versus docetaxel in previously treated patients with advanced non-small-cell lung cancer: two-year outcomes from two randomized, open-label, phase III trials (CheckMate 017 and CheckMate 057). J. Clin. Oncol. JCO2017743062.
Huang, J., Jochems, C., Anderson, A.M., Talaie, T., Jales, A., Madan, R.A., Hodge, J.W., Tsang, K.Y., Liewehr, D.J., Steinberg, S.M., Gulley, J.L., Schlom, J., 2013. Soluble CD27-pool in humans may contribute to T cell activation and tumor immunity. J. Immunol. 190, 6250–6258.
Huang, R.Y., Francois, A., Mcgray, A.R., Miliotto, A., Odunsi, K., 2017. Compensatory upregulation of PD-1, LAG-3, and CTLA-4 limits the efficacy of single-agent checkpoint blockade in metastatic ovarian cancer. OncoImmunology 6, e1249561.
Huard, B., Gaulard, P., Faure, F., Hercend, T., Triebel, F., 1994. Cellular expression and tissue distribution of the human LAG-3-encoded protein, an MHC class II ligand. Immunogenetics 39, 213–217.
Ino, K., Yoshida, N., Kajiyama, H., Shibata, K., Yamamoto, E., Kidokoro, K., Takahashi, N., Terauchi, M., Nawa, A., Nomura, S., Nagasaka, T., Takikawa, O., Kikkawa, F., 2006. Indoleamine 2,3-dioxygenase is a novel prognostic indicator for endometrial cancer. Br. J. Cancer 95, 1555–1561.
Irving, B.A., Weiss, A., 1991. The cytoplasmic domain of the T cell receptor zeta chain is sufficient to couple to receptor-associated signal transduction pathways. Cell 64, 891–901.
Jensen, M.C., Popplewell, L., Cooper, L.J., Digiusto, D., Kalos, M., Ostberg, J.R., Forman, S.J., 2010. Antitransgene rejection responses contribute to attenuated persistence of adoptively transferred CD20/CD19-specific chimeric antigen receptor redirected T cells in humans. Biol. Blood Marrow Transplant. 16, 1245–1256.
Johnston, R.J., Comps-Agrar, L., Hackney, J., Yu, X., Huseni, M., Yang, Y., Park, S., Javinal, V., Chiu, H., Irving, B., Eaton, D.L., Grogan, J.L., 2014. The immunoreceptor TIGIT regulates antitumor and antiviral CD8(+) T cell effector function. Canc. Cell 26, 923–937.
Jones, S., Zhang, X., Parsons, D.W., Lin, J.C., Leary, R.J., Angenendt, P., Mankoo, P., Carter, H., Kamiyama, H., Jimeno, A., Hong, S.M., Fu, B., Lin, M.T., Calhoun, E.S., Kamiyama, M., Walter, K., Nikolskaya, T., Nikolsky, Y., Hartigan, J., Smith, D.R., Hidalgo, M., Leach, S.D., Klein, A.P., Jaffee, E.M., Goggins, M., Maitra, A., Iacobuzio-Donahue, C., Eshleman, J.R., Kern, S.E., Hruban, R.H., Karchin, R., Papadopoulos, N., Parmigiani, G., Vogelstein, B., Velculescu, V.E., Kinzler, K.W., 2008. Core signaling pathways in human pancreatic cancers revealed by global genomic analyses. Science 321, 1801–1806.
Jung, K., Choi, I., 2013. Emerging Co-signaling Networks in T Cell immune regulation. Immune. Netw. 13, 184–193.
Kalos, M., Levine, B.L., Porter, D.L., Katz, S., Grupp, S.A., Bagg, A., June, C.H., 2011. T cells with chimeric antigen receptors have potent antitumor effects and can establish memory in patients with advanced leukemia. Sci. Transl. Med. 3, 95ra73.

Kaufman, H.L., Ruby, C.E., Hughes, T., Slingluff JR., C.L., 2014. Current status of granulocyte-macrophage colony-stimulating factor in the immunotherapy of melanoma. J. Immunother. Cancer 2, 11.

Kaufman, H.L., Russell, J., Hamid, O., Bhatia, S., Terheyden, P., D'angelo, S.P., Shih, K.C., Lebbe, C., Linette, G.P., Milella, M., Brownell, I., Lewis, K.D., Lorch, J.H., Chin, K., Mahnke, L., Von Heydebreck, A., Cuillerot, J.M., Nghiem, P., 2016. Avelumab in patients with chemotherapy-refractory metastatic Merkel cell carcinoma: a multicentre, single-group, open-label, phase 2 trial. Lancet Oncol. 17, 1374–1385.

Keir, M.E., Butte, M.J., Freeman, G.J., Sharpe, A.H., 2008. PD-1 and its ligands in tolerance and immunity. Annu. Rev. Immunol. 26, 677–704.

Keir, M.E., Liang, S.C., Guleria, I., Latchman, Y.E., Qipo, A., Albacker, L.A., Koulmanda, M., Freeman, G.J., Sayegh, M.H., Sharpe, A.H., 2006. Tissue expression of PD-L1 mediates peripheral T cell tolerance. J. Exp. Med. 203, 883–895.

Khalil, D.N., Budhu, S., Gasmi, B., Zappasodi, R., Hirschhorn-Cymerman, D., Plitt, T., De Henau, O., Zamarin, D., Holmgaard, R.B., Murphy, J.T., Wolchok, J.D., Merghoub, T., 2015. The new era of cancer immunotherapy: manipulating t-cell activity to overcome malignancy. Adv. Cancer Res. 128, 1–68.

Kim, E.S., 2017. Avelumab: first global approval. Drugs 77, 929–937.

Kim, I.K., Kim, B.S., Koh, C.H., Seok, J.W., Park, J.S., Shin, K.S., Bae, E.A., Lee, G.E., Jeon, H., Cho, J., Jung, Y., Han, D., Kwon, B.S., Lee, H.Y., Chung, Y., Kang, C.Y., 2015. Glucocorticoid-induced tumor necrosis factor receptor-related protein co-stimulation facilitates tumor regression by inducing IL-9-producing helper T cells. Nat. Med. 21, 1010–1017.

Kim, J.H., Oh, J.Y., Park, B.H., Lee, D.E., Kim, J.S., Park, H.E., Roh, M.S., Je, J.E., Yoon, J.H., Thorne, S.H., Kirn, D., Hwang, T.H., 2006. Systemic armed oncolytic and immunologic therapy for cancer with JX-594, a targeted poxvirus expressing GM-CSF. Mol. Ther. 14, 361–370.

Kim, S., Poursine-Laurent, J., Truscott, S.M., Lybarger, L., Song, Y.J., Yang, L., French, A.R., Sunwoo, J.B., Lemieux, S., Hansen, T.H., Yokoyama, W.M., 2005. Licensing of natural killer cells by host major histocompatibility complex class I molecules. Nature 436, 709–713.

Kleffel, S., Posch, C., Barthel, S.R., Mueller, H., Schlapbach, C., Guenova, E., Elco, C.P., Lee, N., Juneja, V.R., Zhan, Q., Lian, C.G., Thomi, R., Hoetzenecker, W., Cozzio, A., Dummer, R., Mihm Jr., M.C., Flaherty, K.T., Frank, M.H., Murphy, G.F., Sharpe, A.H., Kupper, T.S., Schatton, T., 2015. Melanoma cell-intrinsic PD-1 receptor functions promote tumor growth. Cell 162, 1242–1256.

Ko, K., Yamazaki, S., Nakamura, K., Nishioka, T., Hirota, K., Yamaguchi, T., Shimizu, J., Nomura, T., Chiba, T., Sakaguchi, S., 2005. Treatment of advanced tumors with agonistic anti-GITR mAb and its effects on tumor-infiltrating Foxp3+CD25+CD4+ regulatory T cells. J. Exp. Med. 202, 885–891.

Kochenderfer, J.N., Feldman, S.A., Zhao, Y., Xu, H., Black, M.A., Morgan, R.A., Wilson, W.H., Rosenberg, S.A., 2009. Construction and preclinical evaluation of an anti-CD19 chimeric antigen receptor. J. Immunother. 32, 689–702.

Kochenderfer, J.N., Rosenberg, S.A., 2013. Treating B-cell cancer with T cells expressing anti-CD19 chimeric antigen receptors. Nat. Rev. Clin. Oncol. 10, 267–276.

Kochenderfer, J.N., Wilson, W.H., Janik, J.E., Dudley, M.E., Stetler-Stevenson, M., Feldman, S.A., Maric, I., Raffeld, M., Nathan, D.A., Lanier, B.J., Morgan, R.A., Rosenberg, S.A., 2010. Eradication of B-lineage cells and regression of lymphoma in a patient treated with autologous T cells genetically engineered to recognize CD19. Blood 116, 4099–4102.

Komohara, Y., Morita, T., Annan, D.A., Horlad, H., Ohnishi, K., Yamada, S., Nakayama, T., Kitada, S., Suzu, S., Kinoshita, I., Dosaka-Akita, H., Akashi, K., Takeya, M., Jinushi, M., 2015. The coordinated actions of TIM-3 on cancer and myeloid cells in the regulation of tumorigenicity and clinical prognosis in clear cell renal cell carcinomas. Cancer Immunol. Res. 3, 999–1007.

Kotteas, E.A., Pavlidis, N., 2015. Neuroendocrine Merkel cell nodal carcinoma of unknown primary site: management and outcomes of a rare entity. Crit. Rev. Oncol. Hematol. 94, 116–121.

Kurtulus, S., Sakuishi, K., Ngiow, S.F., Joller, N., Tan, D.J., Teng, M.W., Smyth, M.J., Kuchroo, V.K., Anderson, A.C., 2015. TIGIT predominantly regulates the immune response via regulatory T cells. J. Clin. Invest. 125, 4053–4062.

Kwon, B., Yu, K.Y., Ni, J., Yu, G.L., Jang, I.K., Kim, Y.J., Xing, L., Liu, D., Wang, S.X., Kwon, B.S., 1999. Identification of a novel activation-inducible protein of the tumor necrosis factor receptor superfamily and its ligand. J. Biol. Chem. 274, 6056–6061.

Lalani, A.A., Mcgregor, B.A., Sonpavde, G.P., Choueiri, T.K., 2017. JAVELIN: avelumab another spear to fight urothelial carcinoma. Lancet Oncol.

Langer, C.J., Gadgeel, S.M., Borghaei, H., Papadimitrakopoulou, V.A., Patnaik, A., Powell, S.F., Gentzler, R.D., Martins, R.G., Stevenson, J.P., Jalal, S.I., Panwalkar, A., Yang, J.C., Gubens, M., Sequist, L.V., Awad, M.M., Fiore, J., Ge, Y., Raftopoulos, H., Gandhi, L., Investigators, K., 2016. Carboplatin and pemetrexed with or without pembrolizumab for advanced, non-squamous non-small-cell lung cancer: a randomised, phase 2 cohort of the open-label KEYNOTE-021 study. Lancet Oncol. 17, 1497–1508.

Larkin, J., Chiarion-Sileni, V., Gonzalez, R., Grob, J.J., Cowey, C.L., Lao, C.D., Schadendorf, D., Dummer, R., Smylie, M., Rutkowski, P., Ferrucci, P.F., Hill, A., Wagstaff, J., Carlino, M.S., Haanen, J.B., Maio, M., Marquez-Rodas, I., Mcarthur, G.A., Ascierto, P.A., Long, G.V., Callahan, M.K., Postow, M.A., Grossmann, K., Sznol, M., Dreno, B., Bastholt, L., Yang, A., Rollin, L.M., Horak, C., Hodi, F.S., Wolchok, J.D., 2015. Combined nivolumab and ipilimumab or monotherapy in untreated melanoma. N. Engl. J. Med. 373, 23–34.

Le Boeuf, F., Selman, M., Son, H.H., Bergeron, A., Chen, A., Tsang, J., Butterwick, D., Arulanandam, R., Forbes, N.E., Tzelepis, F., Bell, J.C., Werier, J., Abdelbary, H., Diallo, J.S., 2017. Oncolytic Maraba virus MG1 as a treatment for sarcoma. Int. J. Cancer 141, 1257–1264.

Le Mercier, I., Chen, W., Lines, J.L., Day, M., Li, J., Sergent, P., Noelle, R.J., Wang, L., 2014. VISTA regulates the development of protective antitumor immunity. Cancer Res. 74, 1933–1944.

Lee, D.W., Kochenderfer, J.N., Stetler-Stevenson, M., Cui, Y.K., Delbrook, C., Feldman, S.A., Fry, T.J., Orentas, R., Sabatino, M., Shah, N.N., Steinberg, S.M., Stroncek, D., Tschernia, N., Yuan, C., Zhang, H., Zhang, L., Rosenberg, S.A., Wayne, A.S., Mackall, C.L., 2015. T cells expressing CD19 chimeric antigen receptors for acute lymphoblastic leukaemia in children and young adults: a phase 1 dose-escalation trial. Lancet 385, 517–528.

Lei, F., Song, J., Haque, R., Haque, M., Xiong, X., Fang, D., Croft, M., Song, J., 2013. Regulation of A1 by OX40 contributes to CD8(+) T cell survival and anti-tumor activity. PLoS One 8, e70635.

Leitner, J., Klauser, C., Pickl, W.F., Stockl, J., Majdic, O., Bardet, A.F., Kreil, D.P., Dong, C., Yamazaki, T., Zlabinger, G., Pfistershammer, K., Steinberger, P., 2009. B7-H3 is a potent inhibitor of human T-cell activation: No evidence for B7-H3 and TREML2 interaction. Eur. J. Immunol. 39, 1754–1764.

Levine, B.L., Miskin, J., Wonnacott, K., Keir, C., 2017. Global manufacturing of CAR T cell therapy. Mol. Ther. Methods Clin. Dev. 4, 92–101.

Lewkowich, I.P., Lajoie, S., Stoffers, S.L., Suzuki, Y., Richgels, P.K., Dienger, K., Sproles, A.A., Yagita, H., Hamid, Q., WILLS-Karp, M., 2013. PD-L2 modulates asthma severity by directly decreasing dendritic cell IL-12 production. Mucosal Immunol. 6, 728–739.

Li, Y., Yang, X., Wu, Y., Zhao, K., Ye, Z., Zhu, J., Xu, X., Zhao, X., Xing, C., 2017. B7-H3 promotes gastric cancer cell migration and invasion. Oncotarget 8, 71725–71735.

Liakou, C.I., Kamat, A., Tang, D.N., Chen, H., Sun, J., Troncoso, P., Logothetis, C., Sharma, P., 2008. CTLA-4 blockade increases IFNgamma-producing CD4+ICOShi cells to shift the ratio of effector to regulatory T cells in cancer patients. Proc. Natl. Acad. Sci. U.S.A 105, 14987–14992.

Lines, J.L., Pantazi, E., Mak, J., Sempere, L.F., Wang, L., O'connell, S., Ceeraz, S., Suriawinata, A.A., Yan, S., Ernstoff, M.S., Noelle, R., 2014a. VISTA is an immune checkpoint molecule for human T cells. Cancer Res. 74, 1924–1932.

Lines, J.L., Sempere, L.F., Broughton, T., Wang, L., Noelle, R., 2014b. VISTA is a novel broad-spectrum negative checkpoint regulator for cancer immunotherapy. Cancer Immunol. Res. 2, 510–517.

Lipson, E.J., Vincent, J.G., Loyo, M., Kagohara, L.T., Luber, B.S., Wang, H., Xu, H., Nayar, S.K., Wang, T.S., Sidransky, D., Anders, R.A., Topalian, S.L., Taube, J.M., 2013. PD-L1 expression in the Merkel cell carcinoma microenvironment: association with inflammation, Merkel cell polyomavirus and overall survival. Cancer Immunol. Res. 1, 54–63.

Liu, B.L., Robinson, M., Han, Z.Q., Branston, R.H., English, C., Reay, P., Mcgrath, Y., Thomas, S.K., Thornton, M., Bullock, P., Love, C.A., Coffin, R.S., 2003. ICP34.5 deleted herpes simplex virus with enhanced oncolytic, immune stimulating, and anti-tumour properties. Gene Ther. 10, 292–303.

Liu, J., Yuan, Y., Chen, W., Putra, J., Suriawinata, A.A., Schenk, A.D., Miller, H.E., Guleria, I., Barth, R.J., Huang, Y.H., Wang, L., 2015. Immune-checkpoint proteins VISTA and PD-1 nonredundantly regulate murine T-cell responses. Proc. Natl. Acad. Sci. U.S.A 112, 6682–6687.

Liu, T.C., Galanis, E., Kirn, D., 2007. Clinical trial results with oncolytic virotherapy: a century of promise, a decade of progress. Nat. Clin. Pract. Oncol. 4, 101–117.

Long, E.O., Kim, H.S., Liu, D., Peterson, M.E., Rajagopalan, S., 2013. Controlling natural killer cell responses: integration of signals for activation and inhibition. Annu. Rev. Immunol. 31, 227–258.

Loo, D., Alderson, R.F., Chen, F.Z., Huang, L., Zhang, W., Gorlatov, S., Burke, S., Ciccarone, V., Li, H., Yang, Y., Son, T., Chen, Y., Easton, A.N., Li, J.C., Rillema, J.R., Licea, M., Fieger, C., Liang, T.W., Mather, J.P., Koenig, S., Stewart, S.J., Johnson, S., Bonvini, E., Moore, P.A., 2012. Development of an Fc-enhanced anti-B7-H3 monoclonal antibody with potent antitumor activity. Clin. Cancer Res. 18, 3834–3845.

Loos, M., Hedderich, D.M., Friess, H., Kleeff, J., 2010. B7-h3 and its role in antitumor immunity. Clin. Dev. Immunol. 2010, 683875.

Loos, M., Hedderich, D.M., Ottenhausen, M., Giese, N.A., Laschinger, M., Esposito, I., Kleeff, J., Friess, H., 2009. Expression of the costimulatory molecule B7-H3 is associated with prolonged survival in human pancreatic cancer. BMC Canc. 9, 463.

Maio, M., Grob, J.J., Aamdal, S., Bondarenko, I., Robert, C., Thomas, L., Garbe, C., Chiarion-Sileni, V., Testori, A., Chen, T.T., Tschaika, M., Wolchok, J.D., 2015. Five-year survival rates for treatment-naive patients with advanced melanoma who received ipilimumab plus dacarbazine in a phase III trial. J. Clin. Oncol. 33, 1191–1196.

Maleki Vareki, S., Chen, D., Di Cresce, C., Ferguson, P.J., Figueredo, R., Pampillo, M., Rytelewski, M., Vincent, M., Min, W., Zheng, X., Koropatnick, J., 2015. IDO down-regulation induces sensitivity to pemetrexed, gemcitabine, FK866, and methoxyamine in human cancer cells. PLoS One 10, e0143435.

Maleki Vareki, S., Rytelewski, M., Figueredo, R., Chen, D., Ferguson, P.J., Vincent, M., Min, W., Zheng, X., Koropatnick, J., 2014. Indoleamine 2,3-dioxygenase mediates immune-independent human tumor cell resistance to olaparib, gamma radiation, and cisplatin. Oncotarget 5, 2778–2791.

Maude, S.L., Frey, N., Shaw, P.A., Aplenc, R., Barrett, D.M., Bunin, N.J., Chew, A., Gonzalez, V.E., Zheng, Z., Lacey, S.F., Mahnke, Y.D., Melenhorst, J.J., Rheingold, S.R., Shen, A., Teachey, D.T., Levine, B.L., June, C.H., Porter, D.L., Grupp, S.A., 2014. Chimeric antigen receptor T cells for sustained remissions in leukemia. N. Engl. J. Med. 371, 1507–1517.

Maude, S.L., Teachey, D.T., Porter, D.L., Grupp, S.A., 2015. CD19-targeted chimeric antigen receptor T-cell therapy for acute lymphoblastic leukemia. Blood 125, 4017–4023.

Mcdermott, D., Haanen, J., Chen, T.T., Lorigan, P., O'day, S., Investigators, M.D.X., 2013. Efficacy and safety of ipilimumab in metastatic melanoma patients surviving more than 2 years following treatment in a phase III trial (MDX010-20). Ann. Oncol. 24, 2694–2698.

Melero, I., Murillo, O., Dubrot, J., Hervas-Stubbs, S., Perez-Gracia, J.L., 2008. Multi-layered action mechanisms of CD137 (4-1BB)-targeted immunotherapies. Trends Pharmacol. Sci. 29, 383–390.

Merck, 2018a. Merck's KEYTRUDA® (Pembrolizumab) Significantly Improved Overall Survival and Progression-free Survival as First-line Treatment in Combination with Pemetrexed and Platinum Chemotherapy for Patients with Metastatic Nonsquamous Non-small Cell Lung Cancer (KEYNOTE-189). Merck & Co.

Merck, 2018b. Merck's KEYTRUDA® (Pembrolizumab) Significantly Improved Recurrence-free Survival Compared to Placebo as Adjuvant Therapy in Patients with Stage 3 Resected High-risk Melanoma (EORTC1325/KEYNOTE-054). Merck & Co.

Milone, M.C., Fish, J.D., Carpenito, C., Carroll, R.G., Binder, G.K., Teachey, D., Samanta, M., Lakhal, M., Gloss, B., Danet-Desnoyers, G., Campana, D., Riley, J.L., Grupp, S.A., June, C.H., 2009. Chimeric receptors containing CD137 signal transduction domains mediate enhanced survival of T cells and increased antileukemic efficacy in vivo. Mol. Ther. 17, 1453–1464.

Monney, L., Sabatos, C.A., Gaglia, J.L., Ryu, A., Waldner, H., Chernova, T., Manning, S., Greenfield, E.A., Coyle, A.J., Sobel, R.A., Freeman, G.J., Kuchroo, V.K., 2002. Th1-specific cell surface protein Tim-3 regulates macrophage activation and severity of an autoimmune disease. Nature 415, 536–541.

Motzer, R.J., Tannir, N.M., Mcdermott, D.F., Aren Frontera, O., Melichar, B., Choueiri, T.K., Plimack, E.R., Barthelemy, P., Porta, C., George, S., Powles, T., Donskov, F., Neiman, V., Kollmannsberger, C.K., Salman, P., Gurney, H., Hawkins, R., Ravaud, A., Grimm, M.O., Bracarda, S., Barrios, C.H., Tomita, Y., Castellano, D., Rini, B.I., Chen, A.C., Mekan, S., Mchenry, M.B., Wind-Rotolo, M., Doan, J., Sharma, P., Hammers, H.J., Escudier, B., Checkmate, I., 2018. Nivolumab plus ipilimumab versus sunitinib in advanced renal-cell carcinoma. N. Engl. J. Med. 378, 1277–1290.

Mullen, J.T., Tanabe, K.K., 2002. Viral oncolysis. Oncol. 7, 106–119.

Munn, D.H., Shafizadeh, E., Attwood, J.T., Bondarev, I., Pashine, A., Mellor, A.L., 1999. Inhibition of T cell proliferation by macrophage tryptophan catabolism. J. Exp. Med. 189, 1363–1372.

Nadler, L.M., Anderson, K.C., Marti, G., Bates, M., Park, E., Daley, J.F., Schlossman, S.F., 1983. B4, a human B lymphocyte-associated antigen expressed on normal, mitogen-activated, and malignant B lymphocytes. J. Immunol. 131, 244–250.

Neelapu, S.S., Locke, F.L., Bartlett, N.L., Lekakis, L.J., Miklos, D.B., Jacobson, C.A., Braunschweig, I., Oluwole, O.O., Siddiqi, T., Lin, Y., Timmerman, J.M., Stiff, P.J., Friedberg, J.W., Flinn, I.W., Goy, A., Hill, B.T., Smith, M.R., Deol, A., Farooq, U., Mcsweeney, P., Munoz, J., Avivi, I., Castro, J.E., Westin, J.R., Chavez, J.C., Ghobadi, A., Komanduri, K.V., Levy, R., Jacobsen, E.D., Witzig, T.E., Reagan, P., Bot, A., Rossi, J., Navale, L., Jiang, Y., Aycock, J., Elias, M., Chang, D., Wiezorek, J., Go, W.Y., 2017. Axicabtagene ciloleucel CAR t-cell therapy in refractory large B-Cell lymphoma. N. Engl. J. Med. 377, 2531–2544.

Neelapu, S.S., Tummala, S., Kebriaei, P., Wierda, W., Gutierrez, C., Locke, F.L., Komanduri, K.V., Lin, Y., Jain, N., Daver, N., Westin, J., Gulbis, A.M., Loghin, M.E., De Groot, J.F., Adkins, S., Davis, S.E., Rezvani, K., Hwu, P., Shpall, E.J., 2018. Chimeric antigen receptor T-cell therapy - assessment and management of toxicities. Nat. Rev. Clin. Oncol. 15, 47–62.

Newick, K., O'brien, S., Moon, E., Albelda, S.M., 2017. CAR T cell therapy for solid tumors. Annu. Rev. Med. 68, 139–152.

Ngiow, S.F., Von Scheidt, B., Akiba, H., Yagita, H., Teng, M.W., Smyth, M.J., 2011. Anti-TIM3 antibody promotes T cell IFN-gamma-mediated antitumor immunity and suppresses established tumors. Cancer Res. 71, 3540–3551.

Nivolumab, F.L., 2014. OPDIVO (nivolumab) injection, for intravenous use. In: Administration, F.D. (Ed.).

Nolte, M.A., Van Olffen, R.W., Van Gisbergen, K.P., Van Lier, R.A., 2009. Timing and tuning of CD27-CD70 interactions: the impact of signal strength in setting the balance between adaptive responses and immunopathology. Immunol. Rev. 229, 216–231.

Noonan, A.M., Farren, M.R., Geyer, S.M., Huang, Y., Tahiri, S., Ahn, D., Mikhail, S., Ciombor, K.K., Pant, S., Aparo, S., Sexton, J., Marshall, J.L., Mace, T.A., Wu, C.S., EL-Rayes, B., Timmers, C.D., Zwiebel, J., Lesinski, G.B., Villalona-Calero, M.A., Bekaii-Saab, T.S., 2016. Randomized phase 2 trial of the oncolytic virus pelareorep (Reolysin) in upfront treatment of metastatic pancreatic adenocarcinoma. Mol. Ther. 24, 1150–1158.

Nowak, A.K., Cook, A.M., Mcdonnell, A.M., Millward, M.J., Creaney, J., Francis, R.J., Hasani, A., Segal, A., Musk, A.W., Turlach, B.A., Mccoy, M.J., Robinson, B.W., Lake, R.A., 2015. A phase 1b clinical trial of the CD40-activating antibody CP-870,893 in combination with cisplatin and pemetrexed in malignant pleural mesothelioma. Ann. Oncol. 26, 2483–2490.

Okamoto, A., Nikaido, T., Ochiai, K., Takakura, S., Saito, M., Aoki, Y., Ishii, N., Yanaihara, N., Yamada, K., Takikawa, O., Kawaguchi, R., Isonishi, S., Tanaka, T., Urashima, M., 2005. Indoleamine 2,3-dioxygenase serves as a marker of poor prognosis in gene expression profiles of serous ovarian cancer cells. Clin. Cancer Res. 11, 6030–6039.

Okazaki, T., Maeda, A., Nishimura, H., Kurosaki, T., Honjo, T., 2001. PD-1 immunoreceptor inhibits B cell receptor-mediated signaling by recruiting src homology 2-domain-containing tyrosine phosphatase 2 to phosphotyrosine. Proc. Natl. Acad. Sci. U.S.A 98, 13866–13871.

Oomizu, S., Arikawa, T., Niki, T., Kadowaki, T., Ueno, M., Nishi, N., Yamauchi, A., Hattori, T., Masaki, T., Hirashima, M., 2012. Cell surface galectin-9 expressing Th cells regulate Th17 and Foxp3+ Treg development by galectin-9 secretion. PLoS One 7, e48574.

Pai-Scherf, L., Blumenthal, G.M., Li, H., Subramaniam, S., Mishra-Kalyani, P.S., He, K., Zhao, H., Yu, J., Paciga, M., Goldberg, K.B., Mckee, A.E., Keegan, P., Pazdur, R., 2017. FDA approval summary: pembrolizumab for treatment of metastatic non-small cell lung cancer: first-line therapy and beyond. Oncol. 22, 1392–1399.

Parkhurst, M., Gros, A., Pasetto, A., Prickett, T., Crystal, J.S., Robbins, P., Rosenberg, S.A., 2017. Isolation of t-cell receptors specifically reactive with mutated tumor-associated antigens from tumor-infiltrating lymphocytes based on CD137 expression. Clin. Cancer Res. 23, 2491–2505.

Parry, R.V., Chemnitz, J.M., Frauwirth, K.A., Lanfranco, A.R., Braunstein, I., Kobayashi, S.V., Linsley, P.S., Thompson, C.B., Riley, J.L., 2005. CTLA-4 and PD-1 receptors inhibit T-cell activation by distinct mechanisms. Mol. Cell Biol. 25, 9543–9553.

Passiglia, F., Galvano, A., Rizzo, S., Incorvaia, L., Listi, A., Bazan, V., Russo, A., 2018. Looking for the best immune-checkpoint inhibitor in pre-treated NSCLC patients: an indirect comparison between nivolumab, pembrolizumab and atezolizumab. Int. J. Cancer 142, 1277–1284.

Patel, M.R., Ellerton, J., Infante, J.R., Agrawal, M., Gordon, M., Aljumaily, R., Britten, C.D., Dirix, L., Lee, K.W., Taylor, M., Schoffski, P., Wang, D., Ravaud, A., Gelb, A.B., Xiong, J., Rosen, G., Gulley, J.L., Apolo, A.B., 2017. Avelumab in metastatic urothelial carcinoma after platinum failure (JAVELIN Solid Tumor): pooled results from two expansion cohorts of an open-label, phase 1 trial. Lancet Oncol.

Perez-Gracia, J.L., Loriot, Y., Rosenberg, J.E., Powles, T., Necchi, A., Hussain, S.A., Morales-Barrera, R., Retz, M.M., Niegisch, G., Duran, I., Theodore, C., Grande, E., Shen, X., Wang, J., Nelson, B., Derleth, C.L., Van Der Heijden, M.S., 2017. Atezolizumab in platinum-treated locally advanced or metastatic urothelial carcinoma: outcomes by prior number of regimens. Eur. Urol.

Petricevic, B., Laengle, J., Singer, J., Sachet, M., Fazekas, J., Steger, G., Bartsch, R., Jensen-Jarolim, E., Bergmann, M., 2013. Trastuzumab mediates antibody-dependent cell-mediated cytotoxicity and phagocytosis to the same extent in both adjuvant and metastatic HER2/neu breast cancer patients. J. Transl. Med. 11, 307.

Ping, Y., Liu, C., Zhang, Y., 2017. T-cell receptor-engineered T cells for cancer treatment: current status and future directions. Protein Cell.

Postow, M.A., 2015. Managing immune checkpoint-blocking antibody side effects. Am. Soc. Clin. Oncol. Educ. Book 76–83.

Poulsen, M., 2004. Merkel-cell carcinoma of the skin. Lancet Oncol. 5, 593–599.

Powles, T., Duran, I., Van Der Heijden, M.S., Loriot, Y., Vogelzang, N.J., De Giorgi, U., Oudard, S., Retz, M.M., Castellano, D., Bamias, A., Flechon, A., Gravis, G., Hussain, S., Takano, T., Leng, N., Kadel 3rd, E.E., Banchereau, R., Hegde, P.S., Mariathasan, S., Cui, N., Shen, X., Derleth, C.L., Green, M.C., Ravaud, A., 2017. Atezolizumab versus chemotherapy in patients with platinum-treated locally advanced or metastatic urothelial carcinoma (IMvigor211): a multicentre, open-label, phase 3 randomised controlled trial. Lancet.

Qureshi, O.S., Zheng, Y., Nakamura, K., Attridge, K., Manzotti, C., Schmidt, E.M., Baker, J., Jeffery, L.E., Kaur, S., Briggs, Z., Hou, T.Z., Futter, C.E., Anderson, G., Walker, L.S., Sansom, D.M., 2011. Trans-endocytosis of CD80 and CD86: a molecular basis for the cell-extrinsic function of CTLA-4. Science 332, 600–603.

Rehman, H., Silk, A.W., Kane, M.P., Kaufman, H.L., 2016. Into the clinic: Talimogene laherparepvec (T-VEC), a first-in-class intratumoral oncolytic viral therapy. J. Immunother. Cancer 4, 53.

Ribas, A., Camacho, L.H., Lopez-Berestein, G., Pavlov, D., Bulanhagui, C.A., Millham, R., Comin-Anduix, B., Reuben, J.M., Seja, E., Parker, C.A., Sharma, A., Glaspy, J.A., Gomez-Navarro, J., 2005. Antitumor activity in melanoma and anti-self responses in a phase I trial with the anti-cytotoxic T lymphocyte-associated antigen 4 monoclonal antibody CP-675,206. J. Clin. Oncol. 23, 8968–8977.

Ribas, A., Dummer, R., Puzanov, I., Vanderwalde, A., Andtbacka, R.H.I., Michielin, O., Olszanski, A.J., Malvehy, J., Cebon, J., Fernandez, E., Kirkwood, J.M., Gajewski, T.F., Chen, L., Gorski, K.S., Anderson, A.A., Diede, S.J., Lassman, M.E., Gansert, J., Hodi, F.S., Long, G.V., 2017. Oncolytic virotherapy promotes intratumoral T cell infiltration and improves anti-PD-1 immunotherapy. Cell 170, 1109–1119 e10.

Ribas, A., Kefford, R., Marshall, M.A., Punt, C.J., Haanen, J.B., Marmol, M., Garbe, C., Gogas, H., Schachter, J., Linette, G., Lorigan, P., Kendra, K.L., Maio, M., Trefzer, U., Smylie, M., Mcarthur, G.A., Dreno, B., Nathan, P.D., Mackiewicz, J., Kirkwood, J.M., Gomez-Navarro, J., Huang, B., Pavlov, D., Hauschild, A., 2013. Phase III randomized clinical trial comparing tremelimumab with standard-of-care chemotherapy in patients with advanced melanoma. J. Clin. Oncol. 31, 616–622.

Riether, C., Schurch, C., Ochsenbein, A.F., 2012. Modulating CD27 signaling to treat cancer. OncoImmunology 1, 1604–1606.

Robert, C., Thomas, L., Bondarenko, I., O'day, S., Weber, J., Garbe, C., Lebbe, C., Baurain, J.F., Testori, A., Grob, J.J., Davidson, N., Richards, J., Maio, M., Hauschild, A., Miller Jr., W.H., Gascon, P., Lotem, M., Harmankaya, K., Ibrahim, R., Francis, S., Chen, T.T., Humphrey, R., Hoos, A., Wolchok, J.D., 2011. Ipilimumab plus dacarbazine for previously untreated metastatic melanoma. N. Engl. J. Med. 364, 2517–2526.

Romagne, F., Andre, P., Spee, P., Zahn, S., Anfossi, N., Gauthier, L., Capanni, M., Ruggeri, L., Benson JR., D.M., Blaser, B.W., Della Chiesa, M., Moretta, A., Vivier, E., Caligiuri, M.A., Velardi, A., Wagtmann, N., 2009. Preclinical characterization of 1-7F9, a novel human anti-KIR receptor therapeutic antibody that augments natural killer-mediated killing of tumor cells. Blood 114, 2667–2677.

Romano, E., Kusio-Kobialka, M., Foukas, P.G., Baumgaertner, P., Meyer, C., Ballabeni, P., Michielin, O., Weide, B., Romero, P., Speiser, D.E., 2015. Ipilimumab-dependent cell-mediated cytotoxicity of regulatory T cells ex vivo by nonclassical monocytes in melanoma patients. Proc. Natl. Acad. Sci. U.S.A 112, 6140–6145.

Rosenberg, J.E., Hoffman-Censits, J., Powles, T., Van Der Heijden, M.S., Balar, A.V., Necchi, A., Dawson, N., O'donnell, P.H., Balmanoukian, A., Loriot, Y., Srinivas, S., Retz, M.M., Grivas, P., Joseph, R.W., Galsky, M.D., Fleming, M.T., Petrylak, D.P., Perez-Gracia, J.L., Burris, H.A., Castellano, D., Canil, C., Bellmunt, J., Bajorin, D., Nickles, D., Bourgon, R., Frampton, G.M., Cui, N., Mariathasan, S., Abidoye, O., Fine, G.D., Dreicer, R., 2016. Atezolizumab in patients with locally advanced and metastatic urothelial carcinoma who have progressed following treatment with platinum-based chemotherapy: a single-arm, multicentre, phase 2 trial. Lancet 387, 1909–1920.

Rosenberg, S.A., 1999. A new era for cancer immunotherapy based on the genes that encode cancer antigens. Immunity 10, 281–287.

Rosenberg, S.A., 2011. Cell transfer immunotherapy for metastatic solid cancer—what clinicians need to know. Nat. Rev. Clin. Oncol. 8, 577–585.

Rosenberg, S.A., Packard, B.S., Aebersold, P.M., Solomon, D., Topalian, S.L., Toy, S.T., Simon, P., Lotze, M.T., Yang, J.C., Seipp, C.A., et al., 1988. Use of tumor-infiltrating lymphocytes and interleukin-2 in the immunotherapy of patients with metastatic melanoma. A preliminary report. N. Engl. J. Med. 319, 1676–1680.

Rosenberg, S.A., Restifo, N.P., 2015. Adoptive cell transfer as personalized immunotherapy for human cancer. Science 348, 62–68.
Roth, T.J., Sheinin, Y., Lohse, C.M., Kuntz, S.M., Frigola, X., Inman, B.A., Krambeck, A.E., Mckenney, M.E., Karnes, R.J., Blute, M.L., Cheville, J.C., Sebo, T.J., Kwon, E.D., 2007. B7-H3 ligand expression by prostate cancer: a novel marker of prognosis and potential target for therapy. Cancer Res. 67, 7893–7900.
Rudolph, M.G., Stanfield, R.L., Wilson, I.A., 2006. How TCRs bind MHCs, peptides, and coreceptors. Annu. Rev. Immunol. 24, 419–466.
Sadelain, M., Brentjens, R., Riviere, I., 2009. The promise and potential pitfalls of chimeric antigen receptors. Curr. Opin. Immunol. 21, 215–223.
Sadelain, M., Brentjens, R., Riviere, I., 2013. The basic principles of chimeric antigen receptor design. Cancer Discov. 3, 388–398.
Sakuishi, K., Apetoh, L., Sullivan, J.M., Blazar, B.R., Kuchroo, V.K., Anderson, A.C., 2010. Targeting Tim-3 and PD-1 pathways to reverse T cell exhaustion and restore anti-tumor immunity. J. Exp. Med. 207, 2187–2194.
Sanchez-Paulete, A.R., Labiano, S., Rodriguez-Ruiz, M.E., Azpilikueta, A., Etxeberria, I., Bolanos, E., Lang, V., Rodriguez, M., Aznar, M.A., Jure-Kunkel, M., Melero, I., 2016. Deciphering CD137 (4-1BB) signaling in T-cell costimulation for translation into successful cancer immunotherapy. Eur. J. Immunol. 46, 513–522.
Schadendorf, D., Hodi, F.S., Robert, C., Weber, J.S., Margolin, K., Hamid, O., Patt, D., Chen, T.T., Berman, D.M., Wolchok, J.D., 2015. Pooled analysis of long-term survival data from phase II and phase III trials of ipilimumab in unresectable or metastatic melanoma. J. Clin. Oncol. 33, 1889–1894.
Schaer, D.A., Li, Y., Merghoub, T., Rizzuto, G.A., Shemesh, A., Cohen, A.D., Li, Y., Avogadri, F., Toledo-Crow, R., Houghton, A.N., Wolchok, J.D., 2011. Detection of intra-tumor self antigen recognition during melanoma tumor progression in mice using advanced multimode confocal/two photon microscope. PLoS One 6, e21214.
Scheuermann, R.H., Racila, E., 1995. CD19 antigen in leukemia and lymphoma diagnosis and immunotherapy. Leuk. Lymphoma 18, 385–397.
Schuster, S.J., Svoboda, J., Chong, E.A., Nasta, S.D., Mato, A.R., Anak, O., Brogdon, J.L., Pruteanu-Malinici, I., Bhoj, V., Landsburg, D., Wasik, M., Levine, B.L., Lacey, S.F., Melenhorst, J.J., Porter, D.L., June, C.H., 2017. Chimeric antigen receptor T cells in refractory B-Cell lymphomas. N. Engl. J. Med. 377, 2545–2554.
Scurr, M., Ladell, K., Besneux, M., Christian, A., Hockey, T., Smart, K., Bridgeman, H., Hargest, R., Phillips, S., Davies, M., Price, D., Gallimore, A., Godkin, A., 2014. Highly prevalent colorectal cancer-infiltrating LAP(+) Foxp3(-) T cells exhibit more potent immunosuppressive activity than Foxp3(+) regulatory T cells. Mucosal Immunol. 7, 428–439.
Segal, N.H., Logan, T.F., Hodi, F.S., Mcdermott, D., Melero, I., Hamid, O., Schmidt, H., Robert, C., Chiarion-Sileni, V., Ascierto, P.A., Maio, M., Urba, W.J., Gangadhar, T.C., Suryawanshi, S., Neely, J., Jure-Kunkel, M., Krishnan, S., Kohrt, H., Sznol, M., Levy, R., 2017. Results from an integrated safety analysis of urelumab, an agonist anti-CD137 monoclonal antibody. Clin. Cancer Res. 23, 1929–1936.
Sharma, P., Allison, J.P., 2015a. The future of immune checkpoint therapy. Science 348, 56–61.
Sharma, P., Allison, J.P., 2015b. Immune checkpoint targeting in cancer therapy: toward combination strategies with curative potential. Cell 161, 205–214.
Sharpe, A.H., Abbas, A.K., 2006. T-cell costimulation—biology, therapeutic potential, and challenges. N. Engl. J. Med. 355, 973–975.

Shen, P., Yue, R., Tang, J., Si, H., Shen, L., Guo, C., Zhang, L., Han, H., Song, H.K., Zhao, P., Wang, N., Song, Z., Guo, C., 2016. Preferential Tim-3 expression on Treg and CD8(+) T cells, supported by tumor-associated macrophages, is associated with worse prognosis in gastric cancer. Am. J. Transl. Res. 8, 3419–3428.

Shimizu, J., Yamazaki, S., Takahashi, T., Ishida, Y., Sakaguchi, S., 2002. Stimulation of CD25(+)CD4(+) regulatory T cells through GITR breaks immunological self-tolerance. Nat. Immunol. 3, 135–142.

Soliman, H.H., Jackson, E., Neuger, T., Dees, E.C., Harvey, R.D., Han, H., Ismail-Khan, R., Minton, S., Vahanian, N.N., Link, C., Sullivan, D.M., Antonia, S., 2014. A first in man phase I trial of the oral immunomodulator, indoximod, combined with docetaxel in patients with metastatic solid tumors. Oncotarget 5, 8136–8146.

Soliman, H.H., Minton, S.E., Han, H.S., Ismail-Khan, R., Neuger, A., Khambati, F., Noyes, D., Lush, R., Chiappori, A.A., Roberts, J.D., Link, C., Vahanian, N.N., Mautino, M., Streicher, H., Sullivan, D.M., Antonia, S.J., 2016. A phase I study of indoximod in patients with advanced malignancies. Oncotarget 7, 22928–22938.

Song, J., So, T., Cheng, M., Tang, X., Croft, M., 2005. Sustained survivin expression from OX40 costimulatory signals drives T cell clonal expansion. Immunity 22, 621–631.

Song, J., So, T., Croft, M., 2008. Activation of NF-kappaB1 by OX40 contributes to antigen-driven T cell expansion and survival. J. Immunol. 180, 7240–7248.

Spain, L., Diem, S., Larkin, J., 2016. Management of toxicities of immune checkpoint inhibitors. Cancer Treat Rev. 44, 51–60.

Stanziale, S.F., Petrowsky, H., Adusumilli, P.S., Ben-Porat, L., Gonen, M., Fong, Y., 2004. Infection with oncolytic herpes simplex virus-1 induces apoptosis in neighboring human cancer cells: a potential target to increase anticancer activity. Clin Cancer Res. 10, 3225–3232.

Stewart, R., Morrow, M., Hammond, S.A., Mulgrew, K., Marcus, D., Poon, E., Watkins, A., Mullins, S., Chodorge, M., Andrews, J., Bannister, D., Dick, E., Crawford, N., Parmentier, J., Alimzhanov, M., Babcook, J.S., Foltz, I.N., Buchanan, A., Bedian, V., Wilkinson, R.W., Mccourt, M., 2015. Identification and characterization of MEDI4736, an antagonistic anti-PD-L1 monoclonal antibody. Cancer Immunol. Res. 3, 1052–1062.

Suerth, J.D., Schambach, A., Baum, C., 2012. Genetic modification of lymphocytes by retrovirus-based vectors. Curr. Opin. Immunol. 24, 598–608.

Sugamura, K., Ishii, N., Weinberg, A.D., 2004. Therapeutic targeting of the effector T-cell co-stimulatory molecule OX40. Nat. Rev. Immunol. 4, 420–431.

Sukari, A., Nagasaka, M., Al-Hadidi, A., Lum, L.G., 2016. Cancer immunology and immunotherapy. Anticancer Res. 36, 5593–5606.

Sun, J., Chen, L.J., Zhang, G.B., Jiang, J.T., Zhu, M., Tan, Y., Wang, H.T., Lu, B.F., Zhang, X.G., 2010. Clinical significance and regulation of the costimulatory molecule B7-H3 in human colorectal carcinoma. Cancer Immunol. Immunother. 59, 1163–1171.

Sun, Y., Wang, Y., Zhao, J., Gu, M., Giscombe, R., Lefvert, A.K., Wang, X., 2006. B7-H3 and B7-H4 expression in non-small-cell lung cancer. Lung Canc. 53, 143–151.

Syed, Y.Y., 2017. Durvalumab: first global approval. Drugs 77, 1369–1376.

Tajiri, K., Aonuma, K., Sekine, I., 2018. Immune checkpoint inhibitor-related myocarditis. Jpn. J. Clin. Oncol. 48, 7–12.

Takahashi, T., Tagami, T., Yamazaki, S., Uede, T., Shimizu, J., Sakaguchi, N., Mak, T.W., Sakaguchi, S., 2000. Immunologic self-tolerance maintained by CD25(+)CD4(+) regulatory T cells constitutively expressing cytotoxic T lymphocyte-associated antigen 4. J. Exp. Med. 192, 303–310.

Tang, A.L., Teijaro, J.R., Njau, M.N., Chandran, S.S., Azimzadeh, A., Nadler, S.G., Rothstein, D.M., Farber, D.L., 2008. CTLA4 expression is an indicator and regulator of steady-state CD4+ FoxP3+ T cell homeostasis. J. Immunol. 181, 1806–1813.

Tomkowicz, B., Walsh, E., Cotty, A., Verona, R., Sabins, N., Kaplan, F., Santulli-Marotto, S., Chin, C.N., Mooney, J., Lingham, R.B., Naso, M., Mccabe, T., 2015. TIM-3 suppresses anti-CD3/CD28-induced TCR activation and IL-2 expression through the NFAT signaling pathway. PLoS One 10, e0140694.

Tutt, A.L., O'brien, L., Hussain, A., Crowther, G.R., French, R.R., Glennie, M.J., 2002. T cell immunity to lymphoma following treatment with anti-CD40 monoclonal antibody. J. Immunol. 168, 2720–2728.

Van der Bruggen, P., Traversari, C., Chomez, P., Lurquin, C., De Plaen, E., Van Den Eynde, B., Knuth, A., Boon, T., 1991. A gene encoding an antigen recognized by cytolytic T lymphocytes on a human melanoma. Science 254, 1643–1647.

Van Kooten, C., Banchereau, J., 2000. CD40-CD40 ligand. J. Leukoc. Biol. 67, 2–17.

Vanky, F., Klein, E., Willems, J., Book, K., Ivert, T., Peterffy, A., Nilsonne, U., Kreicbergs, A., Aparisi, T., 1986. Lysis of autologous tumor cells by blood lymphocytes tested at the time of surgery. Correlation with the postsurgical clinical course. Cancer Immunol. Immunother. 21, 69–76.

Vennapusa, B., Baker, B., Kowanetz, M., Boone, J., Menzl, I., Bruey, J.M., Fine, G., Mariathasan, S., McCaffery, I., Mocci, S., Rost, S., Smith, D., Dennis, E., Tang, S.Y., Damadzadeh, B., Walker, E., Hegde, P.S., Williams, J.A., Koeppen, H., Boyd, Z., 2018. Development of a PD-L1 complementary diagnostic immunohistochemistry assay (SP142) for atezolizumab. Appl. Immunohistochem. Mol. Morphol. https://doi.org/10.1097/PAI.0000000000000594.

Vey, N., Dumas, P.Y., Recher, C., Gastaud, L., Lioure, B., Bulabois, C.E., Pautas, C., Marolleau, J.P., Lepretre, S., Raffoux, E., Thomas, X., Hicheri, Y., Bonmati, C., Quesnel, B., Rousselot, P., Castaigne, S., Jourdan, E., Malfuson, J.V., Guillerm, G., Bourhis, J.H., Ojeda-Uribe, M., Hunault, M., Ifrah, N., Gardin, C., Delannoy, A., Beautier, L., Paturel, C., André, P., Zerbib, R., Preudhomme, C., Toubert, A., Dulphy, N., Olive, D., Pigneux, A., Dombret, H., 2017. Randomized phase 2 trial of lirilumab (anti-KIR monoclonal antibody, mAb) as maintenance treatment in Elderly patients (pts) with acute myeloid leukemia (AML): results of the Effikir trial. In: ASH 59th Annual Meeting and Exposition. Atlanta, GA.

Vonderheide, R.H., 2007. Prospect of targeting the CD40 pathway for cancer therapy. Clin. Cancer Res. 13, 1083–1088.

Walunas, T.L., Lenschow, D.J., Bakker, C.Y., Linsley, P.S., Freeman, G.J., Green, J.M., Thompson, C.B., Bluestone, J.A., 1994. CTLA-4 can function as a negative regulator of T cell activation. Immunity 1, 405–413.

Wang, C., Thudium, K.B., Han, M., Wang, X.T., Huang, H., Feingersh, D., Garcia, C., Wu, Y., Kuhne, M., Srinivasan, M., Singh, S., Wong, S., Garner, N., Leblanc, H., Bunch, R.T., Blanset, D., Selby, M.J., Korman, A.J., 2014. In vitro characterization of the anti-PD-1 antibody nivolumab, BMS-936558, and in vivo toxicology in non-human primates. Cancer Immunol. Res. 2, 846–856.

Wang, L., Rubinstein, R., Lines, J.L., Wasiuk, A., Ahonen, C., Guo, Y., Lu, L.F., Gondek, D., Wang, Y., Fava, R.A., Fiser, A., Almo, S., Noelle, R.J., 2011. VISTA, a novel mouse Ig superfamily ligand that negatively regulates T cell responses. J. Exp. Med. 208, 577–592.

Ward, E., Mittereder, N., Kuta, E., Sims, G.P., Bowen, M.A., Dall'acqua, W., Tedder, T., Kiener, P., Coyle, A.J., Wu, H., Jallal, B., Herbst, R., 2011. A glycoengineered anti-CD19 antibody with potent antibody-dependent cellular cytotoxicity activity in vitro and lymphoma growth inhibition in vivo. Br. J. Haematol. 155, 426–437.

Waterhouse, P., Penninger, J.M., Timms, E., Wakeham, A., Shahinian, A., Lee, K.P., Thompson, C.B., Griesser, H., Mak, T.W., 1995. Lymphoproliferative disorders with early lethality in mice deficient in Ctla-4. Science 270, 985–988.
Weber, J., Mandala, M., Del Vecchio, M., Gogas, H.J., Arance, A.M., Cowey, C.L., Dalle, S., Schenker, M., Chiarion-Sileni, V., Marquez-Rodas, I., Grob, J.J., Butler, M.O., Middleton, M.R., Maio, M., Atkinson, V., Queirolo, P., Gonzalez, R., Kudchadkar, R.R., Smylie, M., Meyer, N., Mortier, L., Atkins, M.B., Long, G.V., Bhatia, S., Lebbe, C., Rutkowski, P., Yokota, K., Yamazaki, N., Kim, T.M., De Pril, V., Sabater, J., Qureshi, A., Larkin, J., Ascierto, P.A., Checkmate, C., 2017. Adjuvant nivolumab versus ipilimumab in resected stage III or IV melanoma. N. Engl. J. Med. 377, 1824–1835.
Weber, J.S., Kahler, K.C., Hauschild, A., 2012. Management of immune-related adverse events and kinetics of response with ipilimumab. J. Clin. Oncol. 30, 2691–2697.
Weinberg, A.D., Rivera, M.M., Prell, R., Morris, A., Ramstad, T., Vetto, J.T., Urba, W.J., Alvord, G., Bunce, C., Shields, J., 2000. Engagement of the OX-40 receptor in vivo enhances antitumor immunity. J. Immunol. 164, 2160–2169.
Wherry, E.J., 2011. T cell exhaustion. Nat. Immunol. 12, 492–499.
Wherry, E.J., Kurachi, M., 2015. Molecular and cellular insights into T cell exhaustion. Nat. Rev. Immunol. 15, 486–499.
Williams, J.B., Horton, B.L., Zheng, Y., Duan, Y., Powell, J.D., Gajewski, T.F., 2017. The EGR2 targets LAG-3 and 4-1BB describe and regulate dysfunctional antigen-specific CD8+ T cells in the tumor microenvironment. J. Exp. Med. 214, 381–400.
Wing, K., Onishi, Y., Prieto-Martin, P., Yamaguchi, T., Miyara, M., Fehervari, Z., Nomura, T., Sakaguchi, S., 2008. CTLA-4 control over Foxp3+ regulatory T cell function. Science 322, 271–275.
Wolchok, J.D., Chiarion-Sileni, V., Gonzalez, R., Rutkowski, P., Grob, J.J., Cowey, C.L., Lao, C.D., Wagstaff, J., Schadendorf, D., Ferrucci, P.F., Smylie, M., Dummer, R., Hill, A., Hogg, D., Haanen, J., Carlino, M.S., Bechter, O., Maio, M., Marquez-Rodas, I., Guidoboni, M., Mcarthur, G., Lebbe, C., Ascierto, P.A., Long, G.V., Cebon, J., Sosman, J., Postow, M.A., Callahan, M.K., Walker, D., Rollin, L., Bhore, R., Hodi, F.S., Larkin, J., 2017. Overall survival with combined nivolumab and ipilimumab in advanced melanoma. N. Engl. J. Med. 377, 1345–1356.
Woo, S.R., Turnis, M.E., Goldberg, M.V., Bankoti, J., Selby, M., Nirschl, C.J., Bettini, M.L., Gravano, D.M., Vogel, P., Liu, C.L., Tangsombatvisit, S., Grosso, J.F., Netto, G., Smeltzer, M.P., Chaux, A., Utz, P.J., Workman, C.J., Pardoll, D.M., Korman, A.J., Drake, C.G., Vignali, D.A., 2012. Immune inhibitory molecules LAG-3 and PD-1 synergistically regulate T-cell function to promote tumoral immune escape. Cancer Res. 72, 917–927.
Wu, C.P., Jiang, J.T., Tan, M., Zhu, Y.B., Ji, M., Xu, K.F., Zhao, J.M., Zhang, G.B., Zhang, X.G., 2006. Relationship between co-stimulatory molecule B7-H3 expression and gastric carcinoma histology and prognosis. World J. Gastroenterol. 12, 457–459.
Wu, Y., Shi, H., Jiang, M., Qiu, M., Jia, K., Cao, T., Shang, Y., Shi, L., Jiang, K., Wu, H., 2017. The clinical value of combination of immune checkpoint inhibitors in cancer patients: a meta-analysis of efficacy and safety. Int. J. Cancer 141, 2562–2570.
Yoshinaga, S.K., Whoriskey, J.S., Khare, S.D., Sarmiento, U., Guo, J., Horan, T., Shih, G., Zhang, M., Coccia, M.A., Kohno, T., TAFURI-Bladt, A., Brankow, D., Campbell, P., Chang, D., Chiu, L., Dai, T., Duncan, G., Elliott, G.S., Hui, A., Mccabe, S.M., Scully, S., Shahinian, A., Shaklee, C.L., Van, G., Mak, T.W., Senaldi, G., 1999. T-cell co-stimulation through B7RP-1 and ICOS. Nature 402, 827–832.

Yu, X., Harden, K., Gonzalez, L.C., Francesco, M., Chiang, E., Irving, B., Tom, I., Ivelja, S., Refino, C.J., Clark, H., Eaton, D., Grogan, J.L., 2009. The surface protein TIGIT suppresses T cell activation by promoting the generation of mature immunoregulatory dendritic cells. Nat. Immunol. 10, 48–57.

Yue, E.W., Sparks, R., Polam, P., Modi, D., Douty, B., Wayland, B., Glass, B., Takvorian, A., Glenn, J., Zhu, W., Bower, M., Liu, X., Leffet, L., Wang, Q., Bowman, K.J., Hansbury, M.J., Wei, M., Li, Y., Wynn, R., Burn, T.C., Koblish, H.K., Fridman, J.S., Emm, T., Scherle, P.A., Metcalf, B., Combs, A.P., 2017. INCB24360 (epacadostat), a highly potent and selective indoleamine-2,3-dioxygenase 1 (Ido1) inhibitor for immuno-oncology. ACS Med. Chem. Lett. 8, 486–491.

Zakharia, Y., Mcwilliams, R., Shaheen, M., Grossman, K., Drabick, J., Milhem, M., Rixie, O., Khleif, S., Lott, R., Kennedy, E., Munn, D., Vahanian, N., Link, C., 2017. Abstract CT117: interim analysis of the Phase 2 clinical trial of the Ido pathway inhibitor indoximod in combination with pembrolizumab for patients with advanced melanoma. Canc. Res. 77, CT117.

Zang, X., Thompson, R.H., Al-Ahmadie, H.A., Serio, A.M., Reuter, V.E., Eastham, J.A., Scardino, P.T., Sharma, P., Allison, J.P., 2007. B7-H3 and B7x are highly expressed in human prostate cancer and associated with disease spread and poor outcome. Proc. Natl. Acad. Sci. U.S.A 104, 19458–19463.

Zhang, L., Conejo-Garcia, J.R., Katsaros, D., Gimotty, P.A., Massobrio, M., Regnani, G., Makrigiannakis, A., Gray, H., Schlienger, K., Liebman, M.N., Rubin, S.C., Coukos, G., 2003. Intratumoral T cells, recurrence, and survival in epithelial ovarian cancer. N. Engl. J. Med. 348, 203–213.

Zheng, X., Koropatnick, J., Li, M., Zhang, X., Ling, F., Ren, X., Hao, X., Sun, H., Vladau, C., Franek, J.A., Feng, B., Urquhart, B.L., Zhong, R., Freeman, D.J., Garcia, B., Min, W.P., 2006. Reinstalling antitumor immunity by inhibiting tumor-derived immunosuppressive molecule Ido through RNA interference. J. Immunol. 177, 5639–5646.

Zhu, C., Anderson, A.C., Schubart, A., Xiong, H., Imitola, J., Khoury, S.J., Zheng, X.X., Strom, T.B., Kuchroo, V.K., 2005. The Tim-3 ligand galectin-9 negatively regulates T helper type 1 immunity. Nat. Immunol. 6, 1245–1252.

CHAPTER SIX

Engineering Platforms for T Cell Modulation

John W. Hickey[*,§,¶,a], **Alyssa K. Kosmides**[*,§,¶,a] **and Jonathan P. Schneck**[§,¶,||,1]
[*]Department of Biomedical Engineering, Johns Hopkins University School of Medicine, Baltimore, MD, United States
[§]Institute for NanoBiotechnology, Johns Hopkins University School of Medicine, Baltimore, MD, United States
[¶]Cell Engineering, Johns Hopkins University School of Medicine, Baltimore, MD, United States
[||]Department of Pathology, Johns Hopkins University School of Medicine, Baltimore, MD, United States
[1]Corresponding author: Email: jschnec1@jhmi.edu

Contents

[a] These authors contributed equally to this work.

International Review of Cell and Molecular Biology, Volume 342
ISSN 1937-6448
https://doi.org/10.1016/bs.ircmb.2018.06.003

Abstract

T cells are crucial contributors to mounting an effective immune response and increasingly the focus of therapeutic interventions in cancer, infectious disease, and autoimmunity. Translation of current T cell immunotherapies has been hindered by off-target toxicities, limited efficacy, biological variability, and high costs. As T cell therapeutics continue to develop, the application of engineering concepts to control their delivery and presentation will be critical for their success. Here, we outline the engineer's toolbox and contextualize it with the biology of T cells. We focus on the design principles of T cell modulation platforms regarding size, shape, material, and ligand choice. Furthermore, we review how application of these design principles has already impacted T cell immunotherapies and our understanding of T cell biology. Recent, salient examples from protein engineering, synthetic particles, cellular and genetic engineering, and scaffolds and surfaces are provided to reinforce the importance of design considerations. Our aim is to provide a guide for immunologists, engineers, clinicians, and the pharmaceutical sector for the design of T cell–targeting platforms.

1. INTRODUCTION

The immune system is a complex organization of many different cell types that protect the human body from infection and disease. This cellular network is made up of innate and adaptive immune cells that can detect and eliminate pathogens with extreme precision. The ability of immune cells to provide constant surveillance and protection throughout the entire human body makes it a powerful shield against foreign material. However, deficiencies in this system can expose vulnerabilities or result in disease. In these cases, therapeutics have been developed to modulate the immune response for more favorable outcomes.

Within the adaptive compartment exist T cells, a subset of lymphocytes responsible for cell-mediated immunity. T cells can detect and destroy infected cells while leaving other nearby healthy cells unharmed. Their precision and cytotoxic ability make T cells a popular target for immunotherapies. Because of their complex activation pathways, interaction with other cell types, and biodistribution, many emerging therapeutics utilize engineering concepts to mimic or alter their normal response. Here, we will review these immunoengineering approaches for modulating the T cell response.

1.1 The T Cell Response

Minimally, T cells require two signals for activation—signal 1, a specific peptide presented in the context of the major histocompatibility complex (MHC), and signal 2, a costimulatory signal such as B7/anti-CD28 monoclonal antibody (mAb; Smith-Garvin et al., 2009). The peptide–MHC (pMHC) complex presented by an antigen-presenting cell (APC) interacts with the T cell via its T cell receptor (TCR) and is what confers the specificity of the T cell's activation. Signal 2 provides costimulation and is required for the activation of naïve T cells. B7 on an APC, which interacts with CD28 on the T cell, is thought of as the prototypical costimulatory molecule, although many different costimulatory molecules exist that modulate the T cell expansion, phenotype, and functionality (Debenedette et al., 1997; Resta and Thompson, 1997; Cannons et al., 2001; Zheng et al., 2004; Zhang et al., 2007; Rudolf et al., 2008; Chacon et al., 2013; Zeng et al., 2014; Buchan et al., 2015; Oh et al., 2015). T cells are also exposed to a third soluble signal, signal 3, that modulates their activity, including proinflammatory or anti-inflammatory cytokines, which are released in an autocrine and paracrine fashion (Curtsinger and Mescher, 2011).

Before activation, TCRs are preclustered in small 35–70 nm islands with 7–30 TCRs per island (Lillemeier et al., 2010). Upon recognition of cognate antigen, these nanoclusters begin to concatenate and form what is known as the immune synapse (Hashimoto-Tane and Saito, 2016). The lateral organization of the microscale immune synapse is defined by supramolecular activation clusters (SMACs). TCR and CD28 colocalize within the central SMAC (cSMAC), with a TCR-rich core surrounded by a CD28-rich periphery (Yokosuka et al., 2008; Saito et al., 2010). Larger adhesion molecules, such as Lymphocyte function-associated antigen 1 (LFA-1), localize within the peripheral SMAC (Dustin and Depoil, 2011). This nanoscale colocalization of TCR and CD28 is necessary for activation, as separating signal 1 and signal 2 activation by several microns inhibits T cell activation (Bashour et al., 2014b).

There are two major subsets of T cells—$CD4^+$ and $CD8^+$ T cells, otherwise known as T helper cells and cytotoxic T cells, respectively. Upon activation, $CD4^+$ T cells differentiate into different subtypes and can carry out multiple functions, from activating B lymphocytes to suppressing the immune reaction (Luckheeram et al., 2012; Zhu, 2017). The major subtypes include Th1 and Th2, type 1 and 2 T helper cells (Zhu, 2017), Th17 cells that play roles in allergy and autoimmunity (Tesmer et al., 2008), and Treg cells (Josefowicz et al., 2012) that play important roles in immune response homeostasis. Cytotoxic $CD8^+$ T cells secrete cytokines and kill target cells upon activation (Zhang and Bevan, 2011). This subtype is responsible for selectively eliminating virally infected cells. Upon activation, T cells clonally divide, secrete cytokines, and most importantly, maintain a memory response that provides long-term protection.

In most cases the immune system is a powerful defense against infection and disease. However, dysfunctions or deficiencies in any part of the system can reduce its effectiveness or even create disease itself. Cancer's ability to evade immune recognition, chronic infection, and autoimmune diseases provides opportunities to intercept and correct the immune response.

1.2 The Cancer-Immunity Cycle

Cancer arises from the culmination of a series of mutations that lead to a break in normal cell-regulating functions, including enhanced proliferation or a lack of normal apoptotic processes. This can lead to the expression of mutated neoantigens (Schumacher and Schreiber, 2015; Bobisse et al., 2016), differentiation antigens, viral antigens, or cancer-testis antigens that can be recognized as foreign by the immune system (Vigneron, 2015). These antigens are presented on MHC-I molecules by resident dendritic cells (DCs) and subsequently activate tumor antigen-specific T cells. While a T cell response can often be mounted against these tumor antigens, mechanisms of central and peripheral tolerance and immunosuppressive actions of the tumor microenvironment often hinder an effective immune response (Vasievich and Huang, 2011).

The interplay between cancer and the immune system can be viewed as a potentially self-propagating cycle, termed the Cancer-Immunity Cycle, which has been elegantly reviewed by Chen and colleagues (Fig. 1; Chen and Mellman, 2013). To achieve effective antitumor immunity, antigens released upon cancer cell death are processed and presented by APCs to activate cognate T cells. These T cells may then recognize and kill the tumor cells, leading to a positive feedback loop for immune activation. Cancer occurs when there is a blockade in a step of this cycle that shields the tumor from immune recognition or attack.

In many cases, tumor antigens do not differ sufficiently from self-antigens to be recognized as foreign by the immune system. This is evidenced by the fact that tumors with higher frequencies of neoantigens, tumor-specific antigens derived from tumor mutations, correlate with better prognosis following immunotherapy (Snyder et al., 2014; Schumacher and Schreiber, 2015; Mcgranahan et al., 2016; Germano et al., 2017). Despite mutational load, the immunosuppressive tumor microenvironment also limits the

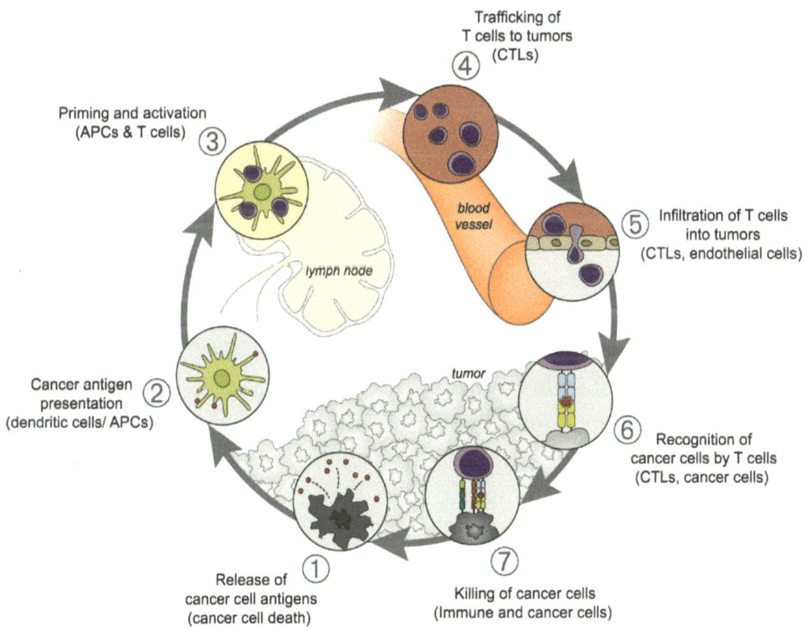

Figure 1 The immune system can recognize and interact with cancer in what is known as the Cancer Immunity Cycle. *Reprinted from Chen, D.S., Mellman, I., 2013. Oncology Meets immunology: the cancer-immunity cycle. Immunity 39 (1), 1–10. https://doi.org/10.1016/j.immuni.2013.07.012, Copyright (2013), with permission from Elsevier.*

immune response. There are three major pathways by which tumor cells evade immune attack—reduced MHC expression of tumor antigens, secretion of immunosuppressive factors, and the upregulation of negative costimulatory pathways within the tumor microenvironment (Rabinovich et al., 2007). Downregulation of MHC expression can protect tumor cell recognition by T cells and is often a method of tumor immune escape (Garrido and Algarra, 2001). Tumor or stromal cells also can secrete soluble factors, including transforming growth factor (TGF)-β, interleukin (IL)-10, and Prostaglandin E2 (PGE_2), or express inhibitory molecules that suppress immune function. Tumor cells that upregulate immunosuppressive T cell immunoglobulin and mucin (TIM) family proteins or programmed death ligand 1 (PD-L1) are correlated with poor prognosis (Baghdadi and Jinushi, 2014; Wang et al., 2017). Immunoengineering therapeutics thus often target these pathways to establish a more robust T cell response against the tumor.

1.3 Chronic Infection

Pathogens are usually rapidly eliminated by an adaptive immune response, to which effector T cells contribute in large numbers but then quickly contract to a smaller subset of memory T cells once the infection has been cleared. However, similar to cancer, chronic infection prolongs T cell exposure to antigen and inflammatory signals. This can cause "exhaustion" of T cells, limiting their effectiveness (Wherry, 2011; Speiser et al., 2014; Wherry and Kurachi, 2015). Exhausted T cells have impaired effector function, lose the ability to secrete cytokines, and upregulate expression of immunoinhibitory molecules like PD-1 and cytotoxic T-lymphocyte-associated protein 4 (CTLA-4) (Fig. 2).

Examples of chronic infection include viral infections like cytomegalovirus (CMV), HIV, and hepatitis C virus. T cell exhaustion prevents effective clearance of pathogenic infection (Almeida et al., 2007). However, this is not a permanent state as small molecule drugs or checkpoint molecules have been shown to reverse T cell functionality (Barber et al., 2006; Chiu et al., 2014). This provides inspiration and hope for engineering approaches to contribute to preventing and rescuing exhausted T cells for fighting chronic infection.

1.4 Autoimmunity

Autoimmunity arises when some level of self-tolerance is broken (Buckner, 2010; Wing and Sakaguchi, 2010; Theofilopoulos et al., 2017). There are

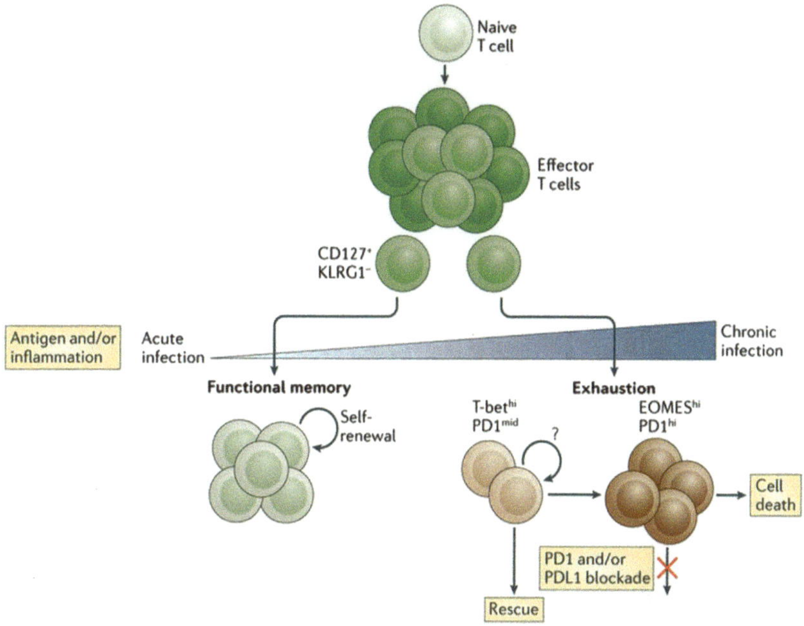

Figure 2 Chronic infection can lead to functional exhaustion of T cells. *Adapted by permission from Macmillan Publishers Ltd.: Wherry, E.J., Kurachi, M., 2015. Molecular and cellular insights into T cell exhaustion. Nat. Rev. Immunol. 15 (8), 486, Copyright (2015).*

several mechanisms of generating immunological tolerance, which many times muddles the elucidation of direct causes of autoimmunity. Main mechanisms of controlling self-reactivity include central tolerance, antigen segregation, peripheral tolerance, regulatory T cells, and activation-induced cell death.

Central tolerance occurs as antigen-specific T cells first go through positive and negative selection in the thymus where T cells with strong-affinity TCRs for self-antigen are eliminated and those with intermediate binding are selected for (Fig. 3). In some cases, strong affinity TCRs for self-antigen are selected for and are caused to differentiate into regulatory T cells. These cells can dampen the response of other self-reactive antigen-specific T cells that escape central tolerance. Additionally, self-reactive T cells that escape central tolerance can be eliminated by anergy or deletion by stimulation of the TCR without costimulatory signals.

Autoimmunity can be triggered by environmental factors, such as an infection, but also requires genetic factors. Examples of autoimmune disorders include type 1 diabetes, rheumatoid arthritis, Chron disease, and

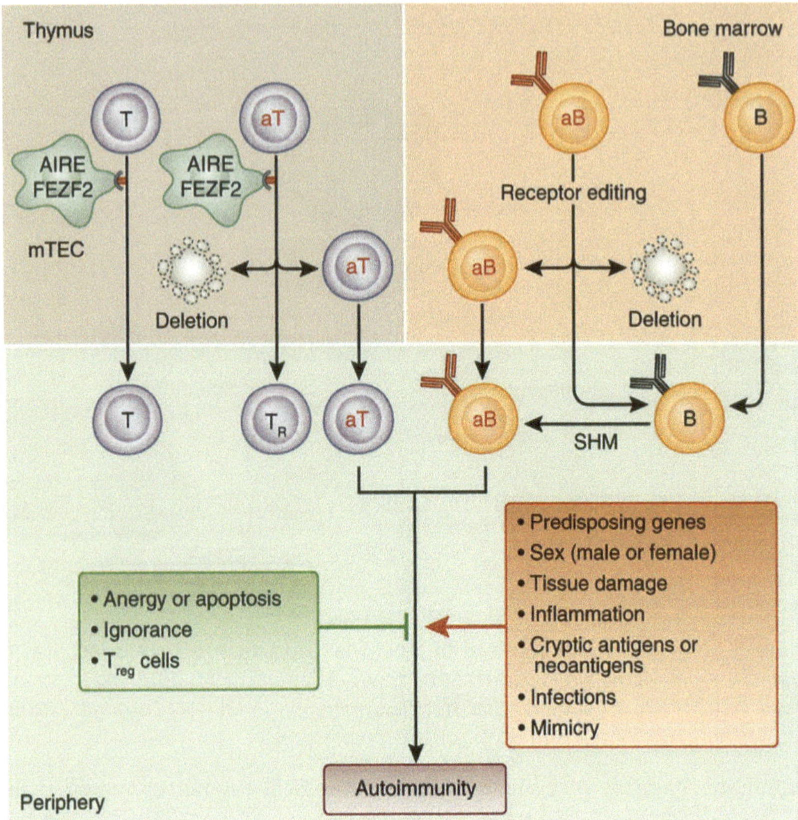

Figure 3 Autoreactive immune cells may escape central tolerance, and risk factors such as genetics or trauma can overcome peripheral tolerance mechanisms such as regulatory T cells to result in autoimmunity. *Reprinted by permission from Macmillan Publishers Ltd.: Theofilopoulos, A.N., Kono, D.H., Baccala, R., 2017. The multiple pathways to autoimmunity. Nat. Immunol. 18 (7), 716, Copyright (2017).*

multiple sclerosis. Severity can be manageable to life-threatening, and, affecting nearly 8% of the population, it is a critical immunological problem to solve, at the root of which are T cells. Many therapies for autoimmune diseases are general immunosuppressants that globally downregulate the immune system. However, new therapies that utilize engineering approaches and can target specific activation receptors on T cells, activate Tregs for adoptive transfer, and engineer cytokine delivery and structure have the potential to work more specifically and eliminate off-target effects (Rosenblum et al., 2012).

2. T CELL MODULATION—PROPERTIES TO CONSIDER WHEN CHOOSING AND DESIGNING A T CELL MODULATING PLATFORM

The dynamics of the immune synapse make T cell activation sensitive to many micro- and nano-scale properties (Perica et al., 2015). The organization and composition of signaling molecules, as well as substrate size, shape, and stiffness, all play intricate roles in modulating T cell activation and resulting phenotype. Immunoengineering approaches allow for these factors to be individually varied and optimized. Here, we will review many of these properties to consider when engineering a platform to modulate T cell activity (Fig. 4, Table 1).

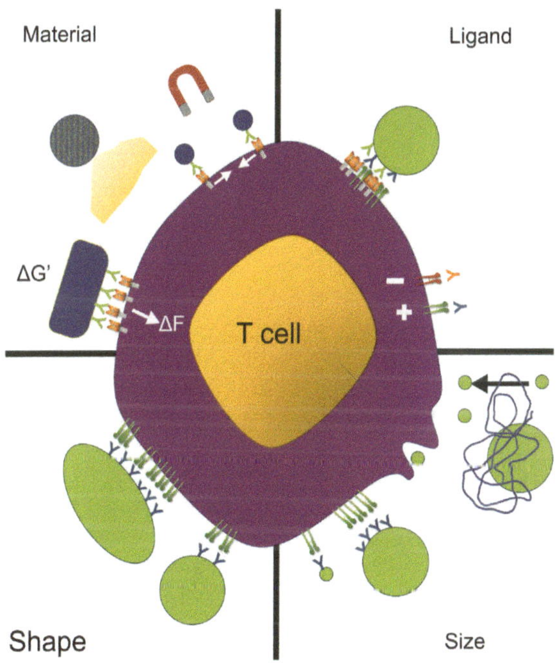

Figure 4 Properties to consider when choosing and designing a platform for modulating T cell activity. Ligand plays a role both in the endogenous density of localized receptors on the T cell surface and the type of receptor which could either inhibit or potentiate T cell responses. Size of the platform could influence its ability to (1) navigate extracellular environment and reach T cells in vivo, (2) be taken up by T cells, or (3) enable multiple receptor interactions from the same particle. Shape of the platform could influence the number of ligand interactions with increased aspect ratios and limit uptake by nonspecific cells. Material can (1) influence the mechanical interactions with the T cell which can lead to beneficial mechanotransduction signaling, (2) be utilized for controlled release of immunomodulatory molecules, or (3) allow external control of materials such as clustering magnetic particles within a magnetic field.

Table 1 Summary of Engineering Design Principles That can Lead to Control Over T Cell Responses

Engineering Design Parameter	Relationship to T Cells	Engineering Example	Impact on T Cells
Size	Engagement with nanometer-sized T cell receptor clusters	Nanoparticle aAPCs	Increased avidity leading to sustained and more effective signaling above 300 nm
	Enhanced draining to lymph node of particles smaller than 150 nm	T cell vaccine particles	Increased response to antigen
	Enhanced tissue retention above 100 nm	Cytokine and costimulatory delivering particles	Decreased off-target effects with local tumor-immune cell activation
	Intracellular particle uptake below 1 μm	In vivo particle transfection of T cells for CAR T cell therapy	T cells were effectively transduced in vivo instead of needing to be removed
	Requirement of multiple costimulatory molecule engagement	Cell-based aAPCs	4-1BB limited apoptosis and surface bound IL-15 increased activation and memory phenotype
Ligand	Choice of costimulatory and coinhibitory molecules affects T cell activation	aAPCs; bispecific antibodies	Costimulatory molecule composition affects T cell proliferation, phenotype, and function
	Signal 1 choice dictates T cell subset to be activated	Bispecific T cell engagers	Targeting CD3 versus peptide-MHC activates all T cells or certain subsets, respectively
	T cell activation is sensitive to signal density	aAPCs	Nanoscale coclustering of signaling molecules is required for robust T cell activation

Shape	Ellipsoidal particles reduce nonspecific RES phagocytcsis	Drug delivery particles	Longer circulation times of particle-based therapeutics yield better outcomes
	Ellipsoidal particles increase T cell—particle contact area	Ellipsoidal aAPCs	Increased contact area improves T cell stimulation by both nano- and micro-particles
	T cell receptor signaling is mechanotransductive	Varying antigen-presenting surfaces	Enhanced T cell activation occurring at stiffness around 100 kPa
Material	Enhanced circulation with "cloaking" materials	Adding CD47 to surface of nanoparticles	Decreased phagocytosis of particle aAPCs in vivo
	Ability to artificially cluster T cell receptors	Magnetic aAPCs	Clustering of receptors initiates effective T cell activation with small aAPCs at a lower dose
	T cells migrate and exist in three-dimensional environments	Artificial lymph nodes	Three-dimensional collagen scaffolds with stromal cells creates lymph node like structures ex vivo and in vivo
	IL-2 necessary as third signal	Biodegradable polymer scaffolds; drug carrier particles	Sustained local release of IL-2 and enables T cell infiltration

aAPCs, artificial antigen-presenting cells; *CAR*, chimeric antigen receptor; *IL*, interleukin; *MHC*, major histocompatibility complex; *RES*, reticuloendothelial system.

2.1 Ligand

Most fundamentally, T cell activation is modulated through different costimulatory and coinhibitory pathways. When developing a therapeutic, one first has to determine which specific signaling pathways to activate or inhibit. In addition to pathway choice, ligand nano- and micro-scale arrangement, density, strength, and duration can have a great impact on T cell activation.

T cells require a signal 1 for activation. This can be provided through stimulating the cognate TCR or CD3, which is constitutively associated with the TCR and is responsible for its intracellular signaling (Birnbaum et al., 2014). pMHC stimulation of TCR provides activation of only cognate antigen-specific T cells, whereas activation through CD3, with an anti-CD3 mAb, for example, activates all T cells regardless of specificity (Smith-Garvin et al., 2009).

While signal 1 is required for activation to occur, the T cell response to that signal is modulated by dozens of types of costimulatory and coinhibitory molecules (Chen and Flies, 2013). In the absence of any costimulation, T cells develop an anergic, under-responsive state. CD28 is the prototypical costimulatory molecule although several other costimulatory molecules exist that modify T cell proliferation and phenotype. For example, 4-1BB costimulation has been shown to preferentially activate $CD8^+$ T cells over $CD4^+$ T cells (Shuford et al., 1997), preferentially activate memory $CD8^+$ T cell populations (Zhang et al., 2007), increase T cell maintenance of CD28 expression (Chacon et al., 2013), and increase $CD8^+$ T cell secretion of IL-2 (Oh et al., 2015). Stimulation through the costimulatory molecules OX40 and CD27 expressed by T cells has also been shown to enhance activation (Zeng et al., 2014; Buchan et al., 2015). Importantly, altering the combinations and ratios of different costimulatory molecules without varying total costimulation can have tremendous impacts on activation (Rudolf et al., 2008; Zeng et al., 2014).

Several coinhibitory pathways exist that diminish T cell activation. Many of these pathways are involved with maintaining homeostasis and establishing a self-regulated immune response following pathogen removal (Okazaki et al., 2013). However, these inhibitory pathways can be utilized by cancer to cause immune evasion or be inhibitory or dysfunctional in autoimmune diseases and chronic infection and are thus often targets of immunotherapeutics. Programmed death 1 (PD-1) and CTLA-4 expressed by T cells are two commonly targeted checkpoint molecules (Callahan and

Wolchok, 2013; Postow et al., 2015). Monoclonal antibodies blocking these checkpoint pathways as well as other immunoinhibitory molecules such as indoleamine 2,3-dioxygenase (IDO) can rescue exhausted T cells (Spranger et al., 2014). Blocking inhibitory pathways along with activating costimulatory pathways can also further augment the T cell response in cancer (Chen et al., 2014; Shindo et al., 2015) and chronic infection(Vezys et al., 2011).

The nano- and micro-scale arrangements of signaling ligands play an important role in modulating the T cell response by engineered platforms as would be speculated from the formation of an immune synapse during endogenous activation. At the microscale, a focal coclustering of signal 1 and signal 2 molecules is necessary to initiate activation. When anti-CD3 mAb is patterned on a planar substrate in focal spots, T cells proliferate and secrete cytokines (Doh and Irvine, 2006). However, T cell activation is reduced when the same anti-CD3 mAb is tethered in an annular pattern that precludes centralized TCR clustering. Signal 2 must also be coclustered with signal 1 during stimulation. Separating anti-CD3 and anti-CD28 by a distance of several microns inhibits IL-2 cytokine production (Bashour et al., 2014b).

In addition to the microscale organization of stimulatory molecules, the TCR intermolecular distance must be sufficiently small to induce activation. When anti-CD3 is patterned on planar substrates at defined densities, only surfaces displaying anti-CD3 with intermolecular distances smaller than approximately 60 nm induce optimal activation (Delcassian et al., 2013; Matic et al., 2013). Interestingly, the minimum intermolecular distance for CD4$^+$ T cell–specific activation with MHC II is close to 115 nm (Deeg et al., 2013), indicating that ligand organization may need to be modulated based on the T cell subset to be affected.

While TCR and CD28 organization have been more extensively studied, some data indicate that signaling through other costimulatory and coinhibitory pathways may also be sensitive to nanoscale arrangement. T cell inhibition through PD-1 signaling has been shown to depend on the formation of microclusters within the immune synapse (Yokosuka et al., 2012; Hui et al., 2017). When the extracellular domain of PD-1 was extended to preclude its colocalization within the immune synapse, inhibition was reversed. 4-1BB has also been shown to localize within the immune synapse along with CD28 during costimulation (Stephan et al., 2007).

2.1.1 Density and Impacts on Signal Strength and Duration

The spatial arrangement of ligands on the surface of both naïve and activated T cells also affects the local density of ligands, which in turn has impacts on molecular and cellular interactions. TCRs are locally arranged on the surface of quiescent T cells in protein islands of about 35–70 nm in diameter and 300 nm in the longest dimension, with 7–20 TCRs found in one island (Lillemeier et al., 2010). Indeed, disruption of T cell lipid rafts by cholesterol depletion disrupts T cell activation further demonstrating the need for multivalent TCR clusters (Tsai et al., 2010). Upon TCR engagement and activation, these TCR nanoislands coalesce to form larger microclusters and eventually develop into the cSMAC of the immune synapse (Fahmy et al., 2001; Varma et al., 2006; Fooksman et al., 2009; Lillemeier et al., 2010). Interestingly, the cognate ligand, MHC molecule, is organized similarly on endogenous APCs. As many as 25–125 MHCs are grouped in 70–600 nm protein patch (Hwang et al., 1998; Bosch et al., 2013). Additionally, it has been shown that even specific protein-MHC complexes are clustered on the surface of APCs (Lu et al., 2012; Ferez et al., 2014).

This nanometer organization is important to modulating the sensitivity of T cell signaling. MHC and TCR binding is a low-affinity interaction with μM Kd compared with nM Kd for antibody-antigen interactions. Kd is a useful description of binding kinetics where it is the equilibrium concentration of ligand where half of the ligand is bound to its receptor, thus a lower concentration such as a nM Kd has much higher affinity for its receptor (Matsui et al., 1991). Consequently, the binding of several MHC and TCR complexes simultaneously is beneficial to initiate engagement of cognate APCs and T cells. Mathematical models of TCR and MHC binding events also show a need for multiple receptors in close proximity (Lever et al., 2014). The models suggest that the reversible binding between MHC and TCR leads to activation when serial binding can occur.

In addition to initiating signaling between APCs and T cells, multivalent interactions would be important to providing sustained signaling (Bousso, 2008). Sustained signaling is essential for effective T cell activation and can influence the fate and function of resultant T cells (Iezzi et al., 1998; Huppa et al., 2003). *In vivo* experiments have shown that once T cells encounter antigen-specific APCs in the lymph node, they form stable connections for several hours (Celli et al., 2007). These studies also demonstrated that this sustained connection is dependent on stable pMHC-TCR interactions.

In addition to these biological findings, researchers have corroborated these findings through directly modifying MHC density. First, researchers have artificially clustered the MHC on APCs and observed enhanced T cell activation with lower levels of MHC (Fooksman et al., 2006). Second, the density of cognate MHC can be modulated by the dose of antigen given to APCs, and as density of surface cognate MHC decreased, so did the in vivo T cell response (Bullock et al., 2003). Third, MHC ligands were patterned on surfaces at varying densities, but the number presented to a T cell was controlled (Manz et al., 2011). Here they found that the density was more important than the number of presented ligands with at least four ligands in close proximity needed to activate a T cell.

In summary, TCRs and MHCs are naturally localized into dense regions, which allow increased cognate recognition and sustained signaling to occur because of serial engagement and multivalent interactions. Though we have focused our analysis how ligand density affects the MHC-TCR interaction, other ligands such as those in the immunological synapse also have distinct spatial locations and densities (Bromley et al., 2001; Fooksman et al., 2009). Therefore T cell—modulating technologies should be designed with consideration given to the density and location of the ligand.

2.2 Size Considerations

In designing cell-modulating technologies, physical size is another important biological constraint. Like with considerations of ligand properties, size design considerations arise from features of T cells. This can arise from molecular size limitations such as distances between ligands on the T cell as described in Section 2.1.1 or even due to the fact that T cells reside within the lymph node. Here we will review some of those interactions and how to design certain technologies based on these scales.

2.2.1 Molecular Interactions

When interacting in biological environments, both ex vivo and in vivo, the number of molecular species in a given environment can be astronomical. And with an equally large number of chemical reactions and molecular binding events occurring, it is hard to achieve the desired interaction and avoid unintended reactions. This poses unique challenges for maintaining the size of synthetic materials created to modulate T cells.

As these materials are usually designed on the micro- to nano-meter scale, the decrease in size increases the free surface energy of the material. As a result of this small size, within minutes of being introduced to a biological media or in vivo, most engineered materials acquire a shell of adsorbed proteins known as a protein corona (Hickey et al., 2015). This protein shell can both determine the fate of the material and cause unintended consequences regardless of T cell targeting. For example, proteins can cause aggregation of the material, changing its size; increase cell-uptake; or even cause negative biological events, such as platelet activation (Nel et al., 2009; Monopoli et al., 2012; Tenzer et al., 2013). Therefore, to control the size, engineers modulate the chemical and biological properties of the surfaces to be inert or covert materials, as described in Section 2.4.

2.2.2 Cellular Interactions

After encountering and interacting with proteins postadministration, therapeutic particles or materials will encounter cells. Cells take up volume around them as a form of communication and defense, and often these materials become subject to this phenomenon. APCs including macrophages phagocytose efficiently to clear antigen and also process antigen to elicit T cell responses. T cells on the other hand are less efficient at phagocytosis. Thus careful design of particle properties can effect cellular and intracellular destinations of particle and cargo. A few notable principles have been able to be distilled from a number of particle size–related studies in cell uptake (Albanese et al., 2012; Shang et al., 2014; Hickey et al., 2015).

First, particle size–dependent cell uptake differences are partially governed by required membrane wrapping energy. Larger particles on the scale of 3 μm and above require more energy to be taken up than smaller nanoparticles, which results in decreased cell uptake (Gratton et al., 2008). In the absence of surface coating the uptake of small particles (<25 nm) is lower than that of larger particles (Shann et al., 2012). Here the surface exposed is insufficient to cause cell receptor–nanoparticle interactions necessary for endocytosis. This results in lower avidity between the cell membrane and particle and makes it more difficult not only for the particle to be "gripped" but also for the cell to initiate signaling for endocytosis machinery (Yuan et al., 2010).

Second, the cellular process for uptake also depends on particle size (Rejman et al., 2004; Lai et al., 2007). Phagocytosis is primarily seen with particles larger than 500 nm and pinocytosis with particles smaller than

500 nm (Kettiger et al., 2013). Even within pinocytosis, cellular uptake processes can differ and have unique vesicle size limits, which then can provide differential intracellular delivery based on particle size. For example, micropinocytosis can engulf particles greater than 1 μm, whereas clathrin-mediated endocytosis can take up particles up to 120 nm, caveolin-mediated up to 60 nm, and clathrin and caveolin-independent endocytosis up to 90 nm (Conner and Schmid, 2003).

Third, cell-uptake parameters differ between different cell types. Cells of the mononuclear phagocytic system will exhibit greater cell uptake than endothelial or even T cells (Zauner et al., 2001). This could affect not only the volume of material that is taken up but also the size dependency trend (Choi et al., 2014).

Understanding how size affects nanoparticle endocytosis will be especially important for the design not only of nanoparticles that can escape cellular uptake but also of effective NP therapeutics such as gene delivery vehicles where endocytosis is a critical component of the delivery process. To increase cell uptake for applications such as gene delivery, nanoparticles can be coated with cell-interacting ligands (Gao et al., 2005; Zhang et al., 2009). This can determine their intracellular destination by methods such as labeling the particle with a microtubule-associated sequence and or nuclear localization signaling sequence to target the nucleus (Narayanan et al., 2013). Additionally, ligands can help to differentiate cell types, such as by coating the particles with antibodies to cell-specific proteins (Jung et al., 2007; Smith et al., 2017).

Besides targeting cells with cell-specific antibodies, particles can be designed to directly provide signaling with ligands (Oelke et al., 2003a). Here size also plays a role as the local density of ligand can be changed by changing the size of a particle. For example, the number of bioavailable ligands on a small 50-nm nanoparticle may only be 2, whereas the number may be closer to 200 on a larger 5-μm particle with equivalent density because the curvature of the particle decreases with increasing size (Hickey et al., 2017). This will affect the local ability to cluster receptors from a single particle and may impact T cell modulation as mentioned in Section 2.1.3, where the spatial relationship among cell surface receptors is critical for T cell signaling.

Finally, when engineering implantable materials (>100 μm) for in vivo applications, then the foreign body response should be considered (Franz et al., 2011). Here since cells cannot uptake the material, then the immune system will become activated based on adsorbed blood proteins. This will cause a cascade of events that may result in either (1) degradation of the material by reactive oxygen species, (2) segregation of the material by formation of scar tissue, or (3) integration of the material into the surrounding tissue. Recently, it has been suggested that the size of the implant can influence this response regardless of material choice, where implants above 1.5 mm in size led to reduced immune response and scarring (Veiseh et al., 2015).

2.2.3 Tissue Interactions

In addition to the cellular destination of materials, material size also affects tissue distribution when injected in vivo (Hickey et al., 2015). T cells are a unique cell type in their mobility and distribution throughout many different compartments (Gong et al., 2014). They arise from within the bone marrow, develop in the thymus, travel from lymph node to lymph node via the lymphatic system, and use the circulatory system to traffic to sources of infection within tissue and can reside there or in lymph nodes for years. Biodistribution will thus affect the ability to target specific T cells with engineered materials.

Broadly speaking, the impact of particle size on biodistribution can be classified by whether particles are introduced to the circulation or to tissues directly. Intravenous administration would be useful for targeting circulating T cells or for reaching T cells systemically. For intravenous administration, particles should be less than 4 μm to prevent occlusion of capillaries. Particles smaller than 6 nm will be cleared by the kidney, and larger particles will mostly be removed by the mononuclear phagocyte system in the liver and spleen. Most particles will end up being cleared from circulation within minutes, but a small percentage will enter different tissue compartments. For example, nanoparticles 20–200 nm in size have shown increased accumulation in tumor tissues based on the increased fenestration of the tumor capillaries—known as the enhanced permeability and retention (EPR) effect (Jain and Stylianopoulos, 2010).

Tissue-based injection would be useful for targeting localized T cells in tissues or for reaching T cells in lymph nodes. Injected particles will diffuse or be convected through the extracellular matrix (ECM). Particles less than 50 nm can move efficiently through spaces in the ECM and be transported via the lymphatics to the lymph node, whereas larger particles are sequestered at the tissue injection site (Reddy et al., 2007; Irvine et al., 2013). Upon reaching the lymph node, low molecular weight proteins (<70 kDa) will be shuttled to fibroblastic reticular cell-lined channels to B cell follicles, while larger proteins and particles will need to be introduced into the lymph node by subcapsular sinus macrophages (Hirosue and Dubrot, 2015). Similar properties govern particle transport in mucosal administration of particles, where particles smaller than 100 nm will have greater diffusion rates and penetration through mucus matrices (Norris et al., 1998). Particles larger than 100 nm may be limited to administration sites and undergo phagocytosis by macrophages or APCs, which subsequently traffic to lymph nodes as mentioned in Section 2.2.2. This property could be harnessed to mimic infection in design of biomaterial vaccines that lead to potent T cell responses.

In conclusion, particle size is a key design parameter that influences whether cell-modulating technologies will reach T cells. Additionally, it impacts how the particles interact with T cells and modulate T cell signaling.

2.3 Shape

When synthesizing synthetic platforms for T cell modulation, particle shape can have surprising impacts on their interaction with T cells and macrophages and their biodistribution. Spherical and ellipsoidal particles have differing interfacial geometries despite their similar length scales and can thus potentially be optimized for T cell interactions. Using shape to vary biodistribution, the half-life and cell and organ targets can also be modified.

Particle shape affects biodistribution in part by modifying phagocytosis by the reticuloendothelial system (RES). In contrast to spherical particles, ellipsoidal particles reduce nonspecific RES phagocytosis, an important consideration for extending the half life of therapeutics injected in vivo (Champion and Mitragotri, 2006; Gratton et al., 2008; Sharma et al., 2010; Yoo and Mitragotri, 2010). In fact, phagocytosis is more dominantly affected by shape than by size in the range of approximately 1–10 μm (Champion and Mitragotri, 2006). However, interestingly, receptor-mediated endocytosis of nonspherical particles is enhanced (Barua et al., 2013). Thus elongating particles for applications where targeted internalization is preferred, such as drug delivery, may also take advantage of longer circulation times.

In addition to affecting biodistribution and half-life, particle geometry plays an important role at the T cell particle interface. While most T cell–interacting particles, such as artificial APCs (aAPC), are spherical in nature, this shape does not recapitulate biological T cell–APC interactions. Spherical particles maximize the curvature and thus minimize the surface area of contact between the T cell and particle. This is especially detrimental at the nanoscale where ligand avidity decreases as particle size decreases, despite biocompatibility incentives to moving to smaller size scales. Ellipsoidal microparticle aAPCs were shown to preferentially interact with T cells along their long axis and significantly increase T cell activation in vitro and in vivo (Sunshine et al., 2014). Ellipsoidal nano-aAPCs also result in a similar increase in T cell activation over spherical particles of similar size while also taking advantage of the increased biocompatibility and superior biodistribution due to small particle size (Meyer et al., 2014).

Particle shape can directly affect T cell activation by influencing avidity. For therapeutics meant to activate a T cell response, ellipsoidal geometries may be preferred due to increased interaction with the T cell surface and superior biodistribution. However, in addition to micro- and nano-particles, shape may be an important factor to consider in larger scaffolds that cannot be directly engulfed by macrophages. The degradation products of implantable T cell–stimulating devices, such as artificial lymph nodes, are often fibers or other nonspherical geometries and may be an important design parameter to limit an unwanted immune reaction (Andorko and Jewell, 2017).

2.4 Material

The choice of material and control of material properties can impart unique functionalities on the T cell–modulating technology. For example, in vivo applications need to consider biocompatible materials with low clearance, efficient targeting, low toxicity, and biodegradability. For ex vivo applications, a broader range of chemistries and materials can be employed to modulate and study T cell responses. Examples include using magnetic materials to cluster signaling components on the surface of the cell and modulating the stiffness of the material to enable study of mechanotransduction pathways.

2.4.1 Material Surfaces

As discussed in Section 2.2, stability in biological fluids and the ability to evade the mononuclear phagocytic system enable therapeutics to reach and target T cells. To this end, reducing the formation of a protein corona on the material is a priority. Engineers have primarily utilized a hydrophilic polymer—polyethylene glycol (PEG)—to create a water-based shell to prevent protein adsorption and particle aggregation (Torchilin and Trubetskoy, 1995; Cheng et al., 2007; Xu et al., 2015). PEG has also been conjugated to protein-based therapeutics to increase solubility, stability, and circulation and reduce immunogenicity (Duncan, 2006).

Often, therapeutics are more than just a delivery vehicle and are coated with either targeting moieties or cell-modulating proteins. In these cases, a PEG layer can help improve stability, but it cannot shield attached protein components from initializing a protein corona or initiating uptake by the mononuclear phagocytic system. Another approach to shielding therapeutics is to use biology as inspiration. For example, liposomes are particles derived from biological lipids that form a bilipid membrane similar to that of cells and can be used as a coating on particles to prevent aggregation and uptake (Safra et al., 2000; Torchilin, 2005).

Similarly, particles have recently been coated with cell membranes, such as red blood cell (RBC) membranes (Hu et al., 2011), tumor cell membranes (Fang et al., 2014), and leukocyte cell membranes (Parodi et al., 2013). Besides increasing stability and decreasing clearance, these membranes can confer additional advantages such as site-specific targeting for targeting inflammation. Lessons learned from these coatings and studying how RBCs avoid uptake have revealed the importance of CD47 as a "don't eat me" signal (Oldenborg et al., 2000). Therefore, instead of coating the particle with entire membranes, researchers have also attached the CD47 protein to nanoparticles, leading to decreased cell uptake and increased circulation (Rodriguez et al., 2013).

As an alternative to coating materials with chemistries or biologics that prevent protein adsorption, these interactions can be taken advantage of intentionally. One example is the design of more efficient vaccines by Liu et al. (2014). They designed their vaccine to have a hydrophobic tail so that it would bind to albumin and traffic to the lymph node. This approach resulted in a more efficient delivery of their vaccine to the lymph node, generating a more robust immune response.

2.4.2 Biointeractive Materials

Beyond avoiding the mononuclear phagocytic system, materials can also be designed to mimic biological properties to engage cells such as T cells. Biomolecules can be attached through adsorption, use of biologic interactions such as streptavidin-biotin, or direct chemical attachment. Already covered in Section 2.1, coating of ligands on materials can mimic existing signals in the body enabling specific T cell modulation. Properties such as ligand choice, arrangement, and density are important factors to consider.

Beyond cell-surface ligands, biomolecules derived from the ECM or cytokines can be coated onto material surfaces. Utilizing such molecules can enable enhanced attachment of cells to synthetic scaffolds (Burdick and Anseth, 2002). Additionally, these molecules such as growth factors can be used to polarize differentiation of cells including stem cells (Discher et al., 2009).

The material stiffness also contributes to the signaling that occurs upon engagement by a cell. This is especially important in mechanotransduction-sensitive pathways. Mechanotransduction plays a role in cell-cell signaling, sensing aspects of the extracellular environment such as fluid flow, and enabling cell attachment and migration through the ECM (Jaalouk and Lammerding, 2009; Humphrey et al., 2014).

T cells are a very active and motile, and mechanotransduction plays a role in similar ways to other cells in areas of cell attachment and migration. More specifically mechanotransduction has also been implicated in TCR signaling (Huse, 2017). As T cells scan the APC, the TCR can engage its cognate MHC, and the resulting force associated with recognition is necessary for T cell activation (Kim et al., 2009). It has been hypothesized that this is due to a catch bond—like behavior of the sensor due to conformational changes in the proteins that result from the force (Liu et al., 2014; Das et al., 2015).

Additionally, once initial TCR signaling has occurred, the immunological synapse develops, which contains integrins and F-actin on the periphery and enables strong adhesion between both cells (Basu and Huse, 2017). Retrograde flow of actin drives engaged TCRs to the center of the immunological synapse, further exerting forces on the TCR-MHC. Additionally, APCs provide rigid surfaces by immobilizing intracellular cell adhesion molecule 1 (ICAM-1) on its surface, which is necessary for T cell activation (Comrie et al., 2015). Therefore mechanical properties are an important consideration for designing materials to directly interact with T cells.

Finally, the actual material can have immunomodulatory effects on immune cells. One noteworthy example is how the polymer poly(beta-amino ester) (PBAE) intrinsically activates the immune response as a vaccine particle. Researchers showed that maximal stimulation of the immune response was achieved with high molecular weight PBAEs or short degradation time points even in the absence of traditional adjuvants (Andorko et al., 2016).

2.4.3 Bioresponsive Materials

A more recent development in material design is to engineer materials that react and respond to biological cues in the body (Lu et al., 2017). The first example is to increase the biodegradation in vivo in targeted locations due to differences in the biological environment, such as changes in pH, redox potential, and enzyme content. Increased acidity in the stomach, tumor microenvironment, or lysosome can be used to cause protonation of the material and trigger to release its cargo (Baeza et al., 2015). Utilizing disulfide linkages, one can exploit the reductive environment of the cytosol or the nucleus compared with the extracellular environment as another method of signaling release (Meng et al., 2009). Finally, enzyme-reactive sequences can be designed into material to take advantage of the increased level of matrix metalloproteinases found in tumors to speed up degradation (Callmann et al., 2015).

Beyond biodegradation, bioresponsive materials can also be designed to respond to mechanical forces, heat, magnetic fields, radiation, and ultrasound to change size, shape, and chemical structure (Mura et al., 2013). Mechanical forces, such as the shear stress induced by fluid flow in circulation, can be utilized to break up larger fragile materials into smaller pieces more amenable to circulation and tissue targeting (Korin et al., 2012). Certain polymers have lower critical solution temperatures close to body temperature, such as N-isopropylacrylamide, which decreases in size due to a phase change (Schmaljohann, 2006). Magnetic materials can be controlled ex vivo by a magnetic field and caused to locally aggregate (Chertok et al., 2008) or generate heat for therapy (Kumar and Mohammad, 2011). Similar control of size change and heat can be generated by gold and light-responsive polymers from activation by specific wavelengths of light (Han et al., 2006; Tong et al., 2012). Like magnetic materials, ultrasound-sensitive materials can be utilized both as imaging agents and to change the size of the material to release cargos locally (Javadi et al., 2012; Wang et al., 2012).

Bioresponsive material strategies can be leveraged to implement more effective modulation of T cells in vivo and ex vivo. First, these material properties could help to target T cells within specific compartments, such as the tumor microenvironment, or even specific T cell intracellular compartments. Second, control over T cell trafficking and response to cargo controlled by temporal external stimuli offers precise control over the timing and level of T cell responses.

2.4.4 Biocompatibility

Material biocompatibility is another important consideration for clinical translation (De Jong and Borm, 2008). Materials already used in U.S. Food and Drug Administration (FDA)-approved devices or applications can make the clinical translation more straightforward, though this may limit the functionalities and engineering design in the therapy (Bobo et al., 2016). Translation becomes increasingly challenging when bioactive proteins and cellular components are added into the material intervention (Aguado et al., 2018).

3. IMMUNOENGINEERING APPROACHES

Engineering concepts have been applied to immunology to develop platforms that alter the T cell response. Scientists and engineers have developed biocompatible platforms that mimic antigen presentation, orchestrate the delivery of immune-modulatory drugs, or prompt a new type of immune interaction. These platforms have been built upon a wide variety of biocompatible platforms, from engineered small molecules and cultured cell lines to biodegradable polymers. Many times, immunoengineering is used to attempt to more accurately mimic biological interactions, while other times, it is used to trigger a new type of interaction that may be more favorable to a patient. Here, we review previously engineered platforms, including protein engineering, synthetic particles, cell and genetic engineering, scaffolds and surfaces, and combination therapies.

3.1 Protein Engineering

At the smallest scale, proteins and small molecules can be engineered to target specific signals involved with T cell activation. This may involve developing a high-affinity antibody to activate or block one or more immune-stimulatory or immune-inhibitory pathways, synthesizing modified cytokines that can more effectively bind their targets, or utilizing antibodies to target the delivery of a drug or cytokine to a specific anatomical site. In this section, we will review various molecular targeting approaches and how engineering principles have been applied to their design. We will focus on how these engineered molecules have been used as monotherapies, although many are used in combination therapies or encapsulated into synthetic particles or scaffolds to further modulate their delivery.

3.1.1 Engineered Cytokines

During T cell activation, T cells receive soluble proinflammatory or anti-inflammatory signals, often referred to as "signal 3." These signals help to drive cell proliferation, differentiation, and chemotaxis, among other functions. Therapeutic administration of proinflammatory cytokines important for immune activation, most commonly IL-2, can activate a robust antitumor immune response, but systemic injection is complicated by life-threatening toxicities (Lurie et al., 1993; Rosenberg, 2014). The clinical use of cytokines is also hindered by their rapid clearance from the bloodstream, leading to a typical serum half-life on the order of minutes. Longer-acting engineered cytokines that can selectively activate immune subsets and antibodies that direct their localization are thus being developed.

In the field of cytokine engineering, scientists and engineers have developed cytokines with altered binding to immune cell subsets to create more clinically favorable therapeutics. IL-2 signaling occurs through either a high affinity ($K_D \sim 10$ pM) trimeric receptor complex, consisting of the IL-2Rα, IL-2Rβ, and IL-2Rγ (γ_c) chains or an intermediate affinity dimeric complex ($K_D \sim 1$ nM) consisting of only the IL-2Rβ and γ_c chains. Different immune cell subsets modulate their sensitivity to IL-2 signaling by altering surface expression of the nonsignaling IL-2Rα chain. Naïve effector T cells only express only the IL-2Rβ and γ_c subunits, whereas activated memory T cells and regulatory T cells also express the high-affinity IL-2Rα and are thus more sensitive to IL-2 activation (Malek and Castro, 2010). Importantly, pulmonary edema, a major toxicity upon systemic IL-2 administration, has been shown to be mainly IL-2Rα-dependent (Krieg et al., 2010). An engineered IL-2 "superkine" was evolved to have 200-fold increased affinity for the IL-2Rβ subunit and thus activated cells in an IL-2Rα–independent fashion. As a result, this IL-2 superkine significantly increased the expansion of $CD8^+$ T cells in vivo but did not alter regulatory T cell expansion compared with the wild-type cytokine. By shifting the cellular response to favor immunostimulatory subsets, the IL-2 superkine significantly reduced pulmonary edema and increase antitumor efficacy (Levin et al., 2012). Further modifications of this engineered IL-2 cytokine have demonstrated how modulating binding activity to the other subunits can result in development of agents that further manipulate the cellular response, such as IL-2 partial agonists or antagonists (Fig. 5A; Mitra et al., 2015).

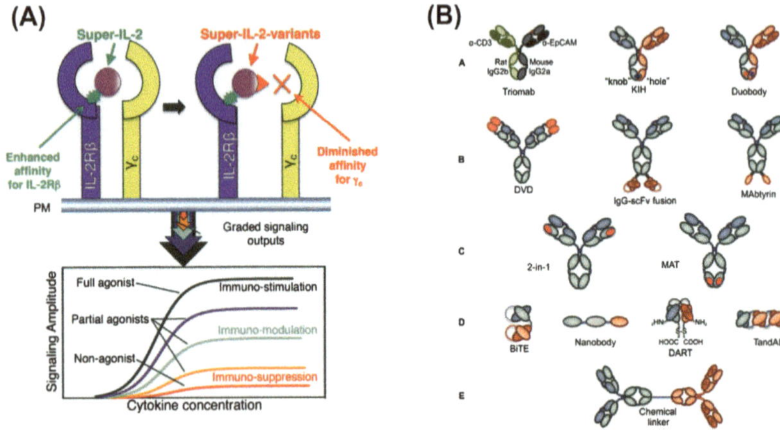

Figure 5 Engineered cytokines and antibodies can alter the T cell response. (A) IL-2 variants can be engineered to have different activity potency and function based on the degree of disruption of γ_C binding. (B) There are five different fundamental formats of bispecific antibodies that can affect how the antibodies interact with cells and their biodistribution in vivo. *IL*, interleukin. *(A) Reprinted by permission from Elsevier: Mitra, S., et al., 2015. Interleukin-2 activity can be fine-tuned with engineered receptor signaling clamps. Immunity 42 (5), 826–838. https://doi.org/10.1126/science.1249098.Sleep, Copyright (2015). (B) Reprinted by permission from Higher Education Press: Strohl, W.R., 2017. Current progress in innovative engineered antibodies. Protein Cell 9 (1), 1–35. https://doi.org/10.1007/s13238-017-0457-8. Higher Education Press, Copyright (2017).*

Superagonists for other cytokine-receptor systems have also been developed. For example, IL-15 is a cytokine that contributes to T cell and NK cell development, proliferation, and activation (Mishra et al., 2014). Endogenously, IL-15 is bound to IL-15Rα and expressed on the cell surface of APCs to T cells and NK cells displaying the intermediate affinity dimeric IL-2 receptor (IL-2Rβ/γ_c). Binding of IL-15 to the IL-15Rα subunit on the surface of APCs induces a conformational change that enhances its interaction with IL-2Rβ/γ_c; thus the administration of soluble IL-15 alone has only a moderate effect on T cell expansion. Administration of preformed complexes of IL-15 with recombinant IL-15Rα was shown to selectively induce robust expansion of memory CD8$^+$ T cells and NK cells (Rubinstein et al., 2006). Building on this work, IL-15 superagonists were engineered by fusing the IL-15 cytokine to a truncated version the IL-15Rα subunit. These engineered fusions exhibited 4- to 5-fold more potent biological activity than the native IL-15 cytokine on IL-2Rβ/γ_c—expressing cells (Zhu et al., 2009), and later iterations of the superagonists were linked to Fc-binding domains to prolong their effects through extension of serum half-life (Bailey et al., 2017).

While high-affinity superagonist cytokines may have increased potency compared with endogenous cytokines, they often do not solve the problem of systemic toxicities. To improve selective targeting, cytokines have been bound to tumor-targeting antibodies to increase their localization to the tumor microenvironment. Over a dozen types of cytokine-antibody fusion proteins (immunocytokines) have been developed. In the simplest terms, cytokines can be linked via the N- or C- terminus of an antibody to leverage the spatial relationship between cytokine delivery and antibody binding (Neri and Sondel, 2016). Additional modifications of the antibody constant region can also affect how the fusion protein interacts with Fcγ receptors or initiates the complement cascade, or they can alter the molecular weight and biodistribution of the cytokine-antibody fusion (Müller, 2014). In addition to targeting cytokines to the tumor microenvironment, the physical constraint of a tumor-targeting antibody with a T or other immune cell-targeting cytokine can be synergistic. For example, immunocytokines that target tumor-expressed antigens have been shown to increase interactions between tumor cells and T cells or natural killer cells much like a bispecific antibody that concurrently targets a tumor-associated antigen and an immune cell surface marker (Neri and Sondel, 2016). Other families of immunocytokines that link cytokines with tumor-associated ECM components are thought to mediate an antitumor effect primarily through localizing cytokine delivery (Carnemolla et al., 2013).

Immunocytokines have been developed that incorporate a wide variety of tumor-associated antigens, but the majority of these constructs utilize the same cytokines: IL-2, IL-12, IL-15, or TNF (Neri and Sondel, 2016). One example of a successful immunocytokine is a fusion between the anti-CD20 antibody rituximab and an IL-15 superagonist, which was shown to exhibit superior antitumor activity to rituximab alone (Vincent et al., 2014; Liu et al., 2016). By linking the superagonist cytokine with a tumor-targeting antibody, the fusion proteins were capable of taking advantages of three mechanisms: specific binding to CD20 on tumor cells; potent immune cell stimulation through the IL-2Rβ/γ_c receptor; and recruitment of antibody-dependent cellular cytotoxicity and phagocytosis through binding to Fcγ receptors on macrophages and natural killer cells. Other immunocytokines that deliver anti-inflammatory cytokines for organ-specific autoimmune disease (Kawalkowska et al., 2016), target cytokine delivery to necrotic areas of tumors through incorporation of anti-DNA antibodies

(Fallon et al., 2017), or enact the blockade of tumor-expressed immune-inhibitory molecules (Chen et al., 2016) have also been developed.

3.1.2 Engineered Antibodies

In addition to the molecular engineering efforts on cytokines, there is great interest in engineering antibodies to block or stimulate cell targets, alter cell-mediated cytotoxicity and biodistribution, or selectively deliver drugs to the site of disease. Monoclonal antibodies can be evolved to selectively bind antigens that may be uniquely expressed or upregulated on tumor cells and also have the advantage of longer half-lives compared with small molecule drugs (Kamath, 2016). Their precision in targeting also mitigates off-target toxicities. However, beyond selecting a molecular target, applying engineering principles to their design can improve their biodistribution, alter their interaction with immune cells, turn them into a vehicle to specifically deliver a drug or nanoparticle to a disease site, or target them to multiple cell types simultaneously. The process of mAb production (Chiu and Gilliland, 2016; Almagro et al., 2018) and the selection of target antigen (Mahoney et al., 2015; Sanchez-Martin et al., 2015) have been reviewed previously. Here, we explore how monoclonal antibodies have been engineered to customize their in vivo effects.

3.1.2.1 Monospecific Antibodies and Mechanisms of Action

Three common classes of therapeutic antibodies include (1) antibodies whose major mechanism of action is through the blockade or stimulation of the targeted pathway; (2) antibodies that are used to deliver a drug payload; and (3) antibodies whose structure has been engineered to engage multiple different cell types simultaneously to redirect functionality. Endogenously, antibodies often bind infected cells or foreign pathogens. The exposed constant region of these antibodies can then mediate antibody-dependent cellular cytotoxicity or complement-dependent cytotoxicity to destroy the targeted entity. Thus most fundamentally, antibody engineering often involves modification of the Fc region to modulate how the bound antibodies will interact with the immune system (Beers et al., 2016). When antibody-mediated cell killing is desired, such as when antibodies target molecules on infected or tumor cells, isotypes such as human IgG1 are often incorporated that initiate these effects (Wang et al., 2015; Taylor and Lindorfer, 2016). In other cases an antibody may be developed to block or stimulate a certain pathway on a healthy cell type, such as the PD-1 inhibitory molecule on T cells. For such cases in which cell death would be detrimental, the antibody is engineered to have an IgG2 or IgG4 isotype that has limited interaction with Fc receptors (Gong et al., 2016). In all cases, antibodies are humanized by replacing the constant region and sometimes parts of the variable regions with the human sequence (Harding et al., 2010).

The interplay between T cells and cancerous cells is multilayered and complex. While cancerous cells may express mutated neoantigens that are recognized by the immune system, they also often upregulate immunosuppressive molecules that inhibit efficient T cell stimulation and lack expression of costimulatory molecules (Chen and Flies, 2013). Many monoclonal antibodies have been developed that block these inhibitory pathways on T cells or tumor cells or that directly stimulate signaling pathways on T cells to jumpstart the cancer immunity cycle. Clinical administration of soluble antagonist antibodies against the inhibitory checkpoint molecules PD-1 (Callahan and Wolchok, 2013; Hamid et al., 2013) and its ligand, PD-L1 (Chen and Han, 2015), and CTLA-4 (Pedicord et al., 2011; Larkin et al., 2015; Postow et al., 2015) have shown tremendous success in recent years. The success of these so-called checkpoint blockade antibodies have been correlated with tumor mutation load—tumors with more mutations and more neoantigens are associated with improved clinical benefit from checkpoint molecule inhibitors (Van Allen et al., 2015; Hugo et al., 2016). Agonistic antibodies have also been developed that stimulate T cell signaling through costimulation, such as anti-CD28, anti-4-1BB, and anti-OX40 antibodies (Buchan et al., 2015; Oh et al., 2015; Tyrsin et al., 2016; Jahan et al., 2017). However, because of the structured arrangement of the signaling molecules during T cell activation, these synthetic costimulatory molecules are often delivered on particle platforms (Rudolf et al., 2008; Zeng et al., 2014) that enable formation of the immune synapse.

The direct function of an antibody is often to affect a signaling pathway. However, antibodies have also been conjugated with various drugs and utilized as carriers to deliver payloads to specific T cells or tumor cells. In cancer immunotherapy, many of these so-called antibody-drug conjugates (ADCs) target tumor-expressed antigens, such as HER2, CD19, or CD22, and novel strategies have been introduced to increase tumor-specificity. For example, an ADC was developed that encodes an additional peptide sequence that shields the antibody's antigen binding site until it is cleaved within the tumor microenvironment (Polu and Lowman, 2014). This method can thus enhance tumor-specificity when the antigen target is not uniquely expressed by cancerous cells. This type of approach could also potentially be used to decrease off-target toxicities of checkpoint inhibitors and costimulatory antibodies as these are not limited to the cancerous tissue. Fusion proteins have also been developed that deliver multiple payloads simultaneously. For example, a trifunctional fusion protein linked an antibody against a tumor-associated antigen with both the costimulatory molecule 4-1BBL and the IL-15 cytokine (Kermer et al., 2014). The fusion protein was more efficient than corresponding bifunctional proteins in activating T cells and inducing an antitumor response in vivo. The approach to treating lymphoid malignancies is slightly different; in this case, antibodies have been developed to deliver cytotoxic drugs to the malignant T cells themselves. An ADC targeting IL-7R, which is expressed on lymphatic cells, demonstrated how this method could be used to treat lymphoid malignancies or autoimmune disorders (Yasunaga et al., 2017).

3.1.2.2 Approaches Antibody Engineering

Many tools are being developed to speed up and diversify the discovery of novel antibody therapies or engineer of antibody variants. One such tool is the development of combinatorial antibody libraries (Lerner, 2016). There are nearly 10^{11} unique antibody sequences, and being able to screen or probe the unique antibody-antigen binding landscape may enable enhanced affinities or unique binding locations which may enable novel mechanisms of therapeutic action, ultimately allowing control over therapeutic antibody design. Nevertheless, the massive libraries would be unexploited if not for high throughput engineering techniques. One such technique is the development of yeast surface display affinity engineering (Packer and Liu, 2015). Here, antibody libraries can be cloned into yeast and fragments can be displayed and allowed to bind to a target protein. Antibodies to both the fragment and bound target protein identify high binders, which can be selected for and further mutated to increase affinity.

There are many variants in structural engineering of the antibody, with particular focus on Fc and Fab portions (Holliger and Hudson, 2005; Owens, 2017). One example that targeted engineering of both ends, looked at eliminating nonspecific thromboembolism events that occurred when the anti-CD40L antibody was administered to treat autoimmune diseases (Xie et al., 2014). Specific targeting sequences of the antibody toward CD40L, termed a domain antibody, were fused to an Fc domain which was engineered to lack any Fc-binding function. Eliminating the Fc, eliminated platelet activation, thus reducing thromboembolism events, and allowing effective treatment of both models of transplantation and lupus.

3.1.2.3 Bispecific Antibodies

One of the most powerful antibody-based approaches in cancer immunotherapy has been the development of bispecific antibodies. Bispecific antibodies are antibodies that have been engineered to comprise two different variable portions, each specific for a different antigen. Their specific structures encompass many formats, and each varies in terms of size, spatial relationship between antigen-binding sites, and synthesis protocol, which may affect their in vivo half-life and interactions with their respective ligands (Fig. 5B; Strohl, 2017). The physical constraint of the two antibody fragments is often central to their function, as bispecific antibodies are used to either physically link two different cell types or dimerize two different molecules on the surface of the same cell. Larger bispecific formats include the individual component antibodies linked by a chemical linker or the a Triomab, which maintains the full IgG shape by linking two half antibodies, each with one heavy and one light chain (Chelius et al., 2016). These families of bispecific antibodies can be easily synthesized by standard protocols to allow for high-throughput manufacturing, and they have extended half-lives compared with smaller proteins. However, they also maintain expression of the Fc region which may be recognized by Fc receptors and lead to unwanted off-target effects. At the opposite end of the spectrum are small bispecific platforms such as tandem single-chain variable fragments (scFvs), in which the heavy and light chains of two different variable domains are linked by a short serine-glycine linker (Huehls et al., 2015). These bispecific antibodies preclude Fc receptor recognition, but their small size (approximately 55 kDa) reduces half-life and necessitates constant intravenous infusion (Klinger et al., 2013).

Likely the most well-known type of bispecific antibodies in the field of cancer immunotherapy are bispecific T cell engagers (BiTEs). BiTEs link an anti-CD3 antibody fragment with an antibody fragment against a tumor-associated antigen, thus simultaneously activating T cells and physically linking them with tumor cells to induce lysis (Yuraszeck et al., 2017). However, because BiTEs incorporate anti-CD3 rather than a pMHC, they nonspecifically activate all T cells and thus are associated with dose-limiting toxicities including cytokine storms (Lee et al., 2016). BiTEs have been developed linking T cell stimulation with various tumor-associated antigens such as CD19 (Mack et al., 1995; Kantarjian et al., 2017), BCMA (Hipp et al., 2017), and EphA2 (Iwahori et al., 2014). However, like many targeted immunotherapies, they require engagement of a tumor-associated antigen and thus antigen escape is a concern. Interestingly, it was recently shown that BiTEs can mediate lysis of bystander tumor cells not expressing the targeted antigen (Ross et al., 2017).

T cell—targeted bispecific antibodies have also been developed that cross-link two molecules on the surface of the same immune cell. A recent bispecific antibody, synthesized by chemically linking agonistic antibodies against the costimulatory receptors CD137 and CD134 on T cells, was more effective at activating an antitumor T cell response than either antibody individually (Ryan et al., 2018). In contrast with monotherapy, the platform uniquely induced preferential expansion of effector T cells. Thus, in addition to redirecting cells, bispecific antibodies can potentially impact T cell activation by physically clustering surface molecules. This approach may be particularly interesting for targeting other T cell signaling molecules since the activation process is highly sensitive to receptor clustering and spatial localization.

With the success of immune-activating bispecifics, significant resources are being dedicated to increase the efficacy of current therapies while limiting toxicity. With nearly 60 preclinical drugs and 30 in clinical trials, knowledge gained from therapeutic efficacy will impact on our fundamental understanding of what constitutes ideal bispecific properties, like pharmacokinetics (Chen and Xu, 2017). Indeed techniques such as developing multiple-cancer epitopes for T cell redirecting antibodies could increase the affinity and sensitivity of the therapy (Sedykh et al., 2018). This would help to reduce nonspecific toxicity such as cytokine storms that result from these immune-activating bispecifics.

In addition to activating the immune system, development of bispecifics is also expanding to other applications such as infectious and autoimmune diseases and diagnostic imaging. Besides T cell redirection, these bispecifics have additional mechanisms of action: T cell recruitment, blockade of T cell acting cytokines, inducing apoptosis, and inactivation of proteins. Indeed, many autoimmune anticytokine therapies demonstrate additive effects when combined justifying the development of bispecific for autoimmune diseases. Just one example includes a bispecific developed toward TNF-α and IL-17 for the treatment of rheumatoid arthritis (Fischer et al., 2015). Here, the bispecific was engineered such that there were an additional two TNF-α and IL-17 Fab portions linked to Fc regions of the antibody, which induced significantly enhanced therapeutic reduction of arthritis potentially due to increased avidity. Similar results were found from a bispecific targeting IL-17A and IL-6 (Lyman et al., 2018).

HIV is an example of bispecific design for infectious disease, where diversity and mutation lend for the need for combinatorial therapies. One example is a neutralizing antibody where one arm of the bispecific is from antibodies toward either CD4 or CCR5, which the HIV virus uses to enter the cell (Huang et al., 2016). The other arm is composed of a Fab that is specific to viral envelope epitopes. These bispecifics reduced viral load and provided complete protection upon viral challenge. Another example to broaden the diversity and increase the avidity of a neutralizing HIV antibody is the development of a trispecific antibody which recognizes the CD4-binding site, V3 glycan patch, and the envelope membrane proximal external region (Steinhardt et al., 2018). The trispecific antibody demonstrated enhanced and broad neutralization in comparison to monospecific antibody therapies.

Another avenue for future bispecific development is improving production and yield of the antibody. Here, principles of directed evolution, cell engineering of hybridomas, and protein engineering of bispecific structure will aim increase yield to decrease high cost associated with these therapies. Alternatively to in vitro production, bispecifics could be produced in vivo through transfection of the bispecific antibody construct. This was recently demonstrated in production of therapeutic levels of an anti-CD3/CD20 bispecific in a mouse for effective antilymphoma therapeutic responses (Pang et al., 2017).

3.2 Synthetic Particles

In contrast to soluble antibodies and small molecules, nanoparticle-based therapies can be engineered to affect biodistribution and drug release profiles. Soluble or particle-conjugated signals can be conjugated to the surface or encapsulated within synthetic particles to directly affect T cell signaling or alter how other cell types interact with T cells. By altering the size, shape, material, or injection route of a particle, drugs can target specific immune subsets in defined timeframes. Here, we will review previous work in this field of immunoengineering and demonstrate how micro- and nanotechnology can be used to deliver extracellular and intracellular signals to modulate T cell responses.

3.2.1 Artificial Antigen-Presenting Cells

aAPC are micro- or nano-sized platforms designed to mimic endogenous APCs for T cell activation. aAPC are three-dimensional cellular or synthetic platforms that minimally present the two necessary signals for T cell activation—pMHC, signal 1 (s1), and a costimulatory molecule, such as B7-1/B7-2 or αCD28 mAb, signal 2 (s2; Oelke and Schneck, 2010; Perica et al., 2015). aAPC have been explored in adoptive cell therapy (ACT), ex vivo activation of autologous lymphocytes with aAPC expressing tumor-specific pMHC, followed by reinfusion of the expanded cells into the patient (Dudley and Rosenberg, 2003). Large numbers of expanded lymphocytes, up to 10^{11}, are needed to treat a single patient with ACT (Rosenberg and Restifo, 2015). Additionally, research has shown the importance of not just quantity but, more importantly, quality of the resultant cells for a long-lived effector population posttransfer (Busch et al., 2016).

Acellular aAPC have grown in importance for a variety of reasons. One reason is cost. Synthetic platforms can decrease costs by creating a more easily manufactured all-in-one off-the-shelf therapy. Another reason is that endogenous APCs can become immunosuppressed and even immunosuppressive in cases such as cancer and chronic infection (Della Bella et al., 2003; Satthaporn et al., 2004; Ye et al., 2010). This poses problems for the achievement of effective innate anticancer responses, therapeutic vaccines, and immunotherapies and ex vivo activation of T cells for adoptive transfer. Thus extensive work has been done to optimize acellular aAPC. Specifically, we will focus on engineering developments of the aAPC that has led to increased functionality: geometry, shape, signaling molecule choice, and material.

3.2.1.1 Size

aAPC size has implications to mimicking surface engagement of endogenous cells, biodistribution, and interaction with the nanoscale arrangement of receptors. Historically, micron-sized aAPC platforms were developed to mimic the endogenous APCs (Mescher, 1992; Curtsinger et al., 1997; Tham et al., 2001; Oelke et al., 2003b; Rudolf et al., 2008). Large, micron-scale particles reduce the curvature at the T cell—aAPC interface and can thus interact with multiple TCRs and induce robust activation. However, spherical microparticles with low MHC signal 1 density cannot activate T cells to the same extent as particles with high density despite increasing particle dose (Mescher, 1992; Motta et al., 1998). Currently, this size is the preferred size for attaching the aCD3 and aC28 molecules used to expand polyclonal populations from tumor infiltrating lymphocyte populations for adoptive immunotherapy (Hinrichs et al., 2011).

Previously it was believed that nanoparticle aAPC could not induce robust T cell activation similar to large aAPC (Mescher, 1992; Steenblock and Fahmy, 2008). In one case, biodegradable aAPCs of 150 nm and 8 μm were formed and showed that microparticles provided enhanced stimulation to antigen-specific T cells, although nanoparticles resulted in some minimal stimulation (Steenblock and Fahmy, 2008). In another report, latex particles of 0.5–5 μm were engineered, and it was shown that larger particles generated greater responses (Mescher, 1992). It was hypothesized that the T cells needed to interact with a contiguous area of stimulatory ligands as opposed to the smaller areas provided by smaller particles.

However, this large size hinders the ability of aAPC to be used in vivo. Particles larger than 2 μm have a chance to cause embolization. Larger particles also have limited abilities to diffuse out into target organs where T cells reside and increase the rate of uptake by the mononuclear phagocytic system. To address this, researchers developed two versions of nanoparticle aAPC from iron oxide and quantum dot materials 30–100 nm in size (Perica et al., 2014). Here, nanoparticles were shown to provide effective antigen-specific stimulation of both murine and human $CD8^+$ T cells and provide tumor prevention. Furthermore, biodistribution studies were done to show that the smaller aAPC drained more efficiently to the lymph node than larger aAPC remained in the subcutaneous injection site.

In an effort to provide mechanistic insight into why these nanoparticles were functional, researchers varied both nanoparticle aAPC size and ligand density (Fig. 6A; Hickey et al., 2017a). Here, particles of 50 nm, 300 nm, 600 nm, and the traditional 4.5 μm were compared head-to-head with controlled densities and quantities of signal. Interestingly, the smallest (50 nm) aAPCs were again found to be less effective; however, aAPCs 300 nm and above were just as effective as large 4.5 μm particles. Mechanistically, this could be due to (1) nearby clustering of signaling molecules was important, and (2) that particles larger than 300 nm aAPC are able to provide effective local engagement of multiple TCRs, while 50 nm aAPC only engage single receptors. This was confirmed by overcoming this limitation by both artificially clustering 50 nm aAPC with a magnetic field and adding saturating doses of 50 nm aAPC.

Ligand density can also be an important parameter, which should be considered in conjunction with size. Increasing the density of the ligand on smaller 50 nm aAPC did not overcome this effect, suggesting a larger particle footprint is needed to engage TCRs in protein nanoislands (Schamel et al., 2005; Lillemeier et al., 2010; Hickey et al., 2017). As the ligand spacing increased to greater than 100 nm, the ability to stimulate $CD8^+$ T cells was abolished even for particles larger than 300 nm. This was consistent

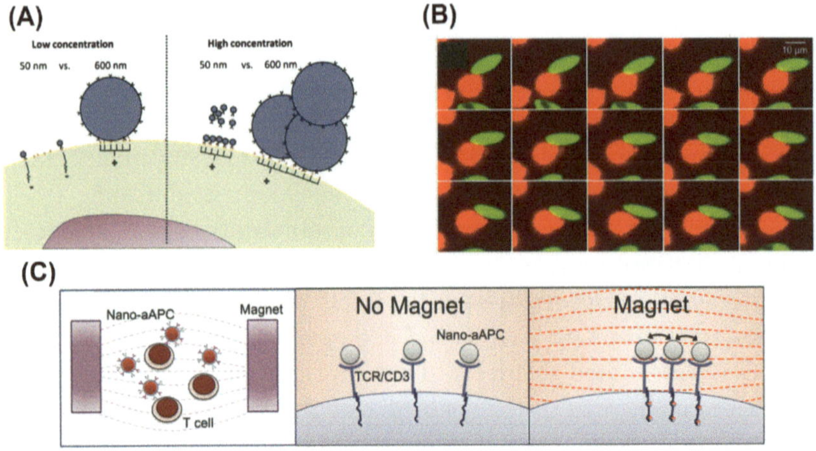

Figure 6 The size, shape, and material of artificial antigen-presenting cells (aAPC) can be engineered to manipulate how they interact with and activate T cells. (A) Nano-aAPC-based T cell activation requires higher concentrations of particles to achieve nanoisland clustering. (B) Ellipsoidal aAPC (green) interact with T cells (red) via their long axis, increasing the length of contact compared with spherical aAPC. (C) Superparamagnetic nano-aAPC can be clustered on the T cell surface using external magnetic fields to induce TCR/CD3 clustering and subsequently T cell activation. *(A) Reprinted with permission from Hickey, J. et al., 2017. Biologically inspired design of nanoparticle artificial antigen-presenting cells for immunomodulation. Nano Lett. 17 (11), 7045–54, Copyright (2017) American Chemical Society. (B) Reprinted from Sunshine, J.C., et al., 2014. Particle shape dependence of CD8$^+$ T cell activation by artificial antigen presenting cells. Biomaterials 35 (1), 269–277. https://doi.org/10.1016/j.biomaterials.2013.09.050. Elsevier Ltd., Copyright (2014), with permission from Elsevier. (C) Reprinted with permission from Perica, K., Tu, A., et al., 2014a. Magnetic field-induced T cell receptor clustering by nanoparticles enhances T cell activation and stimulates antitumor activity. ACS Nano 8 (3), 2252–2260. https://doi.org/10.1021/nn405520d, Copyright (2014) American Chemical Society.*

with the ligand requirements found from planar surface studies (Deeg et al., 2013; Delcassian et al., 2013; Matic et al., 2013) and could explain why nano-aAPC were not stimulatory previously. Therefore both nano- and micro-sized aAPC can be effective with control of the ligand density, as both properties affect the area and number of ligands available to stimulate a CD8$^+$ T cell effectively.

3.2.1.2 Shape

Despite their success in activating T cells at the cell–particle level, microscale particles have poor biodistribution and are thus suboptimal for in vivo applications. Microparticles cannot take advantage of the EPR effect to localize aAPC at the tumor site (Fang et al., 2011) and do not effectively drain to lymph nodes where immune stimulation occurs (Helle et al., 2013; Perica et al., 2014). Thus recent work has explored ways to optimize aAPC on more biocompatible, nano-scale platforms.

The spherical shape of aAPC is popular largely because of the ease of synthesis or fabrication procedures such as emulsion techniques for poly(lactic-co-glycolic acid (PLGA; Astete and Sabliov, 2006; Danhier et al., 2012). However, spheres minimize the contact area at the T cell—aAPC interface despite the importance of multivalent pMHC—TCR interactions. Recent work has demonstrated how altering aAPC geometry can improve the stimulatory capabilities of both micro- and nano-aAPC. Importantly, modifications to aAPC shape have also helped to bring nano-aAPC into the playing field.

Ellipsoidal synthetic aAPC can be developed by physically stretching spherical PLGA particles made by standard emulsion techniques (Sunshine et al., 2014). These oblong particles can be generated with aspect ratios over six and conjugated with equivalent protein densities, allowing for comparison between them and their spherical counterparts (Fig. 6B). Increasing the micro-aAPC aspect ratio (i.e., degree of stretching) positively correlates with $CD8^+$ T cell proliferation and maximally induces a 20-fold stronger T cell response. Interestingly, proliferation did not increase linearly with increased particle stretching but rather resulted in more significant gains at specific aspect ratios. This geometry-driven change in T cell proliferation was shown to be correlated with changes to the T cell—aAPC interface. Ellipsoidal aAPC bound to more T cells, and individual contact areas were larger than those in spherical interactions, thus suggesting that increased avidity and interactions with multiple TCR may drive their improved efficacy.

The geometry effect is also maintained on the nanoscale (Meyer et al., 2014). Ellipsoidal nano-aAPCs with spherical diameters on the order of approximately 200 nm induce a fivefold stronger T cell response than their corresponding spherical nano-aAPC. As shown with other systems (Champion and Mitragotri, 2006; Gratton et al., 2008; Sharma et al., 2010), ellipsoidal aAPC resisted uptake by RES cells and thus had longer circulation times in vivo. This increased ability to stimulate T cells and longer circulation times led to significantly increased T cell proliferation in vivo.

In contrast to methods that manipulate the geometry of the base particle, magnetic fields have also been utilized to alter how nano—aAPC interact with T cells. Superparamagnetic spherical iron-dextran aAPCs with a diameter of 100 nm do not normally activate a robust T cell response. However, when an external magnetic field is applied, the nano-aAPC cluster on the T cell surface, changing the perceived shape and size of the aAPC by the T cell, and likewise, increase proliferation (Perica et al., 2014; Hickey et al., 2017). This method enables nano-aAPC to activate rare endogenous T cell populations for adoptive cell transfer.

3.2.1.3 Ligand Choice

Ligand choice is another critical parameter in the development of aAPC. Traditionally, aAPC are conjugated with the two necessary signals for T cell activation—anti-CD3 mAb or a specific pMHC, signal 1, and B7 or anti-CD28 mAb, signal 2. The choice of anti-CD3 mAb or pMHC dictates whether all T cells or only T cells recognizing a specific target are activated. Often, a specific pMHC will be chosen to target antigen-specific T cell responses, although interesting work has demonstrated how phenotypic markers can be used post-nonspecific activation to select specific T cell subsets (Ye et al., 2014). A great deal of work in antigen discovery and immunogenicity has been conducted and is ongoing in the selection of a specific pMHC targets for different applications.

Despite the fact that signaling through CD28 is the main costimulatory signal provided by synthetic aAPC, the composition of costimulatory signals greatly impacts T cell activation. There are dozens of different types of costimulatory signal 2 molecules, and altering the combinations and ratios of these signals can impact proliferation, phenotype, and survival in vivo. These T cell costimulatory molecules include pathways such as CD70-CD27, CD40-CD40L, 4-1BB-4-1BBL, OX40L-OX40, and others expressed on the APC and T cell, respectively (Chen and Flies, 2013).

Studies have shown that costimulation through 4-1BB during the rapid expansion protocol for adoptive cell transfer improves the expansion and effector function of tumor-specific T cells (Hernandez-chacon et al., 2012). 4-1BB costimulation has also been associated with the preferential expansion of memory T cells, which are often desired for in vivo transfer applications (Zhang et al., 2013). Combining multiple costimulatory signals on the same aAPC also can have significant effects on T cell functionality. A study combining CD28, 4-1BB, and CD27 signaling on K562 aAPC demonstrated that certain combinations of the costimulatory molecules can increase proliferation over 10-fold (Zeng et al., 2014). Even when the same total amount of costimulation is present, simply changing the ratio of signal 2 ligands can impact the resultant T cells. In one study, micron-sized polystyrene aAPC conjugated with different ratios of anti-CD28 and anti-4-1BB mAb affected antigen-specific proliferation as much as fivefold (Rudolf et al., 2008). Other costimulatory molecules, such as activation through OX40 (Jahan et al., 2017) and CD40L (Singh et al., 2017; Tay et al., 2017), which have been shown to have stimulatory effects on T cells, could also be incorporated in next-generation aAPC. A new type of aAPC activation platform where each type of signaling molecule is separated onto distinct, magnetically clustered nanoparticles allow for high throughput customization of signal 2 choice, ratio, and amount by enabling an infinite number of activation conditions from a finite number of functionalized nanoparticles (Kosmides et al., 2018). Using this platform, combinations of anti-CD28, anti-4-1BB, and anti-CD27 signaling were shown to affect T cell proliferation and phenotype. This approach may serve as a useful mechanism to systematically study costimulatory composition.

In addition to the choice of signal 1 and signal 2, aAPC can be conjugated with other molecules to enhance their efficacy. CD47 is a molecule ubiquitously expressed by normal cells that protects self-cells against phagocytosis (Poon et al., 2014). aAPCs that also display CD47 have been shown to also be protected from phagocytosis and induce greater stimulation of T cells in vivo (Bruns et al., 2015).

3.2.1.4 Material

aAPC have been formed from materials that are inorganic, organic, biologically derived, or a combination thereof. Each material choice enables unique functionality and imparts distinct aAPC properties. For example, functions include responsiveness to external stimuli, ligand mobility, biocompatibility, degradability, biodistribution, nanoscale arrangement of receptors, and encapsulation of additional modulators. Therefore the desired function of the aAPC can drive selection of material.

Iron oxide—based aAPC can be effectively used in adoptive immunotherapy expansion protocols as they can be easily removed through magnetic field separation, prior to infusion (Levine et al., 1998; Oelke and Schneck, 2010). Magnetic forces have been used additionally to induce aAPC clustering on the surface of $CD8^+$ T cells and enrich antigen-specific cells (Fig. 6C; Perica et al., 2014; Perica et al., 2015). First, clustering of nanoparticle iron oxide aAPC within a magnetic field lead to much greater stimulation of $CD8^+$ T cells at a much lower dose of aAPC than without the magnetic field (Perica et al., 2014). This increased TCR cluster size on the surface of the T cells has been shown to be important for T cell activation (Varma et al., 2006). The enhanced ex vivo stimulation translated to increased in vivo efficacy of adoptively transferred cells in a tumor treatment model. More recently, magnetic fields have been used to home magnetic aAPC in vivo to tumor sites to mediate effective tumor killing (Zhang et al., 2017).

Beyond clustering, Perica et al. used these antigen-specific nano-aAPC to magnetically enrich for antigen-specific $CD8^+$ T cells (Perica et al., 2015). The approach is to first add the aAPC to bind rare antigen-specific $CD8^+$ T cells, then enrich the antigen-specific cells, and because the aAPC are still bound to the enriched cells, simultaneously stimulate only the enriched cells to proliferate. Impressively, antigen-specific murine and human $CD8^+$ T cells were expanded more than 1000- to 10,000-fold over their precursor frequency. This was used to identify novel neoantigens predicted by tumor sequencing and also used in adoptive transfer model of murine melanomas.

Organic materials are used for improved biocompatibility, biodegradation properties, ability to form shapes other than spheres, and capacity to encapsulate material (Cheng et al., 2015). This is true as well for the aAPC. Particle aAPC have been made from PLGA, which is a component of several FDA-approved devices and therapeutics and has high biocompatibility, allowing aAPC to be used for direct in vivo therapy (Kosmides et al., 2017). Direct in vivo therapy with aAPC is desirable not only because it would allow continued modulation and activation of $CD8^+$ T cells in vivo but also because it would potentially eliminate the need for ex vivo stimulation. These biodegradable aAPC were effective at stimulating $CD8^+$ T cells ex vivo and provided enhanced activation in response to a tumor when combined with checkpoint blockade molecule anti-PD1. This in vitro finding was confirmed in vivo as the combination of aAPC, and anti-PD1 was able to delay tumor growth in an aggressive melanoma cancer treatment model more than either individual modality alone. Additionally, polymers like PLGA are plastic so they can be formed into different shapes as discussed in Section 3.3.1.2, which may enable enhanced biodistribution and T cell contact (Sunshine et al., 2014; Meyer et al., 2015).

Moreover, organic particles formed by emulsion techniques can be utilized for encapsulation of soluble signals. T cell stimulation often requires a third signal from cytokine IL-2 from $CD4^+$ T cells to become fully activated and have the capacity to proliferate (Janssen et al., 2003; Shedlock and Shen, 2003; Malek, 2008). Steenblock et al. developed a PLGA aAPC with IL-2 encapsulated to mediate close paracrine delivery of the cytokine, while the aAPCs are bound to antigen-specific $CD8^+$ T cells to mimic signal 3 delivery in vivo (Steenblock et al., 2011). Having the IL-2 encapsulated improved the $CD8^+$ to $CD4^+$ T cell ratio and enhanced $CD8^+$ T cell activation with upregulation of the IL-2 receptor, CD25. Encapsulation of IL-2 was 100 times more potent than the simple addition of soluble IL-2 to the culture medium, suggesting the advantage of cytokine encapsulation.

As an alternative to organic materials, biological materials are another biocompatible platform to which T cell stimulatory signals can be attached. Beyond biocompatibility, these materials may evade the mononuclear phagocytic system more effectively and thus provide a more effective in vivo therapeutic. Another advantage for biological materials is the ability to have fluid membranes, which may impact the mobility of the signaling receptors and allow for the needed clustering of the receptors upon engagement.

Recently, RBCs have been utilized as a bioinspired method for cargo delivery or as a depot of imaging agents (Yoo et al., 2011). RBCs have "don't-eat-me" signals on their surface, which enables long circulation times. Sun et al. recently attached signaling molecules, including IL-2, to the surface of RBCs (Sun et al., 2017). While there was no in vivo work demonstrated, these provided effective stimulation to antigen-specific cells that were able to mediate cancer-specific killing in vitro.

Liposomes are particles developed from lipids that form a lipid bilayer similar to that of the cell, but the size can be modulated from the nanometers to micrometers range. Liposome-based aAPCs were developed to utilize clustering technology derived from lipid rafts to show that having distinct, locally dense MHC rather than diffusely spread MHC improved T cell activation (Giannoni et al., 2005). Additionally, because it has been shown that endogenous MHCs are clustered in lipid rafts on DCs, another group has taken natural MHCs from DCs and added the molecules to liposomes (Ding et al., 2013). These RAFTsome aAPCs showed effective activation in vitro and tumor protection in an immunization model. This could be potentially used to load patient-derived MHCs preloaded with cancer antigens to induce a polyclonal antigen-specific response without characterizing individual antigenic specificities.

Finally, as each material has unique functionalities, researchers have aimed to combine materials for more elaborate aAPC designs. One example is an aAPC developed from carbon nanotubes, polymer (PLGA) nanoparticles, and magnetite (Fadel et al., 2014). The carbon nanotubes were used to form the particle scaffold, which proves the high surface area needed for effective multivalent aAPC—T cell interactions. The PLGA nanoparticles were loaded with IL-2 to induce paracrine signaling of signal 3. Furthermore, magnetite was embedded in these particles to enable magnetic removal of particles following T cell stimulation for adoptive immunotherapy. These aAPCs provided effective long-term stimulation of both murine and human $CD8^+$ T cells and delayed tumor growth in a melanoma treatment model.

3.2.2 Particles With Ligands for T Cells

In addition to creating aAPCs for activating T cell responses, pMHC molecules have been attached to particles for limiting the immune response (Tsai et al., 2010; Clemente-Casares et al., 2016; Singha et al., 2017). This approach aims to mimic peripheral tolerance where presentation of antigen in the absence of costimulatory molecules leads to anergy, apoptosis, or conversion to a regulatory phenotype. Indeed, this has been done for both attaching pMHC class I and II molecules and inducing both antigen-specific regulatory $CD4^+$ and $CD8^+$ T cells with the effect of reducing autoimmunity in multiple models including diabetes, multiple sclerosis, and arthritis. To be able to administer these particle intravenously, particles less than 20 nm were utilized for engineering studies to identify key parameters necessary for both in vivo function and safety (Singha et al., 2017). Indeed, similar to aAPCs studies, density of the pMHC was a key component of the efficacy of their particles, where less than 3—6 nm intermolecular spacing was most efficacious in eliciting antigen-specific T cell responses. This density may allow enhanced binding with TCR nanoclusters, and clustered binding in macroclusters is observed experimentally and confirmed in silico through modeling.

In addition to particles that interact only with T cells (i.e., aAPC) and particles that interact only with APCs (i.e., particle-based vaccines), particles can be engineered to bind multiple cell types simultaneously, inducing new interactions or changing existing ones. In these instances, a nanoparticle platform is used to increase avidity for the targeted antigens or alter biodistribution. In one study, 50 nm particles were used to redirect a T cell response against a specific tumor antigen, similar in approach to BiTEs (Shutz et al., 2016). These nanoparticles, termed antigen-specific T cell redirectors (ATR), were conjugated with an antibody against the CD19 tumor

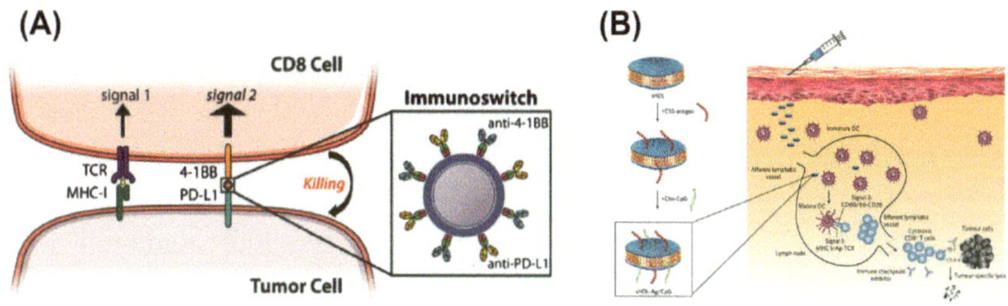

Figure 7 Engineered nanoparticles can be used to affect T cell interaction and stimulation by other cell types. (A) Nanoparticles functionalized with anti-4-1BB and anti-PD-L1 mAbs convert inhibitory signals to costimulatory signals for T cells. (B) A nanodisc platform codelivers antigen peptides and adjuvants to dendritic cells, eliciting robust antigen presentation by dendritic cells and activation of $CD8^+$ T cells. *(A) Reprinted with permission from Kosmides, A.K., et al., 2017. Dual targeting nanoparticle stimulates the immune system to inhibit tumor growth. ACS Nano 11 (6), 5417–5429. https://doi.org/10.1021/acsnano.6b08152, Copyright (2017) American Chemical Society. (B) Reprinted by permission from Springer Nature: Kuai, R., et al., December 2016. Designer vaccine nanodiscs for personalized cancer immunotherapy. Nat. Mater 1, 1–10. https://doi.org/10.1038/nmat4822, Copyright (2016).*

antigen and nontumor pMHC. When MHC-bound ATR were loaded with a human flu peptide, human flu-specific T cells were redirected to kill $CD19^+$ tumor cells.

In contrast to a tumor cell–T cell dual-binding nanoparticle that presents a specific pMHC complex, nanoparticles that bind to other nonspecific cell surface molecules on the two cell types have shown similar efficacy. For example, an 80-nm particle, termed immunoswitch, links antibodies against the costimulatory 4-1BB molecule expressed by T cells with antibodies against the coinhibitory molecule PD-L1 expressed by tumor cells and thus converts an inhibitory signal into a costimulatory signal within the tumor microenvironment (Fig. 7A; Kosmides et al., 2017). These particles were shown to increase effector-target cell conjugation and induce a more robust $CD8^+$ tumor-infiltrating lymphocyte response and delay or eliminate tumor growth in multiple murine tumor models.

3.2.3 Particle Vaccines for T Cell Activation

Synthetic particles can also be engineered to affect which cell types T cells interact with, alter a cell interaction that is already present, or alter the phenotype and activity of other immune cells that modulate T cell activity. DCs play a central role in orchestrating both the innate and adaptive immune response (Steinman, 2012). In an optimal response, DCs express peptide fragments of invading pathogens as well as express costimulatory molecules that enable their efficient activation of T cells. However, in the instance of cancer or chronic infection, this process does not always occur appropriately. One method to initiate a more robust T cell response is thus to enhance antigen presentation or costimulation by DCs which can then mediate the T cell response. Since our analysis is focused on T cells, we will only provide a brief summary of particle properties for vaccines, where more detailed reviews analyzing the density of antigen, size of particle, and material chemistry have already been completed (Bookstaver et al., 2017).

Activation of DCs and other APCs, such as T cell activation, is sensitive to the environment. A stimulatory environment with appropriate stimulation kinetics is crucial to induce activation rather than tolerance. In order to control the stimulatory conditions, DCs can be extracted from a patient, and the signals for activation can be fully controlled ex vivo. Cell-based DC vaccines, such as Sipuleucel-T (Kantoff et al., 2010; Paller and Antonarakis, 2012), involve this ex vivo activation of DCs against a tumor antigen followed by their reinfusion into the patient where they can circulate throughout the body and activate an antitumor immune response. However, this approach is time-consuming and expensive as it involves manipulating cells outside of the human body and often has to be completed at nonlocal sites. Nanoparticles have generated great interest in the field of vaccine delivery due to their ability to be engineered to manipulate in vivo biodistribution and kinetics of vaccine delivery (Gu and Mooney, 2015). Biomaterials can also potentially be used to target vaccination to specific DC subsets which has an effect on vaccine efficacy (Leleux et al., 2015).

Various studies have explored using nanoparticles to effectively delivery antigen to DCs directly in vivo. In the design of nanoparticle-based vaccination, particle size plays an important role in lymph node trafficking. Pluronic copolymer—coated nanoparticles (25 nm) injected intradermally were shown to travel to draining lymph nodes through interstitial flow and remain there for at least 120 h where they could activate a humoral and cellular immune response against a model antigen (Reddy et al., 2007). In contrast, particles just four times larger—100 nm—were approximately 10% as efficient and were undetectable in draining lymph nodes within 24 h. In another study using polystyrene particles ranging in size from 20 nm to 2 μm, 40-nm particles most efficiently trafficked to lymph nodes and activated resident DCs (Fifis et al., 2004). Similar to spherical nanoparticles, nanodiscs have been used for vaccine delivery. Small vaccine-carrying lipoprotein nanodiscs approximately 10 nm in diameter were shown to be more effective than soluble vaccines (Kuai et al., 2016). These nanodiscs were coupled with antigen peptides and adjuvant and were shown to elicit up to 47-fold greater frequency of tumor-specific T cells than soluble vaccines (Fig. 7B). Utilization of a nanodisc carrier allowed antigen and adjuvant to accumulate in lymph nodes, where there is a high density of DCs, more than soluble injected vaccines. Most successful particle-based vaccines are engineered with sub-100 nm particles. Particles larger than approximately 200 nm require cellular transport by DCs in the skin to travel to the lymph node (Manolova et al., 2008).

Particle-antigen conjugates have also been used to generate tolerance in the setting of autoimmune diseases. These approaches focus on delivering auto-antigen similar to particle vaccines, but in the absence of danger signal to generate tolerance. One recent example utilized a quantum dot base particle to enable imaging of trafficked particles within lymph nodes and cellular compartments (Hess et al., 2017). Interestingly, the investigators found that higher numbers of low-density auto-antigen on the particles were more effective than lower numbers of high-density auto-antigen in reducing autoimmunity in a model of multiple sclerosis in mice by increasing regulatory T cells. Potentially, the therapeutic effect could be similar to how peripheral tolerance is developed through high systemic levels of auto-antigen in the absence of danger signal.

In addition to controlling antigen density and dose, codelivery of immunosuppressive agents has been key for mediating effective regulatory immune responses in autoimmune therapies. One example is the use of adjuvants for toll-like receptors (TLRs) such as GpG, which promotes tolerance through TLR9. Because it is electrostatic, it has been complexed with self-antigen to form polyelectrolyte multilayer particles and used to effectively eliminate disease in a mouse model of multiple sclerosis (Tostanoski et al., 2016). Another example utilized different-sized microparticles, where a smaller 0.5- to 2-μm microparticle could be phagocytosed with either vitamin D3 or insulin B peptide, and a larger 30 μm microparticle could be a depot of immune-regulatory signal such as TGF-β or GM-CSF (Lewis et al., 2015). The combination of all four particles enabled prevention of diabetes through controlling DC phenotype and antigen presentation. In addition to the many various strategies to deliver self-antigen with particles (Hunter et al., 2014), researchers are also conjugating self-antigen to the surface of RBCs to utilize a natural pathway for tolerance through the clearing of apoptotic cells (Pishesha et al., 2017). Using this approach, there were benefits in alleviating multiple sclerosis and diabetes in mice both as a preventative and therapeutic treatment.

Various other similar techniques have been utilized that take advantage of nanoparticle trafficking to secondary lymphoid organs by altering size or the biomaterial or incorporating specific receptors. Nanoparticle "backpacks" have been developed that deliver vaccines to lymph node DCs by binding DC receptors, such as CD40 or DEC-205, or by binding endogenous albumin (Lee et al., 2012; Raghuwanshi et al., 2012; Cruz et al., 2014; Liu et al., 2014; Li et al., 2014). In addition to modulating the biodistribution of the vaccine, approaches have been developed to alter how the delivered peptide is presented by the DC. During normal activation, intracellular antigens are presented on MHC-I and activate a cell-mediated $CD8^+$ T cell response, whereas endocytosed antigens are presented on MHC-II and activate a humoral $CD4^+$ T cell—mediated response. To induce presentation by DC on MHC-I, nanoparticle-based vaccines have been developed that are capable of escaping the endosome upon internalization (Scott et al., 2012).

3.2.4 Drug Carrier Particles

Synthetic particles that allow for the engineering of shape, modulation of biodegradation properties, and conjugation with or encapsulation of drugs that affect these various pathways provide a mechanism to direct drug release kinetics with defined distribution throughout the body. In terms of extracellular T cell modulation, conjugating aAPCs with the necessary signals for T cell activation is the main approach. However, biodegradable particles can also be used to selectively deliver soluble signals that affect T cell activation.

During endogenous T cell stimulation, T cells receive soluble "signal 3" molecules that drive proliferation and cytotoxicity (Curtsinger and Mescher, 2011). These soluble mediators are secreted in a localized paracrine or autocrine fashion that increases their local concentration and limits exposure to the desired cells. Because these signals are often broadly immune activating, local delivery ensures that only the target cells are affected to avoid undesirable autoimmune responses. This important aspect of local immune stimulation cannot generally be achieved when immune-activating drugs are administered systemically. To more closely mimic the endogenous process and enable maximum T cell expansion, nanoparticles have been used to selectively deliver stimulatory signals in a paracrine manner to T cells of interest. Biodegradable polymers are often the material of choice because their degradation and intracellular delivery properties can be manipulated.

Local delivery can be crucial especially in the context of autoimmune and allergic disease, where current therapies are systemic immunosuppression, though majority of the disease is not. One example of particle delivery of immunomodulatory agents was the delivery of IL-2, TGF-β1, and Rapamycin biodegradable particles to suppress allergic contact dermatitis (Balmert et al., 2017). This combination mimics and potentiates the signals given by DCs to expand regulatory T cells in vivo. Encapsulation of these agents into particle allows codelivery to eliminate potential immune activating effects of IL-2, local delivery to target inflamed areas, and prevent future allergic responses. Another example is biodegradable particle delivery of CCL22, a chemokine that recruits regulatory T cells, to treat autoimmune-mediated dry eye disease (Ratay et al., 2017), transplant rejection (Jhunjhunwala et al., 2012), and periodontal disease (Glowacki et al., 2013).

Various methods have been developed to target these drug-delivery particles to specific immune subsets. One approach is to combine drug delivery with an aAPC to take advantage of both aAPC-based extracellular stimulation as well as localized drug delivery, as is the case with IL-2–encapsulated aAPC (Steenblock and Fahmy, 2008; Steenblock et al., 2011). Biodegradable particles decorated with specific pMHC have also been used to selectively deliver cytotoxic drugs to cognate T cells for applications in autoimmunity (Fahmy et al., 2007).

Similarly, nanoparticles conjugated with T cell signaling molecules have been used to simply target the nanoparticles to T cells with the goal of intracellular drug delivery. Gelatin nanoparticles approximately 200 nm in diameter conjugated with anti-CD3 mAb have been shown to be selectively internalized by T cells (Dinauer et al., 2005). Over 80% of T cell leukemia cells were observed to take up the particles, demonstrating the potential for nanoparticles for the delivery of intracellular signals. More recently, this approach has demonstrated success in transfecting T cells with genetic material. Anti-CD3 mAb-coated biodegradable nanoparticles approximately 150 nm in diameter were engineered to encapsulate a DNA plasmid encoding for a specific chimeric antigen receptor (CAR; Smith et al., 2017). The particles showed specific transfection of T cells with approximately only 5% of non–T cells showing uptake following intravenous administration. Remarkably, systemic administration of these nanoparticles resulted in the same degree of antileukemia activity when compared with adoptive cell transfer of ex vivo transfected CAR T cells, which will be discussed in the following in greater detail. Thus the nanoparticle approach was able to significantly reduce the cost and time requirements of lymphodepletion and ex vivo T cell manipulation. More specific targeting of T cell subsets is also possible, as anti-CD4 mAb-conjugated lipid nanoparticles have been shown to selectively target $CD4^+$ T cells in vitro and in vivo (Ramishetti et al., 2015). These nanoparticles selectively delivered CD45 siRNA and induced gene silencing up to 20% in $CD4^+$ T cells 5 days post-intravenous administration.

Localized drug delivery to T cells can also be achieved through combination with ACT. For example, nanoparticles can be covalently linked to T cells expanded in vitro prior to reinfusion. In one study, drug-releasing nanoparticles were covalently linked to the surface of tumor-targeting T cells prior to adoptive cell transfer (Stephan et al., 2012). Using nanoparticles that released an inhibitor of immune-inhibitory signals, nanoparticle-functionalized T cells achieved greater expansion at the tumor site compared with systemic administration of the drug, achieving a 14-day survival advantage. Importantly, this study also demonstrated that the nanoparticles were translocated into the immune synapse during T cell activation. To further modulate not just the cell-proximity of drug release but also its temporal profile, T cell–bound nanoparticles have been developed that release drug only upon T cell stimulation (Jones et al., 2017). When $CD8^+$ T cells become activated, they release lytic granules, and this property has recently been used to trigger nanoparticle degradation. These nanoparticles, when bound to HIV-specific $CD8^+$ T cells, were shown to enable the delivery of an immunotherapeutic to a site of high viral replication.

Most T cell–targeting nanoparticles for local drug delivery to date have targeted CD4, CD3, or TCR expression. However, it is important to appropriately select the targeting molecule for the most effective response. Whether or not a targeting molecule is internalized can significantly affect the T cell response to nanoparticle-mediated drug delivery. A study that investigated targeting nanoparticles that released a TGF-β inhibitor to T cell internalizing versus noninternalizing receptors showed differing responses, and this effect changed if T cells were conjugated with nanoparticles prior to adoptive cell transfer or targeted directly by intravenously administered particles (Zheng et al., 2017).

When widespread circulation of a drug is important, smaller particles or molecules are often preferred due to their superior biodistribution. However, in some cases, limiting the circulation of a drug may be desired, such as with nonspecific immune activators meant to target T cells infiltrating a tumor. Systemic administration of stimulatory molecules such as IL-2 and anti-4-1BB mAb is associated with various off-target toxicities due to widespread immune activation (Panelli et al., 2004; Ascierto et al., 2010). Thus a different approach to drug delivery is to anchor drugs to particles and inject them locally at the site of interest for the purpose of prolonging drug retention. Studies have shown that anchoring IL-2 and anti-4-1BB mAb to the surface of liposomes approximately 160 nm in diameter significantly reduces off-target toxicities when injected intratumorally (Kwong et al., 2013; Zhang et al., 2018). Even when injected intratumorally, the soluble antibody and cytokine were detectable at high levels 18 h postinjection and resulted in significant weight loss of treated mice. In contrast, the nanoparticle-anchored drugs limited drug exposure to the tumor site and were shown to treat murine melanoma. Similar results have been observed with locally administered nanoparticles functionalized with anti-4-1BB and anti-PD-L1 mAbs (Kosmides et al., 2017).

3.3 Cellular and Genetic Engineering

The ability to directly engineer cellular function through genetic techniques has increased due to new molecular tools such as CRISPR system (Clustered Regularly Interspaced Short Palindromic Repeats). Additionally, the motivation to engineer cells has increased with the recent success and approvals of CAR T cell therapy. Again, this section will look at engineering principles that enable specificity, control, and enhancement of these cellular engineering approaches and therapies.

3.3.1 Cell-Based aAPCs

In addition to synthetic substrates, cell lines have been used to develop aAPC with the aim of eliminating the burden of using endogenous APCs for stimulating T cells for adoptive immunotherapy (Butler and Hirano, 2014). This cellular approach enables greater biomimicry than synthetic aAPC; however, standardization and stability of the construct can be difficult to control. Additionally, cells used for aAPC are often not APCs and do not express important costimulatory molecules or cytokines needed for effective T cell activation. Therefore cell-based aAPC will need to be efficiently transfected, stably express manipulated genes, and downregulate all other nonspecific human leukocyte antigen (HLA) molecules to avoid activation of allospecific T cells.

A popular choice of cell line to manipulate is the K562 erythromyeloid cell line by retroviral transfection. Researchers have transduced this cell line to stably express Fcγ receptors and 41BB ligand (Maus et al., 2002). Anti-CD3 and Anti-CD28 were added to form the complete aAPC to provide polyclonal T cell expansion. The addition of 41BB ligand induced effective T cell proliferation and limited $CD8^+$ T cell apoptosis (Fig. 8A). Another group also showed the incorporation of the 41BB ligand improved stimulation with other costimulatory ligands, including CD70 and CD80 (Zeng et al., 2014). Costimulation including CD83 was able to stimulate T cells without the addition of exogenous cytokines (Suhoski et al., 2007). Thus the choice of costimulatory molecule(s) for cell-based aAPC impacts the ability to provide effective T cell stimulation.

Cytokines act as a third signal in T cell stimulation and are expensive to add to cultures. Therefore work is being done to incorporate the expression of T cell—stimulating cytokines into cell-based aAPC. IL-21, a cytokine that signals through the IL-2Rγ receptor, has been included into the membrane of aAPC and increased the number of CAR T cells and improved tumor killing in vivo (Singh et al., 2011). Similarly, IL-15 has been stably expressed on the membrane of cell-based aAPC (Hasan et al., 2010). Although soluble IL-15 enhanced T cell activation and cytotoxicity, membrane-bound IL-15 established greater levels of activation, proliferation, and maintenance of the central memory phenotype, as well as nearly three times the cytotoxicity.

Researchers have also used cell-based aAPC to stimulate CMV-specific T cell responses (Papanicolaou et al., 2003). Here they transduced their aAPC to express HLA-A2, CMV peptide, costimulatory marker CD80, and adhesion ligands ICAM-1 and LFA-3. This aAPC produced similar levels of antigen-specific cells compared with endogenous APCs and transfected primary B cells used as APCs. Similarly, this antigen-specific activation has also been applied to $CD4^+$ T cells for immunotherapy (Garnier et al., 2016).

Beyond polyclonal and virus-specific antigens, these aAPC are primarily designed for adoptive immunotherapy in cancer therapy. To this end, cells were engineered to activate cancer-specific T cells recognizing the MART-1 antigen found within melanoma (Butler et al., 2007). Importantly, these cells were developed under Good Manufacturing Practices guidelines, enabling the potential translation to the clinic. Another example of generating tumor-specific T cells is the manipulation of another popular cell line, fibroblast NIH 3T3 cells. Telomerase-specific $CD8^+$ T cells were activated by engineered 3T3 fibroblasts in an effort to target tumors overexpressing the protein (Dupont et al., 2005). T cells stimulated by these aAPC were cytotoxic in a chromium release assay against human telomerase reverse transcriptase expressing tumor cell lines.

3.3.2 CAR T Cells

CAR T cells are T cells that have been genetically engineered to express a receptor with two major functions: (1) antigen recognition, typically through antibody binding, and (2) T cell activation through phosphorylation of canonical intracellular domains (Fig. 8B; Fesnak et al., 2016). These CAR T cells have had tremendous clinical success in treating B cell tumors (Kochenderfer et al., 2010; Porter et al., 2011; Maude et al., 2014). Recently, Tisagenlecleucel became the first FDA-approved CAR T cell therapy for pediatric patients with recurrent B cell acute lymphoblastic leukemia after failing two previous therapies (Prasad, 2018). This has spurred development of CAR T cell construct for applications other than cancer including regulatory CAR T cells for autoimmune disease and tissue transplantation (Jethwa et al., 2014). Despite incredible clinical progress, major hurdles face the translation of this technology, including safety, specificity, and cost (Srivastava and Riddell, 2015; Lim and June, 2017). Here, we will highlight some of the more recent engineering approaches used to address these issues.

One of the main toxicities associated with CAR T cell therapy is the severe cytokine release syndrome or cytokine storm, leading to complications and death (Morgan et al., 2010; Kochenderfer et al., 2012; Davila et al., 2014). Clinically, groups have managed the toxicities using biomarkers, such as cytokine levels, as an indication for early intervention (Davila et al., 2014). More recently, researchers have developed techniques to "turn off" CAR T cells to limit the toxicity. One example is the incorporation of the Caspase-9 gene into the CAR T cell construct (Gargett and Brown, 2014). When the small molecule AP1903 is administered, this dimerizes to cause apoptosis in CAR T cells.

Another method of decreasing the toxicity of CAR T cells is to increase the specificity of the construct. For current therapies, the construct identifies CD19, which is expressed on both healthy and malignant B cells. To target cancers specifically, researchers are targeting antigens like HER2, NY-ESO-1, ROR-1, and MUC16 that are overexpressed within tumor tissues

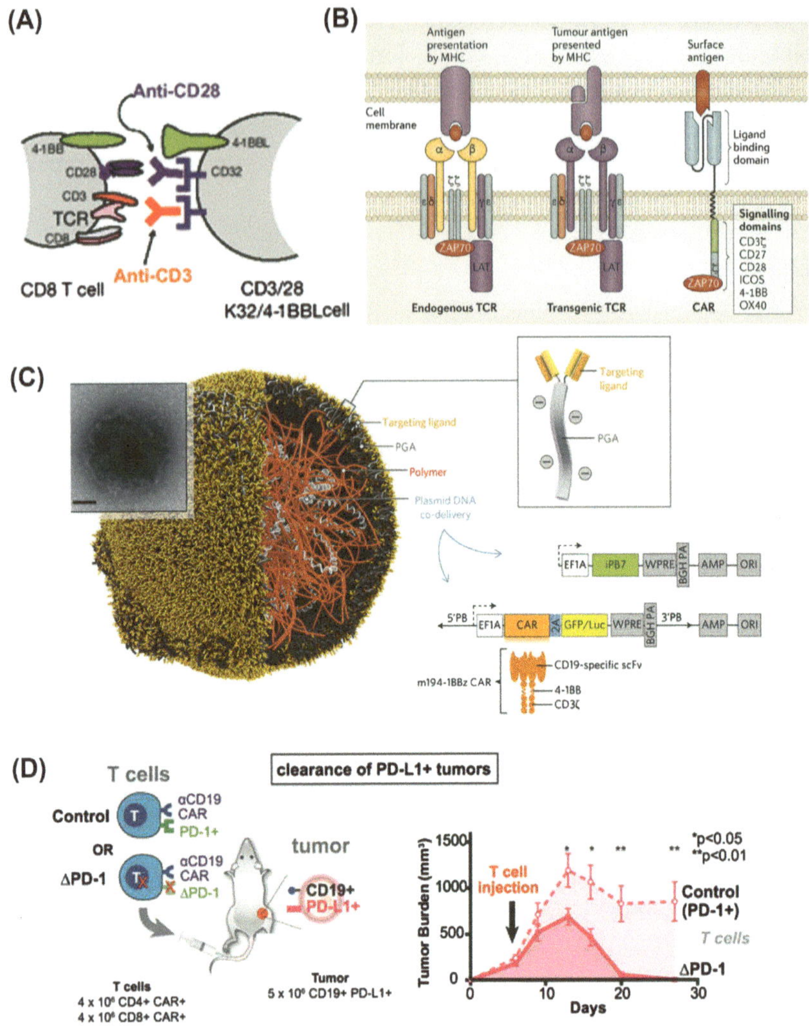

Figure 8 Genetic engineering approaches can be employed to directly or indirectly modulate T cell function. (A) Cell-based artificial antigen-presenting cells can be generated by transfecting a cell line such as K562 cells to express T cell–stimulating signals such 41BB and receptors for anti-CD3 and anti-CD28. (B) T cells can be manipulated to express chimeric antigen receptors which recognize surface antigens extracellularly and contain canonical TCR-activating domains intracellularly. (C) Genetic engineering of T cells in vivo has been accomplished through using DNA-containing, biodegradable nanoparticles that target T cells through an attached targeting ligand. (D) In addition to adding receptors, genetic engineering approaches have demonstrated knock down within T cells with this example of PD1 receptor knockdown in CAR T cells for more effective cancer killing *(A) Reprinted by permission from Macmillan Publishers Ltd.: Maus, M.V., et al., 2002. Ex vivo expansion of polyclonal and antigen-specific cytotoxic T*

(Chekmasova et al., 2010; Hudecek et al., 2010; Morgan et al., 2010; Rapoport et al., 2015).

An additional method to increase the specificity is to engineer the CAR construct to recognize tumor-specific antigens expressed in HLA molecules similar to how TCR molecules recognize antigens. The majority of CAR T cells are developed from scFvs. Early scFvs originated with tumor-specific monoclonal antibodies; however, utilizing scFvs have much higher affinity than TCRs which can cause issues in intracellular signaling and may not recognize intracellular antigens. Indeed scFvs have been made to recognize cancer antigens in the context of HLA, but the affinity of the construct limited efficacy compared with a low-affinity TCR construct recognizing the same HLA antigen construct (Oren et al., 2014). To accomplish scFv design, researchers employed antibody engineering techniques as mentioned in Section 3.1.2 such as phage display to fine tune antibody specificity and affinity toward the desired peptide HLA complex. This limits discovery to affinity and specificity but does not inform on cell activity upon ligation or even expression. Researchers have recently developed another screening technique which includes transduction of T cells with scFv libraries, such that structural design and specificity can be linked back to function and phenotype (Alonso-Camino et al., 2009, 2013).

Further specificity and control can be programmed by development of a "two-signal" approach, similar to canonical T cell stimulation where costimulation is needed. However, in this case, CAR T cell costimulatory constructs are designed to recognize a second antigen to enhance the specificity of tumor killing (Wilkie et al., 2012; Kloss et al., 2013). Similarly, instead of an AND gate, a NOT switch can be developed by adding a second signal that is present on normal cells to act as an inhibitory signal, similar to

lymphocytes by artificial APCs expressing ligands for the T-cell receptor, CD28 and 4-1BB. Nat. Biotechnol. 20 (2), 143–148. https://doi.org/10.1038/nbt0202-143, Copyright (2002). (B) Reprinted by permission from Macmillan Publishers Ltd.: Fesnak, A.D., June, C.H., Levine, B.L., 2016. Engineered T cells: the promise and challenges of cancer immunotherapy. Nat. Rev. Canc. 16 (9), 566–581. https://doi.org/10.1038/nrc.2016.97. Nature Publishing Group, Copyright (2016) (C) Reprinted by permission from Macmillan Publishers Ltd.: Smith, T.T., et al., 2017b. In situ programming of leukaemia-specific T cells using synthetic DNA nanocarriers. Nat. Nanotechnol. 12 (April), 813. https://doi.org/10.1038/nnano.2017.57. Nature Publishing Group, Copyright (2017). (D) Reprinted by permission from Macmillan Publishers Ltd.: Rupp, L.J., et al., 2017. CRISPR/Cas9-mediated PD-1 disruption enhances anti-tumor efficacy of human chimeric antigen receptor T cells. Sci. Rep. 7 (1), 737, Copyright (2017).

canonical PD-1 signaling (Fedorov et al., 2013). An ON switch has also been reported, where CAR T cells can only be activated in the presence of a small-molecule to cross-link signaling domains (Wu et al., 2015). Indeed, the repertoire engineering approaches to regulate CAR signaling through modifying the signal inputs or interpretation is a growing field (Fesnak et al., 2016).

Density can also be used to regulate specificity, with the design of lower affinity CAR constructs. One group decreased the affinity of Epidermal growth factor receptor (EGFR)-targeting CAR T cells to demonstrate lower toxicity to in normal cells that express a lower density of EGFR than cancer cells (Caruso et al., 2015). Additionally, the TCR—MHC interaction is lower affinity than antibody—antigen interactions and is proposed to confer increased sensitivity to lower density of antigens through serially signaling (Lever et al., 2014). This fact may also help in targeting lower density antigens where increased sensitivity is required and has been shown to result in increased efficacy of CAR T cells in comparison to higher affinity antibody-based CAR T cells (Oren et al., 2014).

Ultimate specificity could be achieved by targeting neoantigens. Neoantigens are novel antigens that are produced by mutations in the tumor (Klebanoff et al., 2016; Yarchoan et al., 2017). Adoptive transfer of endogenously activated neoantigen-specific T cells has also shown good clinical responses (Robbins et al., 2013; Tran et al., 2016). However, adopting this approach to CAR T cell therapy is difficult because each patient has unique mutations, identifying the reactivity to mutations is time-consuming, and development of a CAR-T cell with a TCR is less established.

While limiting toxicity is a primary aim of advancing CAR T cell therapy, there are developments being investigated in tandem to increase its efficacy. One main challenge facing all immunotherapies is the tumor immunosuppressive environment. To combat this, CAR T cells can be used in conjunction with checkpoint blockade molecules (Chong et al., 2017) or be programmed to decrease inhibitory markers like PD-1 (Rupp et al., 2017), secrete proinflammatory cytokines (Roybal et al., 2016), and create OR switches to target multiple antigens independently (Grada et al., 2013).

Exhaustion is a challenge that faces CAR T cells in addition to adoptively transferred cells. To overcome exhaustion and increase the functionality of their CAR T cells, researchers have adopted the more recent CRISPR/Cas9 genome editing technology (Eyquem et al., 2017). CRISPR/Cas9 allowed the insertion of the CAR TCR α constant region of the genome to more accurately mimic TCR expression. Indeed, CAR constructs were downregulated following stimulation and then reexpressed similar to TCRs upon T cell activation, which led to a decrease in exhaustion markers. Doing so generated CAR T cells that were much more potent than conventionally retrovirally transduced CAR T cells in treating a mouse model of acute lymphoblastic leukemia.

Additionally, standardizing cell source and maintenance of cell phenotype ex vivo have been challenges for efficacy, particularly for regulatory CAR T cell development. In a recent study, researchers isolated regulatory T cells that were $CD25^+$ $CD45RA^+$, which show more stable maintenance of regulatory phenotype and confirmed the phenotype stability by examining epigenetic markers close to FOXP3 (MacDonald et al., 2016). These cells were effective at preventing graft-versus-host disease in a mouse model; however, development and expansion of these cells still required substantial time, multiple processes, and feeder cells.

Finally, cost and technical challenges are major hurdles to translating this therapy for widespread adoption. One reason is all the ex vivo manipulation of patient cells, limiting this therapy to specific certified locations. To eliminate the need for ex vivo handling of cells, Smith et al. developed nanoparticles that transfected immune cells in vivo by targeting the CD3 receptor on T cells (Fig. 8C; Smith et al., 2017). They demonstrated effective in vivo transfection and tumor treatment in model murine tumors.

3.3.3 Other Genetic Engineering Applications

Genetic engineering of T cells goes beyond CAR T cells in the fields of infectious disease and autoimmunity. The HIV virus targets $CD4^+$ T cells through human chemokine receptor 5 gene (CCR5). To minimize HIV disease spreading and latency in patients, researchers knocked out the CCR5 gene in ex vivo manipulated $CD4^+$ T cells and reintroduced into patients (Tebas et al., 2014). $CD4^+$ T cells without CCR5 persisted longer than untreated cells, and patients who received the therapy had increased levels of $CD4^+$ T cells and lower levels of viremia.

Another example of knocking down a gene takes inspiration from the success that checkpoint blockade therapy has experienced. Researchers used a CRISPR Cas9 approach to eliminate expression of PD-1 by CAR T cells (Fig. 8D; Rupp et al., 2017). Gene editing produced much more functional CAR T cells in completely eliminating established tumors that expressed PD-L1. Finally, engineering of cytokine and chemokine receptors has been used to study cytokine pathways and also to control activation of T cells in vivo (Conklin et al., 2008; Kawahara et al., 2010).

Genetic engineering has been used to convert and maintain T cell phenotype. $CD4^+$ regulatory T cells are an example of this genetic manipulation because the stable expression of regulatory constructs plays a crucial role at establishing and maintaining tolerance; however, when isolated and cultured in vitro often may lose their phenotype or effector T cells will dominate. To solve this, lentiviral transduction of $CD4^+$ T cells with FOXP3 enabled sustained suppressor activity by naïve and memory T cells (Allan et al., 2008). Transduced $CD4^+$ T cells with FOXP3 or IL-10 have demonstrated tolerance induction in models of tissue transplantation (Chai et al., 2005; Andolfi et al., 2012). In addition to FOXP3, $CD4^+$ T cells have been transduced to express class II MHC loaded with autoantigen and FOXP3 and enhanced suppression of autoimmune arthritis in mice (Qian et al., 2013).

3.4 Scaffolds and Surfaces

Similar to particle-based technologies, scaffolds and surfaces are materials that can be engineered to directly modulate T cell responses. What makes scaffolds and surface technologies unique are the (1) ability to directly pattern and coat with proteins in distinct geometric locations, (2) ability to tune unique biophysical properties such as stiffness, (3) size of the material enabling development of multicellular interactions, (4) capacity to move to three-dimensional interactions and potential to create artificial organs. Here, we will focus on both two-dimensional and three-dimensional technologies that can be utilized both in vitro and in vivo to modulate T cell function.

3.4.1 Two-Dimensional Scaffolds and Surfaces

3.4.1.1 Patterning and Coatings

Modern lithographic techniques have enabled precise geometric patterning on the micron scale. This has been utilized by researchers in developing biocoatings and micron-sized structures to further probe T cell biology. Additionally, with the rise of tissue and stem cell engineering, biomaterial engineers have developed materials that can vary in stiffness to probe mechanotransduction pathways. Researchers have also employed these materials to study TCR signaling.

The geometric organization of stimulatory and nonstimulatory molecules during T cell activation is highly regulated as seen in the immune synapse needed for T cell activation (Varma et al., 2006; Fooksman et al., 2009).

Using multiple rounds of microcontact printing, researchers were able to control locations of signal 1 (anti-CD3) and signal 2 (anti-CD28) on planar surfaces (Fig. 9A; Shen et al., 2008). T cells stimulated with the two signals segregated demonstrated enhanced signaling rather than when they are coclustered. The geometry was distinguished by signal 2 activation pathways such as PKC-θ and LcK (Shen et al., 2008; Bashour et al., 2014b). Similar results were demonstrated with signal 1 and adhesion molecule ICAM-1 (Doh and Irvine, 2006). If signal 1 (anti-CD3) was not clustered in a

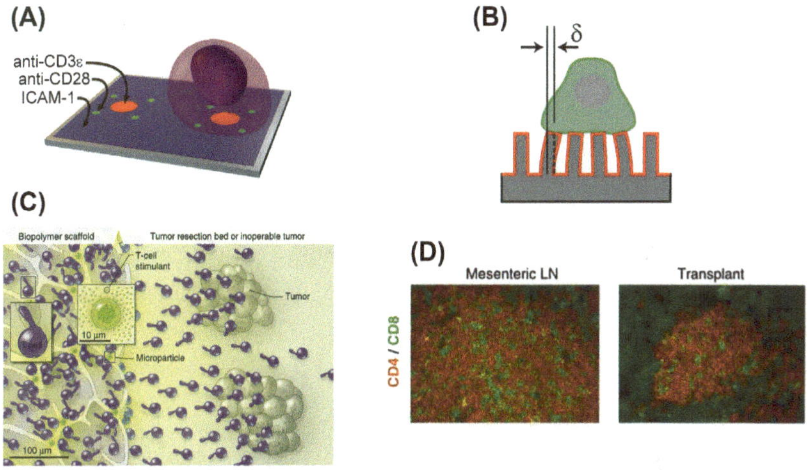

Figure 9 Bulk material properties can be engineered both to study T cell responses and to control T cell response. (A) Lithographic printing technology allows for precise control over T cell—stimulating ligands to interrogate the importance of spacing and location in understanding the immune synapse. (B) To probe forces involved in T cell signaling, researchers have coated micropillar polydimethylsiloxane micropillars with T cell signaling molecules and observe deflection. (C) Hydrogel scaffolds can act as effective delivery vehicles for both T cell—stimulating particles and CAR T cells in cancer immunotherapy. (D) Scaffolds can also be used to generate lymphoid-like organs containing correct conformations and populations of T cells when transferred into mice *(A) Reprinted by permission from National Academy of Sciences: Shen, K., et al., 2008. Micropatterning of costimulatory ligands enhances CD4^{+} T cell function. Proc. Natl. Acad. Sci. USA 105 (22), 7791—7796, Copyright (2008). (B) Reprinted by permission from Bashour, K.T., Gondarenko, A., et al., 2014. CD28 and CD3 have complementary roles in T-cell traction forces. Proc. Natl. Acad. Sci. USA, 111 (6), 2241—2246, Copyright (2014). (C) Reprinted by permission from Macmillan Publishers Ltd.: Stephan, S.B., et al., 2015. Biopolymer implants enhance the efficacy of adoptive T-cell therapy. Nat. Biotechnol. 33 (1), 97, Copyright (2014). (D) Reprinted by permission from Macmillan Publishers Ltd.: Suematsu, S., Watanabe, T., 2004. Generation of a synthetic lymphoid tissue—like organoid in mice. Nat. Biotechnol. 22 (12), 1539bib_Suematsu_and_Watanabe_2004, Copyright (2004).*

centralized format, T cells did not form efficient contacts or fully activate. Microcontact printing is a technology that has elucidated important micro-scale geometric arrangements of receptors.

Another technique is block copolymer micellar nanolithography. This technique allows nanometer resolution of biomolecule printing, where gold nanoparticles are hexagonally packed and coated with a distinct layer of polymer that can be removed, to which biological compounds can be attached (Spatz et al., 2000). A number of researchers have used this technique to probe the question of the influence of nanoscale signal 1 density on T cell activation. Interestingly, the reports found commonly that when using antigen-independent signal 1 with anti-CD3, efficient T cell activation required ligands to be spaced below 60–70 nm apart (Delcassian et al., 2013; Matic et al., 2013). When using antigen-specific signal 1 (peptide-loaded MHC class II), the threshold was even higher at around 100 nm spacing (Deeg et al., 2013). Although these articles did not include costimulatory or adhesion molecules, they demonstrate the importance of ligand density in providing effective T cell stimulation.

In addition to confining biomolecules my micro- or nano-printing, biomolecules can be coated with lipid bilayers onto surfaces to mimic the fluidity of the cell membrane (Groves and Dustin, 2003). Researchers have used this technique to coat silica surfaces with a lipid bilayer with signal 1 (peptide-loaded MHC) and adhesion molecule ICAM-1 (Mossman et al., 2005). However, these molecules were constrained by microfabricating barriers of differing sizes and shapes to observe the effects of restricting ligand mobility. By observing how signaling molecules formed, moved, and clustered helped elucidate processes of the immune synapse formation such as the process of TCR microclusters developing before transfer to the cSMAC.

3.4.1.2 Mechanical Manipulation of Materials

Beyond microfabrication to enable ligand location, this technique can form structures to help probe additional biophysical properties such as creating elastomer micropillars for mechanical signaling processes. TCR signaling has been implicated to be a mechanosensitive process (Huse, 2017); so researchers have attached signal 1 and signal 2 to micropillar arrays to observe the force generation of T cells in engagement with activation signals (Fig. 9B; Bashour et al., 2014a). They found that nearly a 100 pN force was generated through the TCR, though CD28 signaling assisted to increase the forces generated. The localization of intracellular proteins regulating contractility suggested interaction of the TCR with intracellular cytoskeleton similar to focal adhesions.

Additionally, bulk material stiffness has been modified to study the mechanosensitive TCR signaling. T cell–stimulating ligands have been attached to two different surfaces—polyacrylamide and poly(dimethylsiloxane)—were manufactured to have stiffness vary from 10 kPa to 2 mPa (Judokusumo et al., 2012; O'Connor et al., 2012). In both studies, optimal stiffness was found to be around 100 kPa. The stiffness additionally was found to impact T cell differentiation and functionality of the T cells, highlighting this important consideration for T cell–stimulating surfaces.

3.4.2 *Three-Dimensional Scaffolds and Surfaces*

3.4.2.1 Three-Dimensional Scaffolds for T Cell Immunotherapy

Increasing the porosity of materials can lead to cells sensing three-dimensional environment. Dimension can be utilized to create environments which deliver protein or cell products in a more controlled fashion or promote additional cell–cell interactions needed for functional T cell development. Three dimensional in vitro environments can also be used to more accurately mimic interactions that occur within the body. These organ mimics could be further developed to replace resected or dysfunctional immune organs.

Like their particle counterparts, scaffolds can be developed from biocompatible and biodegradable materials that have the potential to release cargo in a controlled fashion. However, when injected, scaffolds are not motile and rather act as a depot delivery system. This is effective in cases where localized delivery is needed.

A specific example is cytokine delivery, as it can be delivered locally to avoid the many off-target effects of systemic administration. IL-2, a cytokine that influences T cell responses and is an FDA-approved therapy but has significant toxicities and side effects (Li et al., 2017). To minimize off-target effects, researchers have encapsulated IL-2 into polymeric biodegradable scaffolds (Hori et al., 2009). This product could be injected, sustained release of IL-2 over days, and enabled cell infiltration to the scaffold.

Besides loading scaffolds with cytokines, cells can be transferred as well. Early on, APCs were the focus of adoptive transfers. DCs have been loaded or recruited to a variety of scaffolds loaded with chemokines, adjuvants, and antigen with the purpose of stimulating and recruiting antigen-specific T cells (Hori et al., 2008; Ali et al., 2009). Beyond DCs, T cells have recently been transferred in vivo in the form of scaffolds (Graham et al., 2013;

Stephan et al., 2015; Smith et al., 2017). The first example demonstrated that loading regulatory T cells (Tregs) onto a PLGA scaffold with islets was effective at protecting grafts from autoimmune rejection in a type 1 diabetes model (Graham et al., 2013). Colocalization and retention of the Tregs and islets was necessary to skew the local microenvironment to immunoinhibitory, thus enabled by the scaffold technology. This further eliminated nonspecific systemic immunosuppressive effects of Treg adoptive transfer.

Another example is the delivery effector T cells to destroy cancer cells (Stephan et al., 2015; Smith et al., 2017). T cells were loaded on an alginate hydrogel along with costimulatory aAPC loaded with IL-15 superagonist (Fig. 9C; Stephan et al., 2015). Here scaffolds were utilized to deliver and further activate T cells in the resection bed or in the tumor local microenvironment. Drastic differences were observed. T cells delivered locally and with additional stimulation were able to cure mice of established tumors, while adoptively transferred T cells did not eliminate tumors. In another study, CAR T cells were loaded into the scaffold and demonstrated similar efficacy in combination with a stimulator of IFN genes (STING agonist; Smith et al., 2017). The controlled release of STING agonists created a vaccine-like environment, which prevented the escape of heterogenous tumor cells not expressing antigens recognized by CAR T cells.

In addition to in vivo delivery and stimulation of T cells for adoptive therapy, three-dimensional environments have been created ex vivo to enhance T cell stimulation (Cheung et al., 2018). High-aspect-ratio mesoporous silica microrods were coated with lipid bilayers, signal 1 and signal 2, and IL-2 to induce T cell stimulation. T cells migrated into the scaffolds and formed more dense clusters of cells than traditional microparticle aAPC and produced higher levels of proliferation and produce antitumor treatment.

3.4.2.2 Tissue Engineering Lymphoid Organs

Three-dimensional scaffolds are utilized frequently to create ex vivo organoids which enable more complex analysis of multiple cell types in a more physiological manner (Fatehullah et al., 2016). However, three-dimensional ex vivo culture is not without challenges including difficulty of working with multiple cell types, large requirements for nutrients, extended culture times, and ability to analyze with current techniques. To tackle some of these issues in developing a human artificial lymph node, Giese et al. compared different bioreactors, source of immune cells, and material for scaffolds (Giese et al., 2006). Based on T cell activation and lymphocyte swarming, they determined nonwoven polyamide fibers and collagen performed best as scaffolds inside disposable bioreactors. Similarly, researchers found that collagen-infused scaffolds provided T cells with the best environment to migrate in three dimension (Stachowiak and Irvine, 2008). These scaffolds were used to study how T cell and DC migration differed in response to the chemokine CCL21.

Collagen-based scaffolds have also been used to develop artificial lymph nodes in vivo (Suematsu and Watanabe, 2004; Okamoto et al., 2007; Kobayashi and Watanabe, 2016). Collagen sponge matrices were loaded with transfected thymus-derived stromal cells to express lymphotoxin α and activated DCs (Suematsu and Watanabe, 2004). Implanted into the renal subcapsular space, these matrices recruited B cells, DCs, T cells, and endothelial cells to form tertiary lymphoid structures in vivo (Fig. 9D). Taken from these mice, these artificial lymph nodes could generate an immune response to a model ovalbumin (OVA) antigen when transferred into severe combined immunodeficient (SCID) mice. The responses were also shown to generate memory T and B cells that migrated to other secondary lymphoid organs such as the spleen (Okamoto et al., 2007). Similar results were recently generated by implanting the scaffold with incorporating lymphotoxin-$\alpha 1\beta 2$, CCL19, CCL21, CXCL12, CXCL13, and soluble RANK ligand in place of the transfected cells (Kobayashi and Watanabe, 2016).

Collagen has also been employed in three-dimensional scaffolds to generate human artificial lymphoid organs such as an artificial thymus (Poznansky et al., 2000; Clark et al., 2005). The collagen coated a tantalum carbon-coated matrix, to which a mixture of keratinocytes and fibroblasts were added ex vivo. Hematopoietic stem cells were added to the matrix, which initiated the maturation of both $CD4^+$ and $CD8^+$ T cells with a diverse repertoire of TCRs similar to a natural thymus. The resultant T cells generated effector-based responses when stimulated yet were tolerant to self-MHC. Development of artificial lymphoid organs can thus elucidate our understanding of how endogenous organs are formed, their function, provide models for testing immunotherapies, and potentially replace dysfunctional immune organs altogether.

3.5 Combination Immunotherapies

Methods of both direct T cell activation and blockade of inhibitory checkpoint pathways have resulted in exciting results in preclinical studies and clinical trials. However, there remains a large population of patients who do not respond to monotherapies. Thus it has become evident that targeting a single stage of the Cancer-Immunity Cycle is not sufficient and that combination therapies that target multiple complementary pathways will have a greater chance of capturing a large response pool. Because there is a diverse array of inhibitory and stimulatory pathways that affect T cell activation, monotherapies often allow for tumor escape through the upregulation of alternative pathways.

Combination therapies against multiple checkpoint molecules can deactivate multiple different T cell inhibitory pathways. Notably, anti-PD-1 mAbs have been coadministered with anti-CTLA-4 mAb in several clinical trials. Because these two pathways are nonredundant, blocking both can synergize T cell activation. The combination therapy significantly enhances the objective response rate to as high as 60% but also increases the amount of adverse toxicities associated with widespread immune activation (Postow et al., 2015). Importantly, the efficacy of combination checkpoint blockade in comparison to monotherapy is stratified by biomarker expression, such as PD-L1, in a patient's tumor and indicates the importance of using biomarkers to guide therapeutic interventions to reduce costs due to unnecessary combinations (Larkin et al., 2015). Additional combinations of inhibitory molecule blockades have also demonstrated improved efficacy. Doublets of anti-CTLA-4, anti-PD-L1, and an IDO inhibitor all enhanced the reactivation of tumor infiltrating lymphocytes in a murine melanoma model (Fig. 10A). These studies demonstrated that the combination therapies were effective by reactivating tumor infiltrating lymphocytes rather than inducing new T cell migration (Spranger et al., 2014). Thus, in tumors lacking T cell infiltration, combination therapies that also induce migration into the tumor may be more effective.

Combination therapies that link T cell stimulation with checkpoint blockade aim to intervene at the two major ways by which T cell activation is modulated. T cell stimulation can be mediated directly through the administration of costimulatory antibodies or indirectly through a DC-based vaccine. Even when an antitumor or antiviral T cell population is present, however, activated T cells upregulate inhibitory molecules such as PD-1 and CTLA-4 in a homeostatic fashion. Tumor cells can also upregulate inhibitory ligands such as PD-L1, especially in response to T cell secretion of IFN-γ, which diminish the activity of these cells.

Checkpoint blockade has now been combined with almost every approach to T cell stimulation to maintain activation. In chronic viral

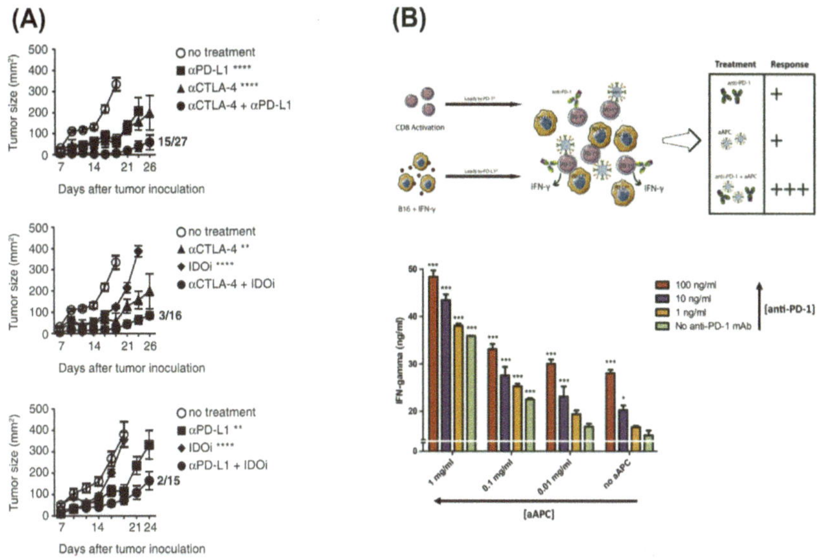

Figure 10 Combination therapies that link multiple immunoengineering approaches can further increase T cell activation for cancer immunotherapy. (A) Doublets of anti-CTLA-4 mAb, anti-PD-L1 mAb, and an IDO inhibitor each enhance T cell activation and delay tumor growth in a B16 murine melanoma model. (B) Antigen-specific stimulation via aAPC synergizes with anti-PD-1 checkpoint blockade to increase T cell activation, measured by IFN-γ secretion. *(A) Reprinted by permission from BioMed Central Ltd.: Spranger, S., et al., 2014. Mechanism of tumor rejection with doublets of CTLA-4, PD-1/PD-L1, or Ido blockade involves restored IL-2 production and proliferation of CD8(+) T cells directly within the tumor microenvironment. J. Immunother. Cancer 2 (1), 3. https://doi.org/10.1186/2051-1426-2-3, Copyright (2014). (B) Reprinted from Kosmides, A.K., et al., 2017. Biomimetic biodegradable artificial antigen presenting cells synergize with PD-1 blockade to treat melanoma. Biomaterials 118, 16–26. https://doi.org/10.1016/j.biomaterials.2016.11.038. Elsevier Ltd., Copyright (2017), with permission from Elsevier.*

infection, coadministration of antagonistic antibodies against PD-L1 and agonistic antibodies against 4 1BB were shown to enhance the expansion of viral-specific CD8$^+$ T cells compared with monotherapy (Vezys et al., 2011). The combination therapy also impacted the kinetics of the T cell response, indicating that combination therapies may be used to modulate how quickly T cells are expanded against a certain antigen. In cancer immunotherapy, anti-PD-1 mAb checkpoint blockade synergizes with in vivo aAPC administration to activate CD8$^+$ T cells against a tumor-expressed antigen (Fig. 10B; Kosmides et al., 2017). The combination therapy, but neither monotherapy, was able to inhibit the growth of palpable murine melanoma and was shown to be mediated by a decrease in PD-1 expression and increase in tumor-specific T cell expansion. Similarly, coadministration of anti-PD-1 and anti-4-1BB can enhance an antitumor response by modulating T cell phenotype and density of tumor infiltrating lymphocytes (Chen et al., 2014; Shindo et al., 2015). Checkpoint blockade has also been shown to increase the efficacy of peptide-pulsed DC vaccination, in this case blocking inhibitory molecules on DCs (Pilon-Thomas et al., 2010).Various other approaches have shown synergy between the combination of anti-PD-1 checkpoint blockade and granulocyte macrophage colony-stimulating factor (Li et al., 2009), OX40 and CD27 costimulation with anti-PD-L1 checkpoint blockade (Buchan et al., 2015), and anti-OX40 costimulation and vaccination (Jahan et al., 2017).

Checkpoint blockade has also been combined with therapeutics that enhance tumor immunogenicity. Treatments that induce tumor cell death or increase mutational load, including oncolytic viruses, radiation, or chemotherapy, can enhance antigen presentation and activation to tumor-specific T cells (Vilgelm et al., 2016). However, the antitumor T cell response is often still limited by the expression of inhibitory molecules by both tumor and immune cells. Radiation has been shown to synergize with anti-CTLA-4 checkpoint blockade, with local radiation leading to the regression of even distant metastases (Chandra et al., 2015; Rudqvist et al., 2017). However, resistance to this dual treatment was shown to be mediated at least partially by tumor upregulation of PD-L1. In response to this observation, combination therapy comprised of radiation, anti-CTLA-4, and anti-PD-L1 was shown to increase the response rate and minimize tumor immune escape (Victor et al., 2015). Checkpoint blockade also can increase the efficacy of oncolytic viruses that preferentially infect and lyses cancerous cells. A study with the oncolytic Newcastle Disease Virus demonstrated that its combination with anti-CTLA-4 could completely eliminate established murine melanoma and even lead to an improved abscopal effect (Zamarin et al., 2014).

While combination therapies often involve the coadministration of multiple therapeutics, another approach to combination therapy is to develop a single therapeutic with multiple mechanisms of action. CD80, the ligand expressed by APCs that induces T cell proliferation upon ligation with T cell expressed CD28, has additional binding partners. CD80 has also been shown to bind PD-L1 and plays a role in inducing apoptosis of activated $CD8^+$ T cells (Butte et al., 2009; Rollins and Gibbons Johnson, 2017). This dual interaction has been taken advantage of in a soluble form of CD80 as a therapeutic (Ostrand-Rosenberg et al., 2015). This

CD80-Fc fusion protein was shown to both neutralize the inhibitory PD-1/PD-L1 interaction while simultaneously stimulating T cells through CD28.

While combining stimulatory and inhibitory molecules in a single all-in-one therapeutic may be desirable from an engineering perspective, the immunological mechanisms behind their success may indicate that sequential delivery may be more effective. In one study, anti-PD-1 administration following anti-OX40 costimulation was effective at delaying tumor growth, but not vise-versa or coinjection—it was necessary for the costimulation to first boost the T cell response to a state where checkpoint molecules played a role in inhibiting the response (Messenheimer et al., 2017). Thus sequential delivery or biodegradable particles that can mediate this type of release may be beneficial.

Most recently, a combination therapy clinical trial including the IDO inhibitor, epacadostat, with an anti-PD1 antibody, pembrolizumab, failed a phase III clinical trial (Mullard, 2018). This combination blocks both immunosuppressive metabolic signals and surface ligands from the tumor microenvironment that diminish T cell activity. However, there was no improvement in progression-free survival when compared with pembrolizumab alone. Consequently, while many combinations do have rational pairings, much is to be learned about how combinatorial pairings should be prioritized and tested as therapies to increase the success rate of costly clinical trials.

4. SUMMARY AND NEXT-GENERATION THERAPIES

Over the past two decades, we have seen a dynamic interaction between our interest in developing new immune-based therapeutics and our understanding of pathogenesis of disease such as cancer and chronic infection. As our understanding grows, so does our capacity to engineer desired T cell responses. This has also provided enthusiasm for developing novel therapies with T cells. Here, we have reviewed general salient engineering principles to control T cells based on basic concepts of T cell biology. We have provided analysis and examples of researchers using these basic principles to control T cells with protein engineering, synthetic particles, cellular and genetic engineering, and scaffolds and surfaces.

Going forward, several areas will continue to be improved using engineering principles. Frequently unintended consequences like autoimmunity, cytotoxicity, or premature cell death result due to a number of unidentified variables activated by engineered inputs. Control over T cell outputs will come from greater understanding of T cell signaling and increased precision in the therapeutic. For example, CAR T cell therapy will benefit from greater systems biology approaches to study complex T cell activation networks and also from development of more sophisticated CAR constructs such as AND, OR, and NOT switches or kill switches.

On the other hand, many T cell—based therapies and modulators have extremely high price tags, and as we increase the precision or functionality, this caveat should be kept in mind. Engineered solutions which decrease cost of therapy will enable widespread translation of therapy that is still personalized. An example may be to engineer methods that successfully manipulate T cells in vivo instead of requiring prohibitively expensive procedures, equipment, personnel, and centers to handle cells ex vivo.

While we have spent time focusing on engineering T cells, progress in T cell therapy will come as other cell types are considered or even simultaneously targeted. T cells are a part of both a broader immune system, spend time traveling through several other organ systems, and require communication and activation from other cells. Additionally, other cell types contribute to the pathogenesis and cure of a disease. For example, the requirement necessary to eliminate large, aggressive, immunoinhibiting tumors required checkpoint blockade, recombinant IL-2, tumor antigen antibody, and a T cell vaccine enabling NK, DC, $CD4^+$, and $CD8^+$ T cell interactions and killing of tumor cells (Moynihan et al., 2016).

Many current advancements have come through cross-disciplinary efforts to tackle important clinical problems. Success has generated clinical, academic, and industrial interest with contributors from immunology and engineering fields. Continued improvements will come with similar engineering approaches as described here but hinge on interdisciplinary collaborations and dissemination of progress. In conclusion, control over T cells will enable further control over disease and enhanced understanding of the immune response.

REFERENCES

Aguado, B.A., et al., 2018. Engineering precision biomaterials for personalized medicine. Sci. Transl. Med. 10 (424), eaam8645.

Albanese, A., Tang, P.S., Chan, W.C.W., 2012. The effect of nanoparticle size, shape, and surface chemistry on biological systems. Annu. Rev. Biomed. Eng. 14, 1—16.

Ali, O.A., et al., 2009. Infection-mimicking materials to program dendritic cells in situ. Nat. Mater. 8 (2), 151—158. https://doi.org/10.1038/nmat2357. Nature Publishing Group.

Allan, S.E., et al., 2008. Generation of potent and stable human CD4$^+$ T regulatory cells by activation-independent expression of FOXP3. Mol. Ther. 16 (1), 194–202. https://doi.org/10.1038/sj.mt.6300341.
Almagro, J.C., et al., 2018. Progress and challenges in the design and clinical development of antibodies for cancer therapy. Front. Immunol. 8 https://doi.org/10.3389/fimmu.2017.01751.
Almeida, J.R., et al., 2007. Superior control of HIV-1 replication by CD8$^+$ T cells is reflected by their avidity, polyfunctionality, and clonal turnover. J. Exp. Med. 204 (10), 2473–2485. https://doi.org/10.1084/jem.20070784.
Alonso-Camino, V., et al., 2009. Lymphocyte display: a novel antibody selection platform based on T cell activation. PLoS One 4 (9), e7174. Public Library of Science.
Alonso-Camino, V., et al., 2013. CARbodies: human antibodies against cell surface tumor antigens selected from repertoires displayed on T cell chimeric antigen receptors. Mol. Ther. Nucleic Acids 2. https://doi.org/10.1038/mtna.2013.19.
Andolfi, G., et al., 2012. Enforced IL-10 expression confers type 1 regulatory T cell (Tr1) phenotype and function to human CD4(+) T cells. Mol. Ther. 20 (9), 1778–1790. https://doi.org/10.1038/mt.2012.71.
Andorko, J.I., et al., 2016. Intrinsic immunogenicity of rapidly-degradable polymers evolves during degradation. Acta Biomater. 32, 24–34. https://doi.org/10.1016/j.actbio.2015.12.026.
Andorko, J.I., Jewell, C.M., 2017. Designing biomaterials with immunomodulatory properties for tissue engineering and regenerative medicine. Bioeng. Transl. Med. 2 (2), 139–155. https://doi.org/10.1002/btm2.10063.
Ascierto, P.A., et al., 2010. Clinical experiences with anti-CD137 and anti-PD1 therapeutic antibodies. Semin. Oncol. 37 (5), 508–516. https://doi.org/10.1053/j.seminoncol.2010.09.008. Elsevier Inc.
Astete, C.E., Sabliov, C.M., 2006. Synthesis and characterization of PLGA nanoparticles. J. Biomater. Sci. Polym. Ed. 17 (3), 247–289. https://doi.org/10.1163/156856206775997322.
Baeza, A., Colilla, M., Vallet-Regí, M., 2015. Advances in mesoporous silica nanoparticles for targeted stimuli-responsive drug delivery. Expet Opin. Drug Deliv. 12 (2), 319–337.
Baghdadi, M., Jinushi, M., 2014. The impact of the TIM gene family on tumor immunity and immunosuppression. Cell. Mol. Immunol. 11 (1), 41–48. https://doi.org/10.1038/cmi.2013.57. Nature Publishing Group.
Bailey, C.P., et al., 2017. New interleukin-15 superagonist (IL-15SA) significantly enhances graft-versus-tumor activity. Oncotarget 8 (27), 44366–44378. https://doi.org/10.18632/oncotarget.17875.
Balmert, S.C., et al., 2017. In vivo induction of regulatory T cells promotes allergen tolerance and suppresses allergic contact dermatitis. J. Contr. Release 261, 223–233. https://doi.org/10.1016/j.jconrel.2017.07.006.
Barber, D.L., et al., 2006. Restoring function in exhausted CD8 T cells during chronic viral infection. Nature 439 (7077), 682–687. https://doi.org/10.1038/nature04444.
Barua, S., et al., 2013. Particle shape enhances specificity of antibody-displaying nanoparticles. Proc. Natl. Acad. Sci. USA 110 (9), 3270–3275. https://doi.org/10.1073/pnas.1216893110.
Bashour, K.T., Gondarenko, A., et al., 2014. CD28 and CD3 have complementary roles in T-cell traction forces. Proc. Natl. Acad. Sci. USA 111 (6), 2241–2246.
Bashour, K.T., Tsai, J., et al., 2014b. Cross talk between CD3 and CD28 is spatially modulated by protein lateral mobility. Mol. Cell. Biol. 34 (6), 955–964. https://doi.org/10.1128/MCB.00842-13.
Basu, R., Huse, M., 2017. Mechanical communication at the immunological synapse. Trends Cell Biol. 27 (4), 241–254.

Beers, S.A., Glennie, M.J., White, A.L., 2016. Influence of immunoglobulin isotype on therapeutic antibody function. Blood 127 (9), 1097–1101. https://doi.org/10.1182/blood-2015-09-625343.

Della Bella, S., et al., 2003. Altered maturation of peripheral blood dendritic cells in patients with breast cancer. Br. J. Cancer 89 (8), 1463.

Birnbaum, M.E., et al., 2014. Molecular architecture of the αβ T cell receptor–CD3 complex. Proc. Natl. Acad. Sci. USA 111 (49), 17576–17581. https://doi.org/10.1073/pnas.1420936111.

Bobisse, S., et al., 2016. Neoantigen-based cancer immunotherapy. Ann. Transl. Med. 4 (14), 262. https://doi.org/10.21037/atm.2016.06.17.

Bobo, D., et al., 2016. Nanoparticle-based medicines: a review of FDA-approved materials and clinical trials to date. Pharm. Res. 33 (10), 2373–2387.

Bookstaver, M.L., et al., 2017. Improving vaccine and immunotherapy design using biomaterials. Trends Immunol. https://doi.org/10.1016/j.it.2017.10.002.

Bosch, B., et al., 2013. Major histocompatibility complex (MHC) class II-peptide complexes arrive at the plasma membrane in cholesterol-rich microclusters. J. Biol. Chem. 288 (19), 13236–13242.

Bousso, P., 2008. T-cell activation by dendritic cells in the lymph node: lessons from the movies. Nat. Rev. Immunol. 8 (9), 675–684.

Bromley, S.K., et al., 2001. The immunological synapse and CD28-CD80 interactions. Nat. Immunol. 2 (12), 1159–1166.

Bruns, H., et al., 2015. CD47 enhances in vivo functionality of artificial antigen-presenting cells. Clin. Canc. Res. 21 (9), 2075–2083. https://doi.org/10.1158/1078-0432.CCR-14-2696.

Buchan, S.L., et al., 2015. OX40- and CD27-mediated costimulation synergizes with anti-PD-L1 blockade by forcing exhausted $CD8^+$ T cells to exit quiescence. J. Immunol. 194 (1), 125–133. https://doi.org/10.4049/jimmunol.1401644.

Buckner, J.H., 2010. Mechanisms of impaired regulation by $CD4^+$ $CD25^+$ $FOXP3^+$ regulatory T cells in human autoimmune diseases. Nat. Rev. Immunol. 10 (12), 849.

Bullock, T.N.J., Mullins, D.W., Engelhard, V.H., 2003. Antigen density presented by dendritic cells in vivo differentially affects the number and avidity of primary, memory, and recall $CD8^+$ T cells. J. Immunol. 170 (4), 1822–1829.

Burdick, J.A., Anseth, K.S., 2002. Photoencapsulation of osteoblasts in injectable RGD-modified PEG hydrogels for bone tissue engineering. Biomaterials 23 (22), 4315–4323.

Busch, D.H., et al., 2016. Role of memory T cell subsets for adoptive immunotherapy. Semin. Immunol. 28 (1), 28–34. https://doi.org/10.1016/j.smim.2016.02.001. Elsevier Ltd.

Butler, M.O., et al., 2007. Long-lived antitumor $CD8^+$ lymphocytes for adoptive therapy generated using an artificial antigen-presenting cell. Clin. Canc. Res. 13 (6), 1857–1867. https://doi.org/10.1158/1078-0432.CCR-06-1905.

Butler, M.O., Hirano, N., 2014. Human cell-based artificial antigen-presenting cells for cancer immunotherapy. Immunol. Rev. 257 (1), 191–209.

Butte, M.J., et al., 2009. PD-L1 interacts specifically with B7-1 to inhibit T cell proliferation. Immunity 27 (1), 111–122. https://doi.org/10.1016/j.immuni.2007.05.016.PD-L1.

Callahan, M.K., Wolchok, J.D., 2013. At the bedside: CTLA-4- and PD-1-blocking antibodies in cancer immunotherapy. J. Leukoc. Biol. 94 (1), 41–53. https://doi.org/10.1189/jlb.1212631.

Callmann, C.E., et al., 2015. Therapeutic enzyme-responsive nanoparticles for targeted delivery and accumulation in tumors. Adv. Mater. 27 (31), 4611–4615.

Cannons, J.L., et al., 2001. 4-1BB ligand induces cell division, sustains survival, and enhances effector function of CD4 and CD8 T cells with similar efficacy. J. Immunol. 167 (3), 1313–1324. https://doi.org/10.4049/jimmunol.167.3.1313.
Carnemolla, B., et al., 2013. Enhancement of the antitumor properties of interleukin-2 by its targeted delivery to the tumor blood vessel extracellular matrix Enhancement of the antitumor properties of interleukin-2 by its targeted delivery to the tumor blood vessel extracellular matrix. Blood 99 (5), 1659–1665. https://doi.org/10.1182/blood.V99.5.1659.
Caruso, H.G., et al., 2015. Tuning sensitivity of CAR to EGFR density limits recognition of normal tissue while maintaining potent antitumor activity. Canc. Res. 75 (17), 3505–3518.
Celli, S., Lemaître, F., Bousso, P., 2007. Real-time manipulation of T cell-dendritic cell interactions in vivo reveals the importance of prolonged contacts for $CD4^+$ T cell activation. Immunity 27 (4), 625–634.
Chacon, J.A., et al., 2013. Co-stimulation through 4-1BB/CD137 improves the expansion and function of CD8(+) melanoma tumor-infiltrating lymphocytes for adoptive T-cell therapy. PLoS One 8 (4), e60031. https://doi.org/10.1371/journal.pone.0060031.
Chai, J.G., et al., 2005. Regulatory T cells, derived from naive $CD4^+$CD25- T cells by in vitro Foxp3 gene transfer, can induce transplantation tolerance. Transplantation 79 (10), 1310–1316 pii:00007890-200505270-00010.
Champion, J.A., Mitragotri, S., 2006. Role of target geometry in phagocytosis. Proc. Natl. Acad. Sci. USA 103 (13), 4930–4934. https://doi.org/10.1073/pnas.0600997103.
Chandra, R.A., et al., 2015. A systematic evaluation of abscopal responses following radiotherapy in patients with metastatic melanoma treated with ipilimumab. Oncoimmunology 4 (11), 1–7. https://doi.org/10.1080/2162402X.2015.1046028.
Chekmasova, A.A., et al., 2010. Successful eradication of established peritoneal ovarian tumors in SCID-Beige mice following adoptive transfer of T cells genetically targeted to the MUC16 antigen. Clin. Canc. Res. 16 (14), 3594–3606.
Chelius, D., et al., May 2016. Structural and functional characterization of the trifunctional antibody catumaxomab. mAbs 0862, 309–319. https://doi.org/10.4161/mabs.2.3.11791.
Chen, D.S., Mellman, I., 2013. Oncology Meets immunology: the cancer-immunity cycle. Immunity 39 (1), 1–10. https://doi.org/10.1016/j.immuni.2013.07.012.
Chen, L., Flies, D.B., 2013. Molecular mechanisms of T cell co-stimulation and co-inhibition. Nat. Rev. Immunol. 13 (4), 227–242. https://doi.org/10.1038/nri3405.
Chen, L., Han, X., 2015. Anti – PD-1/PD-L1 therapy of human cancer : past, present, and future. J. Clin. Invest. 125 (9) https://doi.org/10.1172/JCI80011.by.
Chen, S., et al., 2014. Combination of 4-1BB agonist and PD-1 antagonist promotes antitumor effector/memory CD8 T cells in a poorly immunogenic tumor model. Cancer Immunol. Res. 3 (2), 149–160. https://doi.org/10.1158/2326-6066.CIR-14-0118.
Chen, X., et al., 2016. Therapeutic efficacy of an anti-PD-L1 antibody based immunocytokine in a metastatic mouse model of colorectal cancer. Biochem. Biophys. Res. Commun. 480 (2), 160–165. https://doi.org/10.1016/j.bbrc.2016.10.011. Elsevier Ltd.
Chen, Y., Xu, Y., 2017. Pharmacokinetics of bispecific antibody. Curr. Pharmacol. Rep. 126–137. https://doi.org/10.1007/s40495-017-0090-5.
Cheng, C.J., et al., 2015. A holistic approach to targeting disease with polymeric nanoparticles. Nat. Rev. Drug Discov. 14 (4), 239.
Cheng, J., et al., 2007. Formulation of functionalized PLGA–PEG nanoparticles for in vivo targeted drug delivery. Biomaterials 28 (5), 869–876.
Chertok, B., et al., 2008. Iron oxide nanoparticles as a drug delivery vehicle for MRI monitored magnetic targeting of brain tumors. Biomaterials 29 (4), 487–496.

Cheung, A.S., et al., 2018. Scaffolds that mimic antigen-presenting cells enable ex vivo expansion of primary T cells. Nat. Biotechnol.
Chiu, M.L., Gilliland, G.L., 2016. Engineering antibody therapeutics. Curr. Opin. Struct. Biol. 38, 163–173. https://doi.org/10.1016/j.sbi.2016.07.012.
Chiu, Y.-L.L., et al., 2014. Sprouty-2 regulates HIV-specific T cell polyfunctionality. J. Clin. Invest. 124 (1), 198–208. https://doi.org/10.1172/JCI70510.
Choi, J.-S., et al., 2014. Size-controlled biodegradable nanoparticles: preparation and size-dependent cellular uptake and tumor cell growth inhibition. Colloids Surfaces B Biointerfaces 122, 545–551.
Chong, E.A., et al., 2017. PD-1 blockade modulates chimeric antigen receptor (CAR)–modified T cells: refueling the CAR. Blood 129 (8), 1039–1041.
Clark, R.A., et al., 2005. Human skin cells support thymus-independent T cell development. J. Clin. Invest. 115 (11), 3239–3249.
Clemente-Casares, X., et al., 2016. Expanding antigen-specific regulatory networks to treat autoimmunity. Nature 530 (7591), 434–440. https://doi.org/10.1038/nature16962.
Comrie, W.A., et al., 2015. The dendritic cell cytoskeleton promotes T cell adhesion and activation by constraining ICAM-1 mobility. J. Cell Biol. 208 (4), 457–473.
Conklin, B.R., et al., 2008. Engineering GPCR signaling pathways with RASSLs. Nat. Methods 5 (8), 673.
Conner, S.D., Schmid, S.L., 2003. Regulated portals of entry into the cell. Nature 422 (6927), 37–44.
Cruz, L.J., et al., 2014. Targeting nanoparticles to CD40, DEC-205 or CD11c molecules on dendritic cells for efficient $CD8^+$ T cell response: a comparative study. J. Contr. Release 192, 209–218. https://doi.org/10.1016/j.jconrel.2014.07.040. Elsevier B.V.
Curtsinger, J., et al., 1997. Artificial cell surface constructs for studying receptor-ligand contributions to lymphocyte activation. J. Immunol. Meth. 209 (1), 47–57.
Curtsinger, J.M., Mescher, M.F., 2011. Inflammatory cytokines as a third Signal for T Cell activation Julie. Curr. Opin. Immunol. 22 (3), 333–340. https://doi.org/10.1016/j.coi.2010.02.013.Inflammatory.
Danhier, F., et al., 2012. PLGA-based nanoparticles: an overview of biomedical applications. J. Contr. Release 161 (2), 505–522. https://doi.org/10.1016/j.jconrel.2012.01.043. Elsevier B.V.
Das, D.K., et al., 2015. Force-dependent transition in the T-cell receptor β-subunit allosterically regulates peptide discrimination and pMHC bond lifetime. Proc. Natl. Acad. Sci. USA 112 (5), 1517–1522.
Davila, M.L., et al., 2014. Efficacy and toxicity management of 19-28z CAR T cell therapy in B cell acute lymphoblastic leukemia. Sci. Transl. Med. 6 (224), 224ra25.
Debenedette, M.A., et al., 1997. Costimulation of CD28- T lymphocytes by 4-1BB ligand. J. Immunol. 158 (2), 551–559.
Deeg, J., et al., 2013. T cell activation is determined by the number of presented antigens. Nano Lett. 13 (11), 5619–5626. https://doi.org/10.1021/nl403266t.
Delcassian, D., et al., 2013. Nanoscale ligand spacing influences receptor triggering in T cells and NK cells. Nano Lett. 13 (11), 5608–5614. https://doi.org/10.1021/nl403252x.
Dinauer, N., et al., 2005. Selective targeting of antibody-conjugated nanoparticles to leukemic cells and primary T-lymphocytes. Biomaterials 26 (29), 5898–5906. https://doi.org/10.1016/j.biomaterials.2005.02.038.
Ding, Q., et al., 2013. RAFTsomes containing epitope-MHC-II complexes mediated $CD4^+$ T cell activation and antigen-specific immune responses. Pharm. Res. 30 (1), 60–69.
Discher, D.E., Mooney, D.J., Zandstra, P.W., 2009. Growth factors, matrices, and forces combine and control stem cells. Science 324 (5935), 1673–1677.

Doh, J., Irvine, D.J., 2006. Immunological synapse arrays: patterned protein surfaces that modulate immunological synapse structure formation in T cells. Proc. Natl. Acad. Sci. USA 103 (15), 5700–5705. https://doi.org/10.1073/pnas.0509404103.

Dudley, M.E., Rosenberg, S.A., 2003. Adoptive-cell-transfer therapy for the treatment of patients with cancer. Nat. Rev. Canc. 3 (9), 666–675. https://doi.org/10.1038/nrc1167.

Duncan, R., 2006. Polymer conjugates as anticancer nanomedicines. Nat. Rev. Canc. 6 (9), 688.

Dupont, J., et al., 2005. Artificial antigen-presenting cells transduced with telomerase efficiently expand epitope-specific, human leukocyte antigen-restricted cytotoxic T cells. Canc. Res. 65 (12), 5417–5427. https://doi.org/10.1158/0008-5472.CAN-04-2991.

Dustin, M.L., Depoil, D., 2011. New insights into the T cell synapse from single molecule techniques. Nat. Rev. Immunol. 11 (10), 672–684. https://doi.org/10.1038/nri3066. Nature Publishing Group.

Eyquem, J., et al., 2017. Targeting a CAR to the TRAC locus with CRISPR/Cas9 enhances tumour rejection. Nature 543 (7643), 113–117. https://doi.org/10.1038/nature21405. Nature Publishing Group.

Fadel, T.R., et al., 2014. A carbon nanotube-polymer composite for T-cell therapy. Nat. Nanotechnol. 9 (8), 639–647. https://doi.org/10.1038/nnano.2014.154. Nature Publishing Group.

Fahmy, T.M., et al., 2001. Increased TCR avidity after T cell activation: a mechanism for sensing low-density antigen. Immunity 14 (2), 135–143.

Fahmy, T.M., Schneck, J.P., Saltzman, W.M., 2007. A nanoscopic multivalent antigen-presenting Carrier for sensitive detection and drug delivery to T cells. Nanomed. Nanotechnol. Biol. Med. 3 (1), 75–85. https://doi.org/10.1016/j.nano.2006.11.010.

Fallon, J.K., et al., 2017. Enhanced antitumor effects by combining an IL-12/anti-DNA fusion protein with avelumab, an anti-PD-L1 antibody. Oncotarget 8 (13), 20558–20571. https://doi.org/10.18632/oncotarget.16137.

Fang, J., Nakamura, H., Maeda, H., 2011. The EPR effect: unique features of tumor blood vessels for drug delivery, factors involved, and limitations and augmentation of the effect. Adv. Drug Deliv. Rev. 63 (3), 136–151. https://doi.org/10.1016/j.addr.2010.04.009. Elsevier B.V.

Fang, R.H., et al., 2014. Cancer cell membrane-coated nanoparticles for anticancer vaccination and drug delivery. Nano Lett. 14 (4), 2181–2188.

Fatehullah, A., Tan, S.H., Barker, N., 2016. Organoids as an in vitro model of human development and disease. Nat. Cell Biol. 18 (3), 246.

Fedorov, V.D., Themeli, M., Sadelain, M., 2013. PD-1–and CTLA-4–based inhibitory chimeric antigen receptors (iCARs) divert off-target immunotherapy responses. Sci. Transl. Med. 5 (215), 215ra172.

Ferez, M., et al., 2014. Cognate peptide–MHC complexes are expressed as tightly apposed nanoclusters in virus-infected cells to allow TCR crosslinking. J. Immunol. 192 (1), 52–58.

Fesnak, A.D., June, C.H., Levine, B.L., 2016. Engineered T cells: the promise and challenges of cancer immunotherapy. Nat. Rev. Canc. 16 (9), 566–581. https://doi.org/10.1038/nrc.2016.97. Nature Publishing Group.

Fifis, T., et al., 2004. Size-dependent immunogenicity: therapeutic and protective properties of nano-vaccines against tumors. J. Immunol. 173 (5), 3148–3154. https://doi.org/10.4049/jimmunol.173.5.3148.

Fischer, J.A.A., et al., 2015. Combined inhibition of tumor necrosis factor α and interleukin-17 as a therapeutic opportunity in rheumatoid arthritis: development and characterization of a novel bispecific antibody. Arthritis Rheumatol. 67 (1), 51–62. https://doi.org/10.1002/art.38896.

Fooksman, D.R., et al., 2006. Clustering class I MHC modulates sensitivity of T cell recognition. J. Immunol. 176 (11), 6673–6680.

Fooksman, D.R., et al., 2009. Functional anatomy of T cell activation and synapse formation. Annu. Rev. Immunol. 28, 79–105.

Franz, S., et al., 2011. Immune responses to implants—a review of the implications for the design of immunomodulatory biomaterials. Biomaterials 32 (28), 6692–6709.

Gao, H., Shi, W., Freund, L.B., 2005. Mechanics of receptor-mediated endocytosis. Proc. Natl. Acad. Sci. USA 102 (27), 9469–9474. https://doi.org/10.1073/pnas.0503879102.

Gargett, T., Brown, M.P., 2014. The inducible caspase-9 suicide gene system as a "safety switch" to limit on-target, off-tumor toxicities of chimeric antigen receptor T cells. Front. Pharmacol. 5, 235.

Garnier, A., et al., 2016. Artificial antigen-presenting cells expressing HLA class II molecules as an effective tool for amplifying human specific memory $CD4^+$ T cells. Immunol. Cell Biol. 94 (7), 662.

Garrido, F., Algarra, I., 2001. MHC antigens and tumor escape from immune surveillance. Adv. Canc. Res. 83, 117–158. https://doi.org/10.1016/S0065-230X(01)83005-0.

Germano, G., et al., 2017. Inactivation of DNA repair triggers neoantigen generation and impairs tumour growth. Nature 552 (7683), 116–120. https://doi.org/10.1038/nature24673. Nature Publishing Group.

Giannoni, F., et al., 2005. Clustering of T cell ligands on artificial APC membranes influences T cell activation and protein kinase C θ translocation to the T cell plasma membrane. J. Immunol. 174 (6), 3204–3211.

Giese, C., et al., 2006. A human lymph node in vitro—challenges and progress. Artif. Organs 30 (10), 803–808.

Glowacki, A.J., et al., 2013. Prevention of inflammation-mediated bone loss in murine and canine periodontal disease via recruitment of regulatory lymphocytes. Proc. Natl. Acad. Sci. USA 110 (46), 18525–18530. https://doi.org/10.1073/pnas.1302829110.

Gong, C., Linderman, J.J., Kirschner, D., 2014. Harnessing the heterogeneity of T cell differentiation fate to fine-tune generation of effector and memory T cells. Front. Immunol. 5.

Gong, Q., et al., 2016. Increased in vivo effector function of human IgG4 isotype antibodies through afucosylation. mAbs 8 (6), 1098–1106. https://doi.org/10.1080/19420862.2016.1189049. Taylor & Francis.

Grada, Z., et al., 2013. TanCAR: a novel bispecific chimeric antigen receptor for cancer immunotherapy. Mol. Ther. Nucleic Acids 2.

Graham, J.G., et al., 2013. PLG scaffold delivered antigen-specific regulatory T cells induce systemic tolerance in autoimmune diabetes. Tissue Eng. 19 (11–12), 1465–1475.

Gratton, S.E.A., et al., 2008. The effect of particle design on cellular internalization pathways. Proc. Natl. Acad. Sci. USA 105 (33), 11613–11618. https://doi.org/10.1073/pnas.0801763105.

Groves, J.T., Dustin, M.L., 2003. Supported planar bilayers in studies on immune cell adhesion and communication. J. Immunol. Meth. 278 (1–2), 19–32.

Gu, L., Mooney, D.J., 2015. Biomaterials and emerging anticancer therapeutics: engineering the microenvironment. Nat. Rev. Canc. 16 (1), 56–66. https://doi.org/10.1038/nrc.2015.3. Nature Publishing Group.

Hamid, O., et al., 2013. Safety and tumor responses with lambrolizumab (anti-PD-1) in melanoma. N. Engl. J. Med. 369 (2), 134–144. https://doi.org/10.1056/NEJMoa1305133.

Han, G., et al., 2006. Light-regulated release of DNA and its delivery to nuclei by means of photolabile gold nanoparticles. Angew. Chem. 118 (19), 3237–3241.

Harding, F.A., et al., 2010. The immunogenicity of humanized and fully human antibodies: residual immunogenicity resides in the CDR regions. mAbs 2 (3), 256–265. https://doi.org/10.4161/mabs.2.3.11641.

Hasan, A.N., et al., 2010. IL-15 enhances in-vitro expansion and functional activity of antigen-specific effector memory T cells (TEM) while Co-Expression of IL-15 and IL-15 Rα on antigen presenting cells also promotes enrichment and preferential expansion of central memory t-cells. Biol. Blood Marrow Transplant. 16 (2), S159.

Hashimoto-Tane, A., Saito, T., 2016. Dynamic regulation of TCR-microclusters and the microsynapse for T cell activation. Front. Immunol. 7 (Jun), 1–8. https://doi.org/10.3389/fimmu.2016.00255.

Helle, M., et al., 2013. Surface chemistry architecture of silica nanoparticles determine the efficiency of in vivo fluorescence lymph node mapping. ACS Nano 7 (10), 8645–8657. https://doi.org/10.1021/nn402792a.

Hernandez-chacon, J.A., et al., 2012. Co-stimulation through the CD137/4-1BB pathway protects human melanoma tumor-infiltrating lymphocytes from activation-induced cell death and enhances anti-tumor effector function. J. Immunother. 34 (3), 236–250. https://doi.org/10.1097/CJI.0b013e318209e7ec.Co-stimulation.

Hess, K.L., et al., 2017. Engineering immunological tolerance using quantum dots to tune the density of self-antigen display. Adv. Funct. Mater. 27 (22) https://doi.org/10.1002/adfm.201700290.

Hickey, J.W., et al., 2015. Control of polymeric nanoparticle size to improve therapeutic delivery. J. Contr. Release 219, 536–547.

Hickey, J.W., et al., 2017. Biologically inspired design of nanoparticle artificial antigen-presenting cells for immunomodulation. Nano Lett. 17 (11), 7045–7054. https://doi.org/10.1021/acs.nanolett.7b03734.

Hinrichs, C.S., et al., 2011. Human effector $CD8^+$ T cells derived from naive rather than memory subsets possess superior traits for adoptive immunotherapy. Blood 117 (3), 808–814. https://doi.org/10.1182/blood-2010-05-286286.

Hipp, S., et al., 2017. A novel BCMA/CD3 bispecific T-cell engager for the treatment of multiple myeloma induces selective lysis in vitro and in vivo. Leukemia 31 (8), 1743–1751. https://doi.org/10.1038/leu.2016.388. Nature Publishing Group.

Hirosue, S., Dubrot, J., 2015. Modes of antigen presentation by lymph node stromal cells and their immunological implications. Front. Immunol. 6.

Holliger, P., Hudson, P.J., 2005. Engineered antibody fragments and the rise of single domains. Nat. Biotechnol. 1126–1136. https://doi.org/10.1038/nbt1142.

Hori, Y., et al., 2008. Injectable dendritic cell-carrying alginate gels for immunization and immunotherapy. Biomaterials 29 (27), 3671–3682.

Hori, Y., Winans, A.M., Irvine, D.J., 2009. Modular injectable matrices based on alginate solution/microsphere mixtures that gel in situ and co-deliver immunomodulatory factors. Acta Biomater. 5 (4), 969–982.

Hu, C.-M.J., et al., 2011. Erythrocyte membrane-camouflaged polymeric nanoparticles as a biomimetic delivery platform. Proc. Natl. Acad. Sci. USA 108 (27), 10980–10985.

Huang, Y., et al., 2016. Engineered bispecific antibodies with exquisite HIV-1-neutralizing activity. Cell 165 (7), 1621–1631. https://doi.org/10.1016/j.cell.2016.05.024.

Hudecek, M., et al., 2010. The B-cell tumor–associated antigen ROR1 can be targeted with T cells modified to express a ROR1-specific chimeric antigen receptor. Blood 116 (22), 4532–4541.

Huehls, A.M., Coupet, T.A., Sentman, C.L., 2015. Bispecific T cell engagers for cancer immunotherapy. Immunol. Cell Biol. 93 (3), 290–296. https://doi.org/10.1038/icb.2014.93.Bispecific.

Hugo, W., et al., 2016. Genomic and transcriptomic features of response to anti-PD-1 therapy in metastatic melanoma. Cell 165 (1), 35–44. https://doi.org/10.1016/j.cell.2016.02.065. Elsevier Inc.

Hui, E., et al., 2017. T cell costimulatory receptor CD28 is a primary target for PD-1–mediated inhibition. Science 355 (6332), 1428–1433. https://doi.org/10.1126/science.aaf1292.

Humphrey, J.D., Dufresne, E.R., Schwartz, M.A., 2014. Mechanotransduction and extracellular matrix homeostasis. Nat. Rev. Mol. Cell Biol. 15 (12), 802.

Hunter, Z., et al., 2014. A biodegradable nanoparticle platform for the induction of antigen-specific immune tolerance for treatment of autoimmune disease. ACS Nano 8 (3), 2148–2160. https://doi.org/10.1021/nn405033r.

Huppa, J.B., et al., 2003. Continuous T cell receptor signaling required for synapse maintenance and full effector potential. Nat. Immunol. 4 (8), 749–755.

Huse, M., 2017. Mechanical forces in the immune system. Nat. Rev. Immunol. 17 (11), 679.

Hwang, J., et al., 1998. Domains in cell plasma membranes investigated by near-field scanning optical microscopy. Biophys. J. 74 (5), 2184–2190.

Iezzi, G., Karjalainen, K., Lanzavecchia, A., 1998. The duration of antigenic stimulation determines the fate of naive and effector T cells. Immunity 8 (1), 89–95.

Irvine, D.J., Swartz, M.A., Szeto, G.L., 2013. Engineering synthetic vaccines using cues from natural immunity. Nat. Mater. 12 (11), 978–990.

Iwahori, K., et al., 2014. Engager T cells: a new class of antigen-specific T cells that redirect bystander T cells. Mol. Ther. https://doi.org/10.1038/mt.2014.156.

Jaalouk, D.E., Lammerding, J., 2009. Mechanotransduction gone awry. Nat. Rev. Mol. Cell Biol. 10 (1), 63.

Jahan, N., Talat, H., Curry, W.T., 2017. Agonist OX40 immunotherapy improves survival in glioma-bearing mice and is complementary with vaccination with irradiated GM-CSF–expressing tumor cells. Neuro Oncol. 20 (July 2017), 44–54. https://doi.org/10.1093/neuonc/nox125.

Jain, R.K., Stylianopoulos, T., 2010. Delivering nanomedicine to solid tumors. Nat. Rev. Clin. Oncol. 7 (11), 653–664.

Janssen, E.M., et al., 2003. $CD4^+$ T cells are required for secondary expansion and memory in $CD8^+$ T lymphocytes. Nature 421 (6925), 852.

Javadi, M., et al., 2012. Encapsulating nanoemulsions inside eLiposomes for ultrasonic drug delivery. Langmuir 28 (41), 14720–14729.

Jethwa, H., Adami, A.A., Maher, J., 2014. Use of gene-modified regulatory T-cells to control autoimmune and alloimmune pathology: is now the right time? Clin. Immunol. 51–63. https://doi.org/10.1016/j.clim.2013.11.004.

Jhunjhunwala, S., et al., 2012. Bioinspired controlled release of CCL22 recruits regulatory T cells in vivo. Adv. Mater. 24 (35), 4735–4738. https://doi.org/10.1002/adma.201202513.

Jones, R.B., et al., 2017. Antigen recognition-triggered drug delivery mediated by nanocapsule-functionalized cytotoxic T-cells. Biomaterials 117, 44–53. https://doi.org/10.1016/j.biomaterials.2016.11.048. Elsevier Ltd.

De Jong, W.H., Borm, P.J.A., 2008. Drug delivery and nanoparticles: applications and hazards. Int. J. Nanomed. 3 (2), 133.

Josefowicz, S.Z., Lu, L.-F., Rudensky, A.Y., 2012. Regulatory T cells: mechanisms of differentiation and function. Annu. Rev. Immunol. 30 (1), 531–564. https://doi.org/10.1146/annurev.immunol.25.022106.141623.

Judokusumo, E., et al., 2012. Mechanosensing in T lymphocyte activation. Biophys. J. 102 (2), L5–L7.

Jung, Y., et al., 2007. Retargeting of adenoviral gene delivery via Herceptin–PEG–adenovirus conjugates to breast cancer cells. J. Contr. Release 123 (2), 164–171.

Kamath, A.V., 2016. Translational pharmacokinetics and pharmacodynamics of monoclonal antibodies. Drug Discov. Today Technol. 21–22 (Iv), 75–83. https://doi.org/10.1016/j.ddtec.2016.09.004. The Author(s).

Kantarjian, H., et al., 2017. Blinatumomab versus chemotherapy for advanced acute lymphoblastic leukemia. N. Engl. J. Med. 376 (9), 836–847. https://doi.org/10.1056/NEJMoa1609783.

Kantoff, P.W., et al., 2010. Sipuleucel-t immunotherapy for castration-resistant prostate cancer. N. Engl. J. Med. 363 (5), 2373–2383. https://doi.org/10.1056/NEJMoa1407764.

Kawahara, M., Ueda, H., Nagamune, T., 2010. Engineering cytokine receptors to control cellular functions. Biochem. Eng. J. 48 (3), 283–294.

Kawalkowska, J.Z., et al., 2016. Targeted IL-4 therapy synergizes with dexamethasone to induce a state of tolerance by promoting Treg cells and macrophages in mice with arthritis. Eur. J. Immunol. 46 (5), 1246–1257. https://doi.org/10.1002/eji.201546221.

Kermer, V., et al., 2014. Combining antibody-directed presentation of IL-15 and 4-1BBL in a trifunctional fusion protein for cancer immunotherapy. Mol. Canc. Therapeut. 13 (1), 112–121. https://doi.org/10.1158/1535-7163.MCT-13-0282.

Kettiger, H., et al., 2013. Engineered nanomaterial uptake and tissue distribution: from cell to organism. Int. J. Nanomed. 8, 3255.

Kim, S.T., et al., 2009. The αβ T cell receptor is an anisotropic mechanosensor. J. Biol. Chem. 284 (45), 31028–31037.

Klebanoff, C.A., Rosenberg, S.A., Restifo, N.P., 2016. Prospects for gene-engineered T cell immunotherapy for solid cancers. Nat. Med. 22 (1), 26–36. https://doi.org/10.1038/nm.4015.

Klinger, M., et al., 2013. Immunopharmacologic response of patients with B-lineage acute lymphoblastic leukemia to continuous infusion of T cell – engaging. Blood 119 (26), 6226–6233. https://doi.org/10.1182/blood-2012-01-400515.

Kloss, C.C., et al., 2013. Combinatorial antigen recognition with balanced signaling promotes selective tumor eradication by engineered T cells. Nat. Biotechnol. 31 (1), 71.

Kobayashi, Y., Watanabe, T., 2016. Gel-trapped lymphorganogenic chemokines trigger artificial tertiary lymphoid organs and mount adaptive immune responses in vivo. Front. Immunol. 7, 316.

Kochenderfer, J.N., et al., 2010. Eradication of B-lineage cells and regression of lymphoma in a patient treated with autologous T cells genetically engineered to recognize CD19. Blood 116 (20), 4099–4102.

Kochenderfer, J.N., et al., 2012. B-cell depletion and remissions of malignancy along with cytokine-associated toxicity in a clinical trial of anti-CD19 chimeric-antigen-receptor–transduced T cells. Blood 119 (12), 2709–2720.

Korin, N., et al., 2012. Shear-activated nanotherapeutics for drug targeting to obstructed blood vessels. Science 1217815.

Kosmides, A.K., et al., 2017. Biomimetic biodegradable artificial antigen presenting cells synergize with PD-1 blockade to treat melanoma. Biomaterials 118, 16–26. https://doi.org/10.1016/j.biomaterials.2016.11.038. Elsevier Ltd.

Kosmides, A.K., et al., 2018. Separating T cell targeting components onto magnetically clustered nanoparticles boosts activation. Nano Lett. https://doi.org/10.1021/acs.nanolett.7b05284 p. acs.nanolett.7b05284.

Kosmides, A.K.A.K., et al., 2017. Dual targeting nanoparticle stimulates the immune system to inhibit tumor growth. ACS Nano 11 (6), 5417–5429. https://doi.org/10.1021/acsnano.6b08152.

Krieg, C., et al., 2010. Improved IL-2 immunotherapy by selective stimulation of IL-2 receptors on lymphocytes and endothelial cells. Proc. Natl. Acad. Sci. USA 107 (26), 11906–11911. https://doi.org/10.1073/pnas.1002569107.

Kuai, R., et al., December 2016. Designer vaccine nanodiscs for personalized cancer immunotherapy. Nat. Mater. 1, 1–10. https://doi.org/10.1038/nmat4822.
Kumar, C.S.S.R., Mohammad, F., 2011. Magnetic nanomaterials for hyperthermia-based therapy and controlled drug delivery. Adv. Drug Deliv. Rev. 63 (9), 789–808.
Kwong, B., et al., 2013. Localized immunotherapy via liposome-anchored Anti-CD137 + IL-2 prevents lethal toxicity and elicits local and systemic antitumor immunity. Canc. Res. 73 (5), 1547–1558. https://doi.org/10.1158/0008-5472.CAN-12-3343.
Lai, S.K., et al., 2007. Privileged delivery of polymer nanoparticles to the perinuclear region of live cells via a non-clathrin, non-degradative pathway. Biomaterials 28 (18), 2876–2884.
Larkin, J., et al., 2015. Combined nivolumab and ipilimumab or monotherapy in untreated melanoma. N. Engl. J. Med. 373 (1), 23–34. https://doi.org/10.1056/NEJMoa1504030.
Lee, I.H., et al., 2012. Imageable antigen-presenting gold nanoparticle vaccines for effective cancer immunotherapy in vivo. Angew. Chem. Int. Ed. 51 (35), 8800–8805. https://doi.org/10.1002/anie.201203193.
Lee, K.J., et al., 2016. Clinical use of blinatumomab for B-cell acute lymphoblastic leukemia in adults. Therapeut. Clin. Risk Manag. 12, 1301–1310. https://doi.org/10.2147/TCRM.S84261.
Leleux, J., Alexandra, A., Roy, K., 2015. Engineering immunity: modulating dendritic cell subsets and lymph node response to direct immune-polarization and vaccine efficacy. J. Control Release 219, 610–621. https://doi.org/10.1177/0963721412473755.Surging.
Lerner, R.A., 2016. Combinatorial antibody libraries: new advances, new immunological insights. Nat. Rev. Immunol. 498–508. https://doi.org/10.1038/nri.2016.67.
Lever, M., et al., 2014. Phenotypic models of T cell activation. Nat. Rev. Immunol. 14 (9), 619–629. https://doi.org/10.1038/nri3728. Nature Publishing Group.
Levin, A.M., et al., 2012. Exploiting a natural conformational switch to engineer an interleukin-2 “superkine”. Nature 484 (7395), 529–533. https://doi.org/10.1038/nature10975. Nature Publishing Group.
Levine, B.L., et al., 1998. Large-scale production of $CD4^{+}$ T cells from HIV-1-infected donors after CD3/CD28 costimulation. J. Hematother. 7 (5), 437–448.
Lewis, J.S., et al., 2015. A combination dual-sized microparticle system modulates dendritic cells and prevents type 1 diabetes in prediabetic NOD mice. Clin. Immunol. 160 (1), 90–102. https://doi.org/10.1016/j.clim.2015.03.023.
Li, A.V., et al., 2014. Generation of effector memory T cell–based mucosal and systemic immunity with pulmonary nanoparticle vaccination. Sci. Transl. Med. 5 (204), 204ra130. https://doi.org/10.1126/scitranslmed.3006516.Generation.
Li, B., et al., 2009. Anti-programmed death-1 synergizes with granulocyte macrophage colony-stimulating factor-secreting tumor cell immunotherapy providing therapeutic benefit to mice with established tumors. Clin. Canc. Res. 15 (5), 1623–1634. https://doi.org/10.1158/1078-0432.CCR-08-1825.
Li, Y., et al., 2017. Regulatory T cells control toxicity in a humanized model of IL-2 therapy. Nat. Commun. 8 (1), 1762.
Lillemeier, B.F., et al., 2010. TCR and Lat are expressed on separate protein islands on T cell membranes and concatenate during activation. Nat. Immunol. 11 (1), 90–96. https://doi.org/10.1038/ni0610-543b.
Lim, W.A., June, C.H., 2017. The principles of engineering immune cells to treat cancer. Cell 168 (4), 724–740.
Liu, B., et al., 2014. Accumulation of dynamic catch bonds between TCR and agonist peptide-MHC triggers T cell signaling. Cell 157 (2), 357–368.

Liu, B., et al., 2016. A novel fusion of ALT-803 (interleukin (IL)-15 superagonist) with an antibody demonstrates antigen-specific antitumor responses. J. Biol. Chem. 291 (46), 23869–23881. https://doi.org/10.1074/jbc.M116.733600.

Liu, H., et al., 2014. Structure-based programming of lymph-node targeting in molecular vaccines. Nature 507 (7493), 519–522. https://doi.org/10.1038/nature12978. Nature Publishing Group.

Lu, X., et al., 2012. Endogenous viral antigen processing generates peptide-specific MHC class I cell-surface clusters. Proc. Natl. Acad. Sci. USA 109 (38), 15407–15412. https://doi.org/10.1073/pnas.1208696109.

Lu, Y., et al., 2017. Bioresponsive materials. Nat. Rev. Mater. 2 (1), 16075.

Luckheeram, R.V., et al., 2012. $CD4^+$ T cells: differentiation and functions. Clin. Dev. Immunol. 2012 https://doi.org/10.1155/2012/925135.

Lurie, N., et al., 1993. A progress report on the treatment of 157 patients with advanced cancer using lymphokine-activated killer cells and interleukin-2 of high-dose interleukin-2 alone. N. Engl. J. Med. 329 (7), 478–482. https://doi.org/10.1056/NEJM199308123290707.

Lyman, M., et al., 2018. A bispecific antibody that targets IL-6 receptor and IL-17A for the potential therapy of patients with autoimmune and inflammatory diseases. J. Biol. Chem. ASBMB, p. jbc-M117.

MacDonald, K.G., et al., 2016. Alloantigen-specific regulatory T cells generated with a chimeric antigen receptor. J. Clin. Invest. 126 (4), 1413–1424. https://doi.org/10.1172/JCI82771.

Mack, M., Riethmuller, G., Kufer, P., 1995. A small bispecific antibody construct expressed as a functional single-chain molecule with high tumor cell cytotoxicity. Proc. Natl. Acad. Sci. USA 92 (15), 7021–7025. https://doi.org/10.1073/pnas.92.15.7021.

Mahoney, K.M., Rennert, P.D., Freeman, G.J., 2015. Combination cancer immunotherapy and new immunomodulatory targets. Nat. Rev. Drug Discov. 14 (8), 561–584. https://doi.org/10.1038/nrd4591.

Malek, T.R., 2008. The biology of interleukin-2. Annu. Rev. Immunol. 26, 453–479.

Malek, T.R., Castro, I., 2010. Interleukin-2 receptor signaling: at the interface between tolerance and immunity. Immunity 33 (2), 153–165. https://doi.org/10.1016/j.immuni.2010.08.004. Elsevier Inc.

Manolova, V., et al., 2008. Nanoparticles target distinct dendritic cell populations according to their size. Eur. J. Immunol. 38 (5), 1404–1413. https://doi.org/10.1002/eji.200737984.

Manz, B.N., et al., 2011. T-cell triggering thresholds are modulated by the number of antigen within individual T-cell receptor clusters. Proc. Natl. Acad. Sci. USA 108 (22), 9089–9094.

Matic, J., et al., 2013. Fine tuning and efficient T cell activation with stimulatory aCD3 nanoarrays. Nano Lett. 13 (11), 5090–5097.

Matsui, K., et al., 1991. Low affinity interaction of peptide-MHC complexes with T cell receptors. Science 254 (5039), 1788–1792.

Maude, S.L., et al., 2014. Chimeric antigen receptor T cells for sustained remissions in leukemia. N. Engl. J. Med. 371 (16), 1507–1517.

Maus, M.V., et al., 2002. Ex vivo expansion of polyclonal and antigen-specific cytotoxic T lymphocytes by artificial APCs expressing ligands for the T-cell receptor, CD28 and 4-1BB. Nat. Biotechnol. 20 (2), 143–148. https://doi.org/10.1038/nbt0202-143.

Mcgranahan, N., et al., 2016. Clonal neoantigens elicit T cell immunoreactivity and sensitivity to immune checkpoint blockade. Science 351 (6280), 1463–1469. https://doi.org/10.1126/science.aaf1490.

Meng, F., Hennink, W.E., Zhong, Z., 2009. Reduction-sensitive polymers and bioconjugates for biomedical applications. Biomaterials 30 (12), 2180–2198.
Mescher, M.F., 1992. Surface contact requirements for activation of cytotoxic T lymphocytes. J. Immunol. 149 (7), 2402–2405.
Messenheimer, D.J., et al., 2017. Timing of PD-1 blockade is critical to effective combination immunotherapy with anti-OX40. Clin. Canc. Res. 23 (20), 6165–6177. https://doi.org/10.1158/1078-0432.CCR-16-2677.
Meyer, R.A., et al., 2014. Biodegradable nanoellipsoidal artificial antigen presenting cells for antigen specific t-cell activation. Small 11 (13), 1519–1525. https://doi.org/10.1002/smll.201402369.
Meyer, R.A., et al., 2015. Biodegradable nanoellipsoidal artificial antigen presenting cells for antigen specific T-cell activation. Small 11 (13), 1519–1525.
Mishra, A., Sullivan, L., Caligiuri, M.A., 2014. Molecular pathways: interleukin-15 signaling in health and in cancer. Clin. Cancer Res. 20 (8), 2044–2050. https://doi.org/10.1158/1078-0432.CCR-12-3603.Molecular.
Mitra, S., et al., 2015. Interleukin-2 activity can be fine-tuned with engineered receptor signaling clamps. Immunity 42 (5), 826–838. https://doi.org/10.1126/science.1249098.Sleep.
Monopoli, M.P., et al., 2012. Biomolecular coronas provide the biological identity of nanosized materials. Nat. Nanotechnol. 7 (12), 779–786.
Morgan, R.A., et al., 2010. Case report of a serious adverse event following the administration of T cells transduced with a chimeric antigen receptor recognizing ERBB2. Mol. Ther. 18 (4), 843–851.
Mossman, K.D., et al., 2005. Altered TCR signaling from geometrically repatterned immunological synapses. Science 310 (5751), 1191–1193.
Motta, I., Lone, Y.C., Kourilsky, P., 1998. In vitro induction of naive cytotoxic T lymphocytes with complexes of peptide and recombinant MHC class I molecules coated onto beads: role of TCR/ligand density. Eur. J. Immunol. 28 (11), 3685–3695. https://doi.org/10.1002/(SICI)1521-4141(199811)28:11<3685::AID-IMMU3685>3.0.CO;2-3.
Moynihan, K.D., et al., 2016. Eradication of large established tumors in mice by combination immunotherapy that engages innate and adaptive immune responses. Nat. Med. 22 (12), 1402.
Mullard, A., 2018. IDO takes a blow. Nat. Rev. Drug Discov. Nature Publishing Group.
Müller, D., 2014. Antibody-cytokine fusion proteins for cancer immunotherapy: an update on recent developments. BioDrugs 28 (2), 123–131. https://doi.org/10.1007/s40259-013-0069-7.
Mura, S., Nicolas, J., Couvreur, P., 2013. Stimuli-responsive nanocarriers for drug delivery. Nat. Mater. 12 (11), 991.
Narayanan, K., et al., 2013. Mimicking cellular transport mechanism in stem cells through endosomal escape of new peptide-coated quantum dots. Sci. Rep. 3.
Nel, A.E., et al., 2009. Understanding biophysicochemical interactions at the nano–bio interface. Nat. Mater. 8 (7), 543–557.
Neri, D., Sondel, P.M., 2016. Immunocytokines for cancer treatment: past, present and future. Curr. Opin. Immunol. 40, 96–102. https://doi.org/10.1016/j.coi.2016.03.006.Immunocytokines.
Norris, D.A., Puri, N., Sinko, P.J., 1998. The effect of physical barriers and properties on the oral absorption of particulates. Adv. Drug Deliv. Rev. 34 (2), 135–154.
O'Connor, R.S., et al., 2012. Substrate rigidity regulates human T cell activation and proliferation. J. Immunol. 189 (3), 1330–1339.

Oelke, M., et al., 2003a. Ex vivo induction and expansion of antigen-specific cytotoxic T cells by HLA-Ig-coated artificial antigen-presenting cells. Nat. Med. 9 (5), 619–624. https://doi.org/10.1038/nm869.

Oelke, M., et al., 2003b. Ex vivo induction and expansion of antigen-specific cytotoxic T cells by HLA-Ig—coated artificial antigen-presenting cells. Nat. Med. 9 (5), 619–625.

Oelke, M., Schneck, J.P., 2010. Overview of a HLA-Ig based "Lego-like system" for T cell monitoring, modulation and expansion. Immunol. Res. 47 (1–3), 248–256. https://doi.org/10.1007/s12026-009-8156-z.

Oh, H.S., et al., 2015. 4-1BB signaling enhances primary and secondary population expansion of $CD8^+$ T cells by maximizing autocrine IL-2/IL-2 receptor signaling. PLoS One 10 (5), e0126765. https://doi.org/10.1371/journal.pone.0126765.

Okamoto, N., et al., 2007. Artificial lymph nodes induce potent secondary immune responses in naive and immunodeficient mice. J. Clin. Invest. 117 (4), 997–1007.

Okazaki, T., et al., 2013. A rheostat for immune responses: the unique properties of PD-1 and their advantages for clinical application. Nat. Immunol. 14 (12), 1212–1218. https://doi.org/10.1038/ni.2762. Nature Publishing Group.

Oldenborg, P.-A., et al., 2000. Role of CD47 as a marker of self on red blood cells. Science 288 (5473), 2051–2054.

Oren, R., et al., 2014. Functional comparison of engineered T cells carrying a native TCR versus TCR-like antibody—based chimeric antigen receptors indicates affinity/avidity thresholds. J. Immunol. 193 (11), 5733–5743.

Ostrand-Rosenberg, S., Horn, L.A., Alvarez, J.A., 2015. Novel strategies for inhibiting PD-1 pathway-mediated immune suppression while simultaneously delivering activating signals to tumor-reactive T cells. Canc. Immunol. Immunother. 64 (10), 1287–1293. https://doi.org/10.1007/s00262-015-1677-5. Springer Berlin Heidelberg.

Owens, B., 2017. Faster, deeper, smaller—the rise of antibody-like scaffolds. Nat. Biotechnol. 35 (7), 602–603. https://doi.org/10.1038/nbt0717-602.

Packer, M.S., Liu, D.R., 2015. Methods for the directed evolution of proteins. Nat. Rev. Genet. 379–394. https://doi.org/10.1038/nrg3927.

Paller, C.J., Antonarakis, E.S., 2012. Sipuleucel-T for the treatment of metastatic prostate cancer. Hum. Vaccines Immunother. 8 (4), 509–519.

Panelli, M.C., et al., 2004. Forecasting the cytokine storm following systematic interleukin (IL)-2 administration. J. Transl. Med. 2, 1–14. https://doi.org/10.1186/1479-5876-2-17.

Pang, X., et al., 2017. Treatment of human B-cell lymphomas using minicircle DNA vector expressing anti-CD3/CD20 in a mouse model. Hum. Gene Ther. 28 (2), 216–225. Mary Ann Liebert, Inc. 140 Huguenot Street, 3rd Floor New Rochelle, NY 10801 USA.

Papanicolaou, G.A., et al., 2003. Rapid expansion of cytomegalovirus—specific cytotoxic T lymphocytes by artificial antigen-presenting cells expressing a single HLA allele. Blood 102 (7), 2498–2505.

Parodi, A., et al., 2013. Synthetic nanoparticles functionalized with biomimetic leukocyte membranes possess cell-like functions. Nat. Nanotechnol. 8 (1), 61.

Pedicord, V.A., et al., 2011. Single dose of anti – CTLA-4 enhances $CD8^+$ T-cell memory formation, function, and maintenance. Proc. Natl. Acad. Sci. USA 108 (1), 266–271. https://doi.org/10.1073/pnas.1016791108/-/DCSupplemental. www.pnas.org/cgi/doi/10.1073/pnas.1016791108.

Perica, K., Tu, A., et al., 2014a. Magnetic field-induced T cell receptor clustering by nanoparticles enhances T cell activation and stimulates antitumor activity. ACS Nano 8 (3), 2252–2260. https://doi.org/10.1021/nn405520d.

Perica, K., De León Medero, A., et al., 2014b. Nanoscale artificial antigen presenting cells for T cell immunotherapy. Nanomed. Nanotechnol. Biol. Med. 10 (1), 119–129. https://doi.org/10.1016/j.nano.2013.06.015. Elsevier Inc.

Perica, K., et al., 2015. Enrichment and expansion with nanoscale artificial antigen presenting cells for adoptive immunotherapy. ACS Nano 9 (7), 6861–6871.

Perica, K., Kosmides, A.K., Schneck, J.P., 2015. Linking form to function: biophysical aspects of artificial antigen presenting cell design. Biochim. Biophys. Acta 1853 (4), 781–790. https://doi.org/10.1016/j.bbamcr.2014.09.001. Elsevier B.V.

Pilon-Thomas, S., et al., 2010. Blockade of programmed death ligand 1 enhances the therapeutic efficacy of combination immunotherapy against melanoma. J. Immunol. 184 (7), 3442–3449. https://doi.org/10.4049/jimmunol.0904114.

Pishesha, N., et al., 2017. Engineered erythrocytes covalently linked to antigenic peptides can protect against autoimmune disease. Proc. Natl. Acad. Sci. USA 114 (12), 3157–3162. https://doi.org/10.1073/pnas.1701746114.

Polu, K.R., Lowman, H.B., 2014. Probody therapeutics for targeting antibodies to diseased tissue. Expet Opin. Biol. Ther. 14 (8), 1049–1053. https://doi.org/10.1517/14712598.2014.920814.

Poon, I.K.H., et al., 2014. Apoptotic cell clearance: basic biology and therapeutic potential. Nat. Rev. Immunol. 14 (3), 166–180. https://doi.org/10.1038/nri3607. Nature Publishing Group.

Porter, D.L., et al., 2011. Chimeric antigen receptor–modified T cells in chronic lymphoid leukemia. N. Engl. J. Med. 365 (8), 725–733.

Postow, M.A., et al., 2015. Nivolumab and ipilimumab versus ipilimumab in untreated melanoma. N. Engl. J. Med. 372 (21), 2006–2017. https://doi.org/10.1056/NEJMoa1414428.

Poznansky, M.C., et al., 2000. Efficient generation of human T cells from a tissue-engineered thymic organoid. Nat. Biotechnol. 18 (7), 729.

Prasad, V., 2018. immunotherapy: Tisagenlecleucel—the first approved Car-t-cell therapy: implications for payers and policy makers. Nat. Rev. Clin. Oncol. 15 (1), 11.

Qian, Z., et al., 2013. Engineered regulatory T cells coexpressing MHC class II:peptide complexes are efficient inhibitors of autoimmune T cell function and prevent the development of autoimmune arthritis. J. Immunol. 190 (11), 5382–5391. https://doi.org/10.4049/jimmunol.1300024.

Rabinovich, G.A., Gabrilovich, D., Sotomayor, E.M., 2007. Immunosuppressive strategies that are mediated by tumor cells. Annu. Rev. Immunol. 25, 267–296. https://doi.org/10.1146/annurev.immunol.25.022106.141609.

Raghuwanshi, D., et al., 2012. A simple approach for enhanced immune response using engineered dendritic cell targeted nanoparticles. Vaccine 30 (50), 7292–7299. https://doi.org/10.1016/j.vaccine.2012.09.036. Elsevier Ltd.

Ramishetti, S., et al., 2015. Systemic gene silencing in primary T lymphocytes using targeted lipid nanoparticles. ACS Nano 6706–6716.

Rapoport, A.P., et al., 2015. NY-ESO-1–specific TCR–engineered T cells mediate sustained antigen-specific antitumor effects in myeloma. Nat. Med. 21 (8), 914.

Ratay, M.L., et al., 2017. Treg-recruiting microspheres prevent inflammation in a murine model of dry eye disease. J. Contr. Release 258, 208–217. https://doi.org/10.1016/j.jconrel.2017.05.007.

Reddy, S.T., et al., 2007. Exploiting lymphatic transport and complement activation in nanoparticle vaccines. Nat. Biotechnol. 25 (10), 1159–1164. https://doi.org/10.1038/nbt1332.

Rejman, J., et al., 2004. Size-dependent internalization of particles via the pathways of clathrin- and caveolae-mediated endocytosis. Biochem. J. 377 (Pt 1), 159–169. https://doi.org/10.1042/BJ20031253.

Resta, R., Thompson, L.F., 1997. T cell signalling through CD73. Cell. Signal. 9 (2), 131–139 pii:S0898-6568(96)00132-5.

Robbins, P.F., et al., 2013. Mining exomic sequencing data to identify mutated antigens recognized by adoptively transferred tumor-reactive T cells. Nat. Med. 19 (6), 747.

Rodriguez, P.L., et al., 2013. Minimal" self" peptides that inhibit phagocytic clearance and enhance delivery of nanoparticles. Science 339 (6122), 971–975.

Rollins, M.R., Gibbons Johnson, R.M., 2017. CD80 expressed by $CD8^+$ T cells contributes to PD-L1-induced apoptosis of activated $CD8^+$ T cells. J. Immunol. Res. 2017, 1–6. https://doi.org/10.1155/2017/7659462.

Rosenberg, S.A., 2014. IL-2: the first effective immunotherapy for human cancer. J. Immunol. 192 (12), 5451–5458. https://doi.org/10.4049/jimmunol.1490019.

Rosenberg, S.A., Restifo, N.P., 2015. Adoptive cell transfer as personalized immunotherapy for human cancer. Science 348 (6230), 62–68.

Rosenblum, M.D., et al., 2012. Treating human autoimmunity: current practice and future prospects. Sci. Transl. Med. 4 (125), 125sr1.

Ross, S.L., et al., 2017. Bispecific T cell engager (BiTE®) antibody constructs can mediate bystander tumor cell killing. PLoS One 12 (8), 1–24. https://doi.org/10.1371/journal.pone.0183390.

Roybal, K.T., et al., 2016. Engineering T cells with customized therapeutic response programs using synthetic notch receptors. Cell 167 (2), 419e16–432e16.

Rubinstein, M.P., et al., 2006. Converting IL-15 to a superagonist by binding to soluble IL-15Ra. Proc. Natl. Acad. Sci. USA 103 (24), 9166–9171. https://doi.org/10.1073/pnas.0600240103.

Rudolf, D., et al., 2008. Potent costimulation of human CD8 T cells by anti-4-1BB and anti-CD28 on synthetic artificial antigen presenting cells. Cancer Immunol. Immunother. 57 (2), 175–183. https://doi.org/10.1007/s00262-007-0360-x.

Rudqvist, N.-P., et al., 2017. Radiotherapy and CTLA-4 blockade shape the TCR repertoire of tumor-infiltrating T cells. Cancer Immunol. Res. https://doi.org/10.1158/2326-6066.CIR-17-0134 p. canimm.0134.2017.

Rupp, L.J., et al., 2017. CRISPR/Cas9-mediated PD-1 disruption enhances anti-tumor efficacy of human chimeric antigen receptor T cells. Sci. Rep. 7 (1), 737.

Ryan, J.M., et al., 2018. A novel biologic platform elicits profound T cell costimulatory activity and antitumor immunity in mice. Canc. Immunol. Immunother. 67 (4), 605–613. https://doi.org/10.1007/s00262-018-2116-1. Springer Berlin Heidelberg.

Safra, T., et al., 2000. Pegylated liposomal doxorubicin (doxil): reduced clinical cardiotoxic ity in patients reaching or exceeding cumulative doses of 500 mg/m^2. Ann. Oncol. 11 (8), 1029–1033.

Saito, T., Yokosuka, T., Hashimoto-Tane, A., 2010. Dynamic regulation of T cell activation and co-stimulation through TCR-microclusters. FEBS Letters 584 (24), 4865–4871. https://doi.org/10.1016/j.febslet.2010.11.036.

Sanchez-Martin, D., et al., 2015. Selection strategies for anti-cancer antibody discovery: searching off the beaten path. Trends Biotechnol. 33 (5), 292–301. https://doi.org/10.1038/scientificamerican031616a.

Satthaporn, S., et al., 2004. Dendritic cells are dysfunctional in patients with operable breast cancer. Cancer Immunol. Immunother. 53 (6), 510–518. https://doi.org/10.1007/s00262-003-0485-5.

Schamel, W.W.A., et al., 2005. Coexistence of multivalent and monovalent TCRs explains high sensitivity and wide range of response. J. Exp. Med. 202 (4), 493–503.

Schmaljohann, D., 2006. Thermo-and pH-responsive polymers in drug delivery. Adv. Drug Deliv. Rev. 58 (15), 1655–1670.

Schumacher, T.N., Schreiber, R.D., 2015. Neoantigens in cancer immunotherapy. Science 348 (6230), 69–74. https://doi.org/10.1126/science.aaa4971.

Scott, E.A., et al., 2012. Dendritic cell activation and T cell priming with adjuvant- and antigen-loaded oxidation-sensitive polymersomes. Biomaterials 33 (26), 6211–6219. https://doi.org/10.1016/j.biomaterials.2012.04.060. Elsevier Ltd.

Sedykh, S.E., et al., 2018. Bispecific antibodies: design, therapy, perspectives. Drug Des. Dev. Ther. 195–208. https://doi.org/10.2147/DDDT.S151282.

Shang, L., Nienhaus, K., Nienhaus, G.U., 2014. Engineered nanoparticles interacting with cells: size matters. J. Nanobiotechnol. 12 (1), 5.

Shann, S.Y., et al., 2012. Size-and charge-dependent non-specific uptake of PEGylated nanoparticles by macrophages. Int. J. Nanomed. 7, 799.

Sharma, G., et al., 2010. Polymer particle shape independently influences binding and internalization by macrophages. J. Contr. Release 147 (3), 408–412. https://doi.org/10.1016/j.jconrel.2010.07.116. Elsevier B.V.

Shedlock, D.J., Shen, H., 2003. Requirement for CD4 T cell help in generating functional CD8 T cell memory. Science 300 (5617), 337–339.

Shen, K., et al., 2008. Micropatterning of costimulatory ligands enhances $CD4^+$ T cell function. Proc. Natl. Acad. Sci. USA 105 (22), 7791–7796.

Shindo, Y., et al., 2015. Combination immunotherapy with 4-1BB activation and PD-1 blockade enhances antitumor efficacy in a mouse model of subcutaneous tumor. Anticancer Res. 35, 129–136.

Shuford, W.W., et al., 1997. 4-1BB costimulatory signals preferentially induce $CD8^+$ T cell proliferation and lead to the amplification in vivo of cytotoxic T cell responses. J. Exp. Med. 186 (1), 47–55.

Shutz, C., et al., 2016. Antigen-specific T cell redirectors: a nanoparticle based approach for redirecting T cells. Oncotarget 7 (42), 68503–68512. https://doi.org/10.18632/oncotarget.11785.

Singh, H., et al., 2011. Reprogramming CD19-specific T cells with IL-21 signaling can improve adoptive immunotherapy of B-lineage malignancies. Canc. Res. 71 (10), 3516–3527.

Singh, M., et al., 2017. Intratumoral CD40 activation and checkpoint blockade induces T cell-mediated eradication of melanoma in the brain. Nat. Commun. 8 (1), 1–10. https://doi.org/10.1038/s41467-017-01572-7. Springer US.

Singha, S., et al., 2017. Peptide-MHC-based nanomedicines for autoimmunity function as T-cell receptor microclustering devices. Nat. Nanotechnol. 12 (7), 701–710. https://doi.org/10.1038/nnano.2017.56.

Smith-Garvin, J.E., Koretzky, G.A., Jordan, M.S., 2009. T cell activation. Annu. Rev. Immunol. 27 (1), 591–619. https://doi.org/10.1146/annurev.immunol.021908.132706.

Smith, T.T., et al., 2017a. Biopolymers codelivering engineered T cells and STING agonists can eliminate heterogeneous tumors. J. Clin. Invest. 127 (6), 2176–2191.

Smith, T.T., et al., 2017b. In situ programming of leukaemia-specific T cells using synthetic DNA nanocarriers. Nat. Nanotechnol. 12 (April), 813. https://doi.org/10.1038/nnano.2017.57. Nature Publishing Group.

Snyder, A., et al., 2014. Genetic basis for clinical response to CTLA-4 blockade in melanoma. N. Engl. J. Med. 371 (23), 2189–2199. https://doi.org/10.1056/NEJMoa1406498.

Spatz, J.P., et al., 2000. Ordered deposition of inorganic clusters from micellar block copolymer films. Langmuir 16 (2), 407–415.

Speiser, D.E., et al., 2014. T cell differentiation in chronic infection and cancer: functional adaptation or exhaustion? Nat. Rev. Immunol. 14 (11), 768.

Spranger, S., et al., 2014. Mechanism of tumor rejection with doublets of CTLA-4, PD-1/PD-L1, or Ido blockade involves restored IL-2 production and proliferation of CD8(+) T cells directly within the tumor microenvironment. J. Immunother. Cancer 2 (1), 3. https://doi.org/10.1186/2051-1426-2-3.

Srivastava, S., Riddell, S.R., 2015. Engineering CAR-T cells: design concepts. Trends Immunol. 36 (8), 494–502. https://doi.org/10.1016/j.it.2015.06.004. Elsevier Ltd.
Stachowiak, A.N., Irvine, D.J., 2008. Inverse opal hydrogel-collagen composite scaffolds as a supportive microenvironment for immune cell migration. J. Biomed. Mater. Res. 85 (3), 815–828.
Steenblock, E.R., et al., 2011. An artificial antigen-presenting cell with paracrine delivery of IL-2 impacts the magnitude and direction of the T cell response. J. Biol. Chem. 286 (40), 34883–34892. https://doi.org/10.1074/jbc.M111.276329.
Steenblock, E.R., Fahmy, T.M., 2008. A comprehensive platform for ex vivo T-cell expansion based on biodegradable polymeric artificial antigen-presenting cells. Mol. Ther. 16 (4), 765–772. https://doi.org/10.1038/mt.2008.11.
Steinhardt, J.J., et al., 2018. Rational design of a trispecific antibody targeting the HIV-1 Env with elevated anti-viral activity. Nat. Commun. 9 (1) https://doi.org/10.1038/s41467-018-03335-4.
Steinman, R.M., 2012. Decisions about dendritic cells: past, present, and future. Annu. Rev. Immunol. 30 (1), 1–22. https://doi.org/10.1146/annurev-immunol-100311-102839.
Stephan, M.T., et al., 2007. T cell-encoded CD80 and 4-1BBL induce auto- and transcostimulation, resulting in potent tumor rejection. Nat. Med. 13 (12), 1440–1449. https://doi.org/10.1038/nm1676.
Stephan, M.T., et al., 2012. Synapse-directed delivery of immunomodulators using T-cell-conjugated nanoparticles. Biomaterials 33 (23), 5776–5787. https://doi.org/10.1016/j.biomaterials.2012.04.029. Elsevier Ltd.
Stephan, S.B., et al., 2015. Biopolymer implants enhance the efficacy of adoptive T-cell therapy. Nat. Biotechnol. 33 (1), 97.
Strohl, W.R., 2017. Current progress in innovative engineered antibodies. Protein Cell 9 (1), 1–35. https://doi.org/10.1007/s13238-017-0457-8. Higher Education Press.
Suematsu, S., Watanabe, T., 2004. Generation of a synthetic lymphoid tissue–like organoid in mice. Nat. Biotechnol. 22 (12), 1539.
Suhoski, M.M., et al., 2007. Engineering artificial antigen-presenting cells to express a diverse array of co-stimulatory molecules. Mol. Ther. 15 (5), 981–988. https://doi.org/10.1038/mt.sj.6300134.
Sun, X., et al., 2017. Surface-engineering of red blood cells as artificial antigen presenting cells promising for cancer immunotherapy. Small 13 (40), 1–8. https://doi.org/10.1002/smll.201701864.
Sunshine, J.C., et al., 2014. Particle shape dependence of $CD8^+$ T cell activation by artificial antigen presenting cells. Biomaterials 35 (1), 269–277. https://doi.org/10.1016/j.biomaterials.2013.09.050. Elsevier Ltd.
Tay, N.Q., et al., 2017. CD40L expression allows $CD8^+$ T cells to promote their own expansion and differentiation through dendritic cells. Front. Immunol. 8 (November) https://doi.org/10.3389/fimmu.2017.01484.
Taylor, R.P., Lindorfer, M.A., 2016. Cytotoxic mechanisms of immunotherapy: harnessing complement in the action of anti-tumor monoclonal antibodies. Semin. Immunol. 28 (3), 309–316. https://doi.org/10.1016/j.smim.2016.03.003. Elsevier Ltd.
Tebas, P., et al., 2014. Gene editing of CCR5 in autologous CD4 T cells of persons infected with HIV. N. Engl. J. Med. 370 (10), 901–910.
Tenzer, S., et al., 2013. Rapid formation of plasma protein corona critically affects nanoparticle pathophysiology. Nat. Nanotechnol. 8 (10), 772–781.
Tesmer, L.A., et al., 2008. Th17 cells in human disease. Immunol. Rev. 223, 87–113. https://doi.org/10.1111/j.1600-065X.2008.00628.x.
Tham, E.L., Jensen, P.L., Mescher, M.F., 2001. Activation of antigen-specific T cells by artificial cell constructs having immobilized multimeric peptide-class I complexes and recombinant B7-Fc proteins. J. Immunol. Meth. 249 (1–2), 111–119.

Theofilopoulos, A.N., Kono, D.H., Baccala, R., 2017. The multiple pathways to autoimmunity. Nat. Immunol. 18 (7), 716.
Tong, R., et al., 2012. Photoswitchable nanoparticles for triggered tissue penetration and drug delivery. J. Am. Chem. Soc. 134 (21), 8848–8855.
Torchilin, V.P., 2005. Recent advances with liposomes as pharmaceutical carriers. Nat. Rev. Drug Discov. 4 (2), 145.
Torchilin, V.P., Trubetskoy, V.S., 1995. Which polymers can make nanoparticulate drug carriers long-circulating? Adv. Drug Deliv. Rev. 16 (2–3), 141–155.
Tostanoski, L.H., et al., 2016. Design of polyelectrolyte multilayers to promote immunological tolerance. ACS Nano 10 (10), 9334–9345. https://doi.org/10.1021/acsnano.6b04001.
Tran, E., et al., 2016. T-cell transfer therapy targeting mutant KRAS in cancer. N. Engl. J. Med. 375 (23), 2255–2262.
Tsai, S., et al., 2010. Reversal of autoimmunity by boosting memory-like autoregulatory T cells. Immunity 32 (4), 568–580. https://doi.org/10.1016/j.immuni.2010.03.015. Elsevier Ltd.
Tyrsin, D., et al., 2016. From TGN1412 to TAB08: the return of CD28 superagonist therapy to clinical development for the treatment of rheumatoid arthritis. Clin. Exp. Rheumatol. 34, 45–48.
Van Allen, E.M., et al., 2015. Genomic correlates of response to CTLA-4 blockade in metastatic melanoma. Science 350 (6257), 207–211.
Varma, R., et al., 2006. T cell receptor-proximal signals are sustained in peripheral microclusters and terminated in the central supramolecular activation cluster. Immunity 25 (1), 117–127.
Vasievich, E.A., Huang, L., 2011. The suppressive tumor microenvironment: a challenge in cancer immunotherapy. Mol. Pharm. 8 (3), 635–641. https://doi.org/10.1021/mp1004228.
Veiseh, O., et al., 2015. Size-and shape-dependent foreign body immune response to materials implanted in rodents and non-human primates. Nat. Mater. 14 (6), 643–651.
Vezys, V., et al., 2011. 4-1BB signaling synergizes with programmed death ligand 1 blockade to augment CD8 T cell responses during chronic viral infection. J. Immunol. 187 (4), 1634–1642. https://doi.org/10.4049/jimmunol.1100077.
Victor, C.T.-S., et al., 2015. Radiation and dual checkpoint blockade activate non-redundant immune mechanisms in cancer. Nature 520 (7547), 373–377. https://doi.org/10.1038/nature14292.
Vigneron, N., 2015. Review article human tumor antigens and cancer immunotherapy. Hindawi 2015, 1–17.
Vilgelm, A.E., Johnson, D.B., Richmond, A., August 2016. Combinatorial approach to cancer immunotherapy: strength in numbers. J. Leukoc. Biol. 100 https://doi.org/10.1189/jlb.5RI0116-013RR.
Vincent, M., et al., 2014. Highly potent anti-CD20-RLI immunocytokine targeting established human B lymphoma in SCID mouse. mAbs 6 (4), 1026–1037. https://doi.org/10.4161/mabs.28699.
Wang, C.-H., et al., 2012. Aptamer-conjugated and drug-loaded acoustic droplets for ultrasound theranosis. Biomaterials 33 (6), 1939–1947.
Wang, Q., Liu, F., Liu, L., 2017. Prognostic significance of PD-L1 in solid tumor: an updated meta-analysis. Medicine 96 (18). https://doi.org/10.1097/MD.0000000000006369.
Wang, W., et al., 2015. NK cell-mediated antibody-dependent cellular cytotoxicity in cancer immunotherapy. Front. Immunol. 6 (JUL) https://doi.org/10.3389/fimmu.2015.00368.
Wherry, E.J., 2011. T cell exhaustion. Nat. Immunol. 131 (6), 492–499. https://doi.org/10.1038/ni.2035. Nature Publishing Group.

Wherry, E.J., Kurachi, M., 2015. Molecular and cellular insights into T cell exhaustion. Nat. Rev. Immunol. 15 (8), 486.

Wilkie, S., et al., 2012. Dual targeting of ErbB2 and MUC1 in breast cancer using chimeric antigen receptors engineered to provide complementary signaling. J. Clin. Immunol. 32 (5), 1059–1070.

Wing, K., Sakaguchi, S., 2010. Regulatory T cells exert checks and balances on self tolerance and autoimmunity. Nat. Immunol. 11 (1), 7.

Wu, C.-Y., et al., 2015. Remote control of therapeutic T cells through a small molecule-gated chimeric receptor. Science 350 (6258), aab4077. https://doi.org/10.1097/TP.0000000000001016.

Xie, J.H., et al., 2014. Engineering of a novel anti-CD40L domain antibody for treatment of autoimmune diseases. J. Immunol. 192 (9), 4083–4092. https://doi.org/10.4049/jimmunol.1303239.

Xu, Q., et al., 2015. Impact of surface polyethylene glycol (PEG) density on biodegradable nanoparticle transport in mucus ex vivo and distribution in vivo. ACS Nano 9 (9), 9217–9227.

Yarchoan, M., et al., 2017. Targeting neoantigens to augment antitumour immunity. Nat. Rev. Canc. 17 (4), 209.

Yasunaga, M., Manabe, S., Matsumura, Y., 2017. Immunoregulation by IL-7R-targeting antibody-drug conjugates: overcoming steroid-resistance in cancer and autoimmune disease. Sci. Rep. 7 (1), 1–14. https://doi.org/10.1038/s41598-017-11255-4. Springer US.

Ye, F., et al., 2010. Alterations of dendritic cell subsets in the peripheral circulation of patients with cervical carcinoma. J. Exp. Clin. Cancer Res. 29, 78. https://doi.org/10.1186/1756-9966-29-78.

Ye, Q., et al., 2014. CD137 accurately identifies and enriches for naturally occurring tumor-reactive T cells in tumor. Clin. Canc. Res. 20 (1), 44–55. https://doi.org/10.1158/1078-0432.CCR-13-0945.

Yokosuka, T., et al., 2008. Spatiotemporal Regulation of T Cell costimulation by TCR-CD28 microclusters and protein kinase C translocation. Immunity 29 (4), 589–601. https://doi.org/10.1016/j.immuni.2008.08.011. Elsevier Inc.

Yokosuka, T., et al., 2012. Programmed cell death 1 forms negative costimulatory microclusters that directly inhibit T cell receptor signaling by recruiting phosphatase SHP2. J. Exp. Med. 209 (6), 1201–1217. https://doi.org/10.1084/jem.20112741.

Yoo, J.-W., et al., 2011. Bio-inspired, bioengineered and biomimetic drug delivery carriers. Nat. Rev. Drug Discov. 10 (7), 521.

Yoo, J.-W., Mitragotri, S., 2010. Polymer particles that switch shape in response to a stimulus. Proc. Natl. Acad. Sci. USA 107 (25), 11205–11210. https://doi.org/10.1073/pnas.1000346107.

Yuan, H., et al., 2010. Variable nanoparticle-cell adhesion strength regulates cellular uptake. Phys. Rev. Lett. 105 (13), 138101.

Yuraszeck, T., Kasichayanula, S., Benjamin, J.E., 2017. Translation and clinical development of bispecific t-cell engaging antibodies for cancer treatment. Clin. Pharmacol. Ther. 101 (5), 634–645. https://doi.org/10.1002/cpt.651.

Zamarin, D., et al., 2014. Localized oncolytic virotherapy overcomes systemic tumor resistance to immune checkpoint blockade immunotherapy. Sci. Transl. Med. 6 (226), 226ra32. https://doi.org/10.3174/ajnr.A1256.Functional.

Zauner, W., Farrow, N.A., Haines, A.M.R., 2001. In vitro uptake of polystyrene microspheres: effect of particle size, cell line and cell density. J. Contr. Release 71 (1), 39–51.

Zeng, W., et al., 2014. Artificial antigen-presenting cells expressing CD80, CD70, and 4-1BB ligand efficiently expand functional T cells specific to tumor-associated antigens. Immunobiology 219 (8), 583–592. https://doi.org/10.1016/j.imbio.2014.03.003. Elsevier GmbH.

Zhang, H., et al., 2007. 4-1BB is superior to CD28 costimulation for generating $CD8^+$ cytotoxic lymphocytes for adoptive immunotherapy. J. Immunol. 179 (7), 4910–4918.

Zhang, H., et al., 2013. 4-1BB is superior to CD28 costimulation for generating $CD8^+$ cytotoxic lymphocytes for adoptive immunotherapy. J. Immunol. 179 (7), 4910–4918.

Zhang, N., Bevan, M.J., 2011. $CD8^+$ T cells: foot soldiers of the immune system. Immunity 35 (2), 161–168. https://doi.org/10.1016/j.immuni.2011.07.010. Elsevier Inc.

Zhang, Q., et al., 2017. Biomimetic magnetosomes as versatile artificial antigen-presenting cells to potentiate T-cell-based anticancer therapy. ACS Nano 11 (11), 10724–10732.

Zhang, S., et al., 2009. Size-dependent endocytosis of nanoparticles. Adv. Mater. 21 (4), 419–424.

Zhang, Y., et al., 2018. Nanoparticle anchoring targets immune agonists to tumors enabling anti-cancer immunity without systemic toxicity. Nat. Commun. 9 (1), 6. https://doi.org/10.1038/s41467-017-02251-3.

Zheng, G., Wang, B., Chen, A., 2004. The 4-1BB costimulation augments the proliferation of $CD4^+CD25^+$ regulatory T cells. J. Immunol. 173 (4), 2428–2434.

Zheng, Y., et al., 2017. Enhancing adoptive cell therapy of cancer through targeted delivery of small-molecule immunomodulators to internalizing or non-internalizing receptors. ACS Nano. https://doi.org/10.1021/acsnano.7b00078 p. acsnano.7b00078.

Zhu, J., 2017. T helper cell differentiation, heterogeneity, and plasticity. Cold. Spring Harb. Perspect. Biol. https://doi.org/10.1101/cshperspect.a030338 a030338.

Zhu, X., et al., 2009. Novel human Interleukin-15 agonists. J. Immunol. 183 (6), 1–27. https://doi.org/10.1016/j.jmb.2008.10.054.The.

FSC
www.fsc.org
MIX
Papier | Fördert
gute Waldnutzung
FSC® C083411

Kolibri 360 GmbH
Ferdinand-Jühlke-Straße 7
99095 Erfurt, Deutschland
produktsicherheit@kolibri360.de